Evolutionary Biology – New Perspectives on Its Development

Volume 11

Evolutionary biology has been a remarkably dynamic area since its foundation. Its true complexity, however, has been concealed in the last 50 years under an assumed opposition between the "Extended Evolutionary Synthesis" and an "Alternative to the Evolutionary Synthesis". This multidisciplinary book series aims to move beyond the notion that the development of evolutionary biology is structured around a lasting tension between a Darwinian tradition and a non-Darwinian tradition, once dominated by categories like Darwinian Revolution, Eclipse of Darwinism, Evolutionary Synthesis, and Post-Synthetic Developments.

The monographs and edited volumes of the series propose an alternative to this traditional outlook with the explicit aim of fostering new thinking habits about evolutionary biology, a multifaceted area composed of changing and interacting research entities and explanatory levels. Contributions by biologists and historians/philosophers are welcomed. Topics covered in the series span from (among many other possibilities):

- An Overview of Neutralist Theories in Evolutionary Biology
- Developmental Biology: From Reductionism to Holism and Back
- Selection Theories Beyond Hard and Soft Inheritance
- Divergent, Parallel, and Reticulate Evolution: Competing or Complementary Research Programs?
- The Rise of Molecular Biology: Between Darwinian and Non-Darwinian
- Biologizing Paleontology: A Tradition with Deep Historical Roots
- The Darwinian Revolution and the Eclipse of Darwinism: Blurring the Historiographical Lines
- Darwinism, Lamarckism, Orthogenesis: Can We Really Define Them by Their Hard Explanatory Cores?
- The Evolutionary Synthesis: A Fabricated Concept?
- The Opposition to the Evolutionary Synthesis: Criticizing a Phantom?
- A Reversed Perspective: Approaching Charles Darwin from the Pre-1859 Period
- The Long Development of the Multilevel Paradigm in Evolutionary Biology
- Self-Organization: A Research Tradition from Morphology to Cosmology
- Human Evolution: Sociobiological or Sociocultural?

All chapters are systematically reviewed by the series editor and respective volume editor(s).For monographs, the editor of the book series reach out to two competent reviewers. The editor ensures that reviews are fair and relevant.For edited volumes, the volume's editor selects two competent reviewers for each chapter, ensuring that reviews are fair and relevant. The series editor oversees the whole evaluation process.

* * *

Mathilde Lequin
Juan Manuel Rodríguez Caso
Editors

The Field of Human Evolution: Critical Perspectives from History and Epistemology

Springer

Editors
Mathilde Lequin
Univ. Bordeaux, CNRS, Ministère de la Culture PACEA, UMR 5199
F-33600 Pessac, France

Juan Manuel Rodríguez Caso
División de Ciencias y Artes para el Diseño
Universidad Autónoma Metropolitana
Unidad Xochimilco
Mexico City, Mexico

ISSN 2524-7751 ISSN 2524-776X (electronic)
Evolutionary Biology – New Perspectives on Its Development
ISBN 978-3-032-13924-5 ISBN 978-3-032-13925-2 (eBook)
https://doi.org/10.1007/978-3-032-13925-2

This Springer imprint is published by the registered company Springer Nature Switzerland AG
The registered company address is: Gewerbestrasse 11, 6330 Cham, Switzerland

Contents

Part I

Introduction

Critical Thinking in Human Evolution: Reflexive Approaches to the Human Past

1

Mathilde Lequin

Abstract

This chapter examines the complexity of critical thinking within the field of human evolution and explores the unique contributions that history and philosophy of science can offer. Contrary to a reductive view that portrays the field of human evolution as narrowly positivist—driven solely by paleontological and archeological discoveries and technological advances—I argue that critical reflection is, in fact, central to its development. The chapter pursues two main objectives. First, it seeks to define and map the various dimensions of "criticism" in human evolution, clarifying the multiple meanings of "critical" and the foundations of critical approaches within the field. Second, it emphasizes that history and philosophy do not exercise their critical capacities *against* paleoanthropology and archeology, but *in concert with* them. I explore the ways in which historical and philosophical inquiries can productively engage with and contribute to research on human evolutionary history.

Keywords

Philosophy of biology · Paleoanthropology · Archeology · Reflexivity · Inference · Interpretation

M. Lequin (✉)
Univ. Bordeaux, CNRS, Ministère de la Culture, PACEA, UMR 5199, F-33600 Pessac, France
e-mail: mathilde.lequin@u-bordeaux.fr

M. Lequin, J. M. Rodríguez Caso (eds.), *The Field of Human Evolution: Critical Perspectives from History and Epistemology*, Evolutionary Biology – New Perspectives on Its Development 11, https://doi.org/10.1007/978-3-032-13925-2_1

1.1 Introduction

Paleoanthropologists and archeologists hardly need external prompting to adopt a critical stance toward their own disciplines. The field of human evolution is characterized by frequent and often intense debates that make critical thinking central to its scientific dynamics. These debates span a wide range of issues, including the validity of inferences drawn from skeletal and genetic data (e.g., Cazenave et al. 2024; d'Errico et al. 2024), the reliability of analytical methods used to assess the fossil record (e.g., Robinson and Terhune 2017), and the theoretical frameworks guiding the interpretation of biological and cultural data (e.g., Kissel and Fuentes 2021; Zilhão et al. 2024). They even extend to fundamental questions about the scientificity and limits of the study of human evolution itself (Smith and Wood 2017; Wood and Smith 2022; Villmoare and Kimbel 2024). Far from being mired in endless controversy, the field actively engages in profound reflection on the nature of science, the robustness of its methods, and the criteria for what counts as evidence.

Critical thinking is often celebrated as a hallmark of rationality and, by extension, of the scientific method (Popper 1959; Lakatos 1978). What, then, is distinctive about thinking critically in the field of human evolution? Unlike physicists, paleoanthropologists and archeologists can hardly test their theories and hypotheses through controlled experiments. Instead, research on human evolution is rooted in the historical sciences, which possess their own epistemological framework. These sciences employ specific methodologies for testing hypotheses about the human past, based on the material traces that have come down to us (Cleland 2002). In this context, critical engagement is notable not only for its scope and intensity but also for how deeply it is embedded within the discipline's internal dynamics, advancing debates about the range of evidential claims that can be made from these material traces. Here, criticism is not merely episodic. It does not only emerge when a dominant framework fails to account for new fossil evidence: it is a constitutive feature of the field, shaping its normal functioning. I use the term normal in reference to the "normal science," as conceptualized by Kuhn (1962), to emphasize that criticism, in the field of human evolution, often operates outside a context of paradigm shifts.

Such a persistent critical stance may result from the inherently sensitive and complex nature of the questions the field addresses—questions about our origins and defining traits—combined with the challenge of working with data that are fragmentary, discontinuous, and fundamentally ambiguous. As Chapman and Wylie (2016, p. 3) observe, material traces of the past "rarely ever speak with a single voice, self-warranting and prophetic; what they say is never unmediated by interpretation." The inherently interpretive character of the field renders inferences from the data open to critical scrutiny. Yet, critical thinking in human evolution does not merely consist in opposing one interpretation to another. Rather, it follows its own productive dynamics, with two central features. First, because "the data do not speak" (Chapman and Wylie 2016, p. 203), they must be made to speak—that is, turned into evidence through inferential scaffolding that integrates multiple, diverse lines of evidence.

This interpretive work is necessarily critical: it involves evaluating the robustness of inferences, identifying potential biases, and proposing alternative interpretations. Second, this criticism is productive because it encounters limits: material traces do not support any claim but resist certain interpretations and, by doing so, provide evidential constraints that allow researchers to discriminate between competing hypotheses and to eliminate some of them (the "paradox of interpretation," Chapman and Wylie 2016, p. 206).

Therefore, critical thinking in the field of human evolution should not be reduced to academic polemics or to speculation arising from the underdetermination to which this field is supposedly condemned by the scarcity of its data (for a discussion about underdetermination in historical sciences, see Currie 2021). It encompasses a broader set of epistemic attitudes and methodological approaches that this volume seeks to examine and that this chapter aims to characterize and organize.

In everyday language, criticism often carries a negative connotation, suggesting the search for flaws or the expression of disapproval. Scientific criticism, however, operates differently. The critical approach developed in this volume is not primarily about exposing what is wrong or deficient in the study of human evolution. Instead, it is about examining how scientific practices, concepts, and methodological frameworks can be reflected upon and questioned—particularly at the intersections of biological anthropology, paleontology, primatology, archeology, molecular biology, humanities and social sciences, all of which contribute to our understanding of human evolution.

In this sense, the volume approaches critical thinking through the lens of reflexivity, i.e. by reflecting on what researchers do, and how they do it, when studying human evolution: the epistemic tools they employ, the objects of knowledge they produce, and the broader implications of studying the human past for the humans we are today. More specifically, this introductory chapter distinguishes and explores four dimensions of reflexivity in the field of human evolution: the historical, methodological, epistemic, and (geo)political aspects of knowledge production.

1.2 Historical Reflexivity: Learning from the Field's Past

Historical reflexivity draws on the history of science to encourage scholars in human evolution to consider how their practices and frameworks relate to the discipline's past—whether shaped by it or, conversely, in rupture with it (Delisle 2007, 2012, 2017). It also prompts reflection on how the field engages with its historical depth and defines itself in relation to its conceptual and methodological legacy.

A pivotal focus lies in highlighting the origins and foundations of this scientific field. As Goodrum (Chap. 4) shows, the discipline evolved from racial anthropology, rooted in craniometry and anthropometry, toward the evolutionary interpretation of fossils and the phylogenetic reconstruction of human origins. In the early period of paleoanthropology (from the 1850s to the early 1880s), studies of human fossils were guided not by principles of human evolution but by the aim of identifying distinctive human races. Acknowledging this history and recognizing that the discipline once focused on classifying races rather than understanding human evolution are crucial for grasping how contemporary biological anthropology defines itself in rupture with a form of physical anthropology historically grounded in race.

A critical reappraisal of the history of human evolution can also serve to dismantle the field's founding myths, such as the so-called "Darwinian revolution," which suggests that biological evolution could only be understood through a single, unified doctrine, Darwinism, supposedly forming the theoretical basis of paleoanthropology. By applying a critical approach to historiographical categories, historical scrutiny instead reveals a more complex constellation of naturalists who conceptualized human evolution in diverse ways (Rodríguez Caso, Chap. 3).

Stepping back to explore how current approaches in human evolution are informed by the field's historical legacy, the integration of the Modern Synthesis into paleoanthropologists' and archeologists' theoretical frameworks in the mid-twentieth century represents a pivotal moment. This integration aligned the study of human evolution with evolutionary biology, emphasizing the centrality of natural selection, the primacy of populations over races, and the graduality of evolutionary change. However, it has also profoundly influenced interpretive practices in paleoanthropology and archeology (Delisle 2007; Goodrum 2023), shaping reconstructions of hominin taxonomy and phylogeny (Schwartz 2017) as well as assessments of hominin diversity (Tattersall 2012).

Within this legacy, two assumptions have particularly far-reaching implications on contemporary interpretations of the fossil and archeological record. First, gene-centrism—the idea that genetic change constitutes the primary source of biological novelty—has strongly influenced interpretations of the production, transmission, and selection of variation (Colagè & d'Errico, Chap. 9). Behavior, in this view, is often treated merely as the outcome of a genetic program rather than as a driver of evolution, and human culture is interpreted primarily as the product of genetic innovations. Yet archeological evidence of cultural innovations challenges this perspective, highlighting culture as an active evolutionary force (Colagè & d'Errico, Chap. 9). The second core assumption is phyletic gradualism, which holds that macroscopic evolutionary changes result from the steady accumulation of microscopic innovations under constant, directional natural selection (Pievani, Tattersall, Eldredge, Chap. 7). This view frames species as gradually evolving lineages and has strongly influenced paleoanthropologists' interpretations of hominin evolution and speciation.

Historical reflexivity is not limited to identifying biases; it also considers how alternative theoretical perspectives arise in response to the Modern Synthesis, within an expanding evolutionary framework. For instance, to address the limitations of phyletic gradualism, Eldredge and Gould (1972) proposed the Punctuated Equilibria model, conceptualizing macroevolution as long periods of stasis interrupted by brief, rapid episodes of speciation. Yet, this model is also built on Mayr's theory of allopatric speciation (Pievani, Tattersall, Eldredge, Chap. 7), which distinguishes long-term stasis in parent species from rapid change in isolated descendant populations adapting to new environments. More recently, the Extended

Evolutionary Synthesis has proposed to recognize diverse sources of biological novelty beyond genetic mutation and change. By integrating behavioral innovation, social learning, niche construction, and non-genetic inheritance as evolutionary forces in their own right, it offers a stimulating framework for archeologists (Colagè & d'Errico, Chap. 9).

Critical investigation of the past also involves examining how earlier work addressed what are now considered central challenges, such as bridging nature and culture in the study of human evolution. Sherwood Washburn, for instance, is widely regarded as having played a crucial role in integrating the Modern Synthesis into paleoanthropology (Smocovitis 2012). However, he was also the founder of an anthropology that aimed to think on what it means to be human in all its dimensions, from the biological to the social—an approach resonating strongly with the orientation of pragmatic philosophy (Mikels, Chap. 6).

1.3 Methodological Reflexivity: Data, Interpretations, Representations

Methodological reflexivity examines how evidence in the study of human evolution is produced, integrated, and interpreted across diverse sources of data. This approach is taken up not only by researchers reflecting on the contexts and limits of evidential claims in their field (e.g., Smith and Wood 2017), but also by philosophers of science, investigating the inferences that are drawn from the paleontological and archeological records to past cognition, behavior and culture (Currie et al. 2024). This perspective involves questioning core assumptions in the interpretation of the fossil and archeological record (Lequin 2018; Meneganzin and Currie 2022), analyzing the construction of inferential scaffolding (Chapman and Wylie 2016), and the methodologies of reasoning from material traces (Currie and Killin 2019; Killin and Pain 2023). It also requires situating these practices within the specific conditions of human evolution research: a science that operates under data scarcity (Currie 2021), yet aiming at identifying species (Meneganzin and Bernardi 2023) and at reconstructing the processes and patterns of human evolution and development (Lequin et al. 2025).

More broadly, methodological reflexivity involves stepping back from the immediate task of explaining features or patterns in the fossil or archeological record to reflect on how such explanations are constructed and to scrutinize the causal mechanisms invoked to account for them. For instance, since behavioral and cultural innovations were long regarded as products of genetic evolution, they used to be considered as proximate rather than ultimate causes—that is, as causes without direct evolutionary relevance (Colagè & d'Errico, Chap. 9). However, archeological evidence demonstrates that cultural traits, such as the use of ochre, spread across different populations and geographic areas in ways that reflect cultural transmission strategies. In such cases, culture emerges as an evolutionary force, shaping human history alongside biology, and prompting a revision of causal frameworks in human evolution (Colagè & d'Errico, Chap. 9).

Methodological reflexivity also encompasses the representations that have shaped our evolutionary narratives, from depictions of past hominins to visualizations of phylogeny. For instance, changing portrayals of Neanderthals have been influenced by academic research but also by novels. This interplay between scholarly and imaginative interpretations has contributed to a more inclusive view of Neanderthals (Zwart et al., Chap. 14). Likewise, images of human phylogeny, such as the scale, the tree, or the tree-on-a-map, play a crucial role in representations of human evolution. Sommer (Chap. 8) traces their complex history, showing how particular diagrammatic forms may persist even as new ways of conceptualizing evolutionary relationships emerge.

Epistemic credentials, i.e. the qualities that make a claim or dataset reliable as evidence, are central to methodological reflexivity, particularly in relation to genetic data. How much evidential weight should such data carry? How does molecular evidence compare with morphological or cultural evidence? And how robust are the claims built upon it? Chikhi et al. (Chap. 10) scrutinize the assumptions underlying the models used by geneticists, focusing on how genetic data are mobilized to support hypotheses about population bottlenecks and admixture events. Their analysis raises a broader question: how does one type of data, especially genetic evidence, come to dominate the construction of evolutionary narratives in a field that, in principle, thrives on integrating multiple lines of evidence?

Another salient point here is that genetic sequences are not self-interpreting; as any data, they require theoretical and contextual framing to become evidence (Chapman and Wylie 2016; Leonelli 2016; Currie 2021). This is equally relevant to human behavioral genetics, which seeks genetic underpinnings for traits such as educational attainment. What, then, is the contribution of human behavioral genetics to understanding human evolution? And how might we foster conceptual bridges between its practitioners and those in human evolutionary genetics (Downes, Chap. 11)? In this light, methodological reflexivity emerges as a critical step toward the integration of diverse forms of data and evidence in the study of our evolutionary past.

1.4 Epistemic Reflexivity: Studying Ourselves Through Deep Time

Epistemic reflexivity differs from methodological reflexivity in that it confronts the fact that research on human evolution might be framed as a research about ourselves. Yet the referent of "ourselves" is not self-evident: does it refer to us as members of the species *Homo sapiens*? It potentially extends beyond this group, as the study of human evolution encompasses extinct populations of *Homo sapiens* as well as many other taxa within and beyond the genus *Homo*. These hominins differ from us to varying degrees and are increasingly recognized in their otherness (Lequin 2026). Therefore, framing the study of human evolution as research about ourselves risks implying a universal perspective on "humankind" made from a supposedly transcendent viewpoint, and thus demands careful scrutiny. However, the fact remains that research on human evolution investigates what it means to be human—across taxonomic boundaries and deep timescales—which is both intrinsic and foundational to the field. The field investigates our ancestry, as a biological species, and our relatedness with extinct hominins; it focuses on the evolutionary origins of human biology, behavior, and culture, and this perspective defines what I call here epistemic reflexivity.

Epistemic reflexivity refers to the recognition that, in studying human evolution, the investigator is not identical with but nonetheless related to the object of investigation. In the field of human evolution, humans are the specific focus. However, humans are also special—as, arguably, many biological taxa are, in distinct ways—in the sense that they are a reflexive species. "Humans are universally interested in who they are and where they come from. Sharks, elephants, bats, chimpanzees, and other species are not" (Marks 2015, p. IX). This configuration of the discipline contributes to shaping the field's questions, interpretations, and claims. However, it can also be a source of bias, favoring the overemphasis on traits regarded as uniquely human (Lequin 2026), the projection of attributes considered as defining current humans on past hominins, or the use of data regarding past hominins to explain the humans we are today.

While epistemic reflexivity, then, is a defining feature of the field, its epistemological and methodological consequences need to be explored. Paleoanthropology is usually framed primarily as a branch of biology, rather than recognized as a distinctive form of anthropology. However, as Marks (2015, p. XI) puts it, "what differentiates biological anthropology (the study of human origins and diversity) from biology (the study of life) is reflexivity—the breakdown of the distinction between subject and object that characterizes modern science." While the implications of such breakdowns have long been discussed in the social and human sciences—notably, in *The Order of Things* (Foucault 1970)—they have rarely been applied to paleoanthropology, precisely because this science is seen as anchored in the field of evolutionary biology.

Acknowledging this epistemic reflexivity has far-reaching implications. Consider, for example, paleoanthropological taxonomy. Hominin taxonomy involves arranging fossils into taxa, but what's behind the names of these taxa? Not just a set of specimens. These fossils are conceptualized as *ancestors*—not only as biological ancestors, a category that is itself complex and ambiguous (Love and Wagner 2023)—but as "anthropological ancestors," supporting "a sacred narrative about our past, a story intended to explain who we are and how we came to be here" (Marks, Chap. 12). Paleoanthropology thus engages in a specific kind of scientific narrative-building, explaining descent and relatedness, and thus addressing the field of study that usually defines social anthropology—namely, kinship (Marks, Chap. 12). Here, epistemic reflexivity makes explicit that paleoanthropology is not only about biological classification but also about our relational engagement with the past.

This recognition resonates with Derricourt's notion of a "paleoanthropological exceptionalism" (Chap. 5). Unlike experimental sciences, which privilege reproducibility and replicability, and even unlike archeology, which emphasizes grouped continuities and associated changes in material culture, paleoanthropology has often stressed the singularity of fossil discoveries, using their perceived uniqueness to justify the assignment of new species or even genera. Such specificity brings epistemological obligations, such as situating research within the historical, geographical, and ideological contexts of the scholars who produce it, and examining how these contexts have shaped their scientific practices (Derricourt, Chap. 5).

Another epistemological obligation arises from the recognition that studying past hominins often opens the possibility of thinking about ourselves. It is therefore necessary to examine the kinds of explanatory links we draw between what we know of our distant past and the humans we are today, since the past is frequently invoked to explain, or even justify, features of contemporary individuals and societies. Such explanations involve a shift from "is" to "ought," from a descriptive approach of how humans were, based on the available data, to a proscriptive approach "about how humans truly are and thus should be" (Fuentes, Chap. 13). A critical perspective is all the more necessary that this move is based on a problematic notion of "naturalness," assuming that our deep evolutionary past gives access to "nature," and can reliably indicate what is "natural" for the humans we are—regarding sex, gender, family, or intergroup relationships (Fuentes, Chap. 13). Here we touch on the philosophical debate on "human nature" and how this notion can be reframed in an evolutionary framework, while such biologicized accounts of human nature should be approached with caution (Downes 2017; Machery 2017; Ramsey 2023).

Finally, reframing paleoanthropology as a form of anthropology opens new conceptual and methodological horizons. Such a reframing is mutually reinforcing: new frameworks generate new research programs, and these, in turn, further establish paleoanthropology's anthropological identity. For instance, the recent shift from focusing solely on Neanderthal extinction to exploring their extended history, lifeways, and cultural expressions moves the discourse away from viewing them as "other" or "inferior" toward treating their study as a "cultural anthropology of the past" (Zwart et al., Chap. 14).

1.5 The (Geo)Politics of Knowledge Production: Colonial Legacies and Global Inequalities

The critical perspective developed in this volume is also concerned with the concrete conditions under which the knowledge about human evolution is produced. This requires interrogating the historical contexts in which the field has developed, recognizing how colonial legacies continue to shape scientific practices, and examining the power structures that remain visible today, including the inequalities they produce between the global North and the global South. Haraway (1989) was a pioneer in showing how the issues of race and gender were involved in the scientific understanding of our evolutionary past. Since then, there has been a growing recognition of the urgent need for such critical examination (Liebmann and Rizvi 2008; Lydon and Rizvi 2012; Porr and Matthews 2017; Porr and Matthews 2019a; Ackermann et al. 2018, 2025), especially since paleoanthropology and archeology often "make a range of universalist claims about humanity's past, how 'we' became human, what it means to be human and so on" (Porr and Matthews 2019b, p. 4).

The study of human evolution has long been entangled with Western nationalism, racism, and coloniality (Porr and Matthews 2019a, 2019b; Liebmann and Rizvi 2008). However, these influences are not confined to the past but may still continue to shape current scientific practices, from the distribution of power in research to the curation and accessibility of fossil and archeological collections (Ackermann et al. 2018). While Africa and Asia have often been interpreted through Eurocentric frameworks (Athreya and Ackermann 2019), Menéndez and Yáñez (Chap. 15) show that Latin America remains marginalized in dominant narratives of human evolution. They also expose how Latin American scholarship is systematically silenced within the hegemonic scientific mainstream, revealing dynamics of exclusion that extend across research infrastructure, funding structures, and editorial and publishing policies.

Similarly, the uses of research on human evolution to serve national causes do not belong to the past only, but still need caution. While nationalism has shaped the early research on hominins (Hammond 1988), Chikhi et al. (Chap. 10) warn us that genomic and archeogenomic data may be used in the construction of narratives that may themselves be used to define national or ethnic identities. Therefore, it seems urgent to develop a more critical attitude towards their interpretation. As noted by Fuentes (2021), the challenge of racism persists particularly in the genomic era, with renewed temptations to link genes and traits in ways that risk resurrecting racial divisions within *Homo sapiens*.

Engaging critically with the discipline's past and its present challenges offers an opportunity to build a more inclusive field. As Menéndez and Yáñez (Chap. 15) stress, this requires a "multidisciplinary, multisituated, and polyphonic approach," fostering a form of scientific objectivity that is not an abstract, transcendent ideal but, as Helen Longino (1990) argues, the outcome of a plurality of perspectives sustained by critical dialogue within a scientific community. Such an approach invites us to reframe the field of human evolution as a situated knowledge (Chapman and Wylie 2016), a "view from somewhere" (Smith and Bolnick 2019), attentive to its historical, cultural, and political entanglements. This might help not only to define better practices within the field but also to reshape our understanding of what this discipline can tell us about ourselves as humans.

This is also the ambition of the present volume: to cultivate a reflexivity that opens new ways of thinking about how to build objectivity in a field that works with rare and fragmentary data, belongs both to evolutionary biology and to the social and human sciences, and constitutes a distinctive form of anthropology. In this spirit, the volume aims to advance a constructive critical perspective on the study of human evolution that strengthens scientific practice and encourages sustained engagement with the history and philosophy of science.

Acknowledgements This research benefited from the scientific framework of the University of Bordeaux's IdEx Investments for the Future program / GPR Human Past.

References

Ackermann RR, Athreya S, Black W, Humphreys R, Pickering R, Sahle Y, Schroeder L (2018) Paleoanthropology: decolonizing and building for the future. In: Callan H (ed) The international encyclopedia of anthropology. Wiley, Hoboken, pp 1–9

Ackermann RR, Molopyane K, Athreya S (2025) Unpacking the 'explorer' narrative and its impacts on African paleoanthropology. S. Afr. J. Sci. 121(1/2)

Athreya S, Ackermann RR (2019) Colonialism and narratives of human origins in Asia and Africa. In: Porr M, Matthews JM (eds) Interrogating human origins. Routledge, London, pp 72–95

Cazenave M, Pina M, Hammond A, Böhme M, Begun D, Spassov N, Wood B (2024) Postcranial evidence does not support habitual bipedalism in *Sahelanthropus tchadensis*: a reply to Daver et al. (2022). J Hum Evol 185:103557

Chapman R, Wylie A (2016) Evidential reasoning in archaeology. Bloomsbury Academic, London

Cleland C (2002) Methodological and epistemic differences between historical science and experimental science. Philos Sci 69(3):474–496

Currie A (2021) Stepping forwards by looking back: Underdetermination, epistemic scarcity and legacy data. Perspect Sci 29(1):104–132

Currie A, Killin A (2019) From things to thinking: cognitive archaeology. Mind Lang 34(2): 263–279

Currie A, Killin A, Lequin M, Meneganzin A, Pain R (2024) Past materials, past minds: the philosophy of cognitive paleoanthropology. Philos Compass 19(6):e13001

d'Errico F, Doyon L, Zilhão J, Baker J (2024) Tooth in the spotlight: exploring the integration of archaeological and genetic data to build multidisciplinary narratives of the past. Comptes Rendus Palevol 23(2):31–43

Delisle RG (2007) Debating humankind's place in nature, 1860–2000: the nature of paleoanthropology. Pearson Prentice Hall, Upper Saddle River

Delisle RG (2012) The disciplinary and epistemological structure of paleoanthropology: one hundred and fifty years of development. Hist Philos Life Sci 34(3):283–329

Delisle RG (2017) The deceptive search for "missing links" in human evolution, 1860–2010: do paleoanthropologists always work in the best interests of their discipline? In: Schwartz JH (ed) Rethinking human evolution. MIT Press, Cambridge, MA, pp 1–30

Downes SM (2017) Human nature: an overview. In: Huebner B, Machery E, Ramsey G (eds) The Routledge handbook of evolution and philosophy. Routledge, London, pp 155–166

Eldredge N, Gould SJ (1972) Punctuated equilibria: an alternative to phyletic gradualism. In: Schopf TJM (ed) Models in paleobiology. Freeman, Cooper & Co, San Francisco, pp 82–115

Foucault M (1970) The order of things: an archaeology of the human sciences. Tavistock Publications, London

Fuentes A (2021) Biological anthropology's critical engagement with genomics, evolution, race/racism, and ourselves: opportunities and challenges to making a difference in the academy and the world. Am J Phys Anthropol 175(2):326–338

Goodrum M (2023) Reconceiving paleoanthropology in the era of the modern evolutionary synthesis. PaleoAnthropology 2023(2):246–257

Hammond M (1988) The shadow man paradigm in paleoanthropology, 1911–1945. In: Stocking GW (ed) Bones, bodies and behavior: essays in behavioral anthropology. University of Wisconsin Press, Madison, pp 117–137

Haraway DJ (1989) Primate visions: gender, race, and nature in the world of modern science. Routledge, New York

Killin A, Pain R (2023) Cognitive archaeology and the minimum necessary competence problem. Biol Theory 18(4):269–283

Kissel M, Fuentes A (2021) The ripples of modernity: how we can extend paleoanthropology with the extended evolutionary synthesis. Evol Anthropol 30(1):84–98
Kuhn TS (1962) The structure of scientific revolutions. University of Chicago Press, Chicago
Lakatos I (1978) The methodology of scientific research programmes. Cambridge University Press, Cambridge
Leonelli S (2016) Data-centric biology: a philosophical study. University of Chicago Press, Chicago
Lequin M (2018) Paleoanthropology's uses of the bipedal criterion. Hist Philos Life Sci 40(1):7
Lequin M (2026) Interpreting variability, diversity, and alterity in the hominin fossil record. In: Abadia O, Porr M (eds) Alterity and human evolution. Berghahn, pp 209–228
Lequin M, Colard T, Colombo A, Le Cabec A, Remy F, Schuh A (2025) Investigating development in human evolution: specificities, challenges, and opportunities. Evol Anthropol Issues News Rev 34(1):e70001
Liebmann M, Rizvi UZ (2008) Archaeology and the postcolonial critique. Altamira Press, Lanham
Longino HE (1990) Science as social knowledge: values and objectivity in scientific inquiry. Princeton University Press, Princeton
Love AC, Wagner GP (2023) Telling stories about unseen ancestors. Curr Biol 33(16):R841–R842
Lydon J, Rizvi UZ (eds) (2012) Handbook of postcolonial archaeology. Left Coast Press, Walnut Creek
Machery E (2017) Human nature. In: Smith DL (ed) How biology shapes philosophy. Cambridge University Press, Cambridge, pp 204–226
Marks J (2015) Tales of the ex-apes: how we think about human evolution. University of California Press, Oakland
Meneganzin A, Bernardi M (2023) Were Neanderthals and *Homo sapiens* 'good species'? Quat Sci Rev 303:107975
Meneganzin A, Currie A (2022) Behavioural modernity, investigative disintegration and Rubicon expectation. Synthese 200(1):47
Popper KR (1959) The logic of scientific discovery. Basic Books, New York
Porr M, Matthews JM (2017) Post-colonialism, human origins and the paradox of modernity. Antiquity 91(358):1058–1068
Porr M, Matthews JM (2019a) Interrogating human origins. Routledge, London
Porr M, Matthews JM (2019b) Interrogating and decolonising the deep human past. In: Porr M, Matthews JM (eds) Interrogating human origins. Routledge, London, pp 3–32
Ramsey G (2023) Human nature. Cambridge University Press, Cambridge
Robinson C, Terhune CE (2017) Error in geometric morphometric data collection: combining data from multiple sources. Am J Phys Anthropol 164(1):62–75
Schwartz JH (2017) What's real about human evolution? Received wisdom, assumptions, and scenarios. In: Schwartz JH (ed) Rethinking human evolution. MIT Press, Cambridge, MA, pp 61–91
Smith RJ, Wood B (2017) The principles and practice of human evolution research: are we asking questions that can be answered? Comptes Rendus Palevol 16(5–6):670–679
Smith RWA, Bolnick DA (2019) Situating science: Doing biological anthropology as a view from somewhere. Am. Anthropol. 121: 465–467
Smocovitis VB (2012) Humanizing evolution: anthropology, the evolutionary synthesis, and the prehistory of biological anthropology, 1927–1962. Curr Anthropol 53(S5):S108–S125
Tattersall I (2012) Paleoanthropology and evolutionary theory. Hist Philos Life Sci 34(3):259–281
Villmoare B, Kimbel WH (2024) On the scientific credibility of paleoanthropology. Evol Anthropol Issues News Rev 33(4):e22037
Wood B, Smith RJ (2022) Towards a more realistic interpretation of the human fossil record. Quat Sci Rev 295:107722
Zilhão J, d'Errico F, Banks WE, Teyssandier N (2024) A data-driven paradigm shift for the Middle-to-Upper Palaeolithic transition and the Neandertal debate. Quaternary Environ Hum 2(6): 100037

Why It Is Necessary to Critically Rethink How (Scientific) Knowledge Is Constructed

2

Juan Manuel Rodríguez Caso

Abstract

This introduction reflects on the methodological and epistemological challenges inherent in constructing critical historical narratives of human evolution. It begins with the recognition of the complexity and controversies surrounding this field of study, in order to seek a reconsideration that a priori considers the contributions of different disciplines. The work advocates for developing new methodological frameworks that can effectively translate and incorporate diverse perspectives while addressing linguistic, cultural, and institutional barriers. This critical approach emphasises the urgent need to decolonise evolutionary studies by challenging the monopoly of Western scientific discourse and creating space for multiple epistemological frameworks that can provide more comprehensive understandings of human origins and their development.

Keywords

Critical thinking · Interdisciplinarity · Critical history · Epistemology · Evolution

2.1 Introduction

At the time of writing this introductory essay, it is the 100th anniversary of one of the most transcendent events in the history of the theory of evolution: the legal case *The State of Tennessee v. John Thomas Scopes*, popularly known as "Scopes trial" or "Scopes Monkey Trial". The significance of that event can be assessed in

J. M. Rodríguez Caso (✉)
División de Ciencias y Artes para el Diseño, Universidad Autónoma Metropolitana, Unidad Xochimilco, Mexico City, Mexico
e-mail: jmrodriguez@correo.xoc.uam.mx

M. Lequin, J. M. Rodríguez Caso (eds.), *The Field of Human Evolution: Critical Perspectives from History and Epistemology*, Evolutionary Biology – New Perspectives on Its Development 11, https://doi.org/10.1007/978-3-032-13925-2_2

several ways: despite its short duration, from 10–21 July 1925, the reasons and consequences of what was discussed there are more valid than ever (Larson 2006). Human evolution became a topic of discussion that transcended academic boundaries, and the public had the opportunity not only to observe the discussion but also to participate by sharing their opinions. An event that, for many, can be understood in local terms became, at the time, the first legal case to be broadcast nationally in the United States, in addition to being covered by newspapers in large cities. An event with a media impact that reached the theatre and the cinema, which has resulted, on the one hand, in the reaffirmation of the importance of science in society, but at the same time, in the omnipresence of intolerance. However, from that "local" event, we have today a movement that has crossed borders, to become a worldwide phenomenon: fundamentalist creationism.

Moreover, why starting with a reminder of an event that, for many people, goes against science and its "progress"? First, because the central theme of that discussion was not only evolution, but human evolution in particular. Furthermore, second, because even scientists today seem to take for granted that a quasi-perfect halo of objectivity and neutrality surrounds scientific knowledge. The latter is the main reason why it becomes increasingly necessary to think, rethink, and deconstruct knowledge. Moreover, that path must promote dialogue between the sciences and the humanities.

In fact, studying human evolution is a subject fraught with controversy. The debates that have arisen about the closeness of humans and other primates (Cavalieri and Singer 1993; McGrew 1996; de Waal 2002; Corbey 2005), frauds related to fossil discoveries such as the famous "Piltdown Man" (Booher 1986; Tobias et al. 1992; Gardiner 2003; Stringer 2012), or associations between anthropological practices as forms of legitimisation of racial controversies (Jackson and Weidman 2004; Reardon 2005; Ellison and Goodman 2006; Conklin 2013; Jackson and Depew 2017; Diogo 2024), are part of this. Although it is true that several problems can still be encountered when attempting to construct a critical historical reconstruction of human evolution, even today. Firstly, the fragmentary nature of the fossil record limits the possibilities for developing comprehensive alternative narratives, as paleoanthropological finds remain scarce and geographically concentrated in certain regions, particularly in eastern and southern Africa. This concentration of physical evidence reinforces the dominant interpretations established by Western scientific institutions, while areas with less archeological activity remain underrepresented in the global evolutionary discourse (Barr and Wood 2024; Monarrez et al. 2022).

The interdependence between the scientific disciplines involved in the study of human evolution generates a highly specialised knowledge structure that makes it difficult to question established paradigms. Genetics, paleoanthropology, archeology, and primatology have developed, at least in an ideal sense, mutually reinforcing theoretical and methodological frameworks. However, dialogue can often be confrontational, and data is integrated differently in each discipline, which has implications for the robustness of the evidence. Furthermore, dependence on institutional funding and pressure to publish in prestigious academic journals also encourage adherence to accepted interpretive frameworks, limiting the exploration of alternative approaches that could challenge hegemonic narratives.

Finally, the incorporation of traditional and indigenous knowledge presents particular challenges due to fundamental differences in epistemological systems and conceptions of time between Western scientific traditions and non-Western worldviews. The translation and integration of these alternative perspectives requires methodological frameworks that are still under development, while linguistic, cultural, and institutional barriers continue to limit access to this alternative knowledge for the predominantly Western academic community that dominates evolutionary studies.

2.2 Critical Thinking and Human Evolution

It is worth asking, then, what is a critical perspective? First and foremost, and without a formally established academic field with that name, it is a genealogy that we can intellectually trace back to the Frankfurt School, with the work of Max Horkheimer, Theodor Adorno, and Walter Benjamin, among others. From there, one way to understand critical history is through the inclusion of explanatory frameworks from sociology, philosophy, political science, and literary theory, all in the pursuit not simply of recounting the past, but of using these tools to establish a critique of the past in order to transform the present (Wiggershaus 1994).

In short, a critical history, unlike a "traditional" approach, establishes that all historical writings reflect the perspectives, biases, and interests of those who produce them. From the perspective proposed by Michael Foucault, a "traditional" narrative is a critique of the discourse of a dominant group that legitimises its authority while minimising and rendering invisible alternative proposals (Foucault 2009). What could be a critical view of human evolution?

A critical history of human evolution should challenge the prevailing narratives that have traditionally positioned Western scientific discourse as the primary authority on human origins, while marginalising indigenous knowledge systems and alternative interpretative frameworks. The conventional evolutionary narrative, constructed primarily through European and North American scientific institutions, reflects particular cultural biases about progress, hierarchy, and human exceptionalism that emerged from Enlightenment thinking and colonial contexts. This dominant discourse has historically privileged certain types of evidence, such as fossil records interpreted through Western taxonomic systems, while dismissing or appropriating indigenous oral histories and traditional ecological knowledge that offer different interpretations of human-environment relationships and time frames. Furthermore, the standard evolutionary narrative often reinforces problematic assumptions about technological advancement as synonymous with human progress, thereby legitimising contemporary power structures that position industrialised societies as more "evolved" than traditional communities.

Similarly, from a critical perspective, human evolutionary narratives must be examined to determine how they perpetuate racialised hierarchies and gender biases that reflect the social contexts of their production rather than objective scientific truths. The historical emphasis on "hunter-gatherer" theories and the prioritisation of tool-making as defining characteristics of humankind reveal androcentric assumptions that marginalise the role of women and community care practices in human development (Tanner and Zihlman 1976; Zihlman 1978; Gero and Conkey 1991). Furthermore, the geographical focus on African origins, while scientifically supported, has been interpreted through colonial frameworks that assert Africa as the cradle of humanity while portraying contemporary African societies as "primitive" or "less evolved" (Athreya and Ackermann 2019). A critical approach must question how evolutionary timelines and classifications serve to naturalise contemporary inequalities and, instead, emphasise the multiplicity of human adaptation strategies, the interconnection of biological and cultural evolution, and the need to decolonise scientific methodologies by incorporating diverse epistemological frameworks that challenge the monopoly of Western scientific discourse on the definition of human origins and development.

2.3 Outline of the Book

This volume is organised into four sections. The introductory essays are reflections in which each of the editors reflects on the two theoretical axes from which the case of human evolution is analysed: epistemology and critical history. Furthermore, it is a timely reminder of the increasingly urgent need to recognise the importance of the humanities in scientific practice. Why? The humanities provide scientists with tools to: see how knowledge is produced and limited; learn how concepts, methods, and theories have changed over time; assess the social and ethical consequences of scientific "advances"; and enable more effective communication and collaboration across disciplines.

The first section presents a series of historical reflections on the study of human evolution. The first chapter (Rodriguez Caso, Chap. 3) aims to expand the narrative surrounding authors who are often considered "canonical" in the context of human evolution. In this case, nineteenth-century evolutionary authors such as Jean Baptiste Lamarck and Alfred Russel Wallace stood out for their deep interest in explaining human beings and their changes over time. On the other hand, an author as "well-known" as Charles Darwin deserves a re-reading of the real impact of works such as *The Descent of Man*. Matthew Goodrum (Chap. 4) presents an extensive analysis of different traditions and geographies, highlighting the various approaches to paleoanthropology that were practised in its early years. This reflection significantly broadens the perspective on the methodologies and theories from which paleoanthropology emerged in the nineteenth century, in parallel with the discovery of new human fossils and subsequent theorising in different European countries. Next, Robin Derricourt (Chap. 5) presents us with the growing and diverse phylogeny that studies on human evolution themselves generated in the twentieth century, a period widely recognised by the numerous fossils that were found in various parts of the world, in which fossil evidence alone is not the only element to consider when reconstructing that phylogeny, but instead opens up a space to recognise the role played by the idiosyncrasies of each of the authors participating in that phylogeny. Jessica Mikels-Carrasco (Chap. 6) introduces us to a proposal that revolutionised anthropological studies in the twentieth century, based on Sherwood Washburn's "new physical anthropology". From this perspective, knowledge of human evolutionary history goes beyond fossils by highlighting different biological aspects—such as genes and behaviour—as complementary forms of knowledge, that is, a new epistemological approach. Finally, this section concludes with a way of understanding and visualising human evolution, proposed in the 1970s and now re-evaluated by Niles Eldredge, one of the original proponents, along with Telmo Pievani and Ian Tattersall (Chap. 7). The aim is a critical reconsideration of the scope of punctuated equilibrium in light of both new fossil evidence and other theoretical proposals, in the search for answers from a pluralistic approach.

The second section of the book presents different ways of interpreting and visualising data. One point, however, is that the same evidence can be presented, interpreted, visualised, or translated in different ways, depending on the discipline in question—such as archeology or genetics—and even on the scale. Furthermore, the scope of these interpretations may vary depending on the intended audience. Marianne Sommers' work (Chap. 8) presents a necessary reflection on the role that images play in understanding human evolution, and, above all, the way in which these images have been constructed historically. One point of criticism is the continued validity of nineteenth-century theories in light of the advances made in various aspects of human evolution, leading us to question the validity of the frameworks frequently used today.

On the other hand, Ivan Colagè and Francesco d'Errico (Chap. 9) revisit a classic discussion of the "Modern Synthesis", namely, the notion of "proximate causes" and "ultimate causes", to reflect on the current scope of that proposal, especially about archeology. Their proposal seeks to go beyond reductionism by proposing comprehensive explanatory frameworks in which culture and behaviour can be understood, both of which are fundamental parts of what it means to be human. The joint work of Lounès Chikhi, Jonathan Racine, Rémi Tournebize, and Arnaud Pocheville (Chap. 10) is a provocative piece that reminds us that data produced by scientific work, such as in this case the genetic and genomic evidence of the evolution of the genus *Homo*, are used as the basis for narratives of all kinds, such as those that define national or ethnic identities. It is a clear call to build bridges between scientists and humanists, in order to generate truly critical and informed attitudes towards information. Finally, Stephen Downes (Chap. 11) takes a comprehensive approach, based on the relationship between human behavioural genetics and human evolution, focusing on research into educational performance and height through genome-wide association studies (GWAS). Through comparative analysis of key concepts such as additivity, genetic load, and phenotype, the chapter demonstrates that behavioural geneticists adapt evolutionary concepts in ways that create conceptual tensions. Furthermore, this adaptation, combined with different methodological approaches and theoretical assumptions, undermines the potential of GWAS-based genetics to make meaningful contributions to our understanding of human evolution.

The third section discusses two key concepts for human evolution from different perspectives: diversity and otherness. Thus, the aim is to understand an aspect that, especially today, we cannot overlook: the close relationship—often poorly understood or frankly misinterpreted—between science and politics. First, Jonathan Marks (Chap. 12) presents a critical analysis of hominid taxonomy in paleoanthropology, highlighting the persistent confusion surrounding the classification of extinct human species due to a fundamental misunderstanding of what these taxonomic units represent. Marks argues that hominid species are not biological entities comparable to other zoological species but rather function as "sacred ancestors" in scientific narratives about origins. Drawing parallels with the historical abandonment of racial taxonomy in physical anthropology, Marks suggests that paleoanthropological species, such as *Homo heidelbergensis* and *Homo naledi*, among others, serve as culturally constructed placeholders in stories about human ancestry, rather than representing discrete biological lineages. All this suggests that the search for a definitive biological classification of human ancestors may be inherently flawed, as these taxonomic categories primarily function to organise symbolic narratives about human origins rather than describe actual biological diversity. For his part, Agustín Fuentes (Chap. 13) examines the problematic tendency to derive prescriptive statements about how human beings "should" behave based on interpretations of human evolutionary history. Fuentes argues that many widely accepted statements about "human nature" fundamentally misrepresent the archeological and paleoanthropological evidence from the Pleistocene period. He bases this on two examples: the assumption that humans developed an inherent xenophobia towards outsiders, and the belief that the nuclear family represents the natural and normative human social structure. Fuentes thus argues that human evolution facilitated multiple successful forms of being human, rather than prescribing a single "natural" model. The chapter emphasises that contemporary human diversity reflects this evolutionary heritage of adaptability and warns against the use of selective interpretations of the past to justify social inequalities or current political positions.

The next chapter, an interdisciplinary study by Hub Zwart, Karel Kuipers, Susan Peeters, and Marie Soressi (Chap. 14), examines the evolution of understanding of interactions between Neanderthals and modern humans through the prism of philosophy, archeology, and literature. The authors argue that traditional research has been dominated by a binary "them versus us" paradigm, which portrays Neanderthals as inferior beings whom superior modern humans inevitably displaced during a brief and violent encounter. However, contemporary paleoarcheological evidence reveals a very different picture: Neanderthals and modern humans coexisted in Europe for thousands of years, during which time there was extensive cultural exchange, technological innovation, and genetic interbreeding. The Châtelperronian period demonstrates that Neanderthals possessed sophisticated capacities for creativity and adaptation, challenging assumptions about the exceptional nature of modern humans. This change is reflected not only in scientific discourse, but also in literary representations, from Jack London's early twentieth-century description of primitive cave dwellers conquered by the advanced "people of the fire," to more recent novels that emphasise cooperation, mutual learning, and shared humanity. To conclude this section, we have the manuscript by Lumila Menéndez and Bernardo Yañez (Chap. 15), which presents a critical analysis of the structural biases existing in research on the settlement of the American continent, arguing that this field is dominated by the perspectives of the so-called "Global North," while the contributions of Latin American academics are systematically marginalised. Three main sources of bias are identified: disparities in infrastructure that favour institutions in the "Global North" with greater resources, funding mechanisms that concentrate resources among established researchers in central countries, and editorial practices that privilege publications in English in high-impact journals. These asymmetries create a "vicious circle" in which researchers from the "Global North" maintain their visibility and influence, while academics from the "Global South" face persistent barriers to recognition, despite conducting rigorous and contextualised research. The article advocates for the adoption of a decolonial, multidisciplinary, multisited, and polyphonic approach to address these inequalities. Drawing on the framework of "world anthropologies", the authors argue that meaningful progress requires structural changes, such as more inclusive funding mechanisms, editorial policies that prioritise linguistic and geographical diversity, and collaborative research frameworks that foster balanced partnerships rather than perpetuating extractive practices.

This volume demonstrates the critical importance of integrating humanistic perspectives into the study of human evolution, revealing how disciplinary boundaries have often obscured rather than illuminated our understanding of humanity's deep past. The collected contributions establish that evolutionary narratives are not merely scientific accounts but complex constructions shaped by epistemological frameworks, cultural contexts, and political imperatives. From the reappraisal of canonical figures like Darwin and Wallace to the examination of contemporary genomic interpretations, the evidence presented consistently demonstrates that scientific knowledge about human evolution cannot be divorced from the social, cultural, and ideological contexts in which it is produced. The interdisciplinary approach advocated throughout these chapters offers a more nuanced understanding of how concepts such as diversity, otherness, and human nature have been constructed and contested across different historical periods and geographical contexts.

Moving forward, this work opens several crucial avenues for future inquiry that demand urgent attention from both scientific and humanistic communities. How might we develop more inclusive methodological frameworks that genuinely incorporate perspectives from the so-called "Global South", moving beyond the extractive research practices that currently dominate paleoanthropological and archeological investigations? What new collaborative models could emerge to address the structural inequalities that continue to marginalise non-Western voices in evolutionary discourse? Furthermore, as genomic technologies advance and generate increasingly complex datasets, how can we ensure that these powerful tools serve to illuminate rather than obscure the full spectrum of human diversity and experience? The challenge ahead lies not merely in generating new knowledge but in fundamentally reconceptualising how we approach questions of human origins, ensuring that future research serves the cause of human understanding rather than reinforcing existing hierarchies of knowledge and power.

References

Athreya S, Ackermann RR (2019) Colonialism and narratives of human origins in Asia and Africa. In: Porr M, Matthews JM (eds) Interrogating human origins. Routledge, London, pp 72–95

Barr WA, Wood B (2024) Spatial sampling bias influences our understanding of early hominin evolution in eastern Africa. Nat Ecol Evol 8:2113–2120. https://doi.org/10.1038/s41559-024-02522-5

Booher HR (1986) Science fraud at Piltdown: the amateur and the priest. Antioch Rev 44:389–407

Cavalieri P, Singer P (1993) The great ape project: equality beyond humanity. St. Martin's Press, New York

Conklin AL (2013) In the Museum of man: race, anthropology, and empire in France, 1850–1950. Cornell University Press, Ithaca/London

Corbey R (2005) The metaphysics of apes: negotiating the animal-human boundary. Cambridge University Press, Cambridge

de Waal F (ed) (2002) Tree of origin: what primate behavior can tell us about human social evolution. Harvard University Press, Cambridge/London

Diogo R (2024) Darwin's racism, sexism, and idolization: their tragic societal and scientific repercussions. Springer, Cham

Ellison G, Goodman AH (2006) The nature of difference: science, society and human biology. CRC Press, Boca Raton/London/New York

Foucault M (2009) Archaeology of knowledge., 1st publ., repr. Routledge, London/New York, NY

Gardiner BG (2003) The Piltdown forgery: a re-statement of the case against Hinton. Zool J Linnean Soc 139:315–335

Gero JM, Conkey MW (eds) (1991) Engendering archaeology: women and prehistory. Basil Blackwell, Oxford/Cambridge

Jackson JP, Depew DJ (2017) Darwinism, democracy, and race: American anthropology and evolutionary biology in the twentieth century. Routledge, London/New York

Jackson JP, Weidman NM (2004) Race, racism, and science: social impact and interaction. ABC-CLIO

Larson EJ (2006) Summer for the gods: the scopes trial and America's continuing debate over science and religion. Basic Books

McGrew WC (1996) Chimpanzee material culture: implications for human evolution. Cambridge University Press, Cambridge

Monarrez PM, Zimmt JB, Clement AM et al (2022) Our past creates our present: a brief overview of racism and colonialism in Western paleontology. Paleobiology 48:173–185. https://doi.org/10.1017/pab.2021.28

Reardon J (2005) Race to the finish: identity and governance in an age of genomics. Princeton University Press, Princeton/Oxford

Stringer C (2012) The 100-year mystery of Piltdown man. Nature 492:177–179

Tanner N, Zihlman A (1976) Women in evolution. Part I: innovation and selection in human origins. Signs J Women Cult Soc 1:585–608. https://doi.org/10.1086/493245

Tobias PV, Bowler PJ, Chamberlain AT et al (1992) Piltdown: an appraisal of the case against sir Arthur Keith [and comments and reply]. Curr Anthropol 33:243–293

Wiggershaus R (1994) The Frankfurt school: its history, theories, and political significance. Polity Press, Cambridge

Zihlman AL (1978) Women in evolution, part II: subsistence and social organization among early hominids. Signs J Women Cult Soc 4:4–20. https://doi.org/10.1086/493566

Part II

Historical Perspectives on the Study of Human Evolution

3 Historical Reflections on Human Evolution in the Nineteenth Century: Beyond Darwin and 'Darwinism'

Juan Manuel Rodríguez Caso

Abstract

Charles Darwin's contribution to the history of human evolution is at the centre of the narrative about the evolutionary theory. It is recurrently asserted that from *The Descent of Man* (1871) onwards, a naturalistic view of the development of human beings became the centrepiece of evolutionary anthropological explanations. In general terms, 'Darwinism' is the theory by which human evolution is explained even in the twentieth century and beyond. However, history is always more complex. This chapter aims to provide a broader view of discussions on human evolution in the nineteenth century by drawing on examples of authors and discussions that understood human evolution from perspectives different from Darwin's. The first point to be discussed will be the case of Darwin himself, above all to rethink what Darwin contributed with his reflections on the human being beyond the continuous presentism surrounding his ideas. A second point will serve to present two other cases that exemplify nineteenth-century discussions on human evolution that have either been ignored or are continually downplayed: Jean Baptiste Lamarck's poorly analysed approach in *Système analytique des connaissances positives de l'homme* (1820), and Alfred R. Wallace's anthropological proposal, developed mainly in the 1860s.

Keywords

Critical history · Mythology · Anthropology · Evolution · Biology

J. M. Rodríguez Caso (✉)
División de Ciencias y Artes para el Diseño, Universidad Autónoma Metropolitana, Unidad Xochimilco, Mexico City, Mexico
e-mail: jmrodriguez@correo.xoc.uam.mx

M. Lequin, J. M. Rodríguez Caso (eds.), *The Field of Human Evolution: Critical Perspectives from History and Epistemology*, Evolutionary Biology – New Perspectives on Its Development 11, https://doi.org/10.1007/978-3-032-13925-2_3

3.1 Introduction

As in any discipline or field of knowledge, there are canonical histories. When talking about human evolution, the first thing that comes to mind is anthropology, which is the discipline that focuses explicitly on the study of humans. However, the history of anthropology is not necessarily the disciplinary history of the study of the human. The discipline, like so many others within the scientific endeavour, emerged in different geographical contexts throughout the nineteenth century. The study of human beings, on the other hand, in terms of a research tradition, can be traced back to works such as *A Treatise of Human Nature* (1739–1740) by the Scottish philosopher David Hume, or *Histoire Naturelle, générale et particulière, avec la description du Cabinet du Roi* (1749–1804) by the French naturalist Georges-Louis Leclerc, comte de Buffon. In these two cases, we can appreciate the emergence of a naturalistic perspective to explain the origin of humanity. Also, Buffon's particular case has come to be regarded, erroneously, as a clear forerunner of an evolutionary view of humanity.

As will be seen throughout this chapter, the study of the human, including its evolution, precedes the emergence and consolidation of a specific field of knowledge. From history, it is essential to understand how, regardless of the field of knowledge to which we refer, the study of human evolution has been based on different approaches. Furthermore, beyond the vast diversity of interests found in evolutionary authors, one of the themes they all addressed was that of the human being. It is also important to understand that these interests ranged from the origin of humanity, the diversification of human groups—a subject that is recurrently related to 'races'—and the structural changes in the human body—associated since nineteenth century with ontogenetic processes—the origin and development of cognitive capacities, and culture as a fundamental aspect of what is human, among others.

In historiographical terms, as is frequently appreciated, it is vital to provide a critical approach to historiographical categories, which for various reasons have been consolidated—to the point of being considered 'basic' in the understanding of historical figures and events—as part of a canonical discourse. In the case of this chapter, one of these categories is the so-called 'Darwinian revolution', based on the consideration that a biological phenomenon such as evolution can only be understood based on one theory, Darwinism. In this sense, there are at least two problems that arise with the dogmatic defence of this position: first, to speak of an -ism refers us to a doctrine, or in the best of cases, a school of thought, which forces us to consider that the idea of evolution is subject not only to the scientific evidence of the moment but to a series of epistemic and ontological commitments—which are often not explicit—that border on dogmatism. Moreover, a second problem is that, as James Moore (1991) pointed out that speaking of a single Darwinism has no historical relevance since this claim ignores elements such as interpretation, the circulation of ideas, the construction of narratives, or simply understanding scientific practice as the result of a social and cultural process, in which individuals, even within the same epistemic community, may have different views on the same subject. Especially in historical terms, it is more appropriate to speak of the concrete ideas presented by any author in a particular period or work. That is, we can speak of the ideas presented by Charles Darwin in *Origin*, published in 1859 or the ideas of Alfred R. Wallace in his book *Darwinism* (1889) as examples of different Darwinisms. However, the situation becomes much more complex if we consider how changeable a term can be: one of the first uses of Darwinism was to describe the ideas of Erasmus Darwin, or the defence by Thomas H. Huxley of his perspective on evolution, or even Charles Lyell—who publicly rejected evolution, especially of the human being—or John Stevens Henslow—who defended 'true Darwinism' as an adaptation to the direct impact of the environment.

One of the aims of this work is to remind us that scientific knowledge, as in this case, the study of human evolution, must be subject to continuous revision, both from epistemology and history. For this reason, we take up authors who are often the subject of numerous mythologies, whose ultimate aim always seems to be to defend Darwin's ideas or at least a particular interpretation of them. The idea of myth is taken up here from proposals such as *Darwin Mythology: Debunking Myths, Correcting Falsehoods* (2024), edited by historian Kostas Kampourakis. Mythologies can be those which seek either to overemphasise some element of truth, or to present politically or ideologically charged visions of historical events.

Three authors are discussed in this chapter: Charles R. Darwin (1809–1882), Alfred R. Wallace (1823–1913), and Jean Baptiste Lamarck (1744–1829). The choice of these authors aims to broaden the discussion on how human evolution was conceived throughout the nineteenth century, especially by obviating an idea that might seem evident to many authors: human evolution has been the subject of study and reflection from different areas of knowledge. For example, in the specific case of Darwin and Wallace, it can be said that, regardless of their different academic interests, humans played a central role in developing their respective proposals on natural selection (Rodríguez Caso et al. 2012). Another point to note about the chosen authors is that they were first and foremost evolutionists, transformationists, or transmutationists, depending on the tradition they ascribe to. Humans became the focus of their reflections at different points in their respective careers.

The cases presented in this chapter seek to contribute in two ways: the necessary re-reading—and consequent reinterpretation—of classic authors such as Charles Darwin, and the also necessary recovery of authors usually mythologised, such as Alfred R. Wallace and Jean Baptiste Lamarck. Today, more than ever, there is a need for a critical view of the history of science, that must be part of a continuous reflective and analytical approach not only to scientific discoveries and advances but also to the social, political and cultural context in which they developed.

3.2 Charles Darwin and the Study of Human Species: Re-evaluating *The Descent of Man*

Beginning this chapter's reflections with Darwin is not an arbitrary decision. Charles Darwin is a ubiquitous author in the historical reconstructions of almost any account of nineteenth-century evolutionary ideas. Furthermore, that is precisely the problem to be analysed here.

The idea that Darwin's theory—on the basis that Darwin proposed a general and broad theory of evolution—included human species from his earliest reflections has been passionately debated. As part of the conclusions he draws in his seminal work, a phrase has been quoted profusely to indicate this interest: 'Light will be thrown on the origin of man and his history' (Darwin 1859, p. 488). However, a lively discussion has arisen as to whether, even in the first edition of the *Origin*, Darwin took up the human case—for example, culture, considered a unique feature of the species at the time—to speak broadly about his theory. It is a fact that Darwin did not start from scratch when he publicly set out his ideas about human in *Descent*. We need to remember the comments and notes he made in his notebooks and correspondence about enslaved people in Brazil or about the inhabitants of Tierra del Fuego who were taken back to their homeland after being 'civilised' (Rodríguez Caso et al. 2012). All these claims served as the basis for what many years later would be the specific work to set out his vision of the human being, *The Descent of Man, and Selection in Relation to Sex* (1871), followed by another work, *The Expression of the Emotions in Man and Animals* (1872), which has received little attention from historians of science.

A recurring temptation is to see *The Descent* as a continuation of a grand 'evolutionary' project in which Darwin sought to provide a comprehensive answer, including the human being and all his characteristics. However, this leads to an oversimplification of Darwin's intellectual work. As simple as it may seem, it is enough to review the chronology of his works to appreciate that Darwin should be valued in a broader sense. Furthermore, despite works that have presented him as 'distanced' from evolution, such as his interest in geology and botany, it is expected to think of Darwin as an author almost obsessed with evolution.

A preliminary consideration is the impact of Darwin's work on humans to date. Recently, we have seen the publication of works such as *Sex, Gender, Ethics and the Darwinian Evolution of Mankind: 150 Years of Darwin's 'Descent of Man'* (2024), edited by Michel Veuille. This collection of essays has the particularity of rethinking Darwin's work on the human being, not from the 'classic' themes such as origin or races, but from a proposal that until *Descent* had played a secondary role in Darwin's work: sexual selection. From a historical point of view, Evelleen Richards took up the impact of sexual selection in *Darwin and the Making of Sexual Selection* (2017) to point out the strategic importance of this proposal within the broad context of Darwin's evolutionary proposal, since it explained aspects that natural selection did not—that is, sexual selection—in addition to maintaining a naturalistic proposal on beauty. Another interesting work published recently is *A Most Interesting Problem: What Darwin's Descent of Man Got Right and Wrong about Human*

Evolution (2021), edited by Jeremy DeSilva. Their aim, in their own words, 'is not so much a celebration of Darwin as it is a tribute to how science operates, how scientific ideas are tested, and the role of evidence in helping structure narratives of human origins' (DeSilva 2021, p. xix). Unlike other authors, such as Lamarck or Wallace, this selection of works allows us to see the lasting impact of Darwin's work, both in a more reflective historical sense and from presentist interpretations. The introduction by Darwin's renowned biographer Janet Browne provides a superb account of the context of *Descent*, not least in highlighting not only the work's place in Darwin's thinking but as the culmination of a scholarly life around the human species that had begun with the voyage of the Beagle (1831–1836) (Browne in DeSilva 2021). Another work that has undoubtedly marked the discussion of Darwin and his conception of human evolution is *Darwin's Sacred Cause* (2009) by Darwin's widely recognised biographers, Adrian Desmond and James Moore. It is a work that recognises the importance of the discussions on slavery and its impact on the understanding of the origins of human 'races' and the contribution of that discussion to Darwin's construction of his theory of natural selection.

These works celebrate the anniversary of the publication, or its impact, albeit somewhat in the logic already imposed by numerous historians and scientists with *Origin* (1859) to justify the 'Darwinian revolution'—as a milestone in scientific thought, and from there, to take up in the first instance—sometimes, the only one—Darwin as the reference for discussion of basically any subject. Numerous historians and scientists have used Darwin's work to justify the 'Darwinian revolution' as a milestone in scientific thought. It has also been used as a reference for discussing a wide range of subjects.

In today's terms, it is possible that assessing a book—and the ideas presented therein—such as *Descent* would have to be much more careful, if not honest. Like any man of his time, Darwin has expressions that inevitably today can be considered erroneous or are nevertheless used with unedifying intentions. As Agustín Fuentes points out:

> Reflecting on 'Descent' today one can look to data demonstrating unequivocally that race is not a valid description of human biological variation, that there is no biological coherence to 'male' and 'female' brains or any simplicity in biological patterns related to gender and sex, and that 'survival of the fittest' does not accurately represent the dynamics of evolutionary processes. (Fuentes 2021)

The *Descent* is a book that has features that are seldom emphasised. It is the first book in which Darwin explicitly uses the word 'evolution' on page 2 of the first volume of the first edition, that is, in 1871. Shortly afterwards, with the appearance of the sixth edition of the *Origin*, he would resume its use. A little-known situation is that it was a work in which his daughter Henrietta participated in reading and proofreading.

Importantly, it is crucial to recognise the prior context that existed before Darwin finally decided to speak publicly about 'man', for in the 1860s there were numerous publications that contributed to the heated discussion: *Geological Evidences of the*

Antiquity of Man (1863) by Charles Lyell; *Evidence as to Man's Place in Nature* (1863), by Thomas H. Huxley; *Pre-Historic Times, As Illustrated by Ancient Remains, and the Manners and Customs of Modern Savages* (1865); and *The Origin of Civilisation and the Primitive Condition of Man* (1865), and *The Origin of Man* (1870), both by John Lubbock; and *Researches into the Early History of Mankind and the Development of Civilization* (1865), by Edward B. Tylor. This selection of authors shows, among other things, the interest of British naturalists in human evolution prior to the publication of Darwin. This interest leads us to reconsider, for example, a recurrent claim of the 'Darwinian revolution', namely Darwin's 'fear of criticism', if he dared to view humans as a material product of nature.

In addition, as will also be seen in Wallace's case, at the institutional level, organisations such as the Ethnological Society of London and the Anthropological Society of London were consolidated, in which interest in discussing human species was publicly promoted, or the less recognised case of the negotiations that led to the formal opening of a department dedicated to anthropology in 1866 within the annual meetings of the British Association for the Advancement of Science. In short, Darwin's work came at a time when the subject of humans was more than present in the scientific community. Moreover, beyond the claims that place the book as a watershed in the naturalistic study of human species, Darwin himself makes it clear from the outset that:

> This work contains hardly any original facts in regard to man; but as the conclusions at which I arrived, after drawing up a rough draft, appeared to me interesting, I thought that they might interest others. (Darwin 1871, p. 3)

Does this mean that it is an irrelevant or uninteresting work? No. Instead, it is a recognition—possibly belated in some cases—of the work that preceded it:

> The conclusion that man is the co-descendant with other species of some ancient, lower, and extinct form, is not in any degree new. Lamarck long ago came to this conclusion, which has lately been maintained by several eminent naturalists and philosophers; for instance by Wallace, Huxley, Lyell, Vogt, Lubbock, Büchner, Rolle, &c., and especially by Häckel. (Darwin 1871, p. 3-4)

In terms of structure and argument, Darwin poses three questions to be answered throughout the text:

> The sole object of this work is to consider, firstly, whether man, like every other species, is descended from some pre-existing form; secondly, the manner of his development; and thirdly, the value of the differences between the so-called races of man. As I shall confine myself to these points, it will not be necessary to describe in detail the differences between the several races—an enormous subject which has been fully discussed in many valuable works. (Darwin 1871, p. 2–3)

One question that may be of interest is how Darwin organised the book. Much has been said about the goodness and scope of the 'one long argument' in *Origin*, but *Descent* seems to follow a different logic. It may even seem that we are dealing with two books that differ in their approach, although they have the human being as a common thread. Wallace was among the first to point this out since, at least initially, these topics were not linked. Wallace himself treated sexual selection and the origin of man as separate issues. Even so, it is worth asking why sexual selection has such a prominent place in *Descent*. Furthermore, as Michael Ruse (2015) points out, the answer is simple because it was the way to challenge Wallace's approach.

As Gregory Radick (2013) points out, it is at least three sets of evidence that led Darwin finally to publish a human-centred work: the discussions about the continuity between humans and animals and their evolutionary relationship; the attempts of authors such as William Gregg and Francis Galton to set out the conditions for encouraging both moral and intellectual progress in civilisation; and the growing disagreements with Alfred R. Wallace about the scope of natural selection and sexual selection. On this last point, it is interesting to recall that Wallace was also a significant motivation for encouraging Darwin to go public in the case of *Origin*. Darwin was even aware of these disagreements with Wallace when, shortly before *Descent's* publication, knowing that Wallace would review the work for *Academy*, he said, 'Do not swear at me more than you can help'. This evidence complements what James Moore and Adrian Desmond, in the introduction published in 2004 as part of the reprinting of *Descent*, point out that three fundamental themes allow us to understand its apparent contradictions: race, slavery and sex.

However, as with *Origin*, in *Descent* we can find some subtle differences between the editions since it is worth remembering that in addition to the original 1871 edition in two volumes, a second edition was published in a single volume in 1874, which was subsequently revised and enlarged in 1877 (Darwin 1877. See also Darwin 2004). In Darwin's own words, and of great relevance given the interpretations that have arisen within the 'Darwinian Revolution' that place him as a staunch defender of natural selection, we find the following clarification in the preface of the second edition of *Descent*:

> I may take this opportunity of remarking that my critics frequently assume that I attribute all changes of corporeal structure and mental power exclusively to the natural selection of such variations as are often called spontaneous; whereas, even in the first edition of the 'Origin of Species,' I distinctly stated that great weight must be attributed to the inherited effects of use and disuse, with respect both to the body and mind. I also attributed some amount of modification to the direct and prolonged action of changed conditions of life. (Darwin 1874, p. v)

On the impact of *Descent*, Janet Browne puts it on a par with *Origin* as a memorable book which responded in a timely way to the controversies surrounding the origin and diversity of humanity, which presents the idea of evolution in a 'modern' sense, and which has the power to explain the natural world (Browne in DeSilva 2021). Without detracting from the importance of the above, it is worth looking at Alvar Ellegård's classic work on the reception of Darwin's ideas. As mentioned above, the 1860s can be seen as the 'Darwinian decade' and as a period in which the human was central to naturalistic discussions. Moreover, despite this context, '[i] t did not cause the sensation the Origin produced twelve years earlier. The public was already familiar with the argumentation on the subject. In fact, one detects a slight tone of disappointment in many reviews of the book' (Ellegård 1990, p. 296).

Considering the observation above, it is possible to reflect that, although Darwin had made numerous observations and reflections on human species, their origin and characteristics throughout his life as a naturalist, *Descent* is his public presentation of the human being at a time when his recognition was wide enough to ensure that a potentially controversial publication would have sufficient impact among both the public and academia.

3.3 The Physical and Moral Transformation of Man, According to Jean Baptiste Lamarck

Lamarck is an author whose place in the history of science has been permanently associated with Darwin. One consequence of this association has been, at least in the Anglo-Saxon sphere, to simplify Lamarck to his transformist work, to the publication of *Philosophie zoologique* (1809), and above all, to denigrate him as an author who put forward a 'wrong' vision.

However, Lamarck produced works that have gone almost unnoticed, at least in the history of science. The book in which he undertook to examine humanity from both physical and moral perspectives was *Système analytique des connaissances positives de l'homme, restreintes à celles qui proviennent directement ou indirectement de l'observation* (1820) (see Corsi 1988, p. 205). One of these exceptions is the work of Léon Szyfman (1982), who argues that Lamarck began to consider anthropogenesis within his nascent transformist vision as part of his sometimes little-appreciated works, such as *Hydrogéologie* (1802), when he relates the changes in the Earth's surface to the development and perfection of the bodies of the organisms that inhabit it, including human species. It was in *Philosophie zoologique* (1809) that Lamarck developed his ideas on the origin of humankind. Although he does not delve into a possible monogenist explanation, he does take up a description that suggests the diversity of human groups, varieties of which, in his own words: 'Les naturalistes qui ont considéré l'homme seulement sous le -rapport de l'organisation, en ont formé avec ses six variétés connues, un genre particulier, constituant lui seul une famille à part [...]' (Lamarck 1809, p. 347).[1]

Another relevant work in a historiographical sense is that of Richard Delisle, who points out that Lamarck is not only the first author to propose a coherent and complete evolutionary theory but also the first to apply it fully to humanity (Delisle 2007). In fact, in a similar vein to Szyfman, Delisle proposes that *Système analytique* is the culmination of a continuous reflection that Lamarck began in *Recherches sur l'organisation des corps vivants* (1802), continued with *Philosophie zoologique*

[1]English translation: 'Naturalists who have considered man only from the point of view of organisation, have formed a special genus with its six known varieties, constituting a family in its own right'

(1809), and closed in 1820. In fact, and along the same lines as Delisle, it must be acknowledged that Lamarck's proposal on humanity and evolution is surprisingly modern. His reflection begins with the origin of humans from primates, and Lamarck considers that sophisticated primates—such as chimpanzees—may have gradually changed into a human form, and all this as part of a series of anatomical changes, which will even have cultural implications:

> The replacement of the grasping foot useful in climbing by a supporting foot more stable in walking; the reconfiguration of the leg shape to permit a bipedal gait; the loss of the projecting snout and the large frontal teeth which were no longer required to either fight or tear food apart because of the freed hands. This last feature, in turn, permitted the use of tools. (Delisle 2007)

In this anthropological sense, *Système* has even been appreciated as one of the first antecedents that sought to address the process of hominisation (Stoczkowski 2002). It is striking that one of the criteria used by Stoczkowski to consider what would have been relevant works is the passage from what he calls *histoire raisonnée* to formal anthropology. This claim makes sense if we remember that *Système* focuses on themes such as human intelligence, reason, and imagination, all considered unique to human species. It is also an alternative reading, either to the history of anthropology or to the interpretations of Lamarck's work and figure, which can be seen as a novel approach.

Another aspect of *Système* that has recently been appreciated is its importance in terms of nature classification (Ragan 2023). Based on Ragan's analysis, the following table summarises the criteria used by Lamarck (Table 3.1):

According to Gabriel Gohau (2006), *Système analytique* is the last synthesis for Lamarck. It is a book he would have written in parallel to the completion of his great taxonomic work, *Histoire naturelle des animaux sans vertèbres*. The work is divided into two parts: 'Des objets que l'homme peut considérer hors de lui, et que l'observation peut lui faire connaître',[2] and 'De l'homme et de certains systèmes organiques: observés en lui, les quels concourent à l'exécution de ses actions'.[3]

As part of the development of his anthropological ideas, we return here to his definition of the human being:

> L'homme, véritable produit de la nature, terme absolu de tout ce qu'elle a pu faire exister de plus éminent sur notre globe, est un corps vivant qui fait partie du règne animal, appartient à la classe des mammifères, et tient par ses rapports aux quadrumanes dont il est distingué par diverses modifications, tant dans sa taille, sa forme, sa stature, que dans son organisation intérieure; modifications qu'il doit aux habitudes qu'il a prises et à sa supériorité qui l'a rendu dominant sur tous les êtres de ce globe, et lui a permis de s'y multiplier, de s'y

[2]English translation: 'Objects which man can consider outside himself, and which observation can make known to him'

[3]English translation: 'Man and certain organic systems: observed in him, which contribute to the execution of his actions'

Table 3.1 Characteristics used by Lamarck to classify living beings in *Système*

Features distinguishing living from inorganic bodies	Plants with non-irritable bodies	Animals with irritable bodies
Composition from molecules resulting in an individual	Inability to move suddenly or iteratively	Instant, iterative contractility
Concrete body parts containing fluids	Inability to locomote	Locomotion
Internal vital motion	Inability of solid parts to move in response to external stimuli (although their fluids can so respond)	Repeated movement in response to repeated excitation
Order and state conducive to vital motion	Absence of specialised internal organs	Motion not entirely determined by its cause
Regeneration and ageing	Absence of internal digestion	Both solid and fluid parts participate in vital motion
Acquisition and transformation of food	Inability to circulate fluids, apart from movement of sap	Digestion of composed solids
Ggrowth and development	Growth both upward (shoot) and downward (roots)	Great range of organisation, including some with specialised organs
Reproduction from like	Tendency for upward growth perpendicular to the horizon	Some only irritable, others also sensitive
Common faculties	The body is usually formed of individuals, often with annual generations	No tendency to develop horizontally
A discrete lifespan		

répandre partout, et d'y comprimer la multiplication de celles des autres races d'animaux qui auraient pu lui disputer l'empire de la force. (Lamarck 1820, p. 149)[4]

Lamarck conceived human species as a product of evolution, but above all, as part of a development understood as continuous progress, starting from the development of the body to the consolidation of his place on the planet through intellectual and cultural development. As Szyfman (1982) points out, three characteristics would have given human species superiority over other animals: speech, languages and idioms ('la parole, les langues et les idiomes'). These characteristics result from

[4]English translation: 'Man, a true product of nature, the absolute culmination of all the most eminent things that nature has been able to bring into existence on our globe, is a living body that is part of the animal kingdom, belongs to the class of mammals, and is related to the quadrumans, from which he is distinguished by various modifications, both in his size, shape and stature, and in his internal organisation; modifications which it owes to the habits it has adopted and to its superiority which has made it dominant over all the beings of this globe, and has allowed it to multiply, to spread everywhere, and to compress the multiplication of those other races of animals which could have competed with it for the empire of the forest'

perfecting the senses, and these processes will result in the human being as the master of nature.

In the context of historical analysis, it is essential to consider the historical and cultural context of the work itself. It is also noteworthy that Lamarck did not write the work in the conventional sense; instead, he dictated it to his daughter Cornelie, who helped him write the text, as he had lost his sight by that time. The book was written from articles that Lamarck prepared for the second edition of *Dictionnaire d'histoire naturelle* by the publisher Jean-François Déterville (1766–1842), published in 36 volumes between 1816 and 1819. Lamarck contributed between 1817 and 1818 to volumes 7–21. It is speculated that since Lamarck became blind from July 1818 onwards, he no longer contributed any new writings (Gohau 2006). However, *Système* was well received as an extension of Lamarck's transformist explanation, but above all, reviews such as that of the physician Nicolas-Philibert Adelon (1782–1862) stress the importance of explaining the phenomenon of *sentiment intérieur* as the most important moment in the history of both man and animals (Adelon 1821, p. 266).

In the first section, Lamarck will present a description of the matter, and the initial statement is surprising, especially if we consider that there has been much discussion of his defence of materialism as a fundamental basis for his materialistic explanations. When he describes nature, and in this sense, the place of man, he emphasises that power which gives order to everything, and which leaves man as the most important product of the process:

> La puissance intelligente et sans bornes, à laquelle tout ce qui est doit réellement son existence, qui a, conséquemment, fait exister tous les êtres physiques, les seuls que nous puissions connaître positivement, a-t-elle créé ces derniers immédiatement et sans intermédiaire, ou n'a-t-elle pas établi un ordre de choses, constituant une puissance particulière et dépendante, mais capable de donner lieu successivement à la production de tous les corps physiques, de quelque ordre qu'ils soient? (Lamarck 1820, p. 22–23)[5]

In a methodological sense, the work's most striking aspect is its materialistic explanation of the physical aspects of the human being, an approach taken up by Lamarck as the basis of his transformism (Noguera Solano et al. 2018, 2021). It dedicates more than half of the work to the development of sensation, instinct, reason, intelligence, and imagination, which are all characteristics associated with the cognitive development of human species. It is important to understand that Lamarck posited a difference between matter and nature. In the France of the Enlightenment, to speak of matter was to refer to the physical substance of which all things are made. It is something tangible and observable, the basis of everything that exists, and it was studied through physics and chemistry.

[5] English translation: 'Did the intelligent and boundless power, to which all things really owe their existence, and which consequently brought into existence all physical beings, the only ones we can positively know, create them immediately and without intermediary, or did it not establish an order of things, constituting a particular and dependent power, but capable of successively giving rise to the production of all physical bodies, of whatever order they may be?'

On the other hand, nature is a much broader concept encompassing not only matter but also the processes and phenomena that occur in the natural world. Nature included life, growth, death, and natural cycles. Methodologically, it was related to observing and understanding natural systems and their functioning. Another relevant aspect is the role of positive knowledge. What does Lamarck mean by this? It should be remembered here that what is usually understood by positive thinking is the school developed by the also French Auguste Comte (1798–1857), Henri de Saint-Simon (1760–1825), and Pierre-Simon Laplace (1749–1827), which generally states that proper knowledge is true—positive in this case—as long as it is derived by reason from sensory experience (Burkhardt 1995).

Here, for example, we can see Lamarck's closeness to that statement:

> Qu'est-ce donc, encore une fois, que la nature, puisque ce n'est point une intelligence ? En quoi consiste cet ordre de choses qui a tant de puissance, et qui, lui-même, en établit d'autres? Et, si ce même ordre de choses est immatériel dans toutes ses parties, par quelle voie pouvons-nous parvenir à le connaître, puisque toutes nos connaissances positives proviennent originairement de nos sensations? Par l'exposition suivante, je crois donner la solution de toutes ces questions. (Lamarck 1820, p. 49) [6]

Madeleine Barthélemy-Madaule, one of the most renowned Lamarck's biographers, emphasise both the scholarly significance and political context of *Système*, presenting it as a prime example of his republican ideals, particularly when we consider that:

> Although it was not a conformist work, there can be no doubt that he asserted the existence and omnipotence of God. On what basis would this conclusion be justified? The careful distinction he draws between creation and production is precisely the distinction between God and nature. (Barthélemy-Madaule 1982 p. 11)

Lamarck's work reflects his commitment to empirical 'positive knowledge', aligning with contemporaries like Comte yet distinguishing divine creation from natural production. Lamarck acknowledged a transcendent ordering force despite his compromise to materialism, reconciling scientific observation with metaphysical inquiry. His synthesis of taxonomy, anthropology, and transformism challenges the narrative of his reduction to Darwin's foil. Though long overshadowed, Lamarck emerges not merely as a historical footnote but as a foundational figure whose complexity and foresight warrant renewed recognition in the annals of science.

[6] English translation: 'What then, once again, is nature, since it is not intelligence? What is this order of things which has so much power, and which itself establishes others? And, if this same order of things is immaterial in all its parts, by what means can we come to know it, since all our positive knowledge comes originally from our sensations? In the following exposition, I hope to provide a solution to all these questions'

3.4 Human Evolution and Natural Selection: The Anthropological Contributions of Alfred R. Wallace

One of the recurrent myths surrounding the figure of Alfred R. Wallace and his relationship with Darwinism is his conception of human evolution. Simplistically, it is claimed that Wallace exchanged his 'scientific' view of the human being for the 'quackery' of spiritualism. Here, it is presented as a myth because it is intended to portray Wallace as a 'bad scientist' and insists on a presentist view of the history of science, which distorts the complexity of scientific practices rather than contributing to understanding the past in its context.

However, beyond Wallace's numerous interests throughout his life, the aim here is to show that, from his beginnings as a naturalist, his main interest was the human being in an integral sense. Wallace's aim as a naturalist was to 'historicise' humanity. What do we mean by historicise? It is the process of constructing human histories for various purposes (scientific, religious and socio-political) and their contributions to debates about the origin of human races (Rodríguez Caso 2019). Wallace's anthropological contributions can be assessed in two ways: his role in consolidating the 'new' discipline, anthropology, and his various theorisations of the human being.

The beginnings of British anthropology can also be understood from the individual efforts made by different people who laid the foundations of the institutions that are still a point of reference today, such as the Royal Anthropological Institute (RAI), founded in 1871. Classic works such as George Stocking's *Victorian Anthropology* (1987) suggest that the formation of the RAI was the result of discussions between two groups, the Ethnological Society of London (ESL) and the Anthropological Society of London (ASL). However, in the British institutional context, there was a space that is not often mentioned, but that played a fundamental role in promoting scientific disciplines in the Victorian era: the British Association for the Advancement of Science (BAAS). In the Association, Wallace played a fundamental role in consolidating a unique space for studying the 'sciences of man', as he was the first president of the nascent Department of Anthropology in 1866. Furthermore, why was Wallace appointed president? Because at the time, he was the only one who met the a priori requirements: he was a naturalist who had theorised about physical and cultural aspects of human beings, and he was considered acceptable to both sides, ESL and ASL (Rodríguez Caso 2015, 2018).

However, Wallace's specific interest in the evolution of human species began at the same time as his road to becoming a naturalist, alongside his reading of a controversial work, *Vestiges of the Natural History of the Creation*, originally published anonymously in 1844, although it was later found to have been written by the Scottish journalist Robert Chambers. In a letter dated 28 December 1845, addressed to his best friend, Henry W. Bates, Wallace presents several elements that allow us to appreciate both his knowledge of the 'sciences of man' at that time and his interest in the development of animals, and especially of human beings:

> I would observe that many eminent writers give great support to the theory of the progressive development of species in Animals & plants—There is a very interesting & philosophical

> work bearing directly on the subject[,] 'Lawrence's Lectures on Man' delivered before the Royal Coll[ege]. of Surgeons & which are now published in a cheap form—The great object of these lectures is to illustrate the different races of mankind & the manner in which they probably originated—and he arrives at the conclusion as does Mr. Pritchard[sic] in his work on the Physical history of man, that the varieties of the Human race have not proceeded from any external cause but have been produced by the development of certain distinctive peculiarities in some Individuals which have become propagated through an entire race. —Now I sh[oul]d. say that a permanent peculiarity not depending produced in any way by external causes is a distinction of species & not of mere variety & thus if the theory of the 'vestiges' is carried out the 'Negro' the red Indian & the European are distinct species of the genus Homo—The Albino which presents as striking a difference as the negro, we have modern & not uncommon instances of the production of but the peculiarity is not propagated so extensively as that of the other varieties—Now it appears to me that the 'albino' and 'negro' are very analogous to what are generally considered as 'variety' & 'species' in the animal world. (McKinney 1969)

From this statement, Wallace's interest can be gauged from his experiences in the field, both in England and on his travels to the Amazon and the Malay Archipelago. When noting the importance of the trips made by Wallace, his contributions to subjects such as biogeography usually stand out, a topic for which he has been widely recognised. However, it was only recently that greater attention has been paid to Wallace's ethnographic contributions during his time in the Amazon (Limeira da Silva 2019).

Focusing on Wallace's ethnographic work is an excellent opportunity to show that much of his time travelling was devoted to observing and theorising about human beings. Now, possibly very briefly, to speak of Wallace's anthropological contributions is to go far beyond his institutional involvement in the initial consolidation of anthropology as a unified discipline in Britain. If we refer specifically to the case of physical anthropology, we have at least three themes of importance: the origin of human 'races', the development of so-called distinctive human characteristics, and the evolutionary relationship between humans and primates.

On the origin of the different human groups, his article 'The Origin of Human Races and the Antiquity of Man Deduced from the Theory of "Natural Selection"' was first presented as a lecture to the Anthropological Society of London and subsequently published in the society's journal (Wallace 1864), a paper that has usually gone unnoticed by historians. It is the first paper that formally seeks to explain the origin and development of human species based on natural selection in an era that, as Darwin's case suggests, was replete with authors and discussions about human species. Now, although it may seem too simple a solution, Wallace's merit was undoubtedly to bring together two positions on 'racial difference' that, at the time, were mortal enemies: monogenism and polygenism. The possibility of establishing a process in which both positions could coexist can be seen as a stroke of genius. On the other hand, what is possibly the most recurrent criticism of Wallace arises in this article: accepting that natural selection has limits, especially in explaining the cognitive capacities of human species.

On the other hand, and in agreement with James Costa (2023) and Charles Smith,[7] the 1864 article is not evidence of any 'conversion'—in the most religious sense of the term—of Wallace to spiritualism or a denial of science. On the contrary, it is a further step in Wallace's intellectual process of understanding in a broad and, above all, evolutionary sense, the human being physically and intellectually. Another important point here is to understand—as evident as it may sound—that Wallace's view of natural selection was not the same in 1864 as the one he presented in the famous 'Ternate essay' since, in the latter, he did not address the case of human species (see Wallace 1864).

Regarding the development of the distinctive characteristics of human species, we must begin by making it clear that Wallace's vision, at least from the 1860s onwards, was markedly utilitarian. Natural selection could only preserve what was helpful in the struggle for existence. Wallace's experiences during his travels, which, as mentioned above, occurred in parallel with his interest in historicising the human being, allowed him to get to know numerous human groups first-hand, which allowed him to reflect that, precisely in utilitarian terms, natural selection would not be acting on the cognitive capacities of the human being. This claim is what the linguist Derek Bickerton (2014) has recently called 'Wallace's problem', which in Wallace's own words is reflected in this idea: 'Natural selection could only have endowed the savage with a brain a little superior to that of an ape whereas he possesses one very little inferior to that of an average member of our learned societies' (Wallace 1869, p. 204). The 'traditional' reading—or rather, interpretation—of Wallace's work invariably redounds to the need for a supernatural explanation for the development of these capacities. However, it must be clear that even Wallace's involvement with spiritualism must be understood as part of a search for scientific answers to human evolution. In short, Wallace's view of the evolutionary process would be closer to monism than dualism, as there is no difference between physical and mental/spiritual evolution. Another important aspect is to appreciate that human evolution for Wallace was anthropocentric, which, as Dick (2008, p. 320) has pointed out, would culminate in *Man's Place in the Universe: A Study of the Results of Scientific Research in Relation to the Unity or Plurality of Worlds* (1903), a work that 'sheds unique light on Wallace's anthropocentric world view, and illuminates his entire career, including his belief that the human mind must be set off from animals in the evolutionary process'. Here, we can see that Wallace came to value the continuity between animals and humans as part of evolutionary development, although he would have advocated a form of 'anthropic principle' towards the end of his life. Despite recognising continuity, the human being must be considered an organism with different characteristics.

One specific example worth returning to here is language. Wallace put forward the gestural theory, which defends the importance of facial gestures in explaining the transition from manual to vocal signs. Again, this proposal arose from fieldwork experience, which forced Wallace to learn numerous languages from the different

[7] See comment in https://people.wku.edu/charles.smith/wallace/S093.htm

groups he encountered during his travels. The central point of the argument is that, when comparing different languages, it can be observed that they have a similarity in how they are pronounced, derived from a striking resemblance in the movements produced by the mouth, which goes beyond a simple process of imitation. In the end, for Wallace, there is a marked determinism between the physical development of certain features and the respective sound (or set of sounds) they can display, common to different human groups (Rodríguez Caso 2020a).

Wallace is a prime example of a different interpretation of what he called 'pure Darwinism' (Wallace 1889; Rodríguez Caso 2020b) since he started from a similar view to Darwin, but his interests and influences led Wallace to propose a series of ideas that are sometimes understood as an affront to what Darwin said. This claim should always be understood as part of a normative historical interpretation of scientific endeavour rather than an integral part of its development.

3.5 Final Reflections

One of the most common discussions about humanity and its evolution is the relationship between the physical, the mental, and the environmental. Furthermore, it has been discussed from the most diverse perspectives since the nineteenth century. Wallace, who, as already mentioned, had his main interest in historicising human species, provided the following definition:

> Humanity—the essentially human emotion—has caused us to save the lives of the weak and suffering, of the maimed or imperfect in mind or body. This has to some extent been antagonistic to physical and even intellectual race-improvement; but it has improved us morally by the continuous development of the characteristic and crowning grace of our human, as distinguished from our animal, nature. (Wallace 1890, p. 337)

In conclusion, Darwin's *Descent* represents a pivotal yet complex contribution to evolutionary discourse, emerging within a vibrant scientific milieu where human origins and diversity were already fervently debated. While often framed as a revolutionary text, it built upon prior works by Lyell, Huxley, and others, reflecting Darwin's synthesis of decades of observation and dialogue with contemporaries like Wallace, particularly on sexual selection as a counterpoint to natural selection. The book's significance lies not in the originality of facts but in its systematic argument for human continuity with nature, despite its entanglement with Victorian-era biases on race, gender, and social hierarchy. Modern scholarship acknowledges Darwin's enduring influence while critically reassessing his legacy, emphasising scientific progress's collaborative and iterative nature. By contextualising *Descent* within its historical framework—marked by institutional debates, preexisting anthropological inquiry, and Darwin's cautious public engagement—we appreciate it as both a product of its time and a catalyst for ongoing exploration into human evolution, underscoring the necessity of balancing historical analysis with present-day scientific and ethical scrutiny.

In a nutshell, Jean-Baptiste Lamarck's intellectual legacy extends far beyond the oversimplified caricature of a 'failed evolutionist' juxtaposed against Darwin. While *Philosophie zoologique* cemented his association with transformism, works such as *Système* reveal Lamarck's nuanced integration of human origins into his transformist framework, positing anatomical transitions from primates to humans—such as bipedalism and hand structure—as drivers of cultural evolution. Lamarck's focus on human uniqueness, including language and reason, positioned him as a pioneer in hominisation studies, while his materialist methodology bridged physical and cognitive development, anticipating modern debates on the interplay between biology and culture.

Alfred R. Wallace's contributions to evolutionary theory and anthropology defy the reductive myth of his supposed abandonment of scientific rigour for spiritualism, revealing a lifelong commitment to historicising humanity through an integrative lens of physical and cultural evolution. As a pivotal figure in institutionalising anthropology—exemplified by his leadership in founding the British Association for the Advancement of Science's Department of Anthropology—Wallace sought to reconcile monogenism and polygenism, positing natural selection as a mechanism for human diversity while acknowledging its limitations in explaining advanced cognition, a dilemma termed 'Wallace's problem'. His utilitarian perspective, shaped by ethnographic fieldwork, underscored the adaptive value of traits yet highlighted the paradox of human intellectual capacities exceeding immediate survival needs. Wallace's gestural theory of language, rooted in cross-cultural observations, and his monist view of evolution, which rejected dualism in favour of continuity between physical and mental development, further illustrate his nuanced approach. Despite later anthropocentric leanings, Wallace's work reflects a coherent, evolving inquiry into humanity's place in nature, challenging simplistic narratives that pit him against Darwin and emphasising the need to evaluate his legacy within Victorian science's complex, interdisciplinary context.

The study of human evolution should not be left solely to a disciplinary tradition such as anthropology itself because, beyond the recognition of certain traditions that can be considered canonical, other traditions, other geographies, and other authors are awaiting analysis, which, above all will allow us to recognise as more complex our understanding of how human evolution has been conceived.

References

Adelon NP (1821) Système analytique des connaissances positives de l'homme. Revue encyclopédique 9:257–267

Barthélemy-Madaule M (1982) Lamarck, the mythical precursor: a study of the relations between science and ideology. MIT Press, Cambridge, MA

Bickerton D (2014) More than nature needs. Harvard University Press, Cambridge/London

Burkhardt RW (1995) The spirit of system: lamarck and evolutionary biology: now with "Lamarck in 1995.". Harvard University Press, Harvard

Corsi P (1988) The age of Lamarck: evolutionary theories in France, 1790–1830. University of California Press, Berkeley/Los Angeles/London

Costa JT (2023) Radical by nature: the revolutionary life of Alfred Russel Wallace. Princeton University Press, Princeton/Oxford

Darwin C (1859) On the origin of species by means of natural selection, or the preservation of favoured races in the struggle for life, 1st edn. John Murray, London

Darwin CR (1871) The descent of man, and selection in relation to sex, 1st edn. John Murray, London

Darwin CR (1874) The descent of man, and selection in relation to sex, 2nd edn. John Murray, London

Darwin CR (1877) The descent of man, and selection in relation to sex, 2nd edn, revised and augmented. John Murray, London

Darwin CR (2004) The descent of man. Penguin Publishing Group

de Lamarck JBPA d M (1802) Hydrogéologie, ou, Recherches sur l'influence qu'ont les eaux sur la surface .... Agasse, Maillar, Paris

de Lamarck JBPA d M (1809) Philosophie Zoologique ou exposition des considérations relatives à l'histoire naturelle des animaux. Dentu, Paris

de Lamarck JBPA d M (1820) Système analytique des connaissances positives de l'homme: restreintes à celles qui proviennent directement ou indirectement de l'observation. A. Belin, Paris

Delisle RG (2007) Debating humankind's place in nature, 1860–2000: the nature of paleoanthropology. Pearson Prentice Hall

DeSilva J (2021) A most interesting problem: what Darwin's descent of man got right and wrong about human evolution. Princeton University Press, Princeton/Oxford

Desmond A, Moore J (2009) Darwin's sacred cause: race, slavery and the quest for human origins. Allen Lane, London

Dick SJ (2008) The universe and Alfred Russel Wallace. In: Smith CH, Becca-loni G (eds) Natural selection and be-yond: the intellectual legacy of Alfred Russel Wallace. Oxford Univer-sity Press, Oxford, pp 320–340

Ellegård A (1990) Darwin and the general reader: the reception of Darwin's theory of evolution in the British periodical press, 1859–1872. University of Chicago Press, Chicago

Fuentes A (2021) "The descent of man," 150 years on. Science 372:769–769

Gohau G (2006) Lamarck "Philosophe"? In: Corsi P, Gayon J, Gohau G, Tirard S (eds) Lamarck, philosophe de la nature. Presses Universitaires de France, Paris, pp 9–35

Kampourakis K (ed) (2024) Darwin mythology: debunking myths, correcting falsehoods, 1st edn. Cambridge University Press, Cambridge/New York

Limeira Da Silva VR (2019) History of the "human sciences" and Wallace's scientific voyage in the Amazon: notes on historiographical absences. Estud Hist (Rio de Janeiro) 32:549–563

McKinney HL (1969) Wallace's earliest observations on evolution: 28 December 1845. Isis 60: 370–373

Moore J (1991) Deconstructing Darwinism: the politics of evolution in the 1860s. J Hist Biol 24: 353–408

Noguera Solano R, Rodríguez Caso JM, Bustillo Ramírez R (2018) El programa de Lamarck: una visión materialista de la vida. Metatheoria—Revista de Filosofía e Historia de la. Ciencia 8:147–156

Noguera-Solano R, Rodríguez-Caso JM, Ruiz-Gutiérrez R (2021) The evolutionary thought of Jean-Baptiste Lamarck: why his original ideas should be taught in classrooms. Sci Educ 30:909–929

Radick G (2013) Darwin and humans. In: Ruse M (ed) The Cambridge encyclopedia of Darwin and evolutionary thought. Cambridge University Press, Cambridge, pp 173–181

Ragan MA (2023) Kingdoms, empires, and domains: the history of high-level biological classification. Oxford University Press, New York, NY

Richards E (2017) Darwin and the making of sexual selection. University of Chicago Press, Chicago/London

Rodríguez Caso JM (2015) Estudiar al hombre en sus más variados aspectos: A. R. Wallace, antropólogo. In: Rodríguez JM, Ruiz R, Oliver C, Abarca C (eds) Alfred Russel Wallace, a cien años de su muerte. UNAM, México, pp 95–102

Rodríguez Caso JM (2018) Institucionalización de la antropología en las reuniones de la British Association for the Advancement of science (BAAS), 1863–1870. Cuicuilco Revista de Ciencias Antropológicas 25:167–188

Rodríguez Caso JM (2019) Historización de la naturaleza del ser humano en A. R. Wallace. Saberes. Revista de historia de las ciencias y las humanidades 2:141–171

Rodríguez Caso JM (2020a) Alfred Russel Wallace, antropólogo: contribuciones a la antropología física. Arbor 196:a565–a565

Rodríguez Caso JM (2020b) El "darwinismo puro" de Alfred Russel Wallace: aportaciones a la teoría evolutiva moderna. Asclepio 72:p324

Rodríguez Caso JM, Noguera Solano R, Ruiz Gutierrez R (2012) Fueguinos, Papúas y Europeos en la Construcción de la Selección Natural. In: Martínez-Contreras J, Ponce de León A (eds) Darwin y el evolucionismo contemporáneo. Siglo XXI - Universidad Veracruzana, México, pp 255–269

Ruse M (2015) Sexual selection: why does it play such a large role in the descent of man? In: Hoquet T (ed) Current perspectives on sexual selection what's left after Darwin? Springer, Dordrecht/Heidelberg/New York/London, pp 3–17

Stocking GW (1987) Victorian anthropology. Free Press, New York

Stoczkowski W (2002) Explaining human origins: myth, imagination, and conjecture. Cambridge University Press, New York

Szyfman L (1982) Jean-Baptiste Lamarck et son époque. Masson, Paris

Veuille M (2024) Sex, gender, ethics and the Darwinian evolution of mankind: 150 years of Darwin's 'Descent of Man. Taylor & Francis, Oxford/New York

Wallace AR (1864) The origin of human races and the antiquity of man deduced from the theory of "Natural Selection.". J Anthropol Soc Lond 2:clviii–clxxxvii

Wallace AR (1869) Sir Charles Lyell on geological climates and the origin of species. Q Rev 126: 359–394

Wallace AR (1889) Darwinism: an exposition of the theory of natural selection. Macmillan and Co, London

Wallace AR (1890) Human Selection. Fortnightly Rev 48:325–337

Wallace AR (1903) Man's place in the universe: a study of the results of scientific research in relation to the Unity or plurality of worlds. Chapman & Hall, London

The Methodological and Theoretical Framework of Early Paleoanthropology

4

Matthew R. Goodrum

Abstract

Paleoanthropology emerged as a field of research in the 1860s, in response to the discovery of fossilized human bones. Darwin had already proposed his theory of evolution, and some scientists interpreted the recently discovered human fossils within the context of human evolution. However, influential anthropologists, especially in France, who were among the first to study these human fossils and were instrumental in establishing this new field of paleoanthropology, relied upon pre-existing methods and ideas drawn from the anthropological study of Neolithic and Bronze Age skeletons conducted during the mid-nineteenth century. This chapter argues that during this early phase of paleoanthropology, from the 1850s to the 1880s, a methodological and theoretical framework used by anthropologists to study prehistoric human skeletons was extended to the study of even earlier Paleolithic skeletons from the Pleistocene. This methodological and theoretical framework was based upon the craniometric and anthropometric measurement of Paleolithic skeletons in order to identify distinctive "fossil human races," combined with the widely held idea that during the prehistoric period populations of racially distinct humans had migrated into Europe at different times. For these early paleoanthropologists, their task was to apply anthropological research methods to the analysis of Pleistocene human fossils in order to identify their race and then to trace the changes in the races that inhabited Ice Age Europe. As a result of discoveries and scientific developments occurring during the 1880s and 1890s, paleoanthropology shifted away from this framework and increasingly adopted methods and theories that emphasized the phylogenetic and evolutionary study of extinct hominid species.

M. R. Goodrum (✉)
Department of Science, Technology, and Society, Virginia Tech, Blacksburg, VA, USA
e-mail: mgoodrum@vt.edu

M. Lequin, J. M. Rodríguez Caso (eds.), *The Field of Human Evolution: Critical Perspectives from History and Epistemology*, Evolutionary Biology – New Perspectives on Its Development 11, https://doi.org/10.1007/978-3-032-13925-2_4

Keywords

History of paleoanthropology · Craniometry · Prehistoric anthropology · Paul Broca · Ernest-Théodore Hamy · Cro-Magnon

4.1 Introduction

Evolution is a central feature of paleoanthropology today. Indeed, it is difficult to imagine a science of human origins that did not include the idea of evolution. Paleoanthropology formally emerged as a new scientific discipline during the 1860s, not long after the appearance of Charles Darwin's theory of evolution, although discussions about the origins of humans and about species change existed prior to this. The importance of evolutionary biology in the study of human origins is evident in the history of paleoanthropology. In works that present a broad overview of the history of paleoanthropology, the discovery of hominid fossils and changing ideas about human evolution and evolutionary theory form dominant themes (Delisle 2007; Henke 2015; Tattersall 1995, 2012). The early influence of Charles Darwin's theory on ideas about human origins during the late nineteenth and early twentieth century has been examined of course (Bowler 1986, 1989; McNabb 2012; Wuketits 2015). In recent years, there has also been great interest in the consequences of the integration of the Modern Evolutionary Synthesis into paleoanthropology beginning in the 1950s (Foley 2001; Delisle 1995; Smocovitis 2012). Modern paleoanthropology, as it has emerged since the 1950s, is a complex discipline that relies upon evidence, theories, and methods drawn from hominid paleontology, evolutionary biology, molecular anthropology, geology, archeology, cultural anthropology, and primatology, to name just some of the more prominent ones.

However, when paleoanthropology first emerged as a field of research in the nineteenth century, it operated under a very different methodological and theoretical framework than the science we are familiar with today. This early period in the history of paleoanthropology, which lasted from the 1850s until the early 1880s, has received much less attention than later periods, and even then the emphasis has overwhelmingly been on the debate surrounding the interpretation of the original Feldhofer Neanderthal fossils and the discovery of the original Cro-Magnon specimens. There were a number of Pleistocene human fossils found during this period, though, and they generated great interest since they provided valuable evidence relating to the question of the problem of the geological antiquity of humans (Grayson 1983; Cohen and Hublin 1989; Van Riper 1993).

When a new field of research called paleoanthropology, which was originally defined as the anthropological study of human fossils, emerged during the 1860s (see Goodrum 2014), it was not created out of a vacuum. The anthropologists who studied and described these Ice Age skeletons relied upon methods rooted in craniometry and anthropometry that anthropologists had used to study prehistoric skeletons found in Stone Age, Bronze Age, and Iron Age tombs earlier in the century. These anthropologists examined these ancient skeletons in order to determine the "races" of the different peoples who had inhabited Europe prior to the arrival of modern Europeans. The principles of racial anthropology, not of human evolution, informed many of the studies of Pleistocene human fossils during this first phase of paleoanthropology in the late nineteenth century. Discussions of human evolution did arise with some scientists during discussions of certain fossils, but this was often not the primary concern of many researchers of this first generation of paleoanthropologists who analyzed human fossils.

In addition to these shared methodological techniques used to analyze and describe human fossils, there was also a theoretical framework that anthropologists inherited from their predecessors that was used to make sense of these Pleistocene humans. Rather than a story of human evolution from earlier apelike ancestors, these anthropologists explained Pleistocene human skeletons as belonging to distinctive human races that had inhabited Europe during the Ice Age. This was consistent with what they had already discovered with regard to later prehistoric Europe, where a succession of Neolithic, Bronze Age, and Iron Age races were thought to have migrated into Europe. Thus, Neanderthals and Cro-Magnons were simply earlier races of humans that had inhabited Europe during the Ice Age, which were subsequently replaced by new Neolithic and Bronze Age races that migrated into Europe after the Pleistocene. This was the broad methodological and theoretical framework that informed a large portion of paleoanthropology research during this first phase of the science from the 1860s to the 1880s. An important series of questions arise once we recognize the characteristics of this early phase of paleoanthropology.

By the 1880s, a new phase of paleoanthropology emerged where new hominid fossils were interpreted within an evolutionary framework and new methods were developed to study hominid fossils. One of the most significant historical developments in the early history of paleoanthropology, therefore, was the transition from the methodological and theoretical framework that emphasized craniometry, racial anthropology, and the interpretation of the prehistoric human skeletal record as evidence of successive waves of habitation of Europe by distinct human races, to a methodological and theoretical framework that emphasized extinct species of hominids represented in the fossil record that were interpreted within an evolutionary framework. Craniometry and racial anthropology persisted into the early twentieth century, as did theories of prehistoric migrations. But anthropologists increasingly integrated these techniques and ideas into evolutionary interpretations of the origins of human races, while paleoanthropologists investigated hominid fossils from an evolutionary and phylogenetic perspective (see Sommer 2015, 2024; Delisle 2007).

4.2 Emerging Evidence for Human Prehistory

Knowledge about the remote human past underwent dramatic change during the nineteenth century. Archeologists excavating ancient tombs and examining megalithic monuments in Europe were uncovering evidence that offered new insights about the primeval inhabitants of Europe, but this evidence fundamentally challenged traditional conceptions of human history. As a direct consequence of the discoveries made during such excavations, the Danish archeologist Christian Jürgensen Thomsen proposed the Three Age System (the chronological sequence of Stone, Bronze, and Iron Ages) in the 1830s. Jens Jacob Asmussen Worsaae, the professor of archeology at the University of Copenhagen, extended the Three Age System to other parts of Europe in the 1840s (Rodden 1981; Gräslund 1987). This system offered a way of arranging artifacts and sites chronologically in time, and it gradually became a centerpiece of archeology by the end of the nineteenth century. Yet the Three Age System was provocative and controversial when it was first proposed, and while archeologists increasingly came to adopt this system, there continued to be prominent opponents of it until the 1870s (Rowley-Conwy 2007).

Meanwhile, in Denmark during the 1840s and 1850s, the naturalist Japetus Steenstrup, along with geologist Johan Georg Forchhammer and Worsaae, excavated the kjökkenmöddings that dotted parts of the Danish coastline. In a series of reports to the Royal Danish Academy of Sciences, they argued that these mounds, composed of mollusk shells and animal bones, were middens containing the discarded remains of animals eaten by the earliest inhabitants of Denmark (Forchhammer, Steenstrup and Worsaae 1852). At the same time, archeologists in Switzerland were investigating the lake dwellings (Pfahlbauten) that became visible during the 1850s when drought lowered the water level of Swiss lakes exposing the remains of wooden structures that had previously been submerged. Ferdinand Keller, Frédéric Troyon, and others produced reports on these Stone and Bronze Age settlements, which contained large numbers of artifacts and the remains of pile dwellings built on the lakes (Keller 1866; Jahn et al. 1857; Troyon 1860). An important consequence of this growing body of archeological evidence was the introduction of the terms "prehistoric" and "prehistory" (préhistorique, förhistorisk, vorsgeschichte) in the 1850s and 1860s to refer to this remote period of human history for which there were no written records (Clermont and Smith 1990; Rowley-Conwy 2006).

A critical question that arose as a consequence of this research concerned the identity of these prehistoric peoples. Despite the lack of firm knowledge about the peoples of primeval Europe, the ancient past was not simply a dark abyss. There were familiar sources that scholars could draw upon to seek answers to this question, though. Historians possessed valuable knowledge about the ancient inhabitants of Europe thanks to the accounts of Roman writers who discussed the peoples they encountered during their conquest of Europe. Julius Caesar described the Celtic inhabitants of Gaul in *The Gallic War*, and Tacitus wrote about the Germanic tribes in his *Germania*. This led many scholars to attribute megalithic monuments and stone tombs to the ancient Celts and Germans. A potentially powerful new tool for identifying the prehistoric peoples of Europe and tracing their migrations was philology, the comparative study of languages and their grammar. Scholars studying Persian and Sanskrit, the ancient languages of Iran and India, noted significant parallels between the grammar and vocabulary of these languages and ancient Greek and Latin. Soon philologists such as Friedrich Schlegel, Franz Bopp, and others recognized a close relationship between these languages and the Germanic,

Slavonic, and Celtic languages spoken in Europe. By the middle of the century, scholars were referring to an Indo-Germanic or Indo-European family of languages that shared similarities because they were all related genealogically (Benes 2008; Rabault-Feuerhahn 2008; Olender 1992; Mallory 1989).

Many philologists believed the similarity of these languages could be explained by the migration of successive waves of Indo-European peoples into Europe from some original homeland located in Asia. By the middle of the century, a host of scholars, including Jacob Grimm in Germany, Adolph Pictet in Switzerland, and Max Müller in England, were trying to use philology and ethnological data about the current populations of Europe to reconstruct the racial composition of prehistoric Europe and link it to the migrations of various Indo-European peoples coming from Asia (Grimm 1848; Pictet 1859–63; Müller 1861). While some ethnologists rejected the idea that ethnic identity correlated with language, many scholars saw philology as a way to identify who the early inhabitants of Europe were and the order of their arrival. As a result, migrations and the displacement of one tribe by an invading newcomer became a common model for explaining European prehistory throughout much of the nineteenth century.

The problem confronting archeologists and historians was how to correlate the specific peoples mentioned by Roman authors and the successive waves of invading Indo-European tribes with the tombs and artifacts assigned to the Stone, Bronze, and Iron Ages. By the middle of the nineteenth century, the emerging science of anthropology offered a possible solution to that problem. Archeologists occasionally encountered human skeletons in the tombs they excavated, and these remains became an invaluable source of information about the peoples who inhabited prehistoric Europe. An early attempt to use anthropological methods to identify a prehistoric people appears in Jens Worsaae's *Danmarks oldtid* (1843). From historical documents, Worsaae knew that Finns and Celts had lived in ancient Denmark, but the archeological evidence found in Stone Age Danish tombs led him to doubt that either of these peoples were the original inhabitants of Denmark. He believed instead that the Stone Age inhabitants of Denmark belonged to a hitherto unknown race that had since disappeared, leaving behind only their crude stone monuments and artifacts (Worsaae 1849: 128–133). To test this notion, Worsaae suggested that researchers should compare the morphology of Stone Age skeletons, and especially their skulls, with those of modern humans in order to identify what race they belonged to. This idea was rooted in the notion that human races varied in their morphological features, including their skeletons, and that while members of the same race shared similar features, different human races possessed their own distinctive features. Worsaae did not think it was strange that historical records had preserved no memory of these people, but he did think it was possible to draw inferences from the archeological record about their manner of living and their place in Danish history. He suggested that the Stone Age inhabitants of northern Europe were a nomadic people who lived by hunting and fishing and that this population had eventually been displaced by newcomers who knew metallurgy, used bronze implements, and practiced agriculture (Worsaae 1849: 133–4) (Fig. 4.1).

Fig. 4.1 Stone Age tomb (Cromlech du Tus on the Island of Guernsey) containing two skeletons excavated in 1844. (J. J. A. Worsaae, *The Primeval Antiquities of Denmark*, London 1849, p. 90)

The ideas and methods that lay behind the anthropological endeavor to identify and distinguish human races on the basis of their morphological features can be traced back to the German physician and anthropologist Johann Friedrich Blumenbach at the end of the eighteenth century. In the United States, Samuel Morton advanced this research, and in the 1830s, he established some of the early methods of studying human skulls that were thought to represent different races (Dewbury 2007). Scientists in Europe were also developing techniques that would allow them to determine race on the basis of measurements of the human body (anthropometry) and particularly measurements of the skull (craniometry). One of the early advocates of this technique was the Swedish anatomist Anders Retzius. After studying medicine at the University of Lund, Retzius was appointed professor of anatomy at the Karolinska Institute in Stockholm. There he became intrigued by the relationship between race, brain anatomy, and the level of cultural development and intellectual ability of different ethnic groups. This led him to investigate a possible correlation between race and the shape of the skull in different groups of people. The Dutch anatomist Petrus Camper had already proposed the facial angle, which measured the degree of prognathism among humans, and the technique became widely used among anatomists as a way of scientifically defining racial morphological features (Meijer 1999). Retzius made a major contribution to craniometry by introducing the cephalic index. This is a number produced by dividing the width of the skull by its length and multiplying the resulting number by one hundred. After taking measurements of human skulls from various ethnic populations, Retzius became convinced that skulls from a single race tended to have similar shapes and a similar cephalic index, while skulls from different races tended to have different shapes and a markedly different cephalic index. Retzius identified two distinct skull types derived from the cephalic index. Skulls that were short and round, with a cephalic index greater than 75, he called brachycephalic; skulls that were long and oval, with a cephalic index less than 75, he called dolichocephalic (Retzius 1849; see also Kyllingstad 2014: 1–24).

Anthropologists quickly adopted craniometry and used it in their analysis of prehistoric human skeletons. In France, the influential anthropologist Paul Broca invented a host of instruments to more accurately measure skulls and other bones, and he promulgated rules for what measurements to make and how to make them. He modified Retzius' classification regarding the cephalic index by introducing a mesaticephalic form of skull that was intermediate between the dolichocephalic and brachycephalic (Broca 1861). The German anthropologist Hermann Welcker (1862: 41–46) independently proposed a similar category, which he called orthocephalic. British anthropologists generally adopted these innovations but preferred the term mesocephalic. Craniometry thus became the fundamental technique that anthropologists used to study prehistoric skulls, comparing them with skulls of modern human races, in order to determine the race they belonged to (see Blanckaert 1987, 1989). The primary problem initially facing anthropologists was the meager number of prehistoric skeletons available for study. Marcel Serres, who taught a course in anthropology at the Muséum d'Histoire Naturelle in Paris, remarked how archeologists had studied Celtic monuments and artifacts, but there had been few corresponding anthropological studies of Gallic skeletons. He went further, identifying a problem that others would mention throughout the century. Serres complained that excavators had scarcely paid any attention to the skeletons they found in Gallic tombs. "These precious remains have been thrown to the wind, or if by chance an antiquary has collected a skull, it is not on it that his attention has been directed." But Serres also noted that historical research was beginning to put an end to this carelessness (Serres 1845: 607–8) (Fig. 4.2).

Likewise, the Scottish archeologist Daniel Wilson criticized British archeologists, especially the Wiltshire antiquary Richard Colt Hoare, for not devoting more effort to the study of the crania and skeletons found in prehistoric tombs. He observed that over the years farmers and archeologists had unearthed hundreds of skulls from ancient barrows. Wilson bemoaned the fact that despite this, "Nothing has yet been done by Scottish archeologists with a view to ascertain the physical conformation of the primitive native races." And he had to admit that his own current foray into the subject was "founded on too limited data to be of very great avail, except perhaps in opening up the subject and leading to more extended observation" (Wilson 1851: 160, 164, 178). Similar complaints continued to be expressed throughout the century, with ire often directed toward careless excavators. French archeologist

Fig. 4.2 Goniometer developed by Broca to measure skulls. (Paul Broca, "Nouveau goniomètre." *Bulletin de la Société d'Anthropologie de Paris* 5 (1864): 943–946, p. 944)

Émile Cartailhac (1889: 331) expressed his frustrations on the matter when he remarked: "We ought to have maps giving the distribution and the density of all the ancient races of France. But the thousands of burials which are known have served no purpose. Archeologists alone have destroyed more bone fragments than either natural agencies or field laborers. One is really amazed to find in the Museum, at the museum of Broca, in our public and private collections such a feeble quantity of specimens in good condition and of certain provenance."

French anthropologist Georges Hervé was equally blunt. While discussing the problem of identifying the prehistoric peoples who inhabited the Iberian Peninsula, he pointedly remarked that "one might have had, if one had wanted, materials as precious as abundant. Unfortunately, we can only deplore by denouncing the stupid vandalism of archeologists (rather collectors, in order not to prostitute the name of scientist), who almost always pushed them, in Spain and Portugal, even more to break, to disperse, as unworthy of attention, so many human bones, irremediably lost to ethnological science!" (Hervé 1899: 273–4). He then proceeds to list the numerous sites where human bones were found but were either not preserved or were subsequently misplaced. Despite this grim assessment of the state of prehistoric anthropology offered by some of the leading prehistoric researchers of the century, the fact was that by the 1850s archeologists were preserving more of the skeletons they uncovered and anthropologists were increasingly recognizing the importance of examining them. Their research resulted in a dramatic reconceptualization of how Europe came to be populated in prehistoric times.

4.3 Stone Age Skeletons and the Ethnology of Prehistoric Europe

While agricultural activity, construction work, and enthusiastic excavators searching for artifacts were frequently responsible for the destruction or simply the careless disregard of prehistoric human remains, archeologists did gradually begin to recognize the potential importance of ancient human skeletons for reconstructing the ethnological history of Europe. In 1845, the French geologist Eugène Robert excavated a large stone tomb at Meudon, near Versailles. It contained about two hundred human skeletons buried with flint axes and arrowheads, animal bones, and coarse pottery. Marcel Serres, who was seeking an opportunity to study the skeletons of ancient Gauls, traveled to Meudon to examine the skeletons. He noted that the pottery from the lower layers of the tomb was purely Celtic, while that from the upper layers dated from Gallo-Roman times. Serres (1845: 613–9) concluded that the skeletons in the tomb belonged to two different peoples; those from the lower layers of the tomb represented what the French ethnologist William Edwards called the "Gall type," while the later skeletons belonged to the "Kimri type".[1]

In Scotland, Daniel Wilson also examined ancient skeletons in order to test recent claims about the prehistoric inhabitants of Europe. Wilson was an archeologist and an early adopter of the Three Age System. He was familiar with the research of the Swedish prehistorian Sven Nilsson, who had excavated prehistoric Scandinavian skulls from tombs and given them to Anders Retzius for examination. Nilsson and Retzius (1843) combined the results from these craniological studies with the archeological and philological evidence for successive migrations of Indo-European peoples, to conclude that Scandinavia was originally populated during the Stone Age by a brachycephalic race. Then a dolichocephalic race arrived before the end of the Stone Age, followed by the invasion during the Bronze Age of a race with an even more pronounced dolichocephalic skull that Nilsson identified as the Celts. Finally there came the Indo-European Swedes who initiated the Iron Age in Sweden. Wilson (1851: 163) also cited the opinions of the influential British ethnologist James Cowles Prichard, who promoted the view that an "Allophylian race" had inhabited Europe before the arrival of the Indo-European Celts.

To test these claims, Wilson decided to study the collections of prehistoric Scottish skulls preserved in the museum of the Society of Antiquaries of Scotland and at the Edinburgh Phrenological Society. Among these were a number of "Celtic skulls" excavated from ancient tombs on the island of Ionia, in the Hebrides, in 1833 and donated to the Society of Antiquaries of Scotland by its secretary Donald Gregory (Wilson 1851: 164, 175). The immediate problem that confronted Wilson was his inexperience in craniology. He decided to follow the craniometic methods and to adopt the terminology of the American craniologist Samuel Morton. Wilson soon found that making accurate measurements was exceedingly difficult or even

[1] According to Edwards (1829), the Gall type was an ancient brachycephalic people while the Kimri type was an ancient dolichocephalic people.

impossible because many prehistoric skulls were either damaged or only part of the skull was preserved. To make up for his lack of experience in craniology, Wilson enlisted the aid of John Zaglas, an anatomical assistant at the University of Edinburgh, and of the physician John Alexander Smith. At the beginning, he was even unsure what measurements he should make. In the end, the specific measurements that Wilson made were influenced by the craniological investigations of Walter Adam, a physician who had studied crania from the catacombs of Paris housed at the University of Paris (Wilson 1851: 165, 176–7).

Wilson (1851: 165) defended the fundamental precept of craniometry that human races could be identified from the form of the skull. He argued that

> there can be no question that a general distinctive cranial conformation is clearly discoverable in modern nations, and is even very markedly observable between the different races of the British Isles. Given a sufficient number of examples of each class, the experienced eye would at once discriminate between the modern European Fin, Germanic Teuton, and British Celt. The conclusion appears therefore inevitable, that if we find in the ancient tumuli like variations in physical form, systematically reducible to two or more classes, we are justified in assuming the existence of diverse primitive races, and of seeking in the accompanying relics for indications of their peculiar arts and customs, as well as of their relative positions as contemporary or successive occupants of the country.

When this data was combined with the discoveries made by archeologists, Wilson (1851: 164) argued, one should be able to recognize when a new race appears in some region both from the anthropological evidence manifested in the change over time in the form of the skulls found in tombs as well as the archeological evidence of a change in artifacts or burial practices.

Wilson observed that nine skulls taken from Scottish Stone Age tombs were dolichocephalic. However, several skulls found with bronze artifacts were brachycephalic, which meant they belonged to an entirely different race. Although he stressed that his evidence was too limited to make any strong assertions about the succession of races that inhabited prehistoric Scotland, he did reach some tentative and important conclusions. Historians and antiquaries had long believed that the Celts were the original inhabitants of Britain. The archeological evidence clearly showed that an aboriginal race inhabited Britain for a long period prior to the arrival of the Indo-European Celts. Now the skeletal evidence from Stone Age tombs supported this view. Furthermore, the results of Wilson's craniometric investigations differed from those of Anders Retzius and Sven Nilsson in Scandinavia. There a brachycephalic Stone Age race was succeeded by a dolichocephalic race. In contrast, Wilson (1851: 167–71, 176–7) found that the earliest Stone Age tombs in Scotland contained dolichocephalic skulls while later tombs contained brachycephalic skulls. Then there was the additional problem of the transition from the Stone Age to the Bronze Age. The archeological record indicated a gradual transition from one period to the other, which was inconsistent with the notion promulgated by some scholars that the Bronze Age was introduced by an invading race that exterminated the aboriginal Stone Age inhabitants of Europe. Wilson (1851: 161, 178) noted that the current skeletal evidence was not sufficient to show the abrupt replacement of one race by another as required by the invasion hypothesis.

The craniological investigations of Retzius, Serres, and Wilson reflected the hope that the study of prehistoric skeletons, and especially the examination of skulls, might answer fundamental questions about the racial identity of the prehistoric populations of Europe. A crucial advance in this program came with the work of Paul Broca. Broca studied medicine at the University of Paris, and in addition to his medical and neurological research, he also became a leader of French anthropology. Broca founded the Société d'Anthropologie de Paris in 1859 with the goal of making anthropology an empirical science that would investigate humans racially, zoologically, anatomically, and geologically. Many of its members were polygenists, proponents of the controversial view that the different human races were in fact distinct species that had separate origins. Broca helped to professionalize anthropology and to elevate the status of the science in France by creating a number of institutions. In addition to the Anthropology Society, he also established the Laboratoire d'Anthropologie in 1867 and the École d'Anthropologie in 1876, as well as the journal *Revue d'Anthropologie* which he edited (see Blanckaert 2009; Pineau 1980; Schiller 1979).

Several opportunities arose in the 1860s for Broca to examine prehistoric skeletons. In October 1863, the comte de Lavaulx unearthed a prehistoric stone tomb while removing some large stones that disrupted the cultivation of a field on his estate in Chamant. The count found between a hundred and one hundred and fifty human skeletons along with course pottery, flint arrowheads and knives, polished flint axes, and animal bones (Anon. 1863). He donated these objects to the Société d'Anthroposlogie, which prompted Broca to visit the site. In May 1864, Broca and several colleagues from the Society resumed excavations of the tomb at Chamant. They found many artifacts and human bones, but most of the skulls were badly damaged and fragmentary. From this material, Broca was only able to determine that two skulls were dolichocephalic while another two were mesaticephalic (Anon. 1864b). Not long after the discovery of the tomb at Chamant, a landowner named Mr. Hazard, who lived near the village of Orrouy, found a prehistoric sepulchral cave while conducting work on his property. Laborers conducting the work found a human skeleton near the entrance of the cave, but below this skeleton they unearthed a mass of human bones piled together. The cave contained approximately 50 bodies as well as animal bones, pieces of course pottery, flint knives, and polished flint axes. Mr. Hazard collected a small portion of the bones, including all 21 skulls, and Broca arranged for most of the collection to be sent to the Société d'Anthropologie (1864a). Broca examined the skulls and noted that some were dolichocephalic, others were brachycephalic, and some were intermediate between the two. Broca (1864) interpreted this as evidence that the skeletons belonged to two distinct races, while the intermediate-shaped skulls were evidence of interbreeding between them. This is very typical of the way anthropologists used craniometry to interpret the skeletal evidence from prehistoric tombs.

Meanwhile, similar discoveries were being made in England. Thomas Bateman (1861), a member of the British Archaeological Association, excavated a large number of prehistoric tombs in the north of England during the 1850s that produced artifacts and also many human skeletons. While prehistoric artifacts garnered considerable attention among British archeologists, John Thurnam was among the few researchers who took an anthropological interest in prehistoric skeletons. Thurnam studied medicine at the University of Aberdeen and held several posts before being appointed the medical superintendent of the Wiltshire county asylum at Devizes when it opened in 1851. After his move to Wiltshire, Thurnam began craniological investigations of prehistoric British skeletons. The significance of this research was reflected in his becoming a Fellow of the Society of Antiquaries and a member of both the Ethnological Society of London and of the Anthropological Society of London. While developing his methods for making craniometric measurements, Thurnam drew from a variety of leading figures, including the Dutch zoologist Jan van der Hoeven who published his *Catalogus craniorum diversorum gentium* in 1860, the German anthropologist Ignaz Pruner, as well as Paul Broca and Thomas Huxley. He determined the cranial capacity of skulls by filling the brain case with fine white sand and measured the circumference by passing a tape above the orbits and around the occiput. He measured the length, breadth, and height of the skull, as well as the length and breadth of the face. These measurements allowed him to calculate the all-important cephalic index (it is worth noting that Thurnam (1865: 459–61) supported Broca's addition of the mesaticephalic to the dolichocephalic and the brachycephalic skull types) (Fig. 4.3).

Using these methods, Thurnam studied skulls found in the "chambered long barrows" of North Wiltshire and Gloucestershire. These tombs only contained flint arrowheads and axes and were attributed to the Stone Age. Thurnam (1865: 120–1, 139–40) observed that the skulls from these tombs were dolichocephalic. However, Thurnam found that the skulls from "circular barrows," which were tombs containing bronze and stone artifacts that were attributed to the Bronze Age, were brachycephalic. These results suggested that Britain had originally been inhabited by a dolichocephalic race during the Stone Age and that a tall brachycephalic race inhabited southern England later during the Bronze Age. Excavations that Thurnam conducted in 1863 of a barrow near Stonehenge supported this notion. The barrow contained a secondary burial of six skeletons of a brachycephalic race, but below these later burials, Thurnam (1865: 140–45) found the older primary burial, a male skeleton with a dolichocephalic skull.

Despite this successful application of craniometry in illuminating the prehistoric ethnology of Britain, the question still remained of linking the races identified through craniometry (Stone Age dolichocephalic and Bronze Age brachycephalic) with the peoples mentioned by ancient Roman writers, such as the Belgae or other Celtic tribes. Thurnam encountered difficulties in achieving this. Yet he remained convinced that the anthropological study of prehistoric human skeletons offered great potential in resolving questions of the racial composition of prehistoric Europe. One of the major questions confronting researchers concerned the hypothetical "immigration into Europe, from high Asia, of Indo-European peoples having a dolichocephalic type of skull," and Thurnam (1865: 167) argued that this could "only be satisfactorily determined by long-continued researches on the form of the

Fig. 4.3 Depiction of a prehistoric tomb in Derbyshire excavated by Thomas Bateman in 1848. (Thomas Bateman, *Ten Years' Diggings in Celtic and Saxon Grave Hills, in the Counties of Derby, Stafford, and York, from 1848 to 1858; with Notices of Some Former Discoveries, Hitherto Unpublished, and Remarks on the Crania and Pottery from the Mounds.* London: J. R. Smith, 1861, p. 23)

skulls found in the earliest, or at least in pre-Roman, graves." Thurnam concluded his discussion with an optimistic, but also cautionary, note that acknowledged the challenges of integrating the results derived from the study of prehistoric skeletons with historical knowledge and with the recent discoveries made by philologists—who promulgated the theory of an Indo-European migration.

> I submit the opinion that the accurate observation, description, and measurement of skulls, ancient and modern, is everywhere to be desired, before the solution of the complex problem of the origins of the different European peoples can be satisfactorily attempted. Premature attempts to reconcile the results deduced from scientific craniology, with those arrived at by philological methods, can only result in disappointment (Thurnam 1865: 167)

In Switzerland, Ludwig Rütimeyer and Wilhelm His had greater success in correlating races determined by craniometric methods with peoples known to historians. Rütimeyer was professor of zoology at the University of Basel, and His was professor of anatomy and physiology at the university. Together they examined

76 skulls excavated from prehistoric sites throughout the country. They identified four different skull types (based upon cranial capacity and cephalic index), which they believed represented distinct peoples. These types were named for the localities where they were found. The Disentis type found in prehistoric graves in eastern Switzerland was brachycephalic and represented the population that inhabited Switzerland in the most remote past. The Sion type was mesocephalic and represented the pre-Roman Helvetians, who were Celts. The Hohberg type was very dolichocephalic and probably represented Roman colonists. Finally, the Bel-Air type was also mesocephalic and belonged to the Burgundian period of Swiss history (Rütimeyer and His 1864).

Other researchers interested in prehistory took a more expansive view. During a visit to Caithness, in the northernmost part of Scotland, in the autumn of 1864 Samuel Laing took the opportunity to examine some prehistoric middens and tombs. Laing was a politician who had held a number of government posts including serving as a Minister of Parliament. He was also intensely interested in human prehistory and was a member of both the Geological Society of London and the Anthropological Society of London. Laing (1865: xx–xxi) wondered if Scotland possessed Stone Age sites like the Danish kjökkenmöddings or Swiss lake dwellings, and he was inspired to conduct excavations after reading about the discoveries made in shell mounds located near the Scottish town of Elgin. In Caithness, he excavated a large burial mound and found more than 60 stone cists that contained human skeletons along with stone hammers, axes, and spearheads. He then excavated two large middens composed of mollusk shells and animal bones. In the first midden, he found crude stone and bone artifacts along with fragments of pottery. The second displayed three distinct occupation layers. The lower layers contained only stone and bone artifacts and crude pottery, but the upper layer contained some bronze objects, and the pottery was of finer quality (Laing 1865: xxiv–xxix). In two other tombs, Laing found human bones along with stone and bone artifacts.

Laing argued that these tombs dated from the "early stone period," judging from the lack of finely polished stone artifacts. Thomas Huxley agreed to examine the human bones, which included several skeletons and six skulls. Huxley was a good choice for the task since in addition to being a leading zoologist and supporter of Darwin's theory of evolution he was also an experienced anatomist. Huxley opens his description of the human remains by explaining that he would use the craniological nomenclature developed by Retzius, Broca, and Welcker, but he also introduced new subdivisions of his own (Laing and Huxley 1866: 83–5). Laing had already noted that the skulls recovered were generally dolichocephalic, but Huxley carefully measured the cephalic index of each skull and described its features. On the basis of Huxley's investigations, Laing remarked that while the skulls were dolichocephalic, they differed from the dolichocephalic skulls from British long barrows described by Thurnam. Some of the Caithness skulls possessed strong prognathism (a protruding jaw) like one finds in the "lower races" but other skulls appeared to belong to a "superior race." Laing noted that this was consistent with skeletons found in tombs elsewhere in Europe, which indicated the coexistence of two distinct races during the Stone Age (Laing and Huxley 1866: 68–9).

Laing (1865: xxxiii–xxxiv) believed the skeletons he unearthed in Caithness belonged to the aboriginal race of Scotland and that this race was very different from any modern European race. From Huxley's craniological analysis, Laing arrived at a "most important and interesting fact," namely that these prehistoric people were "so savage and degraded, as to present more resemblance to the lowest Australian than to any historical European race" (Laing and Huxley 1866: 62). When comparing the skeletons from Caithness with those of other modern and prehistoric peoples, it was possible to draw conclusions about their place in the generally accepted hierarchy of human races. "No one can doubt that we have here an extremely savage type. No one can doubt that among the family of savage types the analogies are altogether with the Negro as opposed to the Mongol; and that this tribe of aboriginal Britons must have much more closely resembled the Australian or Tasmanian than the Laplander or Esquimaux" (Laing and Huxley 1866: 63). But who were these people and where had they come from?

Laing had followed recent developments in prehistoric archeology and sought connections between his discoveries in Scotland and discoveries made elsewhere in Europe. The crude artifacts and the animal bones found in the tombs and shell mounds of Caithness resembled those found in the kjökkenmöddings of Denmark. Archeologists generally accepted that the kjökkenmöddings were older than many of the Stone Age tombs found throughout northern Europe. However, geologists and archeologists had also begun to find crudely chipped flint artifacts in geologic deposits that contained the fossil bones of extinct animals such as the mammoth and wooly rhinoceros. Jacques Boucher de Perthes, a French customs official and amateur archeologist, had accumulated a large collection of flint artifacts associated with extinct animal bones in the deposits along the Somme River near Abbeville during the 1840s and 1850s, and geologists soon found similar finds at the site of St. Acheul. In Britain, excavations conducted at Brixham cave in 1858 and 1859 also produced flint tools associated with the bones of extinct animals. These surprising discoveries led most European scientists to accept the notion that humans had lived during the "Glacial Period" when Europe's climate was much cooler than at present and it was inhabited by animals such as mammoths and reindeer (Grayson 1983; Cohen and Hublin 1989; Van Riper 1993). This meant that human prehistory needed to be extended into a deeper past than had hitherto been suspected.

Laing initially suggested in his 1865 paper that the skeletons and artifacts he unearthed at Caithness dated from the "early stone period." While they were not as old as the flint tools found with extinct animal bones, he wondered if it might not be possible that a relic of this "aboriginal population may not have lingered on in this remote corner of Britain, without progress or improvement, down to a comparatively recent period," only to be replaced later by "historical races" who possessed metal. These skeletons and artifacts, he concluded, must belong to the "aboriginal race of North Britain, which is either itself of great antiquity, or represents a race who had lived on little changed from the earliest period of the occupation of the country" (Laing 1865: xxxii). A year later, he was prepared to expand upon these claims. In *Prehistoric Remains of Caithness*, Laing observed that if the material unearthed at Caithness dated from the time of the Danish kjökkenmöddings, which seemed likely, then there would have been ample time for either a succession of races to inhabit the region, or for an original race to have undergone modification over time. The shape of the skull of these savage people could have changed with the passage of time, acquiring the appearance of the prehistoric skulls found later throughout England and Scotland. The subsequent arrival of a more advanced brachycephalic race who interbred with the aboriginal inhabitants of Caithness could have produced a mixed race who gave rise to the "first historical Britons," or perhaps this mixed race was itself overrun by yet another invading race who introduced the use of metals (Laing and Huxley 1866: 64).

Laing even proposed that the aboriginal inhabitants of Caithness had migrated into Britain from continental Europe along with specific animal species. Some anthropologists thought the dolichocephalic Ice Age populations of Europe migrated into Europe from the south accompanied by an African fauna of hippopotamus and elephant, while the brachycephalic Ice Age peoples arrived from the north along with an Arctic fauna consisting of the mammoth and reindeer. Laing suggested that these populations could have met and interbred during the Stone Age producing mixed races, followed later by the invasion of superior Indo-European races. In his opinion, these Stone Age populations probably lived in Britain for an indefinite period of time until they were either exterminated or enslaved and assimilated by superior races that migrated into Britain shortly before the historical period (Laing and Huxley 1866: 63, 70–1). In this way, Laing was able to integrate his specific discoveries of Stone Age human skeletons from Caithness into the prevailing view of human prehistory whereby different "races" identifiable by their anatomical traits had successively migrated into Europe over the course of prehistory, culminating in the invasion of Indo-European peoples such as the Celts and the Germanic tribes. With the growing evidence for the geological antiquity of humans, it also became necessary to explore the possible links between Ice Age human populations and later Europeans.

However, there were scientists and scholars who questioned the series of Aryan migrations proposed by philologists and anthropologists. Thomas Huxley was one of them. He interpreted the human skeletons from Caithness and their place in European prehistory differently than Laing. Huxley remarked that "For anything we know, or have other than philological reasons for believing to the contrary, fair brachycephali ("Belgae", Fins, etc.); fair dolichocephali (Scandinavians); dark but pale-complexioned dolichocephali (Iberians, etc.); and dark and yellow skinned brachycephali (Lapps), may have inhabited some part or other of the area they now respectively occupy in Europe, for long ages before the dawn of history." He also offered this word of caution and an alternative explanation of the linguistic evidence. "Hence, it is worth while to reflect, that the current notions respecting the migrations of races from east to west may be myths developed out of the facts of philology; and that successive waves of language may have spread over Europe by washing over, instead of being carried by, its populations" (Laing and Huxley 1866: 133). There were different ways that prehistorians could interpret the skeletal, archeological, and philological evidence as they tried to identify the various "races" that inhabited Europe during prehistory and to determine how the racial composition of prehistoric Europe may have changed over time. But the anthropologists studying prehistoric skeletons relied upon craniometry as their method for identifying prehistoric races, and they believed in the validity of anthropology's ability to determine race on the basis of skeletal and especially cranial morphology.

4.4 The Caverne de l'Homme Mort and the Tombs of Petit-Morin

The human skeletons recovered from Stone Age tombs acquired a new importance when geologists began to discover fossilized human bones associated with Ice Age animal bones and crude stone tools in cave deposits. In Belgium, Édouard Dupont (1872) found human bones, including a human jaw from Trou de la Naulette, while excavating caves along the Lesse River valley from 1865 to 1867. While the geologic age of these bones was uncertain, there were fewer doubts about the human bones unearthed near the French village of Les Eyzies in 1868. Workmen constructing a railroad noticed bones in a rock shelter in the limestone cliff. The Minister of Public Education invited the geologist Louis Lartet to excavate the site at Cro-Magnon. Lartet found four adult human skeletons and one partial skeleton of a child, along with numerous artifacts and the bones of extinct animals (Lartet 1868b: 346–7). There was little doubt that the skeletons from Cro-Magnon dated to the Ice Age. Paul Broca (1868) examined the skeletons, employing craniometry to measure the skulls and comparing them with other prehistoric and modern skulls. Broca noted that the skulls differed in their morphology from other modern or prehistoric skulls. In his opinion, the troglodytes of Cro-Magnon belonged to a completely new and unknown prehistoric race.

The discovery of human fossils in Pleistocene deposits generated questions about the culture and the racial identify of these people, in much the same way that the recovery of artifacts and skeletons from prehistoric tombs had previously. They also raised questions about the fate of these Ice Age races and their relationship to the later Neolithic populations whose skeletons European anthropologists had been examining over the last few decades. Potential clues came from excavations conducted at the Caverne de l'Homme Mort and in the burial tombs at Petit-Morin. The Caverne de l'Homme Mort is located in the department of Lozère in southern France. Lozère contains numerous prehistoric tombs, and for some years Barthélémy Prunières, a physician living in Marvejols, had excavated these tombs. Prunières was interested in the prehistoric population of Lozère and, as a member of the Société d'Anthropologie de Paris as well as the anthropological societies located in Lille and Bordeaux, he had acquired a reputation as a serious researcher. In 1870 a local farmer gathering stones from the Caverne de l'Homme Mort stumbled across human bones and soon thereafter the vicar of Saint-Pierre-des-Tripiés learned about the discovery. The vicar was a friend of Prunières and occasionally joined him in excavating tombs, so he informed Prunières of the discovery. Despite sending instructions to the local parish priest to preserve the bones until he could inspect the site for himself, by the time Prunières arrived in June a third of the cave had been cleared and a large number of bones lay scattered on the ground outside the cave entrance (Prunières 1872: 430; Broca 1873a: 30). Despite this, Prunières set about collecting the bones and artifacts unearthed during the clearing of the cave, but he also excavated the portion of the cave that had not been disturbed.

Prunières found many human bones but no intact skeletons. In one part of the cave, he encountered nine skulls arranged in a row, although their mandibles were scattered. Prunières was able to recover 20 complete skulls in all, but many others lay broken in the nearby ravine where the farmer had tossed them when clearing the cave. Prunières (1873: 751; 1872: 430) kept five skulls for his private collection but sent the rest, along with a collection of bones and artifacts, to the Société d'Anthropologie de Paris with a note to Broca. Intrigued, Broca traveled to Lozère where he and Prunières inspected the cave, finding hearths and an array of artifacts. Broca also took the opportunity to measure the skulls of several local townspeople, who all turned out to be brachycephalic (Prunières 1873: 751–2). Prunières and Broca were eager to conduct further excavations at the site, but a few days after their visit to the cave, the Franco-Prussian War began, putting an end to those plans. Illness delayed Prunières' plans even further, but eventually in April 1872, he was able to renew his exploration of the cave. In the end, Prunières conducted three excavations at the Caverne de l'Homme-Mort (Prunières 1873: 750, 757).

Prunières collected the bones from about 50 skeletons, including 20 undamaged skulls. Among the artifacts were fragments of coarse pottery, flint tools, and a bone awl. He noted that none of the flint artifacts were polished except for a spearhead made from a fragment of polished axe. This was important because French archeologists distinguished an age of polished stone, which the English archeologist John Lubbock designated the Neolithic, from an earlier age of crudely chipped stone tools, which Lubbock called the Paleolithic. Judging from the artifacts found in the cave, and especially the lack of any Gallo-Roman pottery, Prunières (1872: 431) believed the skeletons in the cave dated from the early part of the "l'époque de la pierre polie" (early Neolithic) (see also Broca 1873a: 1). But it would require an anthropological analysis of the bones to determine who these people were.

Prunières sent his collection of bones to the Société d'Anthropologie de Paris where Broca examined them. He measured and described the skeletons, noting that they represented both men and women of a wide range of ages. The craniometric examination of the skulls showed they were orthognathic (flat-faced, lacking a protruding jaw) and extremely dolichocephalic (Broca 1873a: 12–13, 26). Broca had already determined that the local population was brachycephalic, and when he compared the skulls from the Caverne de l'Homme Mort with skulls Prunières had excavated from the prehistoric Gallic tombs of Lozère, he found the skulls were quite different. In fact, Broca (1873a: 26; 1873b: 763) observed, the skulls from the Caverne de l'Homme Mort did not resemble skulls from any known living population anywhere in Europe. He concluded that these ancient people had completely disappeared in France many centuries ago, probably during the early Neolithic. However, Broca noted that the skeletons from the Caverne de l'Homme Mort did share many features with the Ice Age skeletons from the Vezère valley, especially two skeletons from Cro-Magnon and a skeleton that Édouard Lartet had unearthed in the rock shelter at La Madeleine in 1864. The primary difference between these populations was that the people from the Caverne de l'Homme Mort were smaller in stature and less muscular than the Cro-Magnons. This suggested to Broca that the Neolithic people found in the Caverne de l'Homme Mort had descended from the Cro-Magnon race that inhabited France during the Pleistocene, with the skeleton from La Madeleine representing an intermediate transitional population linking the two. During the "incalculable number of centuries" that separated the Cro-Magnons from the people of the Caverne de l'Homme Mort, the climate, fauna, and food had undergone significant changes, especially the disappearance of the reindeer that had been such an important source of food for Ice Age peoples. Broca (1873a: 12–13, 26; 1873b: 765) thought these environmental changes explained why the Cro-Magnon people had gradually changed into the smaller inhabitants of the Caverne de l'Homme Mort. But he was convinced that both groups belonged to a single race that inhabited France from the time of the mammoths down to the early Neolithic.

For Broca, these ancient skeletons offered crucial insights into the ethnology of prehistoric France and of Europe in general. The excavation of skeletons from caves in France, Belgium, and Germany demonstrated that several distinct human races had inhabited Europe during the Pleistocene. Craniometric examination of Neolithic skeletons indicated that the Neolithic inhabitants of France were a different race from their Paleolithic predecessors. To explain the anatomical similarity (and supposed racial continuity) between the Paleolithic Cro-Magnons and the early Neolithic people of the Caverne de l'Homme Mort, Broca proposed the following scenario. While in many parts of France the Neolithic inhabitants (identified by their polished stone tools) exterminated the local Paleolithic inhabitants, in some remote places such as this gorge in Lozère, the Paleolithic peoples persisted and preserved some of their customs, while they also adopted polished stone tools and pottery. Broca concluded from the archeological and skeletal evidence that the people represented by the skeletons from the Caverne de l'Homme Mort, descendants of the Paleolithic Cro-Magnon race, had coexisted with the Neolithic newcomers represented by the skeletons found in the megalithic tombs located nearby. In Broca's opinion, that Neolithic race either eventually exterminated the older population or racially and culturally absorbed it (Broca 1873a: 10–11, 27–28, 49).

While Broca was examining the skeletons from the Caverne de l'Homme Mort, another series of excavations conducted east of Paris was providing additional information about the prehistoric inhabitants of France. Baron Joseph Berthelot de Baye was a French aristocrat who began collecting prehistoric artifacts when he was a boy. He began pursuing archeology seriously in 1871 under the instruction of Abbé Alphonse Bordé. In 1872 de Baye began excavating a series of caves cut into the chalk cliffs along the Petit-Morin valley located in the department of Marne. The caves appeared to have been artificially created since their walls bore the marks of the flint tools used to make them. Among these, de Baye (1874: 227–32) identified small caves that appeared to be tombs since they usually contained many human skeletons. In one of these caves, for instance, he found 40 skeletons, of different ages and sexes. There were an even greater number of caves that were more spacious, with steps cut into the rock at their entrances, and they often had shelves cut into the walls. De Baye (1874: 232–4, 238, 243) thought these were the dwelling places of these ancient people. In addition to the human skeletons, he also found thousands of polished stone axes along with flint arrowheads, spearheads, and knives. There were also bone tools as well as necklaces made of shells.

De Baye presented his findings at the Congrès Archéologique de France in 1873 and at the Société d'Anthropologie de Paris in 1874. By this time, he had excavated about one hundred and twenty caves. Judging from the artifacts recovered, de Baye argued that these people had lived during the Neolithic period. Paul Broca (1875a: 28) was immediately interested in these new discoveries and traveled to examine the caves for himself. He agreed that the tombs dated from the late Neolithic, and he was eager to examine the human remains. De Baye had collected more than two hundred partial human skeletons, and he sent 44 of the best-preserved skulls, along with a large number of limb bones, to Broca in Paris. Broca measured their cranial capacity and determined their cephalic index, but the results were unusual. There were four brachycephalic skulls, three sub-brachycephalic, ten mesaticephalic, twelve sub-dolichocephalic, and ten dolichocephalic skulls. The maximum cephalic index he recorded was 85.71, but the minimum was 71,65. Since such a wide range did not normally occur within "pure races" (races pures), Broca (1875a: 30–1) concluded that the skeletons from the sepulchral caves of Petit-Morin were the product of a sub-brachycephalic population that had interbred with a dolichocephalic population.

This was consistent with the growing body of anthropological data about the prehistoric peoples that inhabited Europe. Broca noted that the skeletons excavated from Neolithic tombs in France belonged predominantly to a dolichocephalic race. There was also a sub-brachycephalic and mesaticephalic race, the so-called Furfooz race, that the Belgian geologist Édouard Dupont had discovered several years earlier. They appeared to have inhabited Belgium during the late Paleolithic or early Neolithic. Given the similarity in the cephalic index and the shape of some of the Petit-Morin skulls with skulls from the Furfooz caves, it was possible that the latter had come into contact and interbred with the local dolichocephalic race during the early Neolithic, producing a mixed-race population. Broca's measurements of the limb bones supported this hypothesis (Broca 1875a: 30–1). The skeletons from the Caverne de l'Homme Mort and from the sepulchral caves of Petit-Morin became important elements in the anthropological interpretation of the growing number of human "races" that seemed to have inhabited Europe during the Paleolithic and Neolithic periods because of the large number of well-preserved bones recovered from each site. These specimens were combined with previous data obtained from skeletons and archeological artifacts from across Europe in the effort to piece together an idea of who the prehistoric inhabitants of Europe were, what their physical features were, what relation these peoples bore to the archeological record, and how the racial composition of Europe had changed over time.

4.5 Human Fossils

As archeologists were extending knowledge of the human past into a prehistory that reached back into the Stone Age, geologists and paleontologists were beginning to make discoveries that would further extend human prehistory into a remote geological past that no one had suspected. Since the beginning of the nineteenth century, geologists had begun to examine the stratigraphic layers accessible in various regions, including the relatively recent diluvial deposits that were referred to as the diluvium. These deposits contained the bones of extinct animals such as the mammoth and the woolly rhinoceros. Geologists were also beginning to accumulate evidence that the climate had been much colder and that glaciers had covered parts of Europe at that time. The British geologist Charles Lyell initially referred to this period as the Post-Pliocene, but he later named it the Pleistocene period. Others referred to it as the Glacial Period, but everyone recognized that it was the last geologic period before the advent of the present world. Geologists and paleontologists throughout the first half of the century were eagerly exploring these glacial deposits in order to unearth the flora and fauna of this previous world. The precise date of the Ice Age was a point of considerable contention. Those who followed the uniformitarian geology of Lyell argued that the Pleistocene occurred hundreds of thousands of years ago, while other geologists argued for a much shorter timeframe.

The excavation of Pleistocene deposits and the discovery of fossils belonging to extinct animal species took on a new significance when crudely chipped flint artifacts began to be found among these fossil bones. At first, these artifacts were dismissed as much more recent Stone Age artifacts that had accidentally become introduced, through geological forces, into the older Pleistocene strata. However, in addition to these stone implements, human bones also began to appear in association with Pleistocene animal bones. William Buckland, professor of geology at Oxford University, unearthed a partial human skeleton from Goat's Hole cave, near Swansea in Wales, in 1823. The skeleton was found among mammoth, rhinoceros, and hyena bones and was accompanied by a flint artifact and ornaments made from shells and ivory. Buckland noted that the cave deposits "appears to have been disturbed by ancient diggings," and he observed that the artifacts found with the skeleton resembled artifacts that the English antiquary Richard Colt Hoare had found in prehistoric British tombs (Buckland 1823: 85–92). Buckland concluded that the human remains belonged to an ancient Briton who had been buried in the cave just before the Roman occupation of the island, and this conclusion was readily accepted by the scientific community (see Sommer 2007: 21–120).

Jules de Christol and Paul Tournal each encountered human bones and flint artifacts among extinct animal bones in cave deposits in France (Christol 1829; Tournal 1830). Meanwhile, in the early 1830s, the Belgian physician and amateur geologist Phillipe-Charles Schmerling began excavating caves near his home in Liége. He found mammoth, rhinoceros, hyena, cave bear, and other fossils. But in the Engis caves, he also found two human skulls and a variety of postcranial bones along with flint and bone artifacts. They were all associated with Pleistocene animal bones. Similarly, at Engihoul cave Schmerling unearthed more human bones and flint artifacts mixed among extinct animal bones. Schmerling published an account of his remarkable discoveries and stated that he was convinced that the human remains and artifacts had been deposited in the cave at the same time as the extinct animal bones and were not the result of accidental mixing, which meant that humans had lived during this remote geologic period (Schmerling 1833: 31,53–66).

During the 1840s and 1850s, geologists throughout Europe continued to encounter flint artifacts in diluvial and cave deposits in association with extinct animal bones. Historians of science have examined the debates these discoveries generated in the academies of Europe (Grayson 1983; Cohen and Hublin 1989; Trautmann 1992; Van Riper 1993; Sackett 2000). Scholars and scientists were convinced that human history extended roughly six thousand years into the past and that humans appeared in the world only after the period of the extinct diluvial fauna. This conviction was grounded in a complex mixture of the biblical chronology, the archeological record of ancient civilizations, and extant historical texts which were generally considered to be consistent with a relatively short human history. These historians also note that many of the early discoveries of Pleistocene human artifacts were made by amateur excavators or by minor scientists, while the leading geologists and archeologists of the century firmly rejected the scattered and controversial evidence for humans coexisting with extinct animals. They also note that many of these discoveries were made in cave deposits, which were notoriously complex, with many means by which later objects could have become accidentally mixed with older animal fossils. It was only after the extensive discoveries of the amateur French archeologist Jacques Boucher de Perthes in the Somme valley in France during the 1840s and the meticulous excavations of Brixham cave conducted by a respected committee of English geologists in 1858 that the "geological antiquity" of humans began to be widely accepted.

By 1860 geologists and prehistoric archeologists in many parts of Europe were finding further examples of flint artifacts associated with Pleistocene animal fossils. As the evidence for Ice Age humans accumulated a range of new questions emerged. Geologists disagreed over just how long ago the "Glacial Period" had come to an end and whether there had been a single glaciation or several glacial periods separated by warmer periods. It became increasingly important for geologists and archeologists to work out a chronology for the Pleistocene deposits and the stone artifacts found in those deposits. The answers to these questions were important because even if it was necessary to recognize that humans had inhabited the alien world of the Pleistocene, there was a tremendous difference between extending human prehistory by ten thousand years as opposed to a hundred thousand years (for a discussion of these debates see O'Connor 2007; McNabb 2012). Besides these geological questions, there was also the important anthropological question of the identity of these Ice Age humans. Were they the same peoples that were found in the Stone Age tombs of Europe, or did they belong to completely unknown races? The only way to answer this question was to find the fossilized remains of the makers of those enigmatic flint implements.

An overriding concern for the researchers now searching Europe's caves for information about these Ice Age humans was to distinguish artifacts and human remains that genuinely dated from the Pleistocene from artifacts and human remains dating from much more recent times that had become accidentally introduced into Pleistocene deposits, either through intrusive burials or as a result of geologic forces. It was notoriously difficult in some cases, especially in complex cave deposits, to be sure that some artifacts or human bones actually dated from the Pleistocene and some notable cases occurred where errors were made in this regard. One of the most important discoveries of human fossils that illustrates these challenges occurred in the Neander valley, near Düsseldorf in Germany. In 1856, quarry workers removing debris from the Feldhofer grotto unearthed some bones. The workers simply cast them aside, thinking they belonged to some animal, but their discovery came to the attention of a local schoolteacher and amateur geologist named Johann Carl Fuhlrott. Fuhlrott went to the site and retrieved as many of the bones as he could find and he carefully questioned the quarry workers about the precise location of the bones in the cave. Unfortunately, no artifacts or animal bones were present in the cave that could have indicated the antiquity of the human remains, and the skeleton's exact position in the cave's deposits was not recorded. However, Fuhlrott noted that the bones were completely fossilized and appeared to be geologically old. At this time, the idea of the geological antiquity of humans had not yet been accepted, but Fuhlrott recognized the potential significance of the find and brought the bones to Hermann Schaaffhausen, an anatomist and anthropologist at the University of Bonn, to examine.

The fossils consisted of arm bones, two thigh bones, and other parts of the skeleton, but most importantly the upper portion of the cranium. The scientific significance of these fossils depended upon their geologic age, so in his description of the specimen, Schaaffhausen (1858: 454–459) described what was known about their stratigraphic location in the cave as well as the results of experiments that confirmed that the bones were truly fossilized. However, his real expertise lay in anthropology, so Schaaffhausen devoted most of his attention to a meticulous examination and description of the bones. He drew particular attention to the peculiar morphology of the cranium. It was very long, with a low forehead, but it had very prominent eyebrow ridges. Schaaffhausen employed the anthropological and craniometric methods of his contemporaries in his analysis of the Neanderthal specimen. He compared the Neanderthal cranium with modern human skulls and with prehistoric German and Celtic skulls, noting that while it shared some features with prehistoric skulls the Neanderthal cranium also possessed features seen in the skulls of some apes (1858: 461).

In his final assessment of the fossils, Schaaffhausen concluded that the Neanderthal skeleton dated to a period before the arrival of the Celts and Germans in Europe. Thus, they probably belonged to one of the "wild tribes of northwestern Europe" (wilden Stämme des nordwestlichen Europa) that were mentioned by Roman writers. This would make them representatives of the autochthonous population who inhabited Germany before the first Indo-European peoples had migrated into the region. Schaaffhausen suggested that this race had lived just at the end of the "diluvial period" (the Pleistocene), although he admitted that the geological evidence was insufficient to prove such a radical conclusion. He was aware that ancient human skeletons had been unearthed in other parts of Europe and these might represent the original inhabitants of those regions, but he believed the Neanderthal skeleton constituted the "most ancient memorial of the early inhabitants of Europe" (für das älteste Denkmal der früheren Bewohner Europa's gehalten warden). Significantly, he also suggested that prior to the "historical period" many barbarous human races might have existed in Europe, only to later disappear just as many animal species had (1858: 453–454, 460, 477).

Schaaffhausen interpreted the Neanderthal specimen by utilizing the framework established by previous anthropologists confronted with ancient human skeletons. Relying upon anatomical and craniometric methods, he examined the bones and concluded that they belonged to a distinctive human race, remarkable for certain distinctive features that suggested a savage and barbarous race, but fully human. What gave this skeleton particular significance was its apparent geologic age. But if it truly dated to the Glacial Period, it could still be integrated into the chronological sequence of successive human races that had inhabited Europe. The Neanderthal fossils generated much discussion among European scientists (see Zängl-Kumpf 2006; Reader 1988: 1–15; Trinkaus and Shipman 1993: 46–90). One component of that debate focused on the supposed geological antiquity of the bones. The influential German anthropologist Rudolf Virchow strongly rejected the suggestion that the bones were ancient as did August Franz Mayer, Schaaffhausen's colleague at Bonn. Doubts over the geologic age of the fossils were compounded by debate over the strange anatomical characteristics of the bones. Virchow argued that these could be explained by the effects of rickets and severe arthritis. When the British zoologist and evolutionist Thomas Henry Huxley (1863: 149–184) examined casts of the Neanderthal cranium, he was struck by its ape-like features, but the size of the Neanderthal braincase convinced Huxley that this specimen belonged to a primitive human race and not to an extinct pre-human ancestor. Paige Madison (2016, 2020) has examined the scientific discussions that swirled around the interpretation of the Neanderthal specimen in Britain in the first decades after its discovery and how these were informed by the racial (and often racist) anthropology of the time. It is also worth noting that only William King (1864), professor of geology at Queen's College in Galway, Ireland, was willing at this time to propose that the fossils belonged to an extinct human species, which he named *Homo neanderthalensis*, and not simply an ancient human race.

Uncertainties over the geologic age of the Neanderthal fossils limited the influence they exerted in the 1860s, as geologists and archeologists searched for further evidence of Ice Age humans. These new investigations were producing exciting results. In the Périgord region of France, the French paleontologist Édouard Lartet and the English businessman Henry Christy began excavating the caves and rock shelters along the Vézère valley in 1863, which produced an amazing variety of artifacts and Pleistocene animal fossils. Meanwhile, in Belgium, the Académie Royale des Sciences de Belgique directed Édouard Dupont to explore the caves in the province of Namur. Dupont was a respected geologist and director of the Musée

Royal d'Histoire Naturelle in Brussels. In 1864 he began work at the Trou des Nutons in Furfooz and in the winter of 1865 he moved on to explore the caves along the Lesse valley. Over the course of 7 years, Dupont excavated about 60 caves and collected thousands of Pleistocene animal fossils and many artifacts made from flint, bone, and reindeer antler (Dupont 1872: 21–24). In 1866 he unearthed a human mandible along with Pleistocene fossils in the Trou de la Naulette. The mandible was striking in that it was very thick, lacked a chin, and possessed other peculiar morphological features (Dupont 1872: 99–100). Given the potential significance of this discovery, Dupont paid careful attention to the stratigraphy of the cave in order to be sure that the mandible had been deposited at the same time as the Pleistocene fossils.

Dupont (1867a, 1872: 85) suggested that at least four of the caves where he found artifacts and other evidence of human occupation (Trou Magrite, Trou de la Naulette, Trou Balleux, and Trou de Chaleux) dated to the Mammoth Age. Édouard Lartet (1861) had suggested that Pleistocene archeological sites could be arranged in a chronological sequence on the basis of the dominant animal species found at each site, since he believed that as Europe's climate changed during the Ice Age, different animals would have inhabited France. This idea was soon abandoned and was replaced by a chronological system based upon artifact typology devised by the French archeologist Gabriel de Mortillet. However, elements of Lartet's system continued to be used into the 1870s, such as references to a Mammoth Age (early Pleistocene) and to a Reindeer Age (late Pleistocene). Near the village of Furfooz, in the Lesse valley, Dupont found artifacts made from flint and bone in two caves (Trou de Nutons and Trou de Frontal) that he dated to the Reindeer Age on the basis of the predominant animal bones in these caves. Dupont considered the Trou de Frontal to be tremendously important because during his excavations in 1867 he discovered 16 human skeletons along with large numbers of finely made stone tools. Many of the bones were badly broken but among them were several fragmentary skulls, and two skulls were intact (1872: 130–131, 138–139).

Dupont was a geologist, so his expertise lay in excavating the fossils and artifacts and recording the geology and stratigraphy of each cave in order to assure the geological antiquity of any artifacts and human remains. However, he enlisted the assistance of Belgian zoologist Édouard van Beneden to identify the extinct animal bones, while the human bones were examined by anthropologists Armand de Quatrefages and Franz Ignaz Pruner. Dupont (1867b: 4) wanted to use the artifacts and human bones he had unearthed to reconstruct the character of these ancient people, their race, their industry, their manners, and their customs ("je vais essayer de reconstituer les caractères de ces hommes, leur race, leur industrie, leurs mœurs et même quelques-unes de leurs coutumes"). The craniological analysis of the human skulls conducted by Quatrefages and Pruner found that they were brachycephalic and quite different from "Indo-Germanic" skulls. Significantly, Dupont concluded that these inhabitants of Furfooz during the Reindeer Age were a distant and destroyed family of what Pruner believed were the primordial "Mongoloid" population that inhabited Europe before the Indo-European invasions (Dupont 1867b: 29–31; 1872: 142). Dupont had succeeded in finding invaluable new evidence that humans had inhabited Europe during the Pleistocene, but the human skeletons from the Trou de Frontal in Furfooz offer a cautionary example of the challenges faced by excavators during the early decades of paleoanthropology. Dupont was convinced these human skeletons dated to the late Pleistocene (Reindeer Age) but later research demonstrated that the Furfooz skeletons dated to the very early Neolithic or the Mesolithic (Fraipont 1900). The mistaken dating of the Furfooz skeletons, as well as other human remains found in the 1860s, introduced a certain amount of confusion into early paleoanthropological research that is important to acknowledge.

Dupont's discoveries in Belgium generated a great deal of discussion, especially the La Naulette jaw. But these were soon overshadowed by events at Les Eyzies, a village in the Vézère valley in France. Near the end of March in 1868, two contractors named Édouard-Francois Bertou-Meyrou and Justin Delmarès were removing gravel to use in the construction of a nearby road when they exposed a rock shelter. As they dug into the Cro-Magnon rock shelter, they found flint artifacts and animal fossils. They contacted Abel Laganne, who was the lead excavator that had assisted Édouard Lartet and Henry Christy with their investigation of Ice Age archeological sites in the region. Laganne examined the site and retrieved two human skulls and other parts of skeletons, as well as some artifacts. Given the potential scientific significance of the site, Victor Duruy, the Minister of Public Instruction, directed Louis Lartet to undertake a full excavation of the Cro-Magnon rock shelter (Lartet 1875: 66).

Lartet was the son of Édouard Lartet and inherited an interest in geology from his father. In 1864 and 1865, Lartet was a member of the expedition led by Honoré-Théodoric-Paul-Joseph d'Albert de Luynes to explore Palestine and investigate the geological history of the Dead Sea. Lartet had an opportunity to investigate some prehistoric archeological sites along the Dead Sea and in Syria during the expedition. Lartet was completing his doctoral thesis at the University of Paris when he received the invitation to excavate at Cro-Magnon. Throughout the 1860s, Lartet had assisted his father during his excavations of Pleistocene archeological sites in the Vézère valley, so he was well qualified to undertake this work. Lartet found animal bones, including mammoth and reindeer, along with artifacts made from flint and reindeer antler as well as an ornament made from perforated shells. More importantly, he recovered the partial skeletons of four adults and one infant (Lartet 1868a, b). Once again, a primary concern was to definitively show that the human skeletons and artifacts were deposited in the rock shelter at the same time as the Pleistocene animal fossils, so Lartet provided a careful account of the geology and stratigraphy of the site. Édouard Lartet (1868) published a description of the animal fossils that were found with the human remains. Both were convinced that the human remains were from the Pleistocene. The human skeletons generated great excitement because for years Édouard Lartet and Henry Christy had been finding the archeological artifacts produced by the Ice Age inhabitants of France, but little was known about the "ethnic character" of these people. It was hoped that these skeletons could offer answers to this question (Lartet 1868a: 144; 1868b: 336–7, 349) (Fig. 4.4).

Lartet presented a paper on his discoveries at Cro-Magnon at the Société d'Anthropologie de Paris on May 21, 1868. He acknowledged that as a geologist

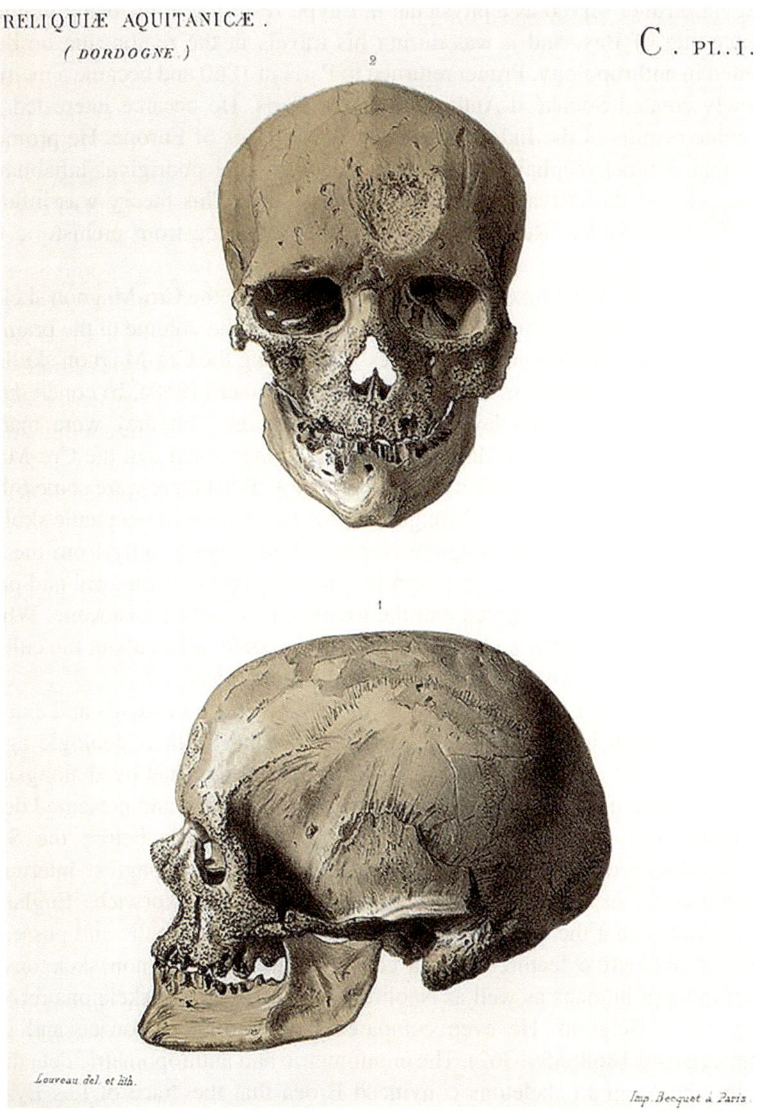

Fig. 4.4 Illustration of one of the skulls found in the Cro-Magnon rock shelter. (Édouard Lartet and Henry Christy, *Reliquæ Aquitanicæ*, London: Williams & Norate, 1875, Appendix "Description of the Plates" C. Skulls and Bones, Plate 1)

he could best describe the geological context of the Cro-Magnon discovery and that only anthropologists could address the question of their race. Among those in the audience for his presentation was the German anthropologist Franz Ignaz Pruner. Pruner studied medicine at the University of Munich, and in 1831, he went to Paris to continue his studies. He joined the expedition led by Karl von Hügels to the Levant and Egypt. Pruner served as a physician in Egypt, resulting in his being granted the honorary title of Bey, and it was during his travels in the region that he became interested in anthropology. Pruner returned to Paris in 1860 and became a member of the newly created Société d'Anthropologie de Paris. He became interested in the geographic origins of the Indo-Europeans and the races of Europe. He promoted a theory that a brachycephalic Mongolian race were the aboriginal inhabitants of Europe, prior to the arrival of Indo-European groups. This theory was influenced by the work of Anders Retzius and the skeletal evidence from prehistoric tombs (Pruner 1864a, b).

Pruner was one of the first anthropologists to examine the Cro-Magnon skeletons. He measured the cephalic index of the skulls as well as the volume of the brain cases and described their morphological features. Comparing the Cro-Magnon skulls with skulls representing various modern human races, Pruner (1868a, b) concluded that the Cro-Magnon crania displayed Mongoloid features. But they were markedly dolichocephalic, unlike most Mongoloid skulls. Pruner noted that the Cro-Magnon skulls resembled modern Estonian skulls, but he noted that there were some Siberian populations that belonged to the Mongoloid race but had dolichocephalic skulls. He also pointed out that the Cro-Magnon crania differed significantly from the skulls that Dupont described from the caves in Belgium. From the geological and paleontological evidence, Pruner agreed that the fossils dated to the Pleistocene. While he was curious about what the archeological evidence could tell us about the culture of these people, he was most interested in their racial identity.

Paul Broca took an immediate interest in the Cro-Magnon fossils and examined them soon after their discovery. Their significance lay in their geologic age and Broca accepted the evidence that the Cro-Magnon skeletons had lived alongside the mammoth during the Ice Age. Broca examined the skeletons and presented detailed descriptions and measurements of the Cro-Magnon fossils before the Société d'Anthropologie de Paris and at the meeting of the Congrès International d'Anthropologie et d'Archéologie Préhistorique held in Norwich, England, in 1868. He found that the Cro-Magnon skulls were dolichocephalic and possessed a number of distinctive features. Broca compared the Cro-Magnon skeletons with those of modern humans as well as Neolithic skeletons and the skeletons recovered by Dupont in Belgium. He even compared them with chimpanzee and gorilla skeletons (Broca 1868: 358–362). The craniometric and anthropometric data derived from the Cro-Magnon skeletons convinced Broca that the "race of Les Eyzies is entirely different from any other race, ancient or modern, that we have ever seen or heard of" (1869: 175). Broca was a leading polygenist in France, promoting the idea that the various human races represented distinct types that did not change over time, and these races had completely separate origins that could not be traced back to a common ancestor. As a consequence, Broca approached the interpretation of the Cro-Magnon skeletons from this polygenist perspective. Broca argued that distinct human races already existed during the Pleistocene, noting the significant anthropological differences between the Cro-Magnon skeletons and the skeletons from the Belgian caves. He also contrasted the Cro-Magnon mandibles with the morphologically different and more "simian" La Naulette mandible (1869: 170–172). As a consequence, he rejected the prevalent theory, promoted by Pruner and others, that a single aboriginal race had inhabited Europe at the end of the Pleistocene. He also argued that human fossils from the Ice Age could help resolve the debate between monogenists and polygenists (1868a, b: 351–352).

Broca also believed that certain cultural and intellectual characteristics of the Cro-Magnon people could be inferred from their anatomy and their archeological artifacts. He observed that

> The great volume of the brain, the development of the frontal region, the fine elliptical profile of the anterior portion of the skull, and the orthognathism of the upper facial region...are incontestable characters of superiority, which are met with usually only in the civilized races. On the other hand, the great breadth of face, the alveolar prognathism, the enormous development of the ascending ramus of the lower jaw, the extent and roughness of the muscular insertions, especially for the masticatory muscles, give rise to the idea of a violent and brutal race... Lastly we may refer to three characters (the excessive breadth of the ascending ramus, the subcoronoidal curve of the ulna, with its very shallow coronoid cavity, and, above all, the flattening of the tibias) as being more or less manifestly Simian; and we shall thus complete the picture of a race which in some of its features attained the highest and noblest stages in human morphology, and in other traits descended below even the lowest of the anthropologic types at present existing. (Broca 1875b: 120)

Broca compared the morphological features of the Cro-Magnon skeletons with those of modern human races, and he interpreted the meaning of those features from the context of nineteenth-century racial anthropology. Many European anthropologists organized the human races into a hierarchy of superior and inferior races, and they believed that a link existed between the physical characteristics of a race and the intellectual and cultural capacity of that race. Working from this framework, Broca argued that the Cro-Magnon skeletons indicated "a race, which, in some of its characteristics, attained the highest and noblest degrees of human morphology, and in others descended even below the most degraded anthropological types of the present day" (1869: 173). For Broca, these morphological traits and the artifacts found with the Cro-Magnon skeletons tell us that these cave dwellers were savages, as all humans of that time were thought to be, and that their skeletons reflected the anatomy of their difficult environmental and social conditions, but their skulls indicated their intelligence and ability to improve their condition of life (1868: 390).

Anthropologists and archeologists were still absorbing the significance of the Cro-Magnon discovery when amateur paleontologist and archeologist Élie Massénat unearthed a complete human skeleton from the rock shelter at Laugerie-Basse, only a few kilometers from the Cro-Magnon site. Massénat began his excavations in 1865 and soon found artifacts made from flint and reindeer antler as well as reindeer antler carved with images of horses and other animals. In the course of these excavations, Massénat unearthed the human skeleton, which had been buried in a crouched position. From the geological context of the burial and the associated artifacts, Massénat and French archeologist Émile Cartailhac argued that the skeleton dated to the Reindeer Age at the end of the Pleistocene (Massénat et al. 1872). These fossils were given to Paul Broca to examine, and as his description of them appeared

(Broca 1873c), news arrived of new discoveries of Ice Age human skeletons from Italy.

Émile Rivière had been studying medicine when an illness prompted him to travel to Cannes in 1868 in the hope that this would improve his health. In April 1869, Rivière and his friend Stanislas Bonfils visited the caves of Balzi Rossi (Baoussé-Roussé), located in the commune of Grimaldi just across the border in Italy. Bonfils operated a natural history museum in the French village of Menton, just across the border, and had previously collected fossils and artifacts from these caves. Filled with a new passion for prehistoric archeology, Rivière obtained permission from the Italian government to pursue research there. Between 1870 and 1875, Rivière conducted a systematic and careful excavation of nine caves at Balzi Rossi. In March 1872, he unearthed a human skeleton in the Grotte du Cavillon. The skeleton was covered in black hematite and red ochre and was found with an ornament made of shells and deer teeth as well as stone artifacts. Rivière found three additional skeletons, one adult and two children, in the Grotte Bausse da Torre in 1873. All of the skeletons were associated with stone artifacts and ornaments made of shells, as well as Pleistocene animal bones belonging to cave bear, hyena, and rhinoceros (Rivière 1887, 1872a, b, 1873, 1874).

As with other such discoveries, Rivière had to make a strong case based on the geological evidence that the human remains had been deposited at the same time as the Pleistocene animal fossils. The animal fossils and the human skeletons were sent to the Muséum National d'Histoire Naturelle in Paris, where paleontologist Albert Gaudry and the zoologist Henri Milne-Edwards helped him identify the animal species from the caves. Rivière was convinced of the geological antiquity of the human skeletons, noting that no polished stone artifacts or pottery was found in the caves, and he described the flint and antler artifacts found with the skeletons. Rivière relied upon his knowledge of anatomy to describe the skeletons. He published measurements for some of the bones and determined that the skulls were dolichocephalic. He particularly compared these skeletons with the Cro-Magnon skeletons and noted the morphological similarity of the Baoussé-Roussé skulls with the Cro-Magnon skulls (Rivière 1872a: 230, 232; 1872b: 166–167; 1874: 96).

By 1873, excavators across Europe had discovered human skeletons from Pleistocene deposits in various parts of Europe, although the geologic age of many of these was disputed. Yet, except for the Neanderthal and the Cro-Magnon skeletons, many of the other specimens had not been fully examined and described. The scientific importance of these purported Pleistocene inhabitants of Europe and the growing public discussion surrounding the nature of these Ice Age people prompted Ernest-Théodore Hamy to undertake a broad and systematic study of the known Pleistocene human fossils. Hamy studied medicine at the Faculty of Medicine in Paris, but while serving as an intern at the Salpêtrière hospital he met Paul Broca. Since he already possessed an interest in anthropology, Broca invited the young Hamy to become an assistant at the Société d'Anthropologie de Paris. Hamy also began taking courses on osteology from Henri Milne-Edwards and on prehistory with Édouard Lartet at the Muséum National d'Histoire Naturelle in 1864. Hamy worked closely with Broca on several projects, and in 1868, Hamy was appointed to the position of préparateur and chef de travaux at the Laboratoire d'Anthropologie at the École Pratique des Hautes Études. However, in 1872 Hamy left this position to become aide-naturaliste (assistant) to Armand de Quatrefages, who held the chair of anthropology at the Muséum National d'Histoire Naturelle. Hamy had already become deeply involved in prehistoric anthropological research by the time he made the move to the museum. In June 1870, Hamy had arranged an agreement with the Paris publisher Jean-Baptiste Baillière to produce a book on the human races, including prehistoric human races. As part of this project, Hamy began working with Quatrefages to examine the known Ice Age human fossils from across Europe.

Hamy (1871: 372) stated that he decided to undertake a systematic investigation of human fossils because he believed there were considerable gaps in the study of the existing human fossil record. Furthermore, he argued that the study of the anatomy of "primitive man" (l'homme primitive) had a more solid foundation than it had in previous decades and, therefore, making an "ethnic determination" (détermination ethnique) of prehistoric bones was easier to accomplish than it had been for his predecessors (1871: 375). He began with a reexamination of the three cranial fragments and two mandibles that Constantin Malaise, professor of natural sciences in Belgium, discovered at Engihoul Cave in 1860 (Malaise 1860). Malaise had provided a brief description of the bones in his account of the find, but Hamy now subjected them to a careful anthropological analysis and compared them with other prehistoric human specimens, including the Furfooz, Neanderthal, and Cro-Magnon fossils as well as Neolithic skeletons. Hamy found that the Engihoul parietal bone differed from the Neanderthal cranium but resembled the Cro-Magnon crania, and the Engihoul occipital bone resembled those from Cro-Magnon. He also noted that the morphology of the Engihoul mandibles differed significantly from the La Naulette mandible but closely resembled the mandibles from Cro-Magnon. Hamy concluded that the Engihoul fossils were very like those from Cro-Magnon, and therefore, the latter were not unique and the Cro-Magnon people had extended from France into Belgium (1871: 376–386).

Hamy turned his attention next to the skeleton that Élie Massénat excavated from the rock shelter at Laugerie-Basse. While the geological and archeological artifacts from the site had garnered much attention, Hamy lamented the fact that no anatomical or anthropological study of the skeleton had yet been conducted. He published a detailed description of the measurements and morphology of the bones and noted the similarity of this skeleton with those from Cro-Magnon (Hamy 1874). By examining the human remains from Laugerie-Basse and comparing them with the skeletons from Cro-Magnon, Hamy (1875: 256, 265) stated that he hoped to determine the "ethnical characteristics" of their "race." This was important in part because these ancient skeletons from the Vézère valley might provide information about the ethnic identity of the people who had inhabited France during the "transition period which unites the age of the Extinct with that of the Retreating Animals," in other words the last phase of the Pleistocene (1875: 265). Hamy also studied the human skeletons that Émile Rivière discovered at Baoussé-Roussé. Again, he carefully measured the bones and described their morphology and subjected the skulls to craniometric analysis. When these skeletons were compared with other Pleistocene human skeletons, it was clear that they closely resembled the skeletons from Cro-Magnon, right down to their mode of burial. For Hamy, it was clear that they belonged to the same "race" (Hamy 1872; Quatrefages and Hamy 1882: 61–62).

Hamy, working with Armand de Quatrefages, examined as many human fossils as they could gain access to, and they compared these fossils with one another as well as with other prehistoric and modern human skeletons. Their systematic survey of the human fossil record resulted in a comprehensive new interpretation of the Pleistocene inhabitants of Europe. In a series of papers published in 1873 and 1874, Hamy and Quatrefages identified several "fossil human races" (races humaines fossiles) represented in the Pleistocene human remains collected across Europe. In their first paper of the series, Hamy and Quatrefages described what they referred to as human fossils belonging to a "Canstadt race," what we today would consider Neanderthals. The name was unfortunate because it derived from a human skull that had been discovered in Germany in 1700 during excavations ordered by Duke Eberhard-Ludwig von Württemberg. It had remained unstudied, but in June 1870 Hamy and Quatrefages were allowed to examine the skull, which was mistakenly believed to be prehistoric. The skull was very dolichocephalic, and when Hamy and Quatrefages compared it with other prehistoric skulls, they found that it shared many characteristics with the famous Neanderthal cranium as well as with a cranium that Charles-Frédéric Faudel unearthed at Eguisheim, in Germany, in 1865 as well as a partial cranium found near the Bohemian town of Brüx in 1871. The La Naulette mandible from Belgium was included because of its morphological similarity with other Neanderthal fossil jaw bones (Quatrefages and Hamy 1873: 520–521). For Hamy and Quatrefages, these fossils represented a distinct human race that lived in Europe during the Pleistocene, which possessed particular morphological characteristics that could be identified through craniometry and the morphological features of the bones. Yet, they noted that similar morphological characteristics could be found in Neolithic skeletons buried in megalithic tombs as well as in skeletons dating from the Medieval Period. They explained this as being the result of interbreeding and the mixing of the Canstadt people with other races (1873: 522).

In addition to this Neanderthal race, Hamy and Quatrefages identified a separate and distinct Cro-Magnon race. Fossils belonging to this race included the skeletons found in the Cro-Magnon rock shelter as well as the human remains that Massenat had found at Laugerie-Basse and the skeletons from Baoussé-Roussé, plus some fossils from several other sites. Skulls of the Cro-Magnon race were dolichocephalic but differed in many ways from skulls belonging to the Neanderthal (Canstadt) race, and the rest of their skeletons differed significantly as well. Hamy and Quatrefages outlined the typical morphological features of these Cro-Magnon people and noted that they seemed to have inhabited the region extending from the Vézère valley in France through northern France and south into Italy during the Pleistocene. As with the Canstadt race, Hamy and Quatrefages recognized Cro-Magnon features in some Neolithic skeletons (Quatrefages and Hamy 1874a, c; 1882: 44–98). Hamy and Quatrefages (1874b) also identified four brachycephalic human races that they thought inhabited Europe during the Pleistocene, represented by the Furfooz crania from Belgium as well as crania found by quarry workers at Grenelle, near Paris, in 1867 and another skull found at Truchère, near Lyon in 1868. All of these fossils were thought to date to the Pleistocene, but by the early twentieth century, they were all recognized as much more recent remains that had gotten interred among Pleistocene animal fossils.

Their survey and interpretation of the human fossil record, which they expanded upon in *Crania ethnica* (1882), is instructive when viewed in the context of other studies of Pleistocene human fossils because it tells us a great deal about the way anthropologists and anatomists investigated and understood these specimens. Hermann Schaaffhausen, Paul Broca, Ernest-Théodore Hamy, and many of the other researchers of this generation who examined these first human fossils to be accepted as authentic Pleistocene remains were trained in an anthropological tradition that sought to identify and characterize the existing human races. The notion that the human species was divided into distinct races, and that these racial distinctions reflected important biological, cultural, and intellectual differences, was widely held by nineteenth-century anthropologists even though monogenists and polygenists disagreed over the origin, nature, and degree of these differences. A consistent theme in most of the descriptions and analyses of the human fossils discussed above is that the anthropologists and anatomists who studied them considered these skeletons to belong to humans, not to some prehuman evolutionary ancestor, although the influence of the new theory of evolution can be seen in occasional references to simian characteristics in some fossils.

The most important question arising from the discovery of Pleistocene human skeletons was to determine what race they belonged to. The established scientific method to answer this question was to examine these fossils using the accepted anthropological methods of craniometry and anthropometry. Equally important, however, and not generally acknowledged, is that the anthropological examination of Neolithic and Bronze Age skeletons that had been occurring for several decades by the time the geological antiquity of humans was accepted played an influential role in the way Pleistocene human remains were studied and interpreted. Human fossil bones were compared with various Neolithic and Bronze Age skeletons, as well as modern human skeletons from various "races" in order to identify similarities and differences. Skills and techniques developed in the process of examining prehistoric human skeletons were applied to the investigation of Ice Age human remains. These formed the methodological foundations of early paleoanthropology.

In addition to providing the accepted methods for how to analyze and interpret Pleistocene human skeletons, the study of Neolithic and Bronze Age skeletons conducted during the middle of the century had also produced a theoretical framework to explain the "racial" differences observed in skeletons from different parts of Europe and from different periods of prehistory. Anthropologists and archeologists argued that there had been successive waves of migration of different races of people into Europe and within Europe. Thus, Stone Age peoples had been succeeded by new races of people during the Bronze Age who in turn were succeeded by other newcomers during the Iron Age. When geologists and archeologists began to discover even more ancient human skeletons from the Pleistocene, these skeletons could quite easily be explained and situated within an existing anthropological and archeological framework. These "fossil human races" simply represented the hitherto unknown peoples that inhabited Europe during the Pleistocene. They now formed the first step in the succession of prehistoric human races that anthropologists had been discovering as they examined human skeletons found in prehistoric tombs. The anthropologists who studied Neolithic (Stone Age) and Bronze Age skeletons during the 1850s and 1860s had established a successful methodological and theoretical framework for explaining and understanding them that was then applied to a new set of even earlier human skeletons from the Pleistocene. The methods for studying these skeletons and the theoretical framework for understanding them remained largely the same. Questions about human evolution certainly existed at this time, but they did not dominate the discussion of the Pleistocene human fossils discovered before the 1880s.

4.6 Conclusion: Establishing Paleoanthropology as a Scientific Discipline

The discovery of fossilized human bones from Pleistocene deposits generated great scientific and public interest. These finds not only contributed to the extension of human prehistory back into the Ice Age, but the fossils themselves offered important evidence pertaining to the identity of these ancient peoples. The first generation of anatomists and anthropologists to examine these human bones, which included Hermann Schaaffhausen, Paul Broca, Ernest-Théodore Hamy, and others, was laying the foundation for a new science that began to be called paleoanthropology. Yet, in doing so they relied on the anthropological interests and methodologies that they had already employed in examining prehistoric human skeletons found in Neolithic, Bronze Age, and Iron Age tombs. The term *paléo-anthropologie* began to be used by French scientists in the late 1860s to refer generally to the study of human paleontology. Clémence Royer (1879) was one of the first people to use the term paléo-anthropologie to specifically refer to the study of the "lost races of humans" through the examination of human fossils. Royer was a French scholar and activist who was particularly interested in anthropology and biology. She produced the first French translation of Darwin's *On the Origin of Species* in 1862 and was the first female member of the Société d'Anthropologie de Paris (Blanckaert 1982; Fraisse 1985; Harvey 2008). Royer recognized that a close relationship existed between paleoanthropology, the study of human fossil remains, and prehistoric archeology, the study of prehistoric artifacts. These two disciplines, she argued, "reveal to us the successive phases of the social evolution of humankind, of its customs, of its instincts and of its rudimentary institutions" (1879: 408). While Royer was one of the first writers to clearly distinguish the discipline of paleoanthropology from prehistoric archeology, the person most responsible for defining and promulgating the objectives of the new science of paleoanthropology was the French anthropologist Paul Topinard.

Topinard studied medicine and obtained his medical degree in 1860. He became increasingly involved in anthropological research after meeting Paul Broca, eventually becoming a professor at the École d'Anthropologie as well as a member of the Société d'Anthropologie de Paris and director of the journal *Revue d'Anthropologie*. In several works published between 1885 and 1891, Topinard clearly defined paleoanthropology and clarified its relationship to other existing sciences (for instance, 1885: 177–178). In a paper presented at the Paris meeting of the International Congress of Prehistoric Anthropology and Archaeology held in 1889, Topinard (1891) carefully distinguished between prehistoric archeology, which studies the artifacts produced by prehistoric humans, and paleoanthropology, which studies the physical characteristics and diverse races of prehistoric humans. In his opinion, prehistoric archeologists should excavate and interpret artifacts, while paleoanthropologists should focus on studying the fossil bones of prehistoric peoples in order to determine their racial characteristics. These distinctions had important implications for the different sets of skills and methods needed by each discipline. Prehistoric archeologists would need to know how to conduct excavations and interpret the geological, stratigraphic, and paleontological context of the artifacts they recovered. However, paleoanthropologists would need to be able to examine and interpret human bones and determine their racial characteristics, which meant they would need to be trained as naturalists or anatomists. The paleoanthropologist, wrote Topinard, "must be a zoologist or a physician, and have the fundamental knowledge that makes an anthropologist" (1891: 177–178).

Thus, paleoanthropology was originally understood to be the study of human fossils, and from the late nineteenth century through the early twentieth century, the individuals who most often examined and described these fossils were either trained as anatomists or anthropologists, with the remainder being primarily vertebrate paleontologists. Their methods and their research questions originated largely from anthropology. Of course, the scientific debate over human evolution was emerging at the same time that Pleistocene human fossils were being discovered in Europe. As paleoanthropology was being established as the anthropological analysis of human fossils, biologists were pursuing ideas about the evolution of humans. Richard Delisle (2007: Chap. 2) has observed that ideas about human evolution proposed during the first few decades after the publication of Charles Darwin's *On the Origin of Species* (1859) heavily relied on the comparative anatomy of monkeys, apes, and humans as well as anthropological data about human races and other notions drawn from zoology. The growing human fossil record was relevant to these discussions, and researchers did sometimes interpret Pleistocene human specimens from an evolutionary perspective. But the fact was that the skeletons so far discovered only offered ambiguous evidence for the evolution of humans from an apelike ancestor, since the Cro-Magnon and even the Neanderthal type skeletons were considered by most scientists to be fully human.

This situation changed after Belgian geologist Maximin Lohest unearthed two human skeletons from the Spy rock shelter in 1886. Lohest invited Julien Fraipont, professor of paleontology at the University of Liège, to study the skeletons, and they published a detailed account of the geology and archeology of the site as well as an anatomical and anthropological description of the skeletons. Fraipont drew upon the work of Broca, Hamy, Schaaffhausen, Thomas Henry Huxley, and others who analyzed the Cro-Magnon and Neanderthal specimens already known and he concluded that the Spy skeletons were Neanderthals (Fraipont and Lohest 1887). However, Fraipont's detailed analysis of these two Neanderthal skeletons clarified many questions about Neanderthal anatomy. Knowledge of Neanderthal anatomy expanded further when the Croatian paleontologist Dragutin Gorjanović-Kramberger assembled a large collection of Neanderthal fossils from a rock shelter near the town of Krapina during excavations conducted from 1899 to 1905 (Gorjanović-Kramberger 1906). The new Neanderthal specimens were influential in convincing some anthropologists, such as Gustav Schwalbe, that the Neanderthals were not a race of modern humans but rather an extinct species that Schwalbe (1901a, b) designated *Homo primigenius*. By the turn of the twentieth century, Schwalbe and other anthropologists increasingly viewed the Neanderthals from an evolutionary perspective and not simply as a barbarous Pleistocene human race.

Dutch anatomist Eugène Dubois' discovery of fossils on the island of Java during the early 1890s, which he attributed to a new species of extinct hominid he called *Pithecanthropus erectus*, also contributed to shifting paleoanthropology to a greater focus on interpreting the human fossil record from an evolutionary perspective (Theunissen 1989). By the turn of the twentieth century, the initial phase of paleoanthropology that had been dominated by anatomists and anthropologists who relied upon craniometry and anthropometry to identify the race of Pleistocene human fossils was coming to a close, and paleoanthropology began to be transformed into a science that studied and interpreted hominid fossils from an evolutionary perspective. The theoretical framework of the first phase, which saw Pleistocene human fossils as the aboriginal races that inhabited Europe during the Paleolithic, who were subsequently succeeded by the arrival of new races of people migrating into Europe first during the Neolithic and then by yet other races during the Bronze and the Iron Ages, was replaced by a new theoretical framework that saw *Pithecanthropus erectus*, *Homo heidelbergensis*, and *Homo neanderthalensis* as extinct hominids who had to be understood within a phylogenetic and evolutionary interpretation of the rise of modern humans. The transition from this first phase of paleoanthropology, with its particular methodological and theoretical framework for interpreting human fossils, to the second phase of paleoanthropology marks an important event in establishing the modern science of human origins research.

References

Anonymous (1863) Fouilles de Chamant; sépultures de l'âge de pierre. Bulletins de la Société d'Anthropologie de Paris 4:652–656

Anonymous (1864a) Sur la caverne sépulcrale d'Orrouy. Bulletins de la Société d'Anthropologie de Paris 5:56–62

Anonymous (1864b) Fouilles de Chamant, (âge de pierre). Bulletins de la Société d'Anthropologie de Paris 5:636–642

Bateman T (1861) Ten years' diggings in Celtic and Saxon grave hills, in the counties of Derby, Stafford, and York, from 1848 to 1858.George Allen & Sons, London
Benes T (2008) In Babel's shadow: language, philology, and the nation in nineteenth-century Germany. Wayne State University Press, Detroit
Blanckaert C (1982) L'anthropologie au féminin: Clémence Royer (1830–1902). Rev Synth 103: 23–38
Blanckaert C (1987) 'Les vicissitudes de l'angle facial' et les débuts de la craniométrie (1765–1875). Rev Synth 108:417–453
Blanckaert C (1989) L'indice céphalique et l'ethnogénie européenne: A. Retzius, P. Broca, F. Pruner-Bey (1840–1870). Bulletin de la Société d'Anthropologie de Paris 1:165–202
Blanckaert C (2009) De la race à l'évolution: Paul Broca et l'anthropologie française (1850–1900). L'Harmattan, Paris
Bowler P (1986) Theories of human evolution: a century of debate, 1844–1944. Johns Hopkins University Press, Baltimore
Bowler P (1989) Holding your head up high: degeneration and orthogenesis in theories of human evolution. In: Moore JR (ed) History, humanity and evolution: essays for John C. Greene. Cambridge University Press, Cambridge, pp 329–353
Broca P (1861) Sur des crânes provenant d'un cimetière de la Cité, antérieur au treizième siècle. Bulletin de la Société d'Anthropologie de Paris 2:501–513
Broca P (1864) Sur les crânes d'Orrouy. Bulletins de la Société d'Anthropologie de Paris 5:718–722
Broca P (1868) Sur les crânes et ossements des Eyzies. Bulletin de la Société d'Anthropologie de Paris 3:350–392
Broca P (1869) On the crania and bones of les Eyzies (Dordogne). Trans Int Congress Prehistoric Archaeol [Norwich] 1868:168–175
Broca P (1873a) Sur les crânes de la caverne de l'Homme-Mort (Lozère). Revue d'Anthropologie 2: 1–53
Broca P (1873b) Sur les crânes et ossements de la caverne de l'Homme-Mort. Association française pour l'avancement des sciences., Comptes-rendus de la 1re session 1872, pp 763–765
Broca P (1873c) Sur trois crânes de l'époque du renne, découverts par M. Elie Massénat à Laugerie-Basse. Bulletin de la Société d'Anthropologie de Paris 8:217–221
Broca P (1875a) Sur les crânes des grottes de Baye. Bulletin de la Société d'Anthropologie de Paris 10(1875):28–32
Broca P (1875b) On the human skulls and bones found in the cave of Cro-Magnon near les Eyzies. In: Lartet E, Christy H (eds) Reliquiae Aquitanicae, being contributions to anthropology and Palaeontology of Périgord and the adjoining provinces of Southern France. Williams and Morgate, London, pp 97–122
Buckland W (1823) Reliquiae Diluvianae; or, observations on the organic remains contained in caves, fissures, and Diluvial gravel, and on other geological phenomena, attesting the action of an Universal Deluge. John Murray, London
Cartailhac É (1889) La France préhistorique d'après les sépultures et les monuments. Baillière, Paris
Clermont N, Smith P (1990) Prehistoric, prehistory, prehistorian, who invented the terms. Antiquity 64:97–102
Cohen C, Hublin JJ (1989) Boucher de Perthes (1788–1868): Les Origines romantiques de la préhistoire. Belin, Paris
de Baye J (1874) Sur les grottes de la vallée de Petit-Morin (Marne). Bulletins de la Société d'Anthropologie de Paris 9:225–244
de Christol J (1829) Notice sur les ossemens humains fossiles des cavernes du département du Gard. J. Martel Ainé, Montpellier
de Quatrefages A, Hamy ET (1873) Races humaines fossiles – races de Canstadt. Bulletin de la Société d'Anthropologie de Paris 8:518–523
de Quatrefages A, Hamy ET (1874a) La race de Cro-Magnon dans l'espace et dans le temps. Bulletin de la Société d'Anthropologie de Paris 9:260–266

de Quatrefages A, Hamy ET (1874b) Races humaines fossiles, mésaticéphales et brachycéphales. Bulletin de la Société d'Anthropologie de Paris 9:819–826

de Quatrefages A, Hamy ET (1874c) Races humaines fossiles, Race de Cro- Magnon. C R Acad Sci Paris 78:861–867

de Quatrefages A, Hamy ET (1882) Crania ethnica. Baillière, Paris

Delisle R (1995) Human palaeontology and the evolutionary synthesis during the decade 1950–1960. In: Corbey R, Theunissen B (eds) Ape, man, Apeman: changing views since 1600. Leiden University, Department of Prehistory, pp 217–228

Delisle R (2007) Debating humankind's place in nature, 1860–2000: the nature of paleoanthropology. Pearson Prentice Hall, Upper Saddle River

Dewbury A (2007) The American school and scientific racism in early American anthropology. Hist Anthropol Annu 3:121–147

Dupont É (1867a) Etude sur cinq cavernes explorées dans la Vallée de la Lesse et le ravin de Falmignoul pendant l'été de 1866. Bulletin de l'Académie Royale des Sciences, Lettres et Beaux-Arts de Belgique 23:244–265

Dupont É (1867b) Étude sur l'ethnographie de l'homme de l'âge du Renne, dans les cavernes de la vallée de la Lesse, ses caractères; sa race; son industrie; ses mœurs. Mémoire de l'Académie Royale des Sciences de Belgique 19:1–76

Dupont É (1872) Les temps préhistoriques en Belgique: L'Homme pendant les Âges de la Pierre dans les environs de Dinant-sur-Meuse, 2nd edn. C. Muquardt, Bruxelles

Edwards W (1829) Des caractères physiologiques des races humaines considérés dans leurs rapports avec l'histoire. Compère Jeune, Paris

Foley RA (2001) In the shadow of the modern synthesis? Alternative perspectives on the last fifty years of paleoanthropology. Evol Anthropol 10:5–14

Forchhammer J, Steenstrup J, Worsaae J (1852) Undersögelser i geologisk-antiqvarisk Retning (Særskilt optryk af Oversigten over det Kongelige Danske Videnskabernes Selskabs Forhandlinger i Aarene 1851), Copenhagen

Fraipont J (1900) Les Néolithiques de la Meuse. Impr. Hayez, Bruxelles

Fraipont J, Lohest M (1887) La race humaine de Néanderthal ou de Canstadt en Belgique. Recherches ethnographiques sur des ossements humains, découverts dans des dépôts quaternaires d'une grotte à Spy et détermination de leur âge géologique. Arch Biol 7:587–757

Fraisse G (1985) Clémence Royer: Philosophe et femme de science. La Découverte, Paris

Goodrum MR (2014) Crafting a new science: defining paleoanthropology and its relationship to prehistoric archaeology. Isis 105:706–733

Gorjanović-Kramberger D (1906) Der Diluviale Mensch von Krapina in Kroatien: ein Beitrag zur Paläoanthropologie. Kneidel, Wiesbaden

Gräslund B (1987) The birth of prehistoric chronology: dating methods and dating systems in nineteenth-century Scandinavian archaeology. Cambridge University Press, Cambridge

Grayson D (1983) The establishment of human antiquity. Academic, New York

Grimm J (1848) Geschichte der deutschen Sprache, vol 2. Weidmannschen Buchhandlung, Leipzig

Hamy ET (1871) Sur quelques ossements humains fossiles de la seconde caverne d'Engihoul, près Liège. Bulletin de la Société d'Anthropologie de Paris 6:370–386

Hamy ET (1872) Observations à propos du squelette humain fossile des cavernes de Baoussé-Roussé, dites grottes de Menton. Bulletins de la Société d'Anthropologie de Paris 7:589–593

Hamy ET (1874) Description d'un squelette humain fossile de Laugerie-Basse. Bulletin de la Société d'Anthropologie de Paris 9:652–658

Hamy ET (1875) Fossil man from La Madelaine and Laugerie Basse. In: Lartet E, Christy H (eds) Reliquiae Aquitanicae, being contributions to anthropology and palaeontology of Périgord and the adjoining provinces of Southern France. Williams and Morgate, London, pp 255–272

Harvey J (2008) Darwin in a French dress: translating, publishing and supporting Darwin in nineteenth-century France. In: Engels EM, Glick TF (eds) The reception of Charles Darwin in Europe, vol 2. Continuum, New York, pp 354–374

Henke W (2015) Historical overview of paleoanthropological research. In: Henke W, Tattersall I (eds) Handbook of paleoanthropology, vol 3, 2nd edn. Springer, New York., vol. 1, pp 4–95
Hervé G (1899) Populations Mésolithiques et Néolithiques de l'Espagne et du Portugal. Revue Mensuelle de L'École d'Anthropologie de Paris 9:265–280
Huxley TH (1863) Evidence as to man's place in nature. Williams & Norgate, London
Jahn A, Uhlmann J, Jahn A (1857) Die Pfahlbau-Alterthümer von Moosseedorf, im Kanton Bern: Ein Beitrag zur ältesten Kultur- und Völkergeschichte Mittel-Europas. Huber & Co., Bern
Keller F (1866) The lake dwellings of Switzerland and other parts of Europe. trans. Lee JE. Longman, Green, London
King W (1864) The reputed fossil man of the Neanderthal. Quart J Sci 1:88–97
Kyllingstad J (2014) Measuring the master race: physical anthropology in Norway, 1890–1945. Open Book Publishers, Cambridge
Laing S (1865) On some ancient shell-mounds and graves, in Caithness. Anthropol Rev 3:xx–xl
Laing S, Huxley TH (1866) Prehistoric remains of Caithness: with notes on the human remains. Williams and Norgate, Edinburgh
Lartet É (1861) Nouvelles recherches sur la coexistence de l'homme et des grands mammifères fossiles. Annales des Sciences Naturelles (Zoologie) 15:177–253
Lartet É (1868) Remarques sur la faune de Cro-Magnon, d'après les débris osseux découverts soit dans la sépulture humaine, soit dans les restes de foyers placés à proximité. Annales des sciences naturelles: Zoologie et paléontologie 10:156–160
Lartet L (1868a) Mémoire sur une sépulture des anciens troglodytes du Périgord. Annales des sciences naturelles: Zoologie et paléontologie 10:133–145
Lartet L (1868b) Une sépulture des troglodytes du Périgord. Bulletins de la Société d'Anthropologie de Paris 3:335–349
Lartet L (1875) A burial-place of the cave-dwellers of Périgord. In: Lartet E, Christy H (eds) Reliquiae Aquitanicae, being contributions to anthropology and Palaeontology of Périgord and the adjoining provinces of southern France. Williams and Morgate, London, pp 62–72
Madison P (2016) The Most brutal of human skulls: measuring and knowing the first Neanderthal. Br J Hist Sci 49:411–432
Madison P (2020) Characterized by darkness: reconsidering the origins of the brutish Neanderthal. J Hist Biol 53:493–519
Malaise C (1860) Note sur quelques ossements humains fossiles et sur quelques silex taillés. Bulletin de l'Académie royale des Sciences, des Lettres et des Beaux-Arts de Belgique 11: 538–546
Mallory J (1989) In search of the indo-Europeans: language, archaeology and myth. Thames and Hudson, London
Massénat É, Lalande P, Cartailhac É (1872) Découverte d'un squelette humain de l'âge du renne, à Laugerie-Basse (Dordogne). Matériaux pour l'histoire primitive et naturelle de l'homme 7:224–228
McNabb J (2012) Dissent with modification: human origins, palaeolithic archaeology and evolutionary anthropology in Britain 1859–1901. Archaeopress, Oxford
Meijer M (1999) Race and aesthetics in the anthropology of Petrus Camper (1722–1789). Rodopi, Amsterdam
Müller M (1861) Lectures on the science of language. Longman, Green, Longman, and Roberts, London
O'Connor A (2007) Finding time for the old stone age: a history of Palaeolithic archaeology and Quaternary geology in Britain, 1860–1960. Oxford University Press, Oxford
Olender M (1992) The languages of paradise: race, religion, and philology in the nineteenth century. trans. A Goldhammer. Harvard University Press, Cambridge
Pictet A (1859–63) Les origines indo-européennes, ou, les aryas primitifs: essai de paléontologie linguistique. 2. J. Cherbuliez, Paris
Pineau H (1980) Paul Broca. L'anthropologiste et le biologiste. Cah Anthropol 1:47–93

Pruner-Bey FI (1864a) Sur l'origine asiatique des européens. Bulletins de la Société d'Anthropologie de Paris 5:223–242
Pruner-Bey FI (1864b) Sur les origines indo-européennes. Bulletins de la Société d'Anthropologie de Paris 5:331–334
Pruner-Bey FI (1868a) Description sommaire des restes humains découverts dans les Grottes de Cro-Magnon, près de la station des Eyzies, arrondissement de Sarlay (Dordogne), en Avril 1868. Annales des sciences naturelles: Zoologie et paléontologie 10:145–155
Pruner-Bey FI (1868b) Sur les ossements humains des Eyzies. Bulletins de la Société d'Anthropologie de Paris 3:416–446
Prunières B (1872) Sur une grotte funéraire dite la Grotte de l'Homme Mort à Saint-Pierre-des-Tripiés (Lozère). Bulletins de la Société d'Anthropologie de Paris 6:428–433
Prunières B (1873) Sur la grotte de Saint-Pierre, dans la Lozère (Caverne de l'Homme Mort). Association française pour l'avancement des sciences., Comptes-rendus de la 1re session 1872, pp 748–760
Rabault-Feuerhahn P (2008) L'archive des origines: Sanskrit, philologie, anthropologie dans l'Allemagne du XIXe siècle. Cerf, Paris
Reader J (1988) Missing links: the hunt for earliest man, 2nd edn. Penguin, London
Retzius A (1843) Om formen af Nordboernes Cranier. P. A. Norstedt, Stockholm
Retzius A (1849) Kraniologisches. Archiv für Anatomie, Physiologie und wissenschaftliche Medizin:554–582
Rivière É (1872a) Sur le squelette humain trouvé dans les cavernes des Baoussé-Roussé (Italie), dites grottes de Menton, le 26 Mars 1872. Matériaux pour l'histoire primitive et naturelle de l'homme 7:228–232
Rivière É (1872b) Note sur l'homme fossile des cavernes de Baoussé Roussé en Italie, dites grottes de Menton. Congrès international d'anthropologie et d'archéologie préhistoriques 1871:164–175
Rivière É (1873) Découverte d'un nouveau squelette humain de l'époque paléolithique dans les cavernes des Baoussé-Roussé (Italie), dites grottes de Menton. C R Hebd Seances Acad Sci 76: 1027–1031
Rivière É (1874) Sur trois nouveaux squelettes humains, découverts dans les grottes de Menton, et sur la disparition des silex taillés et leur remplacement par des instruments en grès et en calcaire. Matériaux pour l'histoire primitive et naturelle de l'homme 5:94–98
Rivière É (1887) De l'antiquité de l'Homme dans les Alpes-Maritimes. J.B. Baillère, Paris
Rodden J (1981) The development of the three age system: archaeology's first paradigm. In: Daniel G (ed) Towards a history of archaeology. Thames and Hudson, London, pp 51–68
Rowley-Conwy P (2006) The concept of prehistory and the invention of the terms 'prehistoric' and 'prehistorian': the Scandinavian origin, 1833–1850. Eur J Archaeol 9:103–130
Rowley-Conwy P (2007) From genesis to prehistory: the archaeological three age system and its contested reception in Denmark, Britain, and Ireland. Oxford University Press, Oxford
Royer C (1879) Les Congrès d'anthropologie, de démographie et d'ethnographie au Trocadéro en 1878. Journal des économistes 5:405–420
Rütimeyer L, His W (1864) Crania helvetica: Sammlung schweizerischer Schädelformen. H. Georg, Basel and Genf
Sackett J (2000) Human antiquity and the old stone age: the nineteenth century background to paleoanthropology. Evol Anthropol 9:37–49
Schaaffhausen H (1858) Zur Kenntniß der ältesten Rassenschädel. Archiv für Anatomie, Physiologie und wissenschaftliche Medicin 5:453–478
Schiller F (1979) Paul Broca: founder of French anthropology, explorer of the brain. University of California Press, Berkeley
Schmerling PC (1833) Recherches sur des ossements fossiles découverts dans les cavernes de la province de Liège, vol 3. Collardin, Liège
Schwalbe G (1901a) Der Neanderthalschädel. C. Georgi, Bonn

Schwalbe G (1901b) Ueber die specifischen Merkmale des Neanderthalschädels. Verh Anat Ges 19: 44–61

Serres M (1845) Sur le monument et les ossements celtiques, découverts à Meudon en juillet 1845. Comptes-rendus des séances de l'Académie des sciences 21:607–619

Smocovitis VB (2012) Humanizing evolution: anthropology, the evolutionary synthesis, and the prehistory of biological anthropology, 1927–1962. Curr Anthropol 53:S108–S125

Sommer M (2007) Bones and ochre: the curious afterlife of the red lady of Paviland. Harvard University Press, Cambridge

Sommer M (2015) Evolutionäre Anthropologie zur Einführung. Junius, Hamburg

Sommer M (2024) The Diagrammatics of 'race': visualizing human relatedness in the history of physical, evolutionary, and genetic anthropology, ca. 1770–2020. Open Book Publishers, Cambridge

Tattersall I (1995) The fossil trail: how we know what we think we know about human evolution. Oxford University Press, Oxford

Tattersall I (2012) Paleoanthropology and evolutionary theory. Hist Philos Life Sci 34:259–281

Theunissen B (1989) Eugène Dubois and the ape-man from Java: the history of the first missing link and its discoverer. Kluwer Academic, Dordrecht/Boston

Thurnam J (1865) On the two principal forms of ancient British and Gaulish skulls. Memoirs of the Anthropological Society of London 1(120–168):459–519

Topinard P (1885) Éléments d'anthropologie générale. Adrien Delahaye et Émile Lecrosnier, Paris

Topinard P (1891) La paléo-anthropologie. Congrès international d'anthropologie et d'archéologie préhistoriques Compte Rendu 1889:382–392

Tournal P (1830) Observations sur les ossements humains et les objets de fabrication humaine confondus avec des ossements de mammifères appartenant à des espèces perdues. Bulletin de la Société Géologique de France 1:195–200

Trautmann T (1992) The Revolution in Ethnological Time. Man 27:379–397

Trinkaus E, Shipman P (1993) The Neandertals: changing the image of mankind. Knopf, New York

Troyon F (1860) Habitations lacustres des temps anciens et modernes. G. Bridel, Lausanne

Van Riper AB (1993) Men among the mammoths: Victorian science and the discovery of human prehistory. University of Chicago Press, Chicago

Welcker H (1862) Untersuchungen über Wachstum und Bau des menschlichen Schädels. Wilhelm Engelmann, Leipzig

Wilson D (1851) The Archæology and prehistoric annals of Scotland. Sutherland and Knox, Edinburgh

Worsaae J (1849) The primeval antiquities of Denmark [Danmarks oldtid oplust ved oldsager og gravhøie]. trans. Thoms WJ. John Henry Parker, London

Wuketits FM (2015) Charles Darwin, paleoanthropology, and the modern synthesis. In: Henke W, Tattersall I (eds) Handbook of paleoanthropology, vol 3, 2nd edn. Springer, New York., vol. 1, pp 97–125

Zängl-Kumpf U (2006) Hermann Schaaffhausen (1816–1893) and the Neanderthal finds of the 19th century. In: Schmitz RW (ed) Neanderthal 1856–2006. Verlag Philipp von Zabern, Mainz am Rhein, pp 45–53

Interpretations, Influences, Personalities, and the Phylogeny of Paleoanthropology

5

Robin Derricourt

Abstract

Much research in biological sciences involves significant samples of materials readily subject to repeat examination and investigation. As the field of paleoanthropology emerged, it was different: reliant on a small sample (or unique specimens) of individual fossil hominin finds. Their description was undertaken by a limited number of specialists whose interpretation was highly influential. One result has been the enthusiasm to name newly studied finds as a new species, or even a new genus, many of which in time would be merged with others. We have listed 104 different hominin names for a genus or species introduced since 1864. The inevitable subjectivity of these research scientists reflected their disciplinary background and training, their national and geographical context, as well as the approaches, prejudices and expectations of the period in which they were active, and their individual personalities. Some were influential mentors and developed their own school of followers. Here, we consider one genealogy or "phylogeny" of individuals active in the narratives of hominin evolution, working across current or former areas of Britain or its dominions: from Grafton Elliot Smith (1871–1937), Davidson Black (1884–1934), Joseph Shellshear (1885–1958), Raymond Dart (1893–1988), Phillip Tobias (1925–2012) to Lee Berger (1965–).

Keywords

Hominins · Paleoanthropologists · Lumpers · Splitters · Grafton Elliot Smith · Raymond Dart · Davidson Black · Joseph Shellshear · Phillip Tobias · Lee Berger · Piltdown

R. Derricourt (✉)
School of Humanities & Languages, University of New South Wales, Sydney, NSW, Australia
e-mail: r.derricourt@unsw.edu.au

M. Lequin, J. M. Rodríguez Caso (eds.), *The Field of Human Evolution: Critical Perspectives from History and Epistemology*, Evolutionary Biology – New Perspectives on Its Development 11, https://doi.org/10.1007/978-3-032-13925-2_5

5.1 Paleoanthropological Exceptionalism

The field of study which developed as paleoanthropology focused on the identification, description, classification and interpretation of hominin physical remains, providing hypotheses and models of evolutionary ancestry and relationships. The character, context and interactions of those who contributed to these studies can provide some insights into how the discipline developed. The nature of paleoanthropology and of the materials it examined have specific characteristics, with the personalities and links of key individuals, alongside national traits and affiliations, reflected in the pattern of assigning (or challenging) distinct species names to newly discovered fossil finds.

While many sciences regard their discipline as unique, the history of paleoanthropology in the twentieth century exhibited unusual and distinct characteristics, many of which may be seen in more recent experience.

Research workers in experimental sciences can report findings which are open to replication and testing by their peers, to confirm, go beyond or question published conclusions. In those scientific areas we could group as *observational*—such as found in geology, botany, zoology or astronomy—it may be typically classes of materials (often with many specimens or examples) which are examined, described, analysed and subject to taxonomic classifications. Others may then locate and study the same range of items. While the discipline of archeology in its formative era may have announced periodic major "discoveries", its development has focused on tracing patterns and trends across time and space, with emphasis on grouped continuities and associated change of material culture, settlement pattern or economic resource exploitation.

By contrast, much of the development of paleoanthropology was marked by the presentation of what were considered unique finds: typically a partial individual cranium or extended skeletal material. The tendency to assign a new species name—or even a new genus—to a new find was a notable feature of fossil hominin studies, resulting (as discussed below) in a somewhat astonishing catalogue of hominins. Over time previously such distinguished species would be combined while attracting criticism from influential senior figures such as Ernst Mayr, Theodosius Dobzhansky and George Gaylord Simpson as they witnessed the accumulation of names.

Given the scarcity of materials in the evolution of paleoanthropology, a large part of activity in the discipline has featured contributions and frequent disputes over the interpretation of previously reported fossils rather than the presentation of new fossil specimens. Especially in the early decades of these discussions, different models of evolutionary theory were in contention to influence interpretations of human ancestry.

All scholarly disciplines have their debates, and behind these may also lie personal, institutional or national rivalries. Episodes in the history of paleoanthropology place particular emphasis on the will, even the need, to disagree sometimes profoundly on the nature of a newly found fossil hominin. As a very recent survey suggests, "But making outlandish claims based on the flimsiest flakes of fossils is all part of the fun of paleoanthropology. The other fun bit is snarking in the pages of *Nature* at other paleoanthropologists for doing this" (Hassett 2022: 247).

An early friendship of Grafton Elliot Smith and Arthur Keith, then the leading human anatomists in London, fell apart initially as they disputed the nature of the Piltdown finds in 1913, and the model was set for what would continue as an energetically disputatious field. A published paper on a Pliocene hominin in 2022 generated the comment from another distinguished colleague: "These critically important fossils deserve better treatment than this shoddy paper provides. The study cherry-picks evidence, ignores recent studies that point to different conclusions than the ones the authors try to defend, and it fails to explore other equally, if not more likely, functional interpretations of these fossils" (Pare and Davis 2022). Social media have been engaged to promote scientific papers with two widely different interpretations of the culture of *Homo naledi*: one with a title beginning: "No scientific evidence that …" (Berger et al. 2023; Martinón-Torres et al. 2023).

The controversial and disputed original paper just mentioned had 37 named authors, reflecting the trend to list all major collaborators, interdisciplinary advisors and sometimes even involved local community members, as authors of a paper they may (or may not) have seen in its various drafts. But there is rarely doubt as to who the lead figure is in a fossil hominin discovery. Another feature of the history of paleoanthropology is the dominance and visibility of key individuals. This is not the god-professor in a hierarchical tradition of academia; it can be more the explorer-hero whose "discovery" may be in the field (like Louis Leakey) or in the laboratory (like Raymond Dart) or both.

Leading scientists in any subject area may expect respectful and admiring obituaries; some may receive the tribute of a conference meeting held in their honour. The importance of their work may be featured in newspaper and magazine articles or included within a television documentary. Many books which propose to present the story of human evolution for a popular general audience tend to focus as much on narratives of discovery and the range of personalities involved as on the nature and interpretation of their finds.

But leading individuals involved in fossil hominin discoveries can go beyond this—with lavish coverage in *National Geographic*, television films devoted to their lives, personalities and work, and dedicated single volumes recounting the narrative of their search and discovery process, whether written by others, by themselves, or in collaboration with a professional writer. Thus inter alia one can read full length books in English alone about Eugène Dubois (Theunissen 1988), G.H.R von Koenigswald (von Koenigswald 1956), Robert Broom (Broom 1950), Raymond Dart (Dart and Craig 1959; Tobias 1984; Wheelhouse and Smithford 2003), Louis Leakey (Leakey 1974; Cole 1975; Morell 1995), Donald Johanson (Johanson and Edey 1981), Richard Leakey (Leakey 1983), Phillip Tobias (Tobias 2005, 2013; Tobias et al. 2008) and Lee Berger (Berger and Hawks 2017, 2023).

5.2 The Speciation of Species (and Genera)

If the century of paleoanthropology from the 1860s to the 1960s was marked by a paucity of specimens to study, it was also marked by the small numbers of specialists undertaking such studies, and many of these could be employed in museums or in fields like human anatomy within medical schools rather than exclusively teaching and researching hominin evolution.

A larger number—especially in North America with its tradition of anthropology schools combining the "four fields" of Physical Anthropology, Cultural Anthropology, Archeology and Linguistics—could be employed to teach the broader area of physical anthropology. For long, this implied relative isolation of those undertaking what we would describe as primary research in paleoanthropology.

A key characteristic of the history of the discipline was the decision to assign a new species name—and often a new genus—to fossil skeletal material under study, leaving to others then or later to group it with other known examples; or sometimes just to rename it. Yet we cannot convincingly explain this just by the isolation of early individual workers, for scientific journals (and monographs) were widely distributed and readily available. Nor does it emerge as an inevitability from the fragmentary nature of fossil finds. It would seem unlikely that pioneer paleoanthropologists at the time were confident that there had been a large number of contemporary hominin species occupying the Pleistocene world of Africa and Eurasia. The dominance of individual scientists with strong personalities making such a choice has to be seen as part of the reason for the extraordinary burst of named species, with egos, institutional pride and nationalist affiliation as a background to the phenomenon.

In a paper I published in *Critique of Anthropology* in 2009, I sought to map the cumulative number of new hominin species and genera between 1890 and 2007 (Derricourt 2009a: 197, fig 1). Since then, I have noted a further nine names for eight hominins have emerged, including *Homo juluensis* from China, said to differ from known species (e.g. Wu and Bae 2025; Bae and Wu 2024). Figure 5.1 in this paper updates this chart with some corrections. It represents 31 genera and 73 hominin species names (listed in Table 5.1), which have been introduced into the scientific literature since 1864 (drawing data primarily from Schwartz and Tattersall 2002–5; Wilson and Reeder 2005 and subsequent hominin descriptions). This is a remarkable assemblage—a "taxonomic inflation" (Lequin 2022: 24)—which some might suggest says more about paleoanthropologists than paleoanthropology. Yet if 104 names were not enough to comprehend, another recent project (Reed et al. 2023), within a less selective if more ambitious framework, has compiled a database of a remarkable 206 classificatory genus and species names for hominin fossil types introduced, suggested or merely floated over 263 years, with more added since.

Reclassification of the fossil finds represented by this panoply of names is more than a simple stylistic or ideological preference between "lumpers" and "splitters", although that is part of the story. The most minimalist approach was probably that of Ernst Mayr who suggested the then known hominins could be grouped into one genus, with three species (*Homo transvaalensis*, *Homo erectus* and *Homo sapiens*);

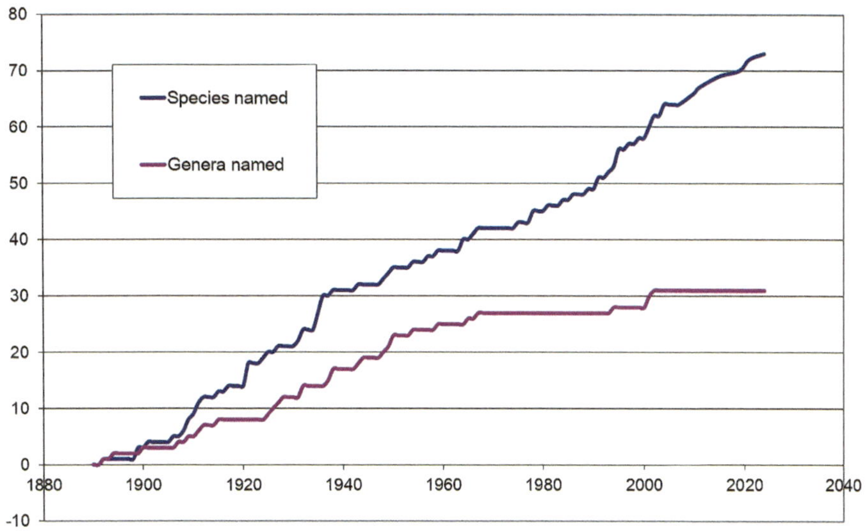

Fig. 5.1 Cumulation of generic and species names for hominins 1890–2024

contrasting the norms of zoology with a pattern in anthropology which was characterised by "the attempt to express every difference of morphology, even the slightest one, by a different name and to do this with the limited number of taxonomic categories that are available" (Mayr 1950: 109).

By the 1960s, many if not most scientific syntheses were satisfied to classify fossil hominin material within just two genera—*Australopithecus* and *Homo*. In part, this was a response to the criticisms by senior scholars as noted above, but it also reflected the perspectives of the post-war "Modern Evolutionary Synthesis", which sought to integrate a contemporary interpretation of Darwinism with the increasing understanding of genetics. Yet despite this, numerous new species names were to be introduced in the succeeding decades.

With the much larger and more diverse range of study material available today, each survey arranges with a different phylogeny those species they consider to include all known hominins. Lumpers may see 10 species, splitters may see 30, or it may be an irrelevant argument (van Holstein and Foley 2022: 167—though this paper suggests the usefulness of a classification with just 10 or alternatively 14 "metapopulations"). Another approach would be to bypass efforts to agree on the number of hominin species and indicate just a "simulum" with classification at genus level only (Lequin 2022: 28). By contrast, others feel confident that tightening of the definitions of what would define a hominin genus can help show what would belong within the *Homo* genus (Wood and Collard 2015). In another generation or two, we might possess so many hominin fossil finds that classification becomes clearer to agree on. Meanwhile, it remains to be seen how much ancient DNA studies will transform the interpretative landscape of the periods for which aDNA samples can be obtained. No less significant are changes and improvements in the way fossil

Table 5.1 Generic and species names applied to hominins

Genus		
Homo	Linnaeus	1758
Anthropopithecus	Dubois	1892
Pithecanthropus	Dubois	1894
Proanthropus	Wilser	1900
Palaeanthropus	Bonarelli	1907
Pseudhomo	Ameghino	1909
Notanthropus	Sergi	1911
Eoanthropus	Woodward	1912
Archanthropus	Arldt	1915
Australopithecus	Dart	1925
Anthropus	Boyd Dawkins	1926
Sinanthropus	Black and Zdansky	1927
Cyphanthropus	Pycraft	1928
Praehomo	von Eickstedt	1932
Javanthropus	Oppenoorth	1932
Plesianthropus	Broom	1937
Paranthropus	Broom	1938
Africanthropus	Weinert	1938
Maueranthropus	Montandon	1943
Meganthropus	Weidenreich	1944
Nipponanthropus	Hasebe	1948
Telanthropus	Broom and Robinson	1949
Pithecanthropus	von Koenigswald	1950
Europanthropus	Wüst	1950
Atlanthropus	Arambourg	1954
Zinjanthropus	Leakey et al.	1959
Tchadanthropus	Coppens	1965
Paraustralopithecus	Coppens	1967
Ardipithecus	White et al.	1994
Orrorin	Senut et al.	2001
Kenyanthropus	Leakey et al.	2001
Sahelanthropus	Brunet et al.	2002
Species		
Homo neanderthalensis	King	1864
Anthropopithecus erectus	Dubois	1892
Homo priscus	Lapouge	1899
Homo spelaeus	Lapouge	1899
Homo primigenius	Schwalbe	1901
Homo grimaldii	Lapouge	1906
Homo heidelbergensis	Schoetensack	1908
Homo spyensis	Krause	1909
Homo mousteriensis	Klaatsch and Hauser	1909

(continued)

Table 5.1 (continued)

Homo aurignacensis	Klaatsch and Hauser	1910
Homo calpicus	Keith	1911
Homo eurafricanus	Sergi	1911
Eoanthropus dawsoni	Woodward	1912
Homo proto-aethiopicus	Giuffrida-Ruggeri	1915
Homo capensis	Broom	1917
Homo wadjakensis	Dubois	1921
Homo cro-magnonensis	Gregory	1921
Homo grimaldiensis	Gregory	1921
Homo rhodesiensis	Woodward	1921
Homo gibraltarensis	Battaglia	1924
Australopithecus africanus	Dart	1925
Sinanthropus pekinensis	Black	1927
Homo drennani	Kleinschmidt	1931
Javanthropus soloensis	Oppenoorth	1932
Homo palestinus	McCown and Keith	1932
Homo neanderthalensis aniensis	Sergi	1935
Homo helmei	Dreyer	1935
Homo kanamensis	Leakey	1935
Australopithecus transvaalensis	Broom	1936
Homo modjokertensis	von Koenigswald	1936
Homo steinheimensis	Berckhemer	1936
Paranthropus robustus	Broom	1938
Homo sapiens protosapiens	Montandon	1943
Pithecanthropus robustus	Weidenreich	1945
Australopithecus prometheus	Dart	1948
Paranthropus crassidens	Broom	1949
Pithecanthropus dubius	von Koenigswald	1950
Atlanthropus mauritanicus	Arambourg	1954
Meganthropus paleojavanicus	Piveteau	1957
Zinjanthropus boisei	Leakey	1959
Sinanthropus lantianensis	Woodward	1964
Homo habilis	Leakey et al.	1964
Homo (erectus or sapiens) paleohungaricus	Thoma	1966
Australopithecus (or Paranthropus) aethiopicus	Coppens	1967
Homo ergaster	Groves and Mazák	1975
Australopithecus afarensis	Johanson	1978
Homo erectus bilzingslebenensis	Vlcek	1978
Homo sapiens daliensis	Wu	1981
Homo louisleakeyi	Kretzoi	1984
Homo rudolfensis	Alexeev	1986
Homo erectus reilingensis	Czarnetzki	1989
Homo erectus narmadensis	Sonakia	1991
Homo erectus wushanensis	Huang et al.	1991

(continued)

Table 5.1 (continued)

Homo cepranensis	Mallegni et al.	1993
Australopithecus anamensis	M Leakey et al.	1995
Australopithecus bahrelghazali	Brunet et al.	1995
Homo microcranous	Ferguson	1995
Homo antecessor	de Castro et al.	1997
Australopithecus garhi	Asfaw et al.	1999
Orrorin tugenensis	Senut et al.	2001
Kenyanthropus platyops	Leakey et al.	2001
Sahelanthropus tchadensis	Brunet et al.	2002
Homo georgicus	Vekua et al.	2002
Ardipithecus kadabba	Haile-Selassie et al.	2004
Homo floresiensis	Brown et al.	2004
Homo denisova	Kraus et al.	2010
Australopithecus sediba	Berger et al.	2010
Homo altaiensis	Derevianko	2011
Homo naledi	Berger et al.	2015
Australopithecus deyiremeda	Haile-Selassie et al.	2015
Homo luzonensis	Détroit et al.	2019
Homo longi	Ji et al.	2021
Homo bodoensis	Roksandic et al.	2021
Homo juluensis	Bae and Wu	2024

remains and their contexts can be dated, which can amend perceptions of the relationships between potentially contemporary species (Grün and Stringer 2023).

Decisions on classificatory nomenclature over time have also reflected individual scientists' loyalties alongside their judgement. Historians of paleoanthropology, and those within the discipline, may choose to consider whether its early history involved active competition between Dutch and German and British scholars; or whether activities led by different 20th century teams in southern Ethiopia were complementary or in rivalry; whether it is realistic to write of "the shark tank of East African paleontology in the 1970s and 1980s" (Berger and Hawks 2017: 51); or whether major egos unafraid of publicity were essential to ensure adequate funding for major projects; or whether scholarship grants for future contributors and leaders in the field from, say, African countries established a beachhead for researchers from the supporting nation; or whether the domestic control of research agendas and projects set different emphases from those of overseas led research.

Paleoanthropology began with scientists from the prosperous nations of the world in the era of colonial powers and competing imperial influence, with national identity and loyalty present alongside commitment to the potential of the emerging field of study. We address here just one narrative in the "phylogeny" of paleoanthropological endeavour: a tradition of mentors that links Grafton Elliot Smith (born 1871) and the most current work a century and a half later at South Africa's Cradle of Humankind.

5.3 A Singular Clade

Many of the major players in the history of anthropology were celebrated for their independence and individualism. Others were linked in a tradition of training, research collaboration, employment and mentorship. Intellectual traditions may be linked by language, nation or institution. The substantial size, and growth, of the university sector in the United States allowed an analysis of a remarkable total of 672 primatologists (Kelley and Sussman 2007) with distinct linear descent in a "genealogy" of training and influence, with a majority of field primatologists tracing such intellectual ancestry through Sherwood Washburn (1911–2000) at the University of Chicago and the University of California Berkeley, back to Earnest Hooton (1887–1954) at Harvard University.

Not all the significant contributors to paleoanthropology from a university base created a cohort of protégés in the discipline. And many others, especially those based in museums, were not in a position to train and direct the research of students. Aleš Hrdlička (1869–1943) was based at the Smithsonian Institution in Washington DC; both Louis Leakey (1903–1972) and Richard Leakey (1944–2022) were primarily associated with the National Museum of Kenya.

One major tradition of anglophone science, based in Britain and its actual or former dominions, links the very recent work on *Homo naledi* back to the early development in Britain inspired and led by the influential if enigmatic figure Grafton Elliot Smith (1871–1937). Notable contemporaries in the emerging field of paleoanthropology were Arthur Keith (1866–1955), Arthur Smith Woodward (1864–1944) and Frederic Wood Jones (1879–1954), but Elliot Smith's impact was particularly significant both directly, and in terms of the phylogeny of the discipline, as his own disciples in turn inspired others who in turn repeated the process.

Elliot Smith was born in the small country town of Grafton in the Australian state of New South Wales in 1871, the year when Charles Darwin's book *The Descent of Man* was published. The speed at which paleoanthropology developed is marked by the fact that by the time of his death, aged 65 in 1937, some 14 of the genera and 30 of the species names listed in this chapter had entered the scientific literature, beginning with that of Eugène Dubois in 1892.

Little stood in the way of Elliot Smith's career in anatomy, which marked him out as a pioneer in a wide range of areas. His interest in the human brain began with a university visit while still at school and stimulated his primary research field; it influenced his later focus on the endocasts of fossil hominins. A bursary helped support his medical degree at Sydney University, where he served as a prosector and (after a year of practical hospital medicine) as university demonstrator in anatomy, while completing a doctorate on the brain of Australia's non-placental mammals and winning the University Gold Medal. He then gained a scholarship to undertake another research degree at Cambridge University. In Cambridge, Elliot Smith's interest remained on the soft tissues of the brain, although he worked under Professor of Anatomy Alexander Macalister, of whom Elliot Smith wrote regretfully he was "intensely devoted to Anthropology of, I am sorry to say, the bone-measuring variety" (Dawson 1938: 21).

Success remained his throughout his life, as attested in numerous tributes (see contributors in Dawson 1938; Elkin and Macintosh 1974). He stayed in Cambridge when awarded a research fellowship after only 3 years and then took up a professorship of anatomy at the Government Medical School in Cairo. During 9 years in Egypt (spending summers based back in Britain), he continued to build up his reputation and research in brain science. Participation in conference presentations and research papers established him as the leading expert of his generation on the organisation of the brain with his development of a model of brain evolution which focused on the neopallium (later called the neocortex), providing the mammalian brain with its functionality and developing to its most sophisticated level in the primates. At the age of 36, he was elected a fellow of the Royal Society of London and 6 years later became its vice-president. Despite initial resistance to any involvement in Egyptology, his presence as an energetic skilled anatomist in Cairo brought him into two additional research fields: reporting on the anatomy, morbidity and supposed racial differences of the large sample of 9000 burials from the explorations ahead of the Nubian dam flooding; and examination of royal mummies in Cairo's Egyptian Museum.

In 1909 Elliot Smith took up the chair of anatomy at the Victoria University of Manchester; serving in due course as the Dean of the Medical School, and during the First World War also undertaking some clinical work and co-authoring a pioneering work on shellshock, what we would now call PTSD (Smith and Pear 1917).

Ten years later, he moved to London as Professor of Anatomy at University College and as head of what became the Institute of Anatomy and Embryology. His arrival there fortuitously coincided with the enthusiasm of the US Rockefeller Foundation to fund the UCL initiative, which saw full-time academic staff in medicine in contrast to a tradition of clinical staff teaching students as an adjunct activity. Funding of £370,000 for the medical school (alongside £835,000 for the associated hospital) underwrote the opportunity Elliot Smith took to support the wide range of his interests, alongside staff and visitors to undertake research there. He would now see anthropology—in its broadest sense—as part of the Institute's brief, including not only the developing fields of physical anthropology and paleoanthropology alongside more traditional and emerging forms of anatomy, but also seeking to bring in social anthropology and his own interest in historical development. It was from his London base that Elliot Smith's existing involvement in fossil hominin studies led to guiding others.

He was already a published author when he moved to the University in 1896. My provisional classification (Fig. 5.2) of his 434 publications (listed in Dawson 1938: 219–253) shows the emphasis of his work on the brain, which continued through his life with 117 publications in that area. The chart indicates his developing interest in aspects of human evolution (90 publications) and the linked topic of racial classification (11 publications). The chart also notes his dedicated and parallel enthusiasm to promote an alternative narrative of human history with 87 publications, including several popular books—facing many critics with what would be described as Egyptocentric hyperdiffusionism (Crook 2012; Derricourt 2015: 104–129).

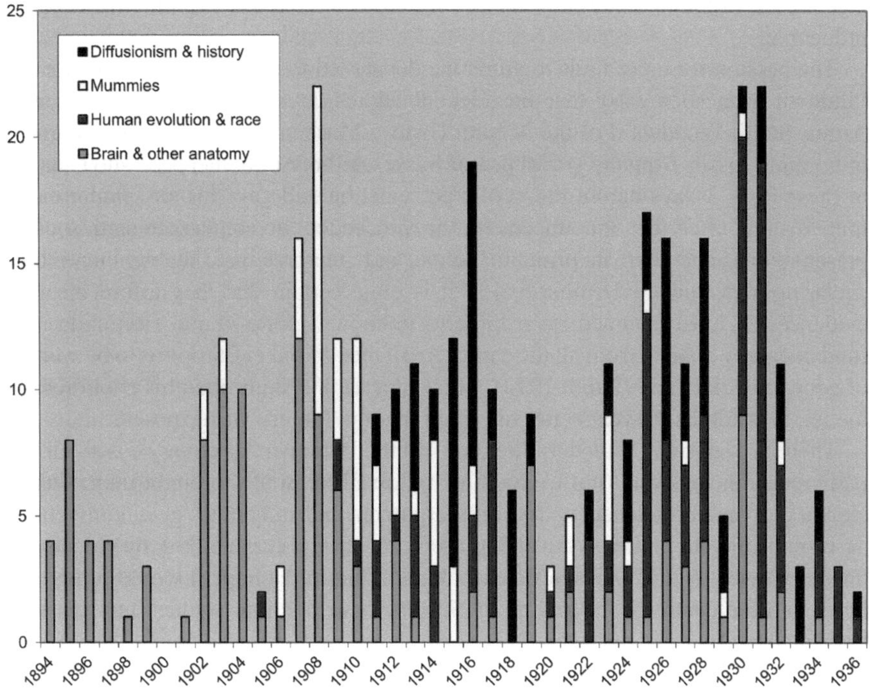

Fig. 5.2 Classification of published works by Grafton Elliot Smith

As a distinguished academic human anatomist, Elliot Smith had been the natural person to approach in Egypt about skeletal remains from excavations or the bodies of Pharaohs in the museum. It was therefore predictable that he would be approached to review publications and comment on finds and interpretations of skeletal finds of fossil hominins as they were uncovered, and became the subject of both scientific and wide popular interest. So by accident, he became an authority in paleoanthropology, a colleague, mentor and guide to others in the new discipline.

His involvement and interest in the topic began when he reviewed a German publication in 1909, and in reviewing another in 1911, he observed that the literature on fossil man had become a torrent. By then, Neanderthals had been joined by *Pithecanthropus* in Java, by the Cro-Magnon finds, by *Homo heidelbergensis* and by claims for hominin ancestors in South America by Argentine zoologist Florentino Ameghino. There was plenty of room for theory, but few physical remains to inform or modify debates. In 1912 as President of the Anthropological Section of the British Association for the Advancement of Science, Elliot Smith's address was on "Human Ancestry" when he focused on the development of the brain structure as the source of developing increased consciousness and learning from experimentation (Smith 1912). Indeed, Elliot Smith's primary research focus on the anatomy of the brain guided his emphasis on the importance and development of the brain (rather than bipedalism) as the measure of hominin evolution, a focus that would remain highly influential.

The passion for more finds to guide the debate led to the unfortunate incident of Piltdown Man soon after that presidential address, when Charles Dawson with Arthur Smith Woodward of the Natural History Museum announced the discovery of hominin fossils from the gravel pits of Sussex. Elliot Smith had been shown casts of these finds 3 days before the public presentation and gave his first impressions immediately after the announcement; he proceeded in conference and society presentations, and then in print, to define and interpret the Piltdown materials, declaring in *Nature* in October 1913 "It is quite certain that they afford the first evidence we have obtained on a hitherto unknown group of the Hominidae, so fundamentally distinct from all the earlier fossil men found in Europe as to be worthy of generic distinction" (Smith 1913). The endocast and skull fitted his evolutionary model; it would be 15 years after his death that the forgery was exposed.

There is a substantial modern literature on the Piltdown "discovery", with all the participants in the event (not excluding Grafton Elliot Smith) brought under suspicion at one time or another (with careful argument and sometimes great conviction) as responsible for undertaking the forgery. A strong current (but by no means universal) view is that Dawson undertook the fraud in the hope it would bring him more into the centre of the emerging elite of fossil hominin studies; but this was prevented by his early death in 1916.

When Elliot Smith visited his homeland of Australia in 1914 for the British Association meeting there, he would mention alongside Piltdown, a cranium from Talgai in Queensland. This earlier discovery was formally announced at the meeting, and "said to be of Pleistocene date". It would be his brother Stewart Smith, another medical academic in anatomy, who would formally publish details of this material.

The Canadian anatomist Davidson Black came to Manchester University for 6 months to work on neuroanatomy under Elliot Smith in 1914. His arrival coincided with the excitement of the Piltdown find and Elliot Smith's involvement in fossil hominin studies. Alongside Elliot Smith, his interests rapidly broadened to physical anthropology (Tobias et al. 2001: 6). Black even spent a day at the Piltdown site. Elliot Smith would recall "At that time I was working on the reconstruction of the Piltdown skull and the study of the endocranial cast obtained from it, and for purpose of comparison had collected casts of all the known fossil human skulls this work aroused a much greater interest in Dr. Davidson Black than the brains of the Dipnoi in which I was trying to engage his interest, and he at once made himself familiar with all of the material I had collected, and informed me that that was the kind of work to which he was determined to devote his life" (Smith 1934). The result, noted below, was his involvement with "Peking Man".

The 1921 find of a fossil then classified as *Homo rhodesiensis* from what was then Broken Hill in Northern Rhodesia, first published by Arthur Smith Woodward, arrived in London where Elliot Smith could examine it, with particular interest again on the endocast and the brain. He suggested in a talk to the Royal Anthropological Institute that the area of the modern human brain associated with speech had not yet developed in this species (Smith 1922).

By then, Elliot Smith had established himself in London and attracted not only medical students and postgraduate students but also scientists who wished to collaborate or work alongside him for a period. In the audience for the 1914 British Association meetings in Australia was Raymond Dart, also a University of Sydney medical school graduate, who would note “I fell under his spell that night and prayed that at some time I would be allowed to work under him” (Dart and Craig 1959: 27).

He followed this dream, moving to Britain where he took a position teaching under Elliot Smith at University College London between 1919 and 1922 (interrupted by a US visit funded by Rockefeller). In London “I had found an exciting new interest—anthropology... Elliot Smith was reconstructing the Piltdown skull and I had gradually been drawn into this branch of my chief’s interest” (Dart and Craig 1959: 31). Elliot Smith recommended him for the position of Professor of Anatomy at the University of the Witwatersrand in Johannesburg, where he would stay from 1923 to his death in 1988.

Dart’s first entry in the field was a paper on the “Boskop man” finds, which Robert Broom had labelled *Homo capensis.* Then came the well-known story of the Taung discovery and Dart’s 1925 description and announcement of *Australopithecus africanus*, but the mentor-pupil relationship had some challenges. Elliot Smith responded with caution to his protégé’s work in *Nature*, suggesting “it would be rash to push the claim in support of the South African anthropoid’s nearer kinship with man” but added “Prof. Dart is probably justified in creating a new species and even a new genus for his interesting fossil” (Smith 1925). In two of his popular books in 1930, Elliot Smith did not accept *Australopithecus* as ancestral to modern humans. Dart took the skull to London in 1931 for Elliot Smith and others to examine (temporarily forgetting it in his taxi) and would now learn from Elliot Smith that his descriptive monograph of the Taung material, whose manuscript he had submitted through him the previous year, had been rejected for publication other than the section on dentition (Dart and Craig 1959: 65); it would not be revived. Elliot Smith said to his London junior associate Solly Zuckerman that it was below the standard required for scientific publication.

Standing in the way of Dart’s confidence in the interpretation of the African find was Elliot Smith’s continuing support for the Eurasian ancestry of *Homo sapiens*. As Robin Dennell (2001) has indicated, for a mixture of political and ideological reasons, the shift from a Eurasian focus to Africa in the interpretation of hominin origins only dated from the late 1950s.

Davidson Black, inspired to physical anthropology by Elliot Smith in Manchester, was now based in China as the Professor of Neurology at the Peking Union Medical College. He was thus ideally placed to study the fossil hominin finds from Zhoukoudian, which in 1927 he would name *Sinanthropus pekinensis*. Elliot Smith was invited to comment in *Nature* on the significance of these finds for the story of human evolution, supporting this with an enthusiasm he had not extended to Dart’s Taung find. Despite this, “it was not surprising that the story circulated that Elliot Smith had sent Dart to South Africa to find the Taung Child, and Black to China to find the Peking Man” (Tobias et al. 2001: 6).

Another of Elliot Smith's protégés and colleagues at University College London was also an Australian, Joseph Shellshear, who knew Davidson Black as well as Raymond Dart with whom he travelled to the USA on Rockefeller funds. He became Professor of Anatomy at the University of Hong Kong in 1922. Elliot Smith travelled to China, where he and Shellshear both examined the Zhoukoudian finds.

Shellshear collaborated with Elliot Smith as the lead author of a detailed study of the interior of the skull of *Sinanthropus* (Shellshear and Smith 1934), reflecting Elliot Smith's continuing focus on endocranial casts rather than the comparative external cranial measurements.

Dart seems to have been somewhat discouraged by Elliot Smith's response to his Taung material, but he remained a loyal admirer, reflected not least in the ideas he published which reflected Elliot Smith's "hyperdiffusionist" alternative historiography (Derricourt 2009b). He would be the recipient of Elliot Smith's library, which his mentor bequeathed to him. But after more than 20 years, Dart resumed his interest and confidence in fossil hominins with his work at Makapan Valley 300 km north of Johannesburg, with finds including hominin fossil material which Dart (confident of its association with fire) would classify in 1948 as *Australopithecus prometheus*.

An inspiration to explore this site came from the activity of Dart's most distinguished protégé, Phillip Tobias. Student expeditions to the Makapan valley led by Tobias raised the possibility that the site could match the dates of the material uncovered by Robert Broom at Sterkfontein, where Tobias had worked briefly as a student (Tobias 2013: 42; Dart and Craig 1959: 89–90). Dart returned to the search for fossil hominins and in his work from 1947 to 1962 was rewarded with the wealth of material from Makapansgat and the opportunity to advance his individualistic claims for the cultural associations of the Australopithecines.

In 1959, with Dart's strongest support, Tobias would succeed Dart as Professor of Anatomy at the University of Witwatersrand at the age of 33.

Tobias in his writing acknowledged the broad range of contemporaries who influenced his work and ideas. Although he stated "When I entered Wits University as a medical student I almost immediately fell under the spell of Raymond Dart", he conceded that as he had learnt little factual anatomy directly from Dart. Nevertheless, his loyalty to Dart and his reputation led him to honour him with "Three honorary publications in tribute to the same man and edited by the same protégé—this is surely unusual. It was a measure of my affection for this father-figure" (Tobias 2013: 219). Tobias' own work would nevertheless contradict some of the published arguments and interpretations of Dart: not least as Tobias' papers and presentations served to undermine the reductive local racial typologies Dart had presented (Morris 2022: 180).

The reputation Tobias developed brought him to the major role he played in the East African work directed by Louis Leakey. In 1955, Tobias had examined some of the Kenyan material from Kanam and Kanjera, which Leakey had assigned to a new species *Homo kanamensis*. Writing of his attendance at the Pan-African Congress of Prehistory in 1959 Tobias noted "While I disproved the evidence on which Leakey had erected a new species *Homo kanamensis*, my analysis largely cleared Leakey of the charges and aspersions which in the late 1930s had threatened to blight his international reputation ... Louis Leakey presented the first account of the important new cranium of Olduvai hominid 5 which Mary Leakey had discovered only 5 weeks earlier. At the same meeting Louis and Mary took me to one side and invited me to undertake the description of this remarkable new cranium" (Tobias 2013: 82). With this find, in due course named *Zinjanthropus boisei* (amended at Tobias' suggestion to *Australopithecus boisei*), Tobias' work with the Leakeys began and would lead to the 1964 announcement and description as a new species of *Homo habilis* (Tobias et al. 2008: 125–142). The actual name "*habilis*" was suggested by Raymond Dart. However, when the Leakeys' son Richard located hominin fossil material in his own excavations in East Africa, he and his colleagues felt it would be inappropriate to involve South African expertise for the paleoanthropological analyses, and they turned elsewhere for studies of this material.

Just as Tobias had been a pupil of Dart and Dart one of the protégés of Elliot Smith, so Tobias in turn trained and inspired a generation of research scientists. He organised and formalised the University of the Witwatersrand's research in 1977 in what would be named the Palaeo-anthropology Research Unit. Ronald Clarke, who undertook his doctorate under Phillip Tobias, secured an appointment at the University and worked at Sterkfontein with finds of both Australopithecines (notably the complete skeleton called "Little Foot") and *Homo* fossils.

Prominent among those working under Tobias at the University of Witwatersrand was the American Lee Berger whose 1994 PhD in paleoanthropology was on the shoulder girdle of hominins. Berger was appointed to a position at the university in 1996 at the age of 33, as director of the paleoanthropology unit, and he has remained at the same university.

He too would use the term "mentor" applied to Tobias, but noting "My instincts for such open collaboration would lead me into conflict with members of my mentor's generation" (Berger and Hawks 2017: 47). This referred to a view that materials should be readily available to a wide range of specialists, reflecting a viewpoint that there was "a clique—a small international coterie of fossil discoverers who involved only a handful of experts in their work and talked mainly to each other" (Berger and Hawks 2017: 208).

Berger then undertook work at Gladysvale in South Africa (with new hominin finds), in Palau in Micronesia, and back in South Africa, where he identified a new species *Australopithecus sediba* from work at Malapa Cave, then another new species *Homo naledi* from the Rising Star cave system. Berger and his work have not been without controversy.

The personal and professional conflict between Clarke and Berger, whether sibling rivalry or interspecific competition, was mentioned in print both by Berger and Hawks (2017: 52–3) and, with apparent sadness, by Tobias et al. (2008: 245–7). No doubt the students of each will develop a further stage in the evolutionary history of paleoanthropology—and add new species (or even generic) names to the discipline, names which different specialists in due course may propose should be lumped with other fossil hominins.

5.4 Conclusions

The characteristics (and idiosyncrasies) of the history of paleoanthropology can best be understood in a framework of the limitation of study material and the frequent isolation of specialists interpreting it. The relative scarcity of hominin fossils has been a catalyst to such problematic approaches, reflecting the witty suggestion that it is "the only branch of science that has more scientists than objects to study" (Berger and Hawks 2017: 15). In that context, national (and in the early era imperial) affiliations and loyalty to mentors have played an important part in the discipline.

It is true that many workers in the field have been self-motivated individualists, and it is true that many fossil finds were, or would remain, isolated examples of a perceived new species. But not all the history of the discipline is of separated workers; a genealogy of influences plays an important role. The phylogeny of paleoanthropologists, like the phylogeny of hominin and broader hominid paleoanthropology, can be described in different ways by different historians of the discipline each with their own phylogenetic tree. Assessing evidence and taxonomy of fossil hominins can be understood alongside assessing also a taxonomy of scientists.

We have described here one highly influential line of distinction in fossil hominin studies: including Lee Berger (born 1965), the most prominent student of Phillip Tobias (born 1925), who was the most prominent student of Raymond Dart (born 1893), who was the most prominent student of Grafton Elliot Smith born in 1871, just as Darwin broke revolutionary new thinking with his *The Descent of Man*.

It is a common perception in humanities and social science disciplines that the conclusions reached by scholarship reflect the context (time, place, ideology) of those presenting that scholarship—not just in the conclusions and interpretations being given, but in the questions being addressed. In the biological and physical sciences there may be a more frequent belief that research reflects an objectivity independent of the affiliations of the researcher. Because paleoanthropology seeks to present, describe and interpret the ancestry and relationships of our own species, it perhaps shares some features with disciplines in the humanities and social sciences.

Personal, institutional and national pride can be found across all sciences. But the rivalries that have accompanied (some would say beset) anthropology over 140 years have been demonstrated in the remarkable number and diversity of species and generic names assigned to fossil remains under examination (and often subsequently abandoned). This alone suggests that paleoanthropology as a discipline has presented some unique and significant characteristics.

References

Bae CJ, Wu X (2024) Making sense of eastern Asian Late Quaternary hominin variability. Nat Commun 15(1):9479

Berger L, Hawks J (2017) Almost human: the astonishing tale of *Homo naledi* and the discovery that changed our human story. National Geographic, Washington, DC

Berger L, Hawks J (2023) Cave of bones: a true story of discovery, adventure, and human origins. National Geographic, Washington, DC

Berger LR et al (2023) Evidence for deliberate burial of the dead by *Homo naledi*. elife 12:RP89106

Broom R (1950) Finding the missing link. Watts, London

Cole S (1975) Leakey's luck: the life of Louis Leakey, 1903–1972. Collins, London

Crook P (2012) Grafton Elliot Smith, Egyptology and the diffusion of culture. Sussex Academic Press, Brighton

Dart RA, Craig D (1959) Adventures with the missing link. Hamish Hamilton, London

Dawson WR (ed) (1938) Sir Grafton Elliot Smith: a biographical record by his colleagues. Jonathan Cape, London

Dennell RW (2001) From Sangiran to Olduvai, 1937–1960: the quest for "centres" of hominid origins in Asia and Africa. In: Corbey R, Roebroeks W (eds) Studying human origins: disciplinary history and epistemology. Amsterdam University Press, Amsterdam, pp 45–66

Derricourt R (2009a) Patenting hominins: taxonomies, fossils and egos. Crit Anthropol 29:193–204

Derricourt R (2009b) The enigma of Raymond Dart. Int J Afr Hist Stud 42:257–282

Derricourt R (2015) Antiquity imagined. IB Tauris, London

Elkin AP, Macintosh NWG (eds) (1974) Grafton Elliot Smith: the man and his work. Sydney University Press, Sydney

Grün R, Stringer S (2023) Direct dating of human fossils and the ever-changing story of human evolution. Quat Sci Rev 322:108379

Hassett B (2022) Growing up human: the evolution of childhood. Bloomsbury, London

Johanson D, Edey MA (1981) Lucy: the beginnings of humankind. Granada, London

Kelley EA, Sussman RW (2007) An academic genealogy on the history of American field primatologists. Am J Phys Anthropol 132:406–425

Leakey LSB (1974) By the evidence. Harcourt Brace Jovanovich, New York

Leakey R (1983) One life: an autobiography. Michael Joseph, London

Lequin M (2022) Comment définir l'humain à partir de sa diversité? Questions épistémologiques et enjeux philosophiques. Bull Mém Soc Anthropol Paris 34:23–30

Martinón-Torres M et al (2023) No scientific evidence that *Homo naledi* buried their dead and produced rock art. J Hum Evol 195:103464

Mayr E (1950) Taxonomic categories in fossil hominids. Cold Spring Harb Symp Quant Biol 15: 109–118

Morell V (1995) Ancestral passions: the Leakey family and the quest for humankind's beginnings. Simon & Schuster, New York

Morris AG (2022) Bones and bodies: how South African scientists studied race. Wits University Press, Johannesburg

Pare S, Davis N (2022) Our oldest known ancestor could probably walk, say researchers. The Guardian, 25 Aug 2022

Reed DN et al (2023) Hominin nomenclature and the importance of information systems for managing complexity in paleoanthropology. J Hum Evol 175:103308

Schwartz JK, Tattersall I (2002–5) The human fossil record (4 volumes). Wiley, New York

Shellshear JL, Smith GE (1934) X. A comparative study of the endocranial cast of *Sinanthropus*. Philos Trans R Soc Lond B 223:469–487

Smith GE (1912) Human ancestry. Br Med J 2:644–646

Smith GE (1913) The Piltdown skull. Nature 92:131

Smith GE (1922) The brain of Rhodesian Man. Nature 109:355–356

Smith GE (1925) The fossil anthropoid ape from Taungs. Nature 115:235

Smith GE (1934) Prof. Davidson Black, F.R.S. Nature 133:521–522
Smith GE, Pear TH (1917) Shell shock and its lessons. Manchester University Press, Manchester
Theunissen LT (1988) Eugène Dubois and the ape-man from Java: the history of the first "missing link" and its discoverer. Kluwer, Dordrecht
Tobias P (1984) Dart, Taung, and the "missing link": an essay on the life and work of Raymond Dart. Witwatersrand University Press, Johannesburg
Tobias P (2005) Into the past: a memoir. Picador, Johannesburg
Tobias P (2013) Into the past: a memoir. Memorial. Picador, Johannesburg
Tobias PV, Wang Q, Cormack JL (2001) The establishment of palaeo-anthropology in South Africa and China: with especial reference to the remarkably similar roles of Raymond A. Dart and Davidson Black. Trans R Soc S Afr 56:1–9
Tobias P, Štrkalj G, Dugard J (2008) Tobias in conversation: genes, fossils and anthropology. Wits University Press, Johannesburg
van Holstein LA, Foley RA (2022) A process-based approach to hominin taxonomy provides new perspectives on hominin speciation. Evol Anthropol 31:166–174
von Koenigswald GHR (1956) Meeting prehistoric man. Thames and Hudson, London
Wheelhouse F, Smithford KS (2003) Dart: scientist and man of grit. Transpareon Press, Sydney
Wilson DE, Reeder DM (eds) (2005) Mammal species of the world. A taxonomic and geographic reference, 3rd edn. Johns Hopkins University Press, Baltimore
Wood MCB, Collard MC (2015) Defining the genus *Homo*. In: Henke I, Tattersall I (eds) Handbook of paleoanthropology, 2nd edn. Springer, Heidelberg, pp 2107–2144
Wu XJ, Bae CJ (2025) Xujiayao *Homo*: a new form of large brained hominin in eastern Asia. PaleoAnthropology 2025:356–369

Sherwood Washburn's Pragmatic Evolutionary Theory

6

Jessica Mikels

Abstract

Sherwood Washburn's "New Physical Anthropology" transformed the study of human evolution within the field of anthropology. This approach shifted the focus from static fossil remains to the dynamic interplay of form, function, genes, and behavior. More than just an expansion of tools and methods, Washburn introduced a new epistemological framework, one that closely resembles Pragmatic Philosophy. This chapter centers on a course Washburn taught late in his career, where he synthesized decades of scholarship and articulated a matured vision of the "New Physical Anthropology." By analyzing these lectures alongside key writings in American and French Pragmatism, I reveal philosophical continuities between Washburn's approach and Pragmatic thought. I argue that this perspective remains highly relevant, and further development of a Pragmatic Evolutionary Theory would deepen our understanding of both our evolutionary past and its implications for the present.

Keywords

Sherwood Washburn · Pragmatic philosophy · Pragmatic epistemology · New physical anthropology · Washburnian anthropology · Pragmatic evolutionary theory

J. Mikels (✉)
University of San Diego, San Diego, CA, USA
e-mail: jmikelscarrasco@sandiego.edu

M. Lequin, J. M. Rodríguez Caso (eds.), *The Field of Human Evolution: Critical Perspectives from History and Epistemology*, Evolutionary Biology – New Perspectives on Its Development 11, https://doi.org/10.1007/978-3-032-13925-2_6

6.1 Introduction

Sherwood Washburn's "New Physical Anthropology" changed the study of human evolution within Anthropology, and this has been well documented (see Strum et al. (1999) for a thorough review). Washburn's New Physical Anthropology (1951) changed the questions and methods in the study of hominid evolution. More than this, Washburn's approach to the study of human evolution also changed the epistemological and philosophical underpinnings of the discipline. I propose that Washburn originated a Pragmatic Evolutionary Theory. Though Washburn did not use the term Pragmatism to describe his work and approach, through an exploration of the overlap between Washburn's thinking and writing on the study of human evolution and Pragmatic Philosophy, I will demonstrate what I mean by Washburn's Pragmatic Evolutionary Theory.

Numerous studies and publications have examined Washburn's influential work, demonstrating the dramatic impact of his ideas of structural complexes (e.g., Morbeck 1997; Strum et al. 1999); his movement away from the traditional static study of fossil remains (e.g., Grand 1999; Mikels-Carrasco 2012); his recognition and support of the study of non-human primates as an important and legitimate course of study in physical anthropology (e.g., Haraway 1989; Washburn 1973); his endorsement of the modern synthesis and the integration of genetic theory into the study of human evolution (e.g., Stini 2010; Zihlman 1999); the movement, along with others working on the modern synthesis, toward population as the unit of analysis (Smocovitis 2012); and his work as a precursor to the biocultural approach, now common throughout biological anthropology (e.g., Yáñez Macías Valadez 2023). This wealth of literature demonstrates the far-reaching impacts of Washburn's work. In interviews conducted in 2004–2005, former students noted that Washburn's impact was so complete, so all-encompassing, that he is often not even given credit because it is just the way physical anthropology is now done (Mikels 2006; Stini 2010; Strum et al. 1999). These impacts and influences on the field are significant. The novel avenues of inquiry and methodologies brought to the study of physical anthropology by Washburn have been explored in depth. This chapter explores the ways of knowing, the philosophy and the epistemology, of the Washburnian approach.

While his published work is a good place to start to understand Washburn's impact, it is in his more private musings and writings where his philosophy really comes through, writings in which he was less constrained by the pressures and conventions of publication. Later in his career, Washburn wrote extensively about his goals of applied physical anthropology and about the vital role he thought the study of evolution had in improving society (Mikels-Carrasco 2010). He also wrote about and created courses demonstrating his approach to physical anthropology and how the science should be done (found throughout the Washburn archives).[1] It is in these writings that I find a strong pragmatic persuasion in his work and thinking.

[1]Washburn's archives are housed at the Bancroft Library, University of California, at Berkeley.

I will examine the culmination of his ideas around how to approach and study physical anthropology and how to use it for the betterment of society, found in his 1977 course titled *Human Sociobiology*.[2] This is the only course fully transcribed in the Washburn archives; something he chose to have done for reasons of posterity, further supporting the legitimacy of focusing on this course for a deeper understanding of Washburn's late career epistemology and philosophy. Importantly, philosophies and epistemologies are tightly intertwined. Deledalle (2002) states that "philosophy is a method, it is the method of method" (p. 8). Pragmatism, it has been argued, is both an epistemology and a philosophy (Kelly and Cordeiro 2020).

I explore what Washburn meant by sociobiology (something vastly different from the work of E.O. Wilson and other scholars, using the same term at the time). Through a critical reading of the course lectures, I explore and demonstrate the overlap between Washburn's approach and the tenets of Pragmatic Philosophy. Biological anthropology[3] today is still shaped by Washburn (Fuentes 2010). His ideas and ideals were formed in collaboration across multiple disciplines and a shared desire with other evolutionary thinkers of his time to create a grand synthesis of human evolution grounded in a secular philosophy of evolutionary humanism (Smocovitis 2012). I begin with a brief look at Washburn's life and career and the influences that brought him to the formulation of the New Physical Anthropology.

6.2 Washburn's Career Overview[4]

Sherwood Washburn was born in 1911 and raised in Cambridge, Massachusetts. Washburn, like many members of his father's side of the family, attended Harvard University for his undergraduate degree (Howell 2003). He completed his Bachelor of Arts *summa* cum *laude* in 1935 (Howell 2003; Zihlman 2001). During his undergraduate studies, he took his first anthropology class taught by Alfred Marston Tozzer. In that class, Washburn became excited about the interdisciplinary potential of anthropology. He stated, "It had the advantage of incorporating biology and social science, history and function theory, all in one" (DeVore 1992: 413).

Washburn then went on to pursue a Doctor of Philosophy in Anthropology at Harvard, under the advisement of Earnest Albert Hooton. Hooton was trained at Oxford in the humanities and sciences and had a significant impact on the field of anthropology, graduating 40 students with advanced degrees in physical anthropology between 1925 and 1950 (Spencer 1982, 1997). Hooton was a traditional

[2] All the course materials are in Carton 4: Folders 33–36 of the Washburn archives. Because there are so many citations from the course throughout this piece, I will provide the date of the lecture being referenced as well as page numbers from the lectures for direct quotes.

[3] For a comprehensive review of the disciplinary shift from physical anthropology to biological anthropology, see Smocovitis (2012) and Yáñez Macías Valadez (2023).

[4] This section draws heavily from the author's 2006 thesis, "Washburn's Continuing Legacy."

physical anthropologist who supported and believed whole-heartedly in the biological basis of race and the inferiority of all races to Western Caucasians (Spencer 1982). Washburn would later reject this sort of racist anthropology and wrote extensively on the equality of the races and the uselessness of racial classification for understanding human evolution (e.g., Washburn 1951, 1963, Washburn archives Carton 3: Folder 8). Hooton was also a committed teacher, and this was something that Washburn embraced and admired in his graduate advisor.

Hooton's students would later head six of the eight new anthropology programs that formed in the United States after World War II (Haraway 1988). Hooton was one of the first anthropologists to recognize the importance of studying nonhuman primates for the understanding of human evolution and human culture (Spencer 1982). Washburn was Hooton's only student whose dissertation concentrated on non-human primates. In it, Washburn analyzed the skeletal proportions of adult langurs and macaques (Haraway 1988).

In 1935–1936, Washburn participated in the Asiatic Expedition, "one of the last of the nineteenth century-style colonial collecting ventures and the first of the new primate behavior field trips conducted in the frame of incipient international conservation politics" (Haraway 1988: 219). Harold J. Coolidge, a zoologist who would later be the first director of the International Union for the Conservation of Nature (IUCN; Barrow 2009), Adolph Hans Schultz, a highly influential physical anthropologist and anatomist in his own right, and the first physical anthropologist to be appointed a position in a medical school at the Carnegie Institute (Erikson 1981), and Clarence Ray Carpenter, a student of Robert Yerkes who conducted one of the first naturalistic studies of primate behavior in the early 1930s, led the Asiatic Expedition (Howell 2003; Spencer 1982; Zihlman 2001). It was on this trip that Washburn gained his first field experience observing non-human primates. The contours of what he would come to refer to as the New Physical Anthropology would begin to form on this expedition. The collaboration of anatomists and behaviorists demonstrated for Washburn the need to link form and function; later, this link would be central to the New Physical Anthropology.

Washburn first took a position in anatomy at Columbia University's College of Physicians and Surgeons, where he worked from 1940 to 1947. While in New York, Washburn began to work with geneticists at the forefront of the modern synthesis (Zihlman 2000). Collaborations with Theodosius Dobzhansky, the geneticist, and William King Gregory, a functional anatomist and paleontologist, furthered Washburn's vision for a multi-disciplinary approach to the study of evolution. The 1950 Cold Springs Harbor Symposium, organized by Washburn and Dobzhansky, marked "the occasion that the study of fossil man was integrated into the evolutionary synthesis" (Smocovitis 2012: S116, quoting Mayr).

"In coming to the modern synthesis, the bridge builders transcended their disciplinary methods and reconstituted the totality of species change. They recognized the necessity of embracing many methods in order to comprehensively address evolutionary problems" (Zihlman 2000: 60). The synthesis established the importance of the link between traditional genetics, field biology, and population genetics. "The new synthesis brought to the study of human evolution an emphasis on process that recognized several biological levels from genes to organisms, to populations, all as important components" (Zihlman 2000: 58). Washburn is considered by many to be the primary catalyst for incorporation of the new synthesis into the field of physical anthropology (Zihlman 2001).

Washburn then headed to the Department of Anthropology at the University of Chicago from 1947 to 1958. It was at the University of Chicago that Washburn worked with Sol Tax as they sought to create a shared vision for the field. Together with others, Washburn and Tax created a course on human origins which integrated the study of human cultures and evolution in the field of anthropology (Smocovitis 2012). Sol Tax was the originator of action anthropology, with its emphasis on application (1975), something Washburn would also integrate into his work. Washburn's time at the University of Chicago is noteworthy because that university was central to the development of American Pragmatism, often linked tightly with the Chicago school of Sociology (Kaushik and Walsh 2019).[5] There is no compelling evidence in the archives or citation records that Washburn read John Dewey or George Herbert Mead (both of whom were at Chicago 20–50 years before Washburn). But their influence and the ubiquity of some of their philosophical musings may have influenced a young Washburn while at Chicago.

Washburn came to the University of California at Berkeley in 1958. He remained an active member of the Berkeley anthropology department until his retirement in 1979. In 1978, he was awarded the honor of faculty emeritus. The course at the center of this analysis, taught in 1977, represents the culmination of Washburn's career and thinking—thinking that emerged from a lifetime of experience, influence, and reflection.

Washburn's career spanned three decades and between 1950 and 1970, he was one of the most influential physical anthropologists in the United States (Strum et al. 1999). Washburn died April 16, 2000, at 88 (Marks 2000). Washburn's legacy is that of a reflective scientist and a humanist wanting to right the wrongs in society through the application of good science, something he shared with the other synthesizers (Smocovitis 2012). In an obituary for the National Academy of Sciences, F. Clark Howell stated, "He was neither shy nor deferential; the postures he assumed and the stances and beliefs he opposed were always serious, even pressing matters and thus warranted outright proselytism on his part" (2003: 17). Washburn was unwavering in his belief in his own ideas and ideals. He would later say there was no reason for him to write an autobiography, because it would simply be an account of how "he was right, and they were wrong" (Washburn 2000).

The examination of this course and reflections on Washburn's epistemology and philosophy highlight areas where his insights remain compelling. At the same time, Washburn's own commitment to introspection and critique of science creates space to reconsider or even set aside some of his specific ideas, without undermining the significance of his broader contributions. This need to philosophize the study of

[5] And indeed, Sol Tax as well. Though it is beyond the scope of this chapter to consider the intellectual influences on Sol Tax's anthropological thinking.

human evolution has been noted before (e.g., Haraway 1988; Fuentes 2010). Perhaps when a science grapples with questions so intimately tied to our understanding of ourselves, such reflection becomes both inevitable and necessary. The convergence of Washburn's theory and method with Pragmatism necessitates an initial examination of the fundamental principles of Pragmatic philosophy.

6.3 Pragmatic Philosophy[6]

American and French Pragmatism are two schools of thought with similar ontologies. The American Pragmatists primarily trace their lineage to scholars such as Charles Sanders Peirce, William James, and John Dewey and are tightly linked to the Chicago School of Sociology (Kaushik and Walsh 2019). The French school comes primarily from scholars such as Bruno Latour and critical history and philosophy of science studies (Blok and Meilvang 2015). Ontologically, both schools of thought are committed to a sort of empirical realism: one that is emergent, contingent, and co-constructed through the interactions of human and non-human actors, or what Latour described as the assemblages that inhabit and shape our lived experience (Latour 2004). The philosophical approach of the Pragmatists is less concerned with ontological debates between realism and existentialism and more concerned with an epistemology of action as being shaped by what works for a given condition in a given time (Dewey 1908; Latour 2004). Importantly, early Pragmatists such as James and Peirce focused on ideas of natural realism and recognized that our minds both filter the world and help us to understand the world in tandem (Putnam 1996); a position that will be further expanded upon in analysis of Washburn's lectures on the human brain. The goal of both American and French Pragmatic schools of thought is not to reduce all to social construction but instead to build up and understand the interactional nature inherent in the idea of social construction. In this way, Pragmatism bridges positivism and constructivism. Demeritt (2005) calls this artifactual construction and notes that ontologically this view holds that while "truth" may be elusive, "facts" can still be helpful, and there is no way to distinguish, ontologically, between some outside truth and human created facts. While Demeritt's 2005 piece places Donna Haraway and Bruno Latour in this line of thinking, he does not cite the Pragmatists. Latour's (2004; 2008) later work demonstrated the link between his approach to science studies and American Pragmatism.

American Pragmatic philosophy orients researchers towards practice, towards action in the world. It begins with several foundational premises—the rejection of Cartesian dualisms, the contingent nature of reality, the formation of knowledge as collective and social, which is closely related to the need for methodological pluralism, and the application of scholarship to real world issues (Kaushik and Walsh 2019). French Pragmatism is a label that has only recently been applied to

[6]For my analysis of Washburnian Pragmatism, I focus primarily on Dewey, Peirce, and Latour.

work primarily in history and philosophy of science led by the research and writings of Bruno Latour (Salinas 2016). French Pragmatism recognizes the contingency of knowledge in both time and space and seeks to uncover the roles of power in the construction of that knowledge (Ward et al. 2018). Like American Pragmatism, it recognizes that knowledge is true in so much as it works and is useful, and that as variables change so too can the understandings and approaches of a given area of inquiry (Ward et al. 2018). In this piece, I explore each of the Pragmatic tenets in turn and demonstrate the relationship to Washburn's approach.

At the heart of both American and French pragmatism is the need to break down Cartesian dualisms—the divides between nature and culture, or socially constructed objects and real-world things in the language of Latour (2004). To divide the world into human artifacts and natural facts (or whatever terminology one chooses) is to make many items incomprehensible, particularly complex assemblages of actors and objects that are at the heart of science and most pressing real-world problems (Latour 1991; 1998). "Since the seventeenth century this conception of experience as the equivalent of subjective private consciousness set over against nature, which consists wholly of physical objects, has wrought havoc in philosophy" (Dewey 1929 (1958): 13). This divide not only has wrought havoc in philosophy but the sciences and philosophy of science as well. Washburn, too, sought to bridge this divide as unnecessary and not useful when attempting to account for the entirety of human experience and evolution.

In his 2020 book, Trevor Pearce argues, "that Pragmatism—the most famous movement in American philosophy in the early 1900s—was the outgrowth of a rich conversation between late nineteenth century philosophers, biologists and social scientists" (p. 3) about the inseparable bond between organisms and their environments, including humans and their environment. Dewey himself authored an article entitled, *The Influence of Darwin on Philosophy* (1910). In that piece, Dewey argued that what troubled evolutionists most about Darwinian evolution was the abandonment of earlier evolutionary ideas that framed species as progressing toward perfection (perfection typically embodied by the European white male). If instead, evolution is change, in directions that can be unpredictable and not directional to perfection, then the rejection of Darwinism was due to the challenges it posed to ideas of superiority and the ranking of species. Dewey took this idea and applied it to society and, most important for Dewey, to the betterment of society. If Darwin is right, according to Dewey, and evolutionary change does not leave us with a structuring of species from least to greatest, then so too the structure of society, with all the disparities and inequalities that go along with that, is not a foregone conclusion of birth. Instead, everyone is born with the same biological basis and the same rights to achieve and live a life free from repression (Pearce 2020). The Pragmatists made this link between the biological and the social in a non-reductionist, liberating way long before the work of Washburn, but these ideas are clearly echoed in Washburn's work.

This connection between the social and the biological, where both are taken seriously, was the goal of Washburn's course in sociobiology. While it could be argued that earlier in his career, Washburn was primarily focused on bringing the biological to bear on the social, by the end of his career Washburn sought to find a way to understand both the findings of genetics in evolution alongside social constructions of the world, a truly non-reductive approach that in many ways looks very similar to the work of the Pragmatists.

Furthermore, for Pragmatism, reality and knowledge are emergent properties of inquiry and interaction (Baggio and Parravicini 2019). It is through collective inquiry that these emergent properties can be realized; reality is not ontologically distinct from the act of inquiry itself (El-Hani and Pihlström 2002). Through interaction (both between humans and non-humans or assemblages (Latour 2004)) new realities can be discovered. These realities are simultaneously socially constructed and ontologically real, leaving behind debates of constructivism vs. realism (Latour 2004). This means that reality, or the item of inquiry, itself is plural, contingent, and real at the same time. This approach to knowledge (or in Latour's (2004) terms, objects and things) is necessarily pluralistic and social in that it must account for multiple perspectives to avoid the biases of a few in defining problem definitions and areas of inquiry (Dewey 1923).

Due to this pluralistic, emergent nature of research and knowledge creation, Pragmatism bridges induction (beginning with observation) and deduction (beginning with theory) with Peirce's idea of abduction (the creation of explanatory hypotheses based on what is known at the time) (Rochberg-Halton 1983; Kaushik and Walsh 2019). This approach allows for an iterative process of inquiry that fluctuates between the three approaches to knowledge creation. For the Pragmatic method, it is not enough to propose philosophical musings; rather concrete problems must be brought to bear in order to understand the real-world consequences of the philosophical ideations and methods, which are good in so much as they are useful for real-world application (Glasgow 2013). For this same reason, Pragmatism is not tightly bound to any one method of inquiry. In a 2005 discussion with Theodore Grand, a former Washburn student, he noted that as discussions in philosophy of science began to focus on deduction vs induction, Washburn did not pay much mind. Instead, he had a "if it works, use it" mentality towards methods and approaches to science. This same methodological principle has often been used to describe Pragmatism.[7]

Finally, due to the concern with the consequences of inquiry, Pragmatism, beginning with early American and later French philosophers, is concerned with real-world problems. It is not enough to theorize or philosophize; rather only through application can we understand the meaning of our thinking and knowledge. For a democratic, liberating philosophy to be democratic and liberating, it must leave academic institutions and be applied. This idea of application was also central to Washburn's later writings (Mikels-Carrasco 2010).

[7]Though this loose approach to methods is certainly one of the critiques of the philosophy. But it is beyond the purview of this paper to dive into the philosophical debates around Pragmatism.

6.4 The Course: Human Sociobiology According to Washburn

> The analysis of any social system requires the consideration of biological, social, and historical factors. Our intellectual traditions place different causal factors in different departments—or even in different divisions of the university. But the problems of human behavior do not fall into traditional academic categories. This course will be built around a series of cases showing the interrelations of biological and social science and history. The aim is to suggest patterns of discovery which will be useful in a wide variety of problems, not just those limited to the cases discussed in the course. (course description from syllabus, Washburn archives Carton 4: Folder 33)

Washburn's course, *Human Sociobiology*, was a 3-month course, offered in the Winter of 1977 (just prior to Washburn being awarded emeritus in 1978 and retiring in 1979) (Mikels 2006). The course itself represents, in its most complete form, Washburn's evolutionary epistemology and philosophy. It came at the end of his career; a point when he was trying to create a complete story of his life's work.

Importantly, Washburn's use of the term sociobiology is nothing like the contemporary sociobiology of E.O. Wilson and others. When Washburn first heard the term sociobiology, he expressed excitement around the work, feeling that it was in line with what he had been doing in his approach to physical anthropology all along. However, he became dissatisfied with the work going on under that label. In a 1976 letter to Marshall Sahlins, of the University of Chicago, Washburn stated, "I think sociobiology has gone mad. Much of it harks back to the mistakes of the nineteenth century. When people describe human marriage as 'pair bonding' and speak of 'pastoral genes,' 'I go up the wall!'" (Washburn archives, Carton 4 Folder 65). Why he used the term for the title of the class is not clear, as much of the class is a rebuke of Wilson's sociobiology. It was a term that Washburn wanted to embrace—blending of the social and the biological for a more complete understanding of humans and human evolution. But Washburn sought to do this in a non-reductionist way, quite different from other sociobiologists at the time.

Washburn began the course with a close analysis of how the human brain works. He noted that if the human brain is the primary tool of the social sciences, we should understand how it works. Early in the course, Washburn stated, "One notion in sociobiology is that we don't know what is going on in our own heads, and what we think is going on in our own heads is a very small portion of what is really happening there, and what we think about it may be entirely wrong" (SLW lecture, January 13: 2). By understanding the human brain, within the context of evolution, we can begin to understand how the brain filters the world without our even knowing it (SLW lecture, January 13: 3). For example, Washburn noted that if you put a human in a rat maze they would fair terribly because our human brains do not have strong olfactory zones, this is something the hominids lost in evolution. When we attempt to understand situations, these biological underpinnings in the brain should be understood and considered before erroneous conclusions are drawn.

In the first few classes, Washburn made clear that without sufficient understanding of the biological *and* the social we end up with flawed applications of both in ways that can have meaningful impact in people's lives. He used the example of the belief in the biological basis of IQ, a still widely held belief throughout much of the US education system and popular psychology. That IQ results are culturally contingent is well documented, and yet the idea persists that IQ tests are testing something innate and biological (SLW lecture, January 11).

Another persistent biological reductionism is the grandmother hypothesis, which posits that women live beyond menopause to aid in caring for their grandchildren. Washburn contended that women's extended post-menopausal lifespan is a result of increased longevity rather than a trait specifically selected for. He pointed out the discrepancy in how such theories often neglect to apply similar evolutionary reasoning to males, noting, "it's interesting that nobody applies this to males; most sociobiology doesn't. Most are biologists and most are male. So they all talk about females and none of them say that it's adaptive for males to have prostatic enlargement"—another biological byproduct of aging rather than a selected trait (SLW lecture, March 8: 18).

Washburn also critiqued the over interpretation of culture without consideration of the biological and used the example of the Sapir-Whorf hypothesis. This hypothesis, held as one of the ultimate examples of the cultural construction of reality, holds that language shapes how we understand the world and in the extreme, "that thought is impossible without words" (SLW lecture, January 20: 2). Washburn saw merit in this idea that language shapes how we understand the world (and later in the class pontificated on the constraints of language in our understanding of ourselves). But taken to its extreme, those species without complex human language then must lack perception and interaction with the world. Non-human animals undoubtedly engage in "thinking" and "planning" that do not rely on language. While the complexity of human symbolic language may be unique, "thinking" still occurs in non-human animals. Washburn, like others, postulated complex language is what first set us on a new evolutionary trajectory (SLW lecture, January 18: 16).

Understanding the filtering of the world by our brains must be the first step in understanding the world. "The point I am trying to make is that the world we see isn't real, except to us. The world that we are seeing out there has all kinds of characteristics that the world in our brain doesn't have. There isn't an objective world out there" (SLW lecture, January 13: 4). Washburn did not reject the Sapir-Whorf hypothesis, he sought to integrate it with biological understanding. The subjective nature of our understanding of the world, of the way we think about the world, has a whole lot of bases in the biology of our brains not just our cultural and language particularities. This sort of co-construction of the social and biological in tandem is revisited throughout the course.

After the first several lectures on the brain, Washburn then turned to looking at non-human primates and modern hunter-gatherer societies for our understanding of the evolution of humanity. He was cautious in this course about the use of either for understanding early humans, noting that there may be too much structural variation between humans and non-human primates (or even between different non-human primate species) to draw neat comparisons for understanding early hominids (SLW lecture, February 1 & 3). He also cautioned using modern hunter-gatherers for comparison, because there is no reason to believe that they are living in the same way that our ancestors lived before farming (SLW lecture, February 8 & 10).

Washburn provided several case studies to better explain what he meant by sociobiology. The topics focused on issues of sleep, sex, aggression, and human nature itself. Throughout the course, Washburn used lines of evidence from the social and the biological with a constant eye toward the history of science and history of thought in these areas. The course ends with ruminations on what we mean by human nature and if we can ever truly comprehend ourselves.

6.5 Washburn's Pragmatic Evolutionary Theory

Using the 1977 Sociobiology course as representative of the culmination of Washburn's thinking in physical anthropology, this section explores the parallels between his approach and the principles of Pragmatic philosophy. I argue that Washburn's work lays the foundation for a Pragmatic Evolutionary Theory —a theory that challenges reductionist frameworks, recognizes the context-dependence of human evolutionary studies, and emphasizes the practical implications for how these ideas relate to lives lived today.

6.5.1 Bridging the Social and Biological: A Non-dualistic, Non-reductive Approach to Science

One of the key tenets of Pragmatism is the rejection of Cartesian dualism (Deledalle 2002), and related to that, the aim of avoiding reductionism (Braun and Castree 1998). Washburn's sociobiology did not seek to reduce the social to the biological, to look for biological explanations of social pursuits. Rather, he looked to bridge the social and the biological in a way that could be beneficial for understanding the totality of human evolution. He stated, "The problem is how to keep a useful relationship between biology and social science" (SLW lecture, January 27: 1). Washburn's idea of sociobiology was dramatically different from scholars of the day, and he spent a significant amount of the class demonstrating those differences. Importantly, it was the reductionism on both sides—those posing purely biological explanations of behavior and those posing purely social explanations of behavior—which Washburn sought to counter.

Early nineteenth-century social scientists were often evolutionists, meaning they saw societies as a gradual progression from less evolved to more evolved. This was a common view held in both sociology and anthropology departments (Bock 1979). Washburn viewed some of the problems with these scientists, such as Herbert Spencer, Lewis Henry Morgan, Auguste Comte, and others, as coming from a desire to emulate biology.

> So, the greatest success of biological science was hitting on a small number of experiments which revealed critical facts, critical relationships which would apply practically regardless

> of time and place and so on. Now, when people started studying human behavior, the beginnings of social science, whether they were consciously doing this or not, searching for facts which would have the same quality as these biological facts. (SLW lecture, March 8: 7)

This desire for laws, or what often ends up being reductionism, holds strong in the sciences in general and in the social sciences in particular. Washburn recognized the same mistake, reducing the social to the biological, both in early evolutionist thinking and later in sociobiology, which began with biological scientists and was later adopted by some anthropologists. The reason for this, according to Washburn, was that "people make tremendous efforts to find such laws. The most recent statement is in Wilson's sociobiology where he says the purpose of sociobiology is to find the laws which govern animal behavior, including human behavior" (SLW lecture, March 8: 8). This desire to generalize across complex systems continues today. It is pervasive not only in the sciences but in pop-psychology and the general zeitgeist. These simplified narratives of human behavior and evolution are easy to digest and understand and often seem so apparent that they are difficult to refute.

The primary problem for Washburn, with both nineteenth-century evolutionism and the sociobiology of his day, was that:

> Once the theory was accepted, this is a point I want to repeat, people that thought the theory was so powerful that they could arrange facts without any regard to the facts. ... It doesn't really matter what the facts were, they all ended up in the same order because the theory was so powerful that you didn't really have to worry that much about the facts. (SLW lecture, March 8: 3)

Later in the same lecture, he continued:

> If you think that genetic theory is so powerful that you can predict human behavior without any information lying between your genetic theory and human behavior, you're in exactly the same situation that the evolutionists were when they could arrange everything according to their theory without doing a thorough job on the intermediate stages. (SLW lecture, March 8: 5)

For Washburn, it was not just the reduction of the social to the biological but the simplicity of being able to go from theory to conclusion without any concern for all the work required in between that was wrong with the sociobiological hypotheses. "... what sociobiology tries to do in the sense of Wilson's socio-biology is to shortcut... there would be nothing on the brain, nothing on hormones, nothing on structure, nothing on locomotion, nothing on emotions. You go straight from a genetic calculus to behavior" (SLW lecture, March 8: 9).

Washburn particularly struggled with the calculus of reciprocal altruism and noted with the calculus of sociobiology came the idea that all customs must be adaptive, and this just does not make sense. "If your brain is emotional as well as rational and you're ignorant, as we all are, one should not expect that everything that one sees is functional. There ought to be a lot of noise in the system" (SLW lecture, March 8: 17). Furthering his critique of reciprocal altruism, in the March 1, 1977, lecture, Washburn stated that another major problem is the calculus involved requires an understanding of the outcomes by the actor. The interpretation of behaviors as altruistic in order to pass on some percentage of one's genes demonstrated Washburn's critique that the theory itself does not require an analysis of the data, it provides a foregone conclusion. Washburn stated, "Now, I have a low opinion of the altruistic motives of the baboon. My guess is that no baboon ever protected if there was no alternative mode of behavior" (SLW lecture, March 1: 6). Instead, Washburn provides room for animal mind in his interpretation of behavior, stating: "Now, the only point I'm making is that it's the animal's appraisal of the situation, not altruism, that makes the difference in what actually happens" (SLW lecture, March 1: 7).

Also, part of Washburn's critique was that individuals in a social group (human primates, non-human primates, or other animals) share most of their DNA with one another. In a letter to Washburn from Richard Currier in 1977, Currier stated, "I was greatly struck by your observation that virtually all genetic material is shared within a population, and that the sociobiological model in which non-kin are depicted as not sharing genotypes at all is, in fact, a highly misleading distortion" (Washburn archives Carton 4: Folder 65). The idea that behavior is calculated based on DNA that is not shared with most of the group means decisions are made based on a tiny fraction of one's DNA. Washburn thought it absurd to think that calculations are being made for that small bit of DNA. The whole genetic game is played out within a fraction of 1% of the genes. "Another thing that happens when you do this, and that is you see that you will preserve almost all your own genes by helping any other member of your own species…" (SLW lecture, March 8: 19).

So then why was sociobiology so attractive during Washburn's time and continues to be today? Washburn found several reasons why this approach was (and continues to be) so appealing—firstly, it provided a simple mechanism for making sense of most behaviors. The theory can postulate a role for genes, even in the absence of evidence supporting their involvement in a particular behavior. Secondly, it is easy to learn and apply because it blurs the line between genetic determination from social learning—leaving little room for learning or the mind and instead reinforcing a deterministic perspective. And thirdly, conclusions can be drawn neatly between theory and behavior, without any consideration of history, anatomy, physiology, culture, etc. (SLW lectures, throughout). "…it's the neatest trick of the century. Because with practically no intellectual effort you predict all kinds of things about animal behavior and human behavior" (SLW lectures, January 27: 20).

Washburn does not leave his critique of the study of human evolution and behavior to the ideas coming out of sociobiology. He also hoped to avoid the reductionism of the social sciences that ignore the biological. In the March lectures, he explained:

> The roots of modern social science, come [from] the French moral philosophers who claimed that social facts form an independent system that can only be understood by the understanding of other social facts. And what they did then, starting to build social science in this way, was to cut all roots with biology. In other words, they saw the mistakes of 19th century evolution, and many of those you've seen we now regard as biological mistakes, and they

> wanted to start social science according to independent social principles which were absolutely separate from biology. (SLW lecture, March 10: 7)

He also stated, in reference to social reductionism: "another way of short cutting this kind of system is to leave out all the biology and just study social behavior. There'd be no biology" (SLW lecture, March 8: 10).

Importantly, Washburn noted, those French moral philosophers were not rejecting modern-day biology, they were rejecting the evolutionism of their time (SLW lecture, March 10: 9). Washburn sought to bring biology back to the social sciences without doing away with the social. In the lecture from January 27, 1977, he spent a fair amount of time discussing Emile Durkheim. Washburn postulated that when Durkheim proposed the role of social facts in understanding human behavior, something that is just accepted in the social sciences today, he was making a major break with the social scientists of his time who were evolutionists. Durkheim demonstrated the social nature of something like suicide by showing how it could vary based on time and place (Durkheim 1897). Washburn posed to the class to consider why it is that so many people in the San Francisco Bay area who die by suicide do so by jumping off the Golden Gate Bridge. There are plenty of other ways to die by suicide, and yet that is the way it is done in that area. This approach to social science is foundational to sociology and cultural anthropology. Washburn did not wish to discard this understanding of social behavior and social systems as socially constructed, but he also did not want to ignore the biological in order to accept Durkheim's perspective.

The approach to what Washburn still wished to call sociobiology brings him neatly in line with the Pragmatists, searching for ways to make sense of the biological alongside, and interactional with, the social. "For Dewey thought comes from within, it prolongs the biological movement: society is in continuity with the world, it is not an addition to it" (Deledalle 2002: 9). Actions of human actors build on the biology (and physics and chemistry) of their bodies and the world; actors are not separable from the world that is. For Washburn, understanding human action required more than just biology, but ignoring biology would mean missing the complexity inherent in human behavior. In one of the early lectures, Washburn asked, "How can you bring biology into this picture and keep your social facts? That it seems to me is the critical thing if there's to be sociobiology. What you cannot do is to take genetic theory and simply say this negates the social facts" (SLW lecture, January 27: 9). I am reminded of an interaction I had as a graduate student (with a fair amount of biological training) in sociology. A fellow student was questioning the role of the senses and why it would be that we do not have a lot of words or discussion around smells. I mentioned that our human olfactory senses have been much reduced in the process of evolution—the other student found this interesting and surprising, not having thought about the biological basis for which senses we focus on. This is exactly the sort of thing Washburn sought to address. "Biology can come into the picture in quite different ways, without worrying about the social fact and without denying the notion of the social fact. One way biology comes in is with interest in general human biological features that lie behind all societies. This does not deny social facts. This does not deny differences" (SLW lecture, January 27: 9).

Department divides and hierarchies in science further the problem of erroneous conclusions and simplification of the social to only biological explanations (or social explanations with no accounting for the biological). Washburn critiqued the disregard that biologists seem to show for social scientists when he stated: "...you want to apply genetic theory to customs. The people that know the customs say it doesn't fit. One way to get out of this dilemma is to say well they say it doesn't fit, but they're only social scientists so it couldn't amount to very much anyway" (SLW lecture, March 12: 20). Working across disciplines is perhaps the only way to avoid the reductionism inherent in a single discipline approach.

6.5.2 The Contingent and Emergent Nature of Reality: A Historical View of Science

While it is beyond the purview of this chapter to review the metaphysics of the different Pragmatic Philosophers, it is fair to say that an understanding of truth and of reality in Pragmatism is emergent from experience. This does not negate the biological or physical but rather understands human truths as historically situated. This is seen most clearly in the French Pragmatists and their study of science. Latour (2010) began his book *On the Modern Cult of Factish God* with a story of how he was accused of trying to diminish science with his science studies, when in fact, he viewed his work as strengthening science. "The question was never to get *away* from facts but *closer* to them, not fighting empiricism but, on the contrary, renewing empiricism" (Latour 2004: 231). For both the French and early American Pragmatists, reflections on the act of doing science should work to build science into a better, more complete, and more attuned agenda.

Throughout the sociobiology course and other writings in the archives, Washburn continually noted the contingent nature of understanding, that research and findings are only as good as the tools of the time. Science is a useful pursuit. It is, nonetheless, incomplete and often wrong. In the January 18, 1977, lecture, Washburn discussed the past medical use of "bleeding" for curing diseases, and that we now know this is not a sound method for curing illnesses. Continuing from this example, he stated, "This kind of thing is very common in science. Where what everybody thinks is correct turns out to be a mistake" (p. 9). The problem, Washburn noted, is that the English language lacks the ability to accurately take into account these changes—it is not a misbelief or a rejection of obvious evidence, rather the evidence itself pointed to things that we now know are not true—the tools of the time and the thinking of the time created facts that are no longer facts.

This is potentially a slippery slope into complete relativism, but Washburn wished to maintain a commitment to empirical science—its ways of thinking and of knowing as still bringing us ever closer to greater understanding. The exact same point that Latour was making. As just one example, when discussing the molecular evidence of the human relationship to chimpanzees, Washburn's added the caveat,

"...within the limits of the method that most modern molecular biologists use..." (SLW lecture, February 1: 1). This reflective approach to science is remarkably similar to those who wish to understand science in its historic context and still be able to maintain that there is a growth in knowledge over time. It is for this reason that Washburn believed that interdisciplinary understanding should include a solid dose of history and philosophy of science as well.

Importantly, this appreciation for the history of science also allows students and scientists to appreciate the role of emotion, of being human, that is inherently part of scientific discovery. Washburn discussed the role of emotional cerebration in scientific theories and our general approach to the world. Our brains evolved to be emotional. Since the brain is the primary tool of scientists, the role of emotion in the support or refutation of different theories cannot be ignored. These strong emotions can stand on either side of a theory just as they stand on either side of political issues. In reference to one of the class readings, Washburn stated, "McLean [the author] further adds that scientific theories rarely die before their authors" (SLW lecture, January 11: 8). This falls neatly in line with the seminal work of Thomas Kuhn (1962) who provided the framework through which many in history and philosophy of science understand the role of scientific theory and the refutation of those theories over time. Evidence fits theories until there is too much evidence that does not fit. And certainly, the originator of the theory is less likely to accept the evidence in refutation of the given theory.

Nineteenth-century evolutionism used the idea of superior/inferior in a way that fit with people's (including scientists') already held views—white, male Europeans were more evolved because they were superior, and other peoples were less evolved because they were inferior. This was something that most did not struggle to accept. However, when Darwin proposed the role of continual change in evolution, this was harder to accept. "One could argue that the notion that Europeans are better than everybody else got into the evolutionary theory even prior to the biological part," Washburn stated in the January 27, 1977 lecture (p. 5).

It is clear from the lectures that Washburn was reading historians of science, though he does not say whom.[8] He noted that, "the most important thing that you do as a scientist is to ask the right questions" (SLW lecture, February 15: 3). And, as noted in the introduction to the course, if most of our thinking about social issues comes from our brain, we should also understand how our brains work. And we also need to understand how science works.

> So the point I'm trying to make is that the theory determines the way you look at the structures which you find... (SLW lecture, March 10: 12) ...it took a long time to convert traditional anatomy over into adaptation, and that the interpretation went from stressing non-adaptive characters to stressing that everything must be adaptive and forgetting that stating that everything must be adaptive is a guide to research. It's not a conclusion. (SLW lecture, March 10: 14)

---

[8] Though the archives contain a correspondence between Washburn and Donna Haraway, a noted feminist philosopher of science.

Furthermore, selection itself is exceedingly complicated, and it is not just one structure that is being selected on, rather it is functional-structural complexes, which makes analysis even more complicated (SLW lecture, March 12: 22). By understanding how our minds work, how reality comes to be through social interaction, and the history of how these things impact science, we can create better theories and hopefully move away from theories of the past that simplify and often recreate societal inequalities.

6.5.3 Knowledge as Collective and Social: Advocating for Methodological Pluralism

Any good theory should help guide the researcher toward methods that fit within the confines of the ways of knowing that the theory proposes. Washburn's Pragmatic Evolutionary Theory also points to the methods that make sense for the questions posed. Out of Pragmatism comes the Pragmatic Method, which clarifies that methods of understanding are evaluated based on their "practical consequences" (this is from Peirce but also in line with Dewey's Pragmatism) (Peirce 1878; Hill 1940). This issue of practical consequences is one Washburn addressed head on when one of his former students (who was later also a professor at Berkeley) became embroiled in controversy for teaching there were biological differences of the races. Berkeley, in support of academic freedom, allowed the course to be taught point/counterpoint. Washburn himself wrote about his concerns with how this type of theorizing could be used off campus to support racism (Mikels 2006).

Similarly, in the critique of theories that lead to conclusions without methods in between, Washburn was searching for those methods that fit with his non-reductionist approach to evolutionary theory. Washburn primarily relied on multi-method approaches, linking the biological and the social, while being careful not to reduce complexity in ways that make for easy answers that ignore the structure, function, history, and culture needed for answering questions in human evolution.

Washburn recognized that knowledge is situated and contingent. But he did not leave science to be merely deconstructed. Instead, by breaking down the nature of knowledge, he then builds it back up by the interdisciplinarity of the science he practiced. This is in line with the Pragmatists, who offer a critique of power and contingency of knowledge yet still seek to recognize and understand the world that is (Latour 1998; 2004). In a critique of the deconstructionists Washburn stated, "what I'm trying to say is that there's been a steady erosion of this kind of a point of view [referring to science] without any other point of view taking its place" (SLW lecture, January 13: 3). This is the exact concern Latour expressed in his 2004 piece, "Why has critique run out of steam? From matters of fact to matters of concern." Washburn recognized that science is contingent but still offers promise to make sense of the world that is. Washburn's and the Pragmatist's point is that science provides us with the ability to test our views and to change them as the evidence changes. Washburn navigated a delicate balance, aiming not to discard science through situatedness and deconstruction, but to achieve a deeper understanding through enhanced empiricism, strengthening and expanding scientific explanations (e.g., James 1906 (1976); Latour 2004).

So, how does Washburn move away from purely deconstruction and ideally reconstruct science? He finds his way through the application of multiple methods and an approach that is neither deductive nor inductive (Grand, personal communication 2005). Peirce's Pragmatic method also moves away from induction and deduction alone to abduction as a third step in the formation of new knowledge (Minnameier 2004). Induction and deduction play a vital role in the interpretation of observation by the application of theories and in the testing of those theories. However, there is a third phase, abduction, where Peirce claimed:

> Abduction is the process of forming an explanatory hypothesis. It is the only logical operation which introduces any new idea; for induction does nothing but determine a value and deduction merely evolves the necessary consequences of a pure hypothesis. Deduction proves that something *must* be; Induction shows that something *actually is* operative; Abduction merely suggests that something *may be*. (Peirce 1903, quoted from Minnameier 2004)[9]

Similarly, Washburn employed a reflective and iterative approach, emphasizing that when iterated across multiple methods and multiple disciplines, findings are more in line with the world that is.

Another reason Washburn sought interdisciplinary methods in science was to recognize the issues of single-lens views on knowledge. It is very human to assume that the way one views an issue is the right way, the way one lives is the right way, and this is also true in science. However, by looking across disciplines and considering multiple lines of evidence, the comparative method becomes more rigorous, thereby clarifying our understanding of the world. Another example from the 1977 class demonstrates this when Washburn stated, "So in each one of these comparisons, what we are trying to do is compare a social behavior, see some of the variety of the social behavior, and then look at the underlying biology and see if this helps us to understand whether the comparison is a valid comparison" (SLW lecture, February 10: 22). It is not until the behavior is looked at in a variety of settings and the biology is considered that any conclusions should be drawn.

In the January 20, 1977, lecture, Washburn addressed the question of the evolution of human language. He demonstrated that the ability of human language to communicate incredibly complex things is what is adaptive about human language. Sometime after the split from chimpanzees, this must have been selected for. In the lecture, Washburn discussed several areas of inquiry into the evolution of language. In particular, he looked at research on the structure and function of the human brain, demonstrating the complex structures that are needed for language. By then turning

[9] It is not my intention to do a complete review of Peirce's triadic theory of knowledge production in science, but rather to briefly summarize that a way of knowing proposed by the Pragmatists is similar to Washburn's multi-method, comparative approach.

to comparative studies, he established how there are fundamental problems with the study of language across species because the underlying structure and function are so fundamentally different.

> The whole biological basis for sounds in the nonhuman primates is fundamentally different from the basis for the production of word in human beings. This, I think, is one of the fundamental reasons why people are constantly confused in comparing the sounds of nonhuman primates to human words. Because the assumption was that one was dealing with the same system and that man just had more of it. (SLW lecture, January 20: 14)

In the February 1, 1977 lecture, in discussing what is useful to compare in the study of non-human primates, Washburn demonstrated that we must understand the biological differences before we compare the social behaviors—if not, we may be comparing things (as with language) that are operating from completely different mechanisms.

> My own bias is that the recent study shows that language-based behavior and cognitive differences are far greater than people thought a short time ago, and at least at the present, they (human and non-human primates) are very, very different with no particular way of jumping across them, from one to the other. (SLW lecture, March 8: 15)

By looking across multiple disciplines and combining multiple methods, the complexity of language and other behaviors becomes all the clearer, but it also begins to help focus on what makes sense for study. This is a crucial point coming from Washburn as he is the progenitor of much of the field of primatology (Haraway 1989; Kelley and Sussman 2007; Mikels 2006). Comparisons only make sense with an understanding of the underlying structure and function, the adaptive complexes that Washburn so often focused on, and understanding these can only come by looking across the fields of science to see where the information that one needs may reside. Elaborating on this idea, Washburn stated:

> You see that this shows that that way of looking at the problem is entirely wrong (that is comparing human language to non-human primate sounds). There is a new system which is based on the cortex, it's based on a sound code which is based on the articulatory control and reception in the brain for the articulations which is really very different from anything in the nonhuman primates (SLW lecture, January 20: 14–15). There were thousands of papers on the origin of language, none of which until very recently, contained the information that could possibly solve the problem. I think that this is a very common situation in the history of science. (SLW lecture, January 20: 15)

But Washburn wanted more. It is not enough to understand the structural and functional differences for understanding the evolution of language, we must account for the social and how the social and biological are interactive.

> So the social fact in this sense is the different languages that you learn that you're told how to say them is a measure of coercion. The biological problem is what lies behind the different social-linguistic social facts. That's of interest if you're interested in the evolution of language. (SLW lecture, January 27: 9)

Later lectures focused on other issues that, like language, could only be understood in their evolutionary context when looked at across multiple methods and perspectives. Washburn's February 1, 1977, lecture focused on the overemphasis on mating patterns as a way to understand social dynamics—when in fact vastly different social dynamics can have similar sexual and mating patterns. This, like dominance, has been overemphasized in the study of other animals and the comparison to humans (Strum 1987 (2001)). Washburn gave the example of human female continual receptivity as often being postulated as the main driver in human social systems, helping to explain the large size and complexity of those systems. And yet, orangutans are continually receptive and solitary, indicating that this single variable alone is insufficient to draw broad conclusions. The overemphasis on human female receptivity led to erroneous conclusions (SLW lecture, February 1).

Washburn believed the family unit primarily functions through economic reciprocity, noting that this dynamic has changed in the modern era. He critiqued researchers for overly focusing on the role of sex within the social structure, arguing that it is not the predominant factor. The idea that primate and human societies are held together by female receptivity does not hold when considering the diverse functional arrangements observed across different primate groups, and that this diversity of social arrangements exists and functions often with the same sexual arrangements and mating patterns. But reasonable scientists came to these conclusions, to which Washburn noted, "Again, we've learned that one person will look and see something different from what the other person will be seeing" (SLW lecture, February 24: 17).

While a focus on sex and mating systems does not do well in accounting for different social structures in human and non-human primates, Washburn did see the lack of complex language as explaining the size limits on non-human primate societies. He proposed that without complex symbolic language non-human primate group size is constrained (SLW lecture, February 1: 11).

There are other things that Washburn did think were useful to compare across human and non-human primate societies. For example, Washburn noted that play behavior across species looks strikingly similar. At the time, play had received little attention—particularly when compared to sexual behaviors, aggression, and dominance—and yet, play is one area that does look similar and provides a glimpse into an important social behavior with biological bases. Furthermore, mother-infant interactions and behaviors are strikingly similar across primates, perhaps because these are the behaviors needed to keep the infant alive (SLW lecture, February 1).

Washburn, like the Pragmatists, believed there is a world to understand (Salinas 2016). However, he allowed for the role of social construction in how we view and understand this world. Our biology is deeply intertwined with social construction, shaping our perception of the world, while our experiences, in turn, influence our biology—often through selective processes. If we are to understand the evolution of humans in its entirety, it is through the application of multiple methods from multiple fields that this may be done. Importantly, if we constrain ourselves to the testing of hypotheses or even the formation of hypotheses within a single line of inquiry, we will often be wrong. Washburn saw this early in his career with the inclusion of the synthesis of genetic theory and evolutionary studies, *in addition* to the analysis of fossil evidence and evaluation of form and function (Mikels-Carrasco 2012; Smocovitis 2012; Zihlman 2000).

It is here, late in his career, that we see Washburn synthesizing his own approach to bridge the social with the biological in a meaningful way. Washburn's approach mirrors Pragmatic methods, utilizing multiple lines of evidence, recognizing creative steps beyond just deduction or induction in scientific inquiry. He supported using methods that are suited for the specific task at hand, grounded in practice; an approach to science that those who have worked in laboratories or the field will likely recognize. This reflects the deeply human pursuit of science, aimed at revealing the nature of reality.

6.5.4 Applied Physical Anthropology: Connecting Scholarship to Real-World Issues

Pragmatism is focused on the production of actionable knowledge (Kelly and Cordeiro 2020). Dewey is perhaps most remembered in this regard with the creation of the University of Chicago Laboratory School, a lasting and well-known application of his thinking. Again, like the Pragmatists, Washburn believed that it was the application of evolutionary theory that was his most important contribution. Later in his career, he hoped that what he termed his applied anthropology would be "regarded as the most useful part of my work" (Washburn archives Carton 1: Folder 34). Specifically, Washburn believed the most promising arena for applying his ideas was the education of young people (Mikels-Carrasco 2010).

Based on his evolutionary theory and understanding, Washburn was a harsh critic of the US education system. In the January 11, 1977, lecture, he stated: "Our educational tradition is an extraordinary tradition. It is not necessary; it is irrational and inefficient" (p. 5). How similar this idea was to those of John Dewey, who stated, "If we teach today's students as we taught yesterday's, we rob them of tomorrow" (Dewey 1923 (1947): 167). Instead, Washburn believed that "education should be fun, it should be practical, and it should be based on an understanding of the nature of the world and the nature of man" (SLW lecture, January 11: 12). Washburn used the sociobiology course as a chance to try out his ideas in teaching. In the March 8, 1977, lecture, Washburn made clear that the progression through the class assignments was to be left up to the students. He put into practice his belief in the freedom of the learner to choose their learning path as an important precept in the teaching of human primates (Mikels-Carrasco 2010).

Washburn believed that social facts and their biological basis could coexist and understanding these things would be fruitful in addressing real-world issues. But he noted the need for caution: "This is tricky. I'm sure people will make mistakes doing this" (SLW lecture, January 27: 10). He then went on to demonstrate a mistake he had made by telling the class he had originally believed that the car accident rate between males and females had a biological basis. But further research and findings demonstrated that if you controlled for driving under the influence, these differences dropped away, meaning there is no biological difference. It is a social difference in behavior. He cautioned that without careful examination things may look biological that are actually social.

Later primatological research within biological anthropology, based on new understandings that often came from Washburn's own students, demonstrated that a focus on aggression as central to the structure of primate (and subsequently human) societies had often been overplayed in the past (Strum 1987; Strum and Fedigan 2000). The idea that much of human behavior comes from adaptations to hunting is not as common in anthropology today (Hart 2018). These were not Washburn's views; he maintained a focus on the role of hunting in understanding human evolution. Nonetheless, by practicing his approach, some of Washburn's students would upset the focus on aggression, dominance, and hunting and instead emphasize friendships and social bonds for understanding primate societies. Though it is important to note, ideas about dominance and aggression are still quite common in the popular zeitgeist. Like non-Washburnian sociobiology, those simple explanations take hold and are often hard to counter. "This is part of the trickiness with biological and social facts because frequently it's very difficult to tell how much of the biology is in fact involved and how much is totally a social fact at an entirely different sort of level" (SLW lecture, February 15: 10).

Getting it right is crucial if we are to apply scholarship to the world in which we live. But we apply incomplete knowledge all the time. We constantly operate on partial knowledge; it is simply a matter of whether we admit that all our knowledge is incomplete. Washburn and the Pragmatists may provide an avenue for applying science to societal problems—one that maintains the recognition that we will sometimes get it wrong yet affirms that it remains a necessary human pursuit.

6.6 Legacy: Washburn's Pragmatic Evolutionary Theory

The Pragmatic lens allows for the understanding of science not through deconstructionism, but as a method to enrich scientific inquiry, where new knowledge continually challenges old theories. Washburn recognized the overemphasis on sexual behaviors as a driving force for the structure of primate societies. He did, however, maintain a focus on aggression and viewed primate structure as being the "anatomy of aggression" (SLW lecture, March 1: 12). Even within that focus on aggression, Washburn recognized the tendency to overinterpret behaviors through that framework. For instance, the assumption that yawning invariably signifies a threat overlooks the fact that animals, including humans, often yawn in response to tension, stress, or fatigue. Furthermore, Washburn stated: "So there's no question, I think, that aggression can be due to biology, but I think also, there's no question that most common aggression isn't due to biology" (SLW lecture, March 12: 15).

Washburn's own students demonstrated that this overemphasis on aggression led to the neglect of the understanding of social bonds and friendships (e.g., Strum 1987). As with all scientists, Washburn demonstrated emotional cerebration and did not necessarily accept these new interpretations (Mikels 2006). But, as with all of science, new methods and new ways of knowing lead to new interpretations of the evidence.

There is the constant challenge of overly simplistic ideas taking hold and not being put to rest. This is the case with much of the reductionism of sociobiology in science and popular culture—the idea that all behaviors ultimately have a selfish underpinning. The reduction of complex actors to selfish genes is a story that is easy to understand; it reduces complexity to a simple mechanism that is easily digestible and fits well within ideals of individualism and capitalist pursuits.

In contrast to the sociobiology of his day, which continues to this day,[10] Washburn's sociobiology offers us something different, something more complex and harder to wrap our heads around. Yet, as we live and work and play in a world that is simultaneously and inextricably physical and constructed, his ideas provide a way of thinking that does not reduce that complexity but rather that celebrates and enriches our understanding of ourselves and other creatures.

> The point I want to make, then, is that although they're biological, although they're genetic, even in a society as simple as a monkey society, it requires learning and appreciation of the social situation before you can predict what kind of actual behavior will take place. So, we don't want to get the feeling, just because there's a great deal of anatomy in there, and a great ancient structure and very complicated physiological structure, that it is not highly modified by experience. (SLW lecture, March 1: 22)

6.7 Conclusion: Pragmatic Evolutionary Theory Today

In his lecture on March 8, 1977, Washburn posed the question of whether we are incomprehensible to ourselves and examined Philip H. Rhinelander's (1973) book of a similar title.

> So the point that I'm trying to make is that it has proved exceedingly difficult to state what human nature is. And what human nature appears to be is the nature of the person who's writing about human nature, talking about human nature (SLW lecture, March 8: 4). Most of the time, the conclusions on "human nature" make no sense across cultures. (SLW lecture, March 8: 5)

Washburn later continued in the same lecture "…it's been exceedingly difficult to free the biological kind of view of human nature from the influence of culture…" (SLW lecture, March 8: 11). But we may still be comprehensible. For the Pragmatists, there is an empirical world we can understand—not separated from our cultural constructions but intertwined and interactive with both biology and culture.

Washburn did not wish to leave humans as incomprehensible to ourselves. He wanted a reflective science, cognizant of its assumptions, to help uncover the story of human evolution. "So theory is important in guiding the research, but theory does not lead you to any particular conclusion. The conclusion depends on doing

---

[10] A quick Google Scholar search finds a plethora of articles still applying things like reciprocal altruism to all sorts of human and non-human behaviors.

something else, doing some more work" (SLW lecture, March 12: 14). Pragmatic method, Peirce's abduction, and Washburn's methodological pluralism share similarities in their recognition of the interactive nature of research and their acceptance of what works as the correct method at a given time.

Washburn wanted a new way of doing, or at least thinking, about science. "So we need, I think, to look for some different sort of a model of science if we're going to get ahead in this question of what human behavior is really all about, where maybe we'll get to the kind of rules someday, but the rules are going to be more complicated and they're a fairly long way off" (SLW lecture, March 8: 8). He was seeking an epistemological pluralism that still eludes us. He stated, "I wish we could leave the word science out of this altogether because it's a loaded word" (SLW lecture, March 15: 7).

Like the Pragmatists and the evolutionary synthesizers (Smocovitis 2012), Washburn was attempting to find a moral philosophy for evolutionary theory, one that recognized that moral phenomena are natural phenomena (Welchman and Cochran 2010). He believed that most of the problems we face are biological, social, and moral. "There isn't any way you can decide what is desirable to do, what's worthwhile to do, without taking a moral stand" (SLW lecture, March 15: 6). As Howell (2003) stated, Washburn did not view his work as just interesting, it was imperative.

Perhaps it is through a deeper examination of the overlap between Pragmatic Philosophy and Washburn's New Physical Anthropology that new ways of knowing and new rules for science can continue to emerge. In a science that seeks to understand what it is to be human, the need to philosophize about the work is imperative (Fuentes 2010).

It has been argued that Washburn (along with other evolutionary synthesizers) was advancing a humanistic vision of evolution rooted in the positivist worldview of the Enlightenment (Smocovitis 2012). However, in the later stages of his career, it was not positivism Washburn promoted but Pragmatism—a philosophical perspective that bridges the gap between positivism and constructivism.

I propose that Washburn laid the groundwork for what can be called a Pragmatic Evolutionary Theory within physical anthropology (and what we now call biological anthropology). Revisiting the work of the Pragmatists offers a valuable opportunity to refine this philosophical and epistemological framework as we continue to grapple with questions about our origins and the implications for our future. This piece aims to open a conversation about the deep connections between Washburn's evolutionary thinking and Pragmatism. Much work remains in developing what a Pragmatic Evolutionary Theory might look like today but recognizing its roots is a crucial first step.

The study of human evolution is not merely a scientific pursuit; it is a deeply human endeavor. In that spirit, it feels fitting to close with a limerick attributed to Washburn himself:

> Pandas, pongos, beasts galore
> Be sure to see more, more, more!
> – Washburn

References

Baggio G, Parravicini A (2019) Introduction to pragmatism and theories of emergency. European Journal of Pragmatism and American Philosophy 11: XI–2

Barrow MV (2009) Nature's ghosts: confronting extinction from the age of Jefferson to the age of ecology. University of Chicago Press, Chicago

Blok A, Meilvang ML (2015) Picturing urban green attachments: civic activists moving between familiar and public engagements in the city. Sociology 49(1):19–37

Bock K (1979) Theories of progress, development, evolution. In: Bottomore T, Nisbet R (eds) A history of sociological analysis. Heinemann Educational Books (online by Cambridge University Press), London, pp 39–79

Braun B, Castree N (eds) (1998) Remaking reality: nature at the millenium, 1st edn. Routledge, New York

Deledalle G (2002) French sociology and American pragmatism: the sociology of Durkheim and the pragmatism of John Dewey. Trans Charles S Peirce Soc 38(1/2):7–11

Demeritt D (2005) Science, social constructivism and nature. In: Bruce B, Castree N (eds) Remaking reality. Routledge, New York, pp 184–204

DeVore I (1992) An interview with Sherwood Washburn. Curr Anthropol 33(4):411–423

Dewey J (1908) What does pragmatism mean by practical? J Philos Psychol Sci Methods 5(4): 85–99

Dewey J (1910) The influence of Darwin on philosophy, and other essays in contemporary thought. Henry Holt, New York, pp 1–19

Dewey J (1923 (1947)) Democracy and education: an introduction to the philosophy of education. Macmillan, New York

Dewey J (1929 (1958)) Experience and nature. Dover Publications, Inc, New York

Durkheim E (1897 (2005)) Suicide: a study in sociology. Routledge, New York

El-Hani CN, Pihlström S (2002) Emergence theories and pragmatic realism. Essays Philos 3(2): 143–176

Erikson GF (1981) Adolph Hans Schultz, 1891-1976. Am J Phys Anthropol 56:365–371

Fuentes A (2010) The new biological anthropology: bringing Washburn's new physical anthropology into 2010 and beyond—the 2008 AAPA luncheon lecture. Am J Phys Anthropol 143(S51): 2–12

Glasgow RE (2013) What does it mean to be pragmatic? Pragmatic methods, measures, and models to facilitate research translation. Health Educ Behav 40(3):257–265

Grand T (1999) Sherry Washburn and the revolution in functional anatomy. In: Strum SC, Lindburg DG, Hamburg D (eds) The new physical anthropology. Simon and Schuster, Delran, pp 228–232

Haraway DJ (1988) Remodeling the human way of life: Sherwood Washburn and the new physical anthropology. In: George J, Stocking W (eds) Bones, bodies, behavior: essays on biological anthropology. The University of Wisconsin Press, Madison, pp 206–259

Haraway DJ (1989) Primate visions: gender, race, and nature in the world of modern science. Routledge, New York

Hart D (2018) Man the hunted: primates, predators, and human evolution. Routledge, New York

Hill WH (1940) Peirce's "pragmatic" method. Philos Sci 7(2):168–181

Howell FC (2003) Sherwood Larned Washburn, 1911-2000. Biogr Mem Natl Acad Sci 84:2–24

James W (1906 (1976)) Essays in radical empiricism. Harvard University Press, Massachusetts

Kaushik V, Walsh CA (2019) Pragmatism as a research paradigm and its implications for social work research. Soc Sci 8(9):255

Kelley EA, Sussman RW (2007) An academic genealogy on the history of American field primatologists. Am J Phys Anthropol 132(3):406–425

Kelly LM, Cordeiro M (2020) Three principles of pragmatism for research on organizational processes. Method Innov 13(2):1–10

Kuhn TS (1962 (2012)) The structure of scientific revolutions, 4th edn. The University of Chicago Press, Chicago
Latour B (1998) To modernize or ecologize? That is the question. In: Braun B, Castree N (eds) Remaking reality: nature at the millennium. Routledge, London
Latour B (1991) We have never been modern. Harvard University Press, Cambridge, MA
Latour B (2004) Why has critique run out of steam? From matters of fact to matters of concern. Crit Inq 30(2):225–248
Latour B (2008) "It's development, stupid!" or: how to modernize modernization. In: Proctor J (ed) Postenvironmentalism. MIT Press, Boston, pp 1–13
Latour B (2010) On the modern cult of the factish god. Duke University Press, Chapel Hill, NC
Marks J (2000) Sherwood Washburn, 1911-2000. Evol Anthropol News 9:225–226
Mikels JD (2006) Sherwood Washburn's continuing legacy. Master's thesis, University of Oklahoma
Mikels-Carrasco J (2010) Teaching human primates: education from an evolutionary perspective, the contributions of Sherwood Washburn. Gen Anthropol 17(1):1–10
Mikels-Carrasco J (2012) Sherwood Washburn's new physical anthropology: rejecting the "religion of taxonomy". Hist Philos Life Sci 34:79–101
Minnameier G (2004) Peirce-suit of truth: why inference to the best explanation and abduction ought not to be confused. Erkenntnis 60(1):75–105
Morbeck ME (1997) Life history, the individual, and evolution. In: Morbek ME, Galloway A, Zihlman AL (eds) The evolving female: a life-history perspective. Princeton University Press, Princeton, pp 3–14
Pearce T (2020) Pragmatism's evolution: organism and environment in American philosophy. University of Chicago Press, Chicago
Peirce CS (1878) How to make our ideas clear. Pop Sci Mon 12:286–302
Putnam H (1996) Pragmatism and realism. Cardozo Law Rev 18:153–170
Rhinelander PH (1973) Is man incomprehensible to man? Scribner, New York
Rochberg-Halton E (1983) The real nature of pragmatism and Chicago sociology. Symb Interact 6(1):139–153
Salinas FJ (2016) Bruno Latour's pragmatic realism: an ontological inquiry. Glob Discourse 6(1–2):8–21
Smocovitis VB (2012) Humanizing evolution: anthropology, the evolutionary synthesis, and the prehistory of biological anthropology, 1927–1962. Curr Anthropol 53(S5):S108–S125
Spencer F (ed) (1982) A history of American physical anthropology: 1930–1980. New York Academic Press, New York
Stini WA (2010) Sherwood L. Washburn and "the new physical anthropology". In: Little MA (ed) Histories of American physical anthropology in the twentieth century. Lexington Books, Lanham, pp 173–185
Spencer F (ed) (1997) History of physical anthropology: an encyclopedia. Garland, New York
Strum SC (1987 (2001)) Almost human: a journey into the world of baboons. The University of Chicago Press, Chicago
Strum SC, Fedigan LM (2000) Changing view of primate society: a situated North American view. In: Strum SC, Fedigan LM (eds) Primate encounters: models of science, gender and society. The University of Chicago Press, Chicago, pp 3–49
Strum SC, Lindburg DG, Hamburg D (eds) (1999) The new physical anthropology. Simon and Schuster, Delran
Tax S (1975) Action anthropology. Curr Anthropol 16(4):514–517
Ward K, Jonas AE, Miller B, Wilson D (2018) The Routledge handbook on spaces of urban politics. Routledge, New York
Washburn S (2000) Memorial to Sherwood L. Washburn, May 21
Washburn SL (papers) BANC MSS 98/132 c, Phonotape 2950, Motion Picture 842, Phonodisc 738. The Bancroft Library University of California Berkeley
(A) Carton 4: Folders 33-36 (1977) Human Sociobiology lectures

(B) Carton 3: Folder 8 (1974) Comment on: AAUP statement on studies of intelligence and race
(C) Carton 4: Folder 64 & 65 (1976-1977) Professional letters
(D) Carton 4: Folder 34 (1979-1982) L.S.B. Leakey Foundation
(E) Carton 5: Folder 29 (1985-1991) Sarich, Vincent

Washburn SL (1951) The new physical anthropology. Trans N Y Acad Sci Ser II 13:298–304

Washburn SL (1963) The study of race. Am Anthropol 65(3):521–531

Washburn SL (1973) The promise of primatology. Am J Phys Anthropol 38:177–182

Welchman J, Cochran M (2010) Dewey's moral philosophy. In: The Cambridge companion to Dewey, Cambridge companions to philosophy. Cambridge University Press, Cambridge, pp 166–186

Yáñez Macías Valadez B (2023) A historiographical perspective on contemporary biological anthropology. Almagest 14(2):64–72

Zihlman A (1999) Fashions and models in human evolution: contributions of Sherwood Washburn. In: Strum SC, Lindburg DG, Hamburg D (eds) The new physical anthropology. Simon and Schuster, Delran, pp 151–161

Zihlman A (2000) A skeletal survey of physical anthropology in the US. Riv Antropol 78:57–66

Zihlman A (2001) In memoriam: Sherwood Washburn, 1911-2000. Am J Phys Anthropol 116:181–183

Punctuated Equilibrium and Human Evolution, 50 Years Later

7

Niles Eldredge, Telmo Pievani, and Ian Tattersall

Abstract

The concept of punctuated equilibrium (PE) had three significant impacts on evolutionary thinking. The first was on the history of biological thought: the critique of a strict phyletic gradualism in paleontology and paleo-anthropology. The second was the theoretical impact for an extension and revision of the neo-Darwinian structure of the theory of evolution, although the organism-centered "Extended Evolutionary Synthesis" shows its blindness in considering the role of macroevolutionary patterns. The acknowledgment not only of the validity of the theory but also of its empirical fecundity is still weakened in philosophy of biology by intentional misrepresentations: for example, the inappropriate conflation of PE with Saltationism. PE concerns both the patterns and the processes of evolution, not only the time but also the mode of speciation. After 50 years of updating, a review of the scientific literature clearly shows that PE has introduced a pluralist view on the rhythms and modes of speciation. The most recent phylogenetic research (integrating molecular, paleontological, morphological data) assigns a central role to "punctuations," with the possibility to calculate their relative frequency group by group. In paleoanthropology, the complex hominin phylogeny—up to the branching diversification of the genus Homo and cohabitation of different recent human species—can find some clarification in the light of allopatric speciations and biogeographic factors. PE

N. Eldredge
Division of Paleontology, American Museum of Natural History, New York, NY, USA

T. Pievani (✉)
Department of Biology, University of Padua, Padua, Italy
e-mail: dietelmo.pievani@unipd.it

I. Tattersall
Division of Anthropology, American Museum of Natural History, New York, NY, USA

M. Lequin, J. M. Rodríguez Caso (eds.), *The Field of Human Evolution: Critical Perspectives from History and Epistemology*, Evolutionary Biology – New Perspectives on Its Development 11, https://doi.org/10.1007/978-3-032-13925-2_7

has given an essential contribution in the recognition of the branching feature of the hominin phylogeny against any scenario of progressive hominization.

Keywords

Punctuated equilibrium · Allopatric speciation · Stasis · Tempo and mode of evolution · Hominin phylogeny · Evolutionary pluralism · Hierarchy theory

7.1 Introduction

The year 2022 marked the passage of half a century since Niles Eldredge and Stephen J. Gould (1972) published their pathbreaking paper on Punctuated Equilibria, and of 40 years since the appearance of "The Myths of Human Evolution" by Niles Eldredge and Ian Tattersall (1982). In this chapter, we look back over those decades and assess the impact that the new perspective, laid out in the first of those contributions and embodied in the second, has had on evolutionary biology in general, and on paleoanthropology in particular. The proposal of punctuated equilibria criticized phyletic gradualism, which had long been influential in paleoanthropology, focusing on speciation processes and ecological factors of physical separation of populations. It also highlighted the role of morphological stasis in paleontology and the opportunity for a multi-level vision of the evolutionary process. We argue that in other paleontological fields, there has been a recognition that punctuated changes do occur—particularly in response to environmental changes and geographic isolation. We show the need to consider mechanisms beyond genetic gradualism (as supported by the Modern Synthesis) as a way to make sense of our complex family tree. This approach would mean recognizing evolutionary processes at multiple scales and a pluralistic approach to evolutionary theory.

7.2 Gradualism: An Influential Image of Evolution

One of the most heated scientific controversies around Charles Darwin's legacy has centered on the great naturalist's gradualist leanings, an issue about which his friend Thomas Henry Huxley expressed misgivings as early as 1859. Although in his youthful Notebooks on Transmutation (1836–1844), Darwin clearly sensed the role of geographic isolation in the emergence of new species (and showed interest in the idea of the Italian geologist Giovanbattista Brocchi that species were to be regarded as individuals with births, histories, and deaths), for various reasons the English naturalist in his later years came to embrace a gradualist view of the evolutionary process (Barrett et al. 1989). Darwin's experience in the study of artificial selection; his observations of gradual changes in some species of birds and barnacles; his training as a geologist; his inability to imagine how isolation could occur over vast stretches of space (e.g., in southern South America); and, above all, his reverence for the geologist Charles Lyell (a firm believer in the uniformity of causes acting in nature in the past as in the present), all combined to reinforce his gradualist standpoint. Further, his Notebook B shows that Darwin's biological gradualism also came directly from Lamarck's gradualism. Darwin consistently opposed Lyell's rejection of Lamarck, starting on the *Beagle* and continuing through the Red and Transmutation Notebooks (Eldredge 2015). Because of this theoretical gradualistic filter, Darwin was forced to deny plausibility to much of the paleontological evidence that already attested to the operation of such influences as relatively rapid speciation, the sudden explosions of forms, and mass extinctions.

The theory of Punctuated Equilibria (PE) was proposed in 1972, some 90 years after Darwin's death, by Niles Eldredge and Stephen J. Gould, paleontologists at the American Museum of Natural History in New York and Harvard University, respectively. Their contribution immediately perturbed the waters of evolutionary biology, which had remained calm for decades under the influence of the scientific research program known as the "Modern Synthesis" (MS), and which encapsulated what two philosophers of biology have described as "the received view" of evolution at that time (Sterelny and Griffiths 1999).

Darwin was well aware of the central problem of gradual evolution, candidly admitting it in *The Origin of Species*: the geological strata do not record all the intermediate links that evolution by small, imperceptible steps would require. The epistemological dismissal of this problem necessarily had to lie in imperfect and deficient documentation had gradualism actually been the norm, the dominant paleontological signal would have been one of directional change. Instead, what is preserved abundantly is the signal of stasis, despite those very same gaps of preservation. Darwin was well aware of stasis, and the only data he personally had on evolution argued strongly for the importance of isolation. However, while in his youthful notebooks the theme of geographical isolation is present (and is even connected to reproductive isolation), in his mature works this aspect loses importance.

Gradualism became an influential image of evolution for a century, even as the theoretical framework it provided clearly required adjusting the data in its favor, while any view denying it (from paleontological catastrophism to the macromutationism of the early geneticists) was considered pseudo-scientific. The paradigmatic assumption of the population geneticists who supported the neodarwinian Modern Synthesis was that all observable changes at the macroscopic level (i.e., the great differences between different species, genera, and families of organisms in the Darwinian tree of life) were traceable to small innovations that accumulated at the microscopic level (i.e., in the genetic makeup), under the constant and directional pressure of natural selection. The slogan of the "phyletic gradualism" of the Modern Synthesis was that macroevolution should be totally traceable (and reducible) to microevolution. Consequently, any macroevolutionary phenomena that denied a strict gradualism (among them such important and well-documented features of the record as rapid speciations, turnovers of species, adaptive radiations, mass-extinctions—all already highlighted in the literature) were either denied or underemphasized.

Cracks in this vision quickly appeared. In 1944, in his *Tempo and Mode in Evolution*, the paleontologist George Gaylord Simpson noted that in many evolutionary trees, notably that of the mammals, it was not always possible to find a sequence of intermediate forms that reflected a gradual passage from ancestral species to current species. In explaining this, Simpson preferred to follow a different path from that of devaluing paleontological data. When the expected intermediate forms were not found, it was because over the course of evolution the pace of diversification had not been uniform. That, he claimed, was due to accelerated evolutionary change among small, geographically isolated populations (a process compatible with a neodarwinian explanation of evolution), and not to the incompleteness of the fossil record. A decade later, however, Simpson had completely abandoned his idea of "quantum evolution" (Eldredge 1995; Gould 2002; Sepkoski 2012). An alternative interpretation of the observed rhythms of organic change was suggested by the processes involved in speciation, which are undervalued in the perspective of neodarwinian gradualism. Accordingly, in 1972 Eldredge and Gould noted the basic contradiction between the paleontological approach to the idea of species in the context of a long-term time continuum ("paleo-species" or "chrono-species"), and the biological approach whereby species are seen as discrete entities endowed with defined reproductive boundaries.

Although the idea that speciation was in some way related to peripheral populations had already been touched upon by Darwin in his discussion of geographical varieties of island species, it was not until the late 1930s (Dobzhansky 1937) that the topics of evolutionary rhythm and diversification among species began to come together. Soon, Ernst Mayr (1942, 1959, 1970) noticed that, when a new species emerged, it was often in association with some geographical or climatic phenomenon. Quickly, the biological notion of species, centered on the set of reproductive relationships among individuals in a population, replaced both Linnaeus' essentialist morphological definition and Darwin's nominalist definition. The species is a reproductively closed system (albeit with certain exceptions), and membership in a species is recognized by the absence of interbreeding with individuals of other species. Thanks to the biological concept of species that emerged within the MS, the species became a real entity in nature, no longer a label to be filled in using resemblances of various kinds among organisms.

It should not be forgotten, however, that the biological notion of species also involves some not insignificant difficulties. The chances of interbreeding among individuals decrease with lessening proximity, and it is often arbitrary to fix a precise moment of "speciation" (as in the case of ring species). In some cases, the absence of interbreeding does not reflect species separation; and, conversely, members of two separate species may occasionally hybridize: a phenomenon very common in plants, and also present in animals when their biogeographical distributions overlap. This is particularly well documented among felines, but it has also apparently occurred among recent species of the genus *Homo*, such as *H. sapiens*, *H. neanderthalensis*, and the Denisovans. It is as if in evolution there is a generation-by-generation dialectic between continuous change (where "continuous" indicates an absence of gaps at the biological level, not a directional process), and the effects of the discrete separation that comes about when an ecological barrier allows a threshold to be crossed, creating an irreversible reproductive barrier.

Nevertheless, thanks to the biological concept of species, the focus in evolutionary biology eventually shifted from the general idea of species to possible mechanisms of speciation, henceforth understood as the establishment of "reproductive isolation." In this sense, the explanatory hypothesis that has won the greatest fealty among biologists since the 1940s is the theory of "allopatric speciation" proposed by Mayr: a potential new species can form in a geographic and environmental context different from that typical of the parent species. This is particularly likely to happen when a small local population remains isolated at the edge of the parent species' geographic range. In such "peripheral isolates," developing isolation mechanisms are capable of preventing the re-establishment of "gene flow" between the isolated population and the still-adjacent parent population. Genetic drift, natural selection, migration, and other factors operating differentially in the two populations cause them to diverge from each other and to evolve intrinsic isolation mechanisms such that, were members of the two populations to meet, they would no longer interbreed. In this way, two new species, two new reproductive entities are born, variably different from the parent.

7.3 Tempo and Mode of Speciation

The speed of the divergence process may be more or less directly proportional to the size of the population involved, which might explain the periods of apparent evolutionary acceleration often noticed by paleontologists. If so, while it may remain a neo-Darwinian mechanism of continuous accumulation of genetic mutations under the effect of natural selection and genetic drift, the relative speed of the speciation event may vary according to the number of individuals involved. The action of selection may indeed be continuous and cumulative, but on the paleontological timescale speciation can appear to be abrupt.

The model suggested by the allopatric theory of speciation favors geography and ecological conditions as crucial evolutionary factors. It also introduces the possibility of the coexistence of two different evolutionary rhythms: the mild rhythm of effective stasis by the parent species (equilibria), and the "accelerated rhythm" of accommodation to a new microenvironment by the descendant species (punctuation). Most importantly, going in the opposite direction from gradualism, under allopatric speciation theory the apparent discontinuities of rhythm found in the fossil record are real, predictable, and intelligible; they are no longer an annoying "paradigm anomaly." In the punctuated equilibria perspective, speciation becomes an ecological, geographic, and sporadic process. And paleontology, freed from the onerous commitment to which it had been bound by neo-Darwinism, has the potential to no longer be an intrinsically imperfect science: the "gaps" in the fossil record encountered by paleontologists are actually real and contain information about evolutionary histories in both temporal and geographical contexts.

Among other things, PE thus represents an extension of Mayr's speciation theory to the general tendency for macro-evolutionary novelties to emerge in nature. It is by no means an anti-Darwinian theory, and it does not contradict the core mechanisms of the Darwinian explanation. It has nothing to do with the saltationism of the early geneticists (something that continues to be widely misunderstood among biologists); instead, it proposes an alternative to the phyletic gradualism of the MS in the light of new knowledge. Re-read in the light of PE, Huxley's admonition regarding the non-necessity of generalized gradualism turns out to have been correct: thanks to the theory of allopatric speciation, normal Darwinian processes can indeed produce what we see in the fossil record, provided we give up the rigid and binding gradualist assumption espoused by both Darwin himself and by neodarwinists. The MS was an unnecessary restraint on the theory. In this sense, we know well that the warning "the absence of evidence is not the evidence of absence" must be kept in mind. However, we do not believe that it should lead to a forced search for gradual processes when they are not there. Thus, PE provides paleontology with a more parsimonious view of the data.

Eldredge and Gould then extended their new view to a more general critique of evolutionary progressionism, starting with human evolution where the temptation to use the past to justify the present (our emergence as *Homo sapiens*, our current solitude as a hominin species, and our evolutionary success) is particularly strong. There is nothing special about hominins, and PE therefore helps to redefine the tempo and mode of speciation in our own subfamily, as we will see in more detail later.

In the last part of their 1972 paper, Eldredge and Gould proposed a pattern of macroevolution characterized by long periods of stasis that are punctuated by rapid speciation events in isolated subpopulations. This is where the idea of "punctuated equilibria" arises as a metaphor for the lives of biological species. Diversity and time, proportional to each other in phyletic gradualism, may actually be disjoint, because speciation tends to be rapid, episodic, and irregular, as well as dependent on ecological and geographic contingency (as the young Darwin had already noted in his Notebooks, using the metaphor not only of the tree, but also of the "coral" of life). In the evolution of multicellular life, most speciations may well occur in physical, environmental turnover events—as seen early on by Alcide d'Orbigny (1842), and as clearly articulated by Vrba (1985).

A classic extrapolation of phyletic gradualism to macroevolution is provided by the so-called biostratigraphic "trends" that are found within long graded stratigraphic series. From the elaboration of the marine shells studied by Geerat J. Vermeij, to the expansion of brain size in the genus *Homo* (Tattersall 2023), such trends are widespread in paleontology and seem to support continuity in macroevolution. However, a series of speciations, together with the differential survival and sorting of branching species, can give the impression of a gradual and directional shift of phenotype in a certain direction. We can see, therefore, that PE does not prejudicially rule out the possibility of gradual evolutionary processes occurring in nature but instead explains them in population and species-level terms rather than in strictly phyletic ones. Punctuated equilibria allows us to identify "species sorting" as a possible cause of perceived evolutionary trends.

In the PE framework, macroevolution is not completely reducible to microevolution, since "vertical" genetic factors are integrated with "horizontal" ecological and geographic ones. In this way, PE has also helped to shift the explanatory weights of genetics and ecology. As recently shown by Sean Anderson and Jason Weir (2022), most speciation events require at least an allopatric phase, and speciation in allopatry does not generally require lineages to exploit new resources or otherwise adapt to distinct ecological pressures, relying instead on their prolonged geographic separation. Geographic isolation thus emerges as the main driver of speciation. When a speciation occurs, the ecological trigger comes first, and the genes follow. Nevertheless, rather than being diminished, the role of genes is integrated into an explanatory context in which ecological factors have equal standing.

Sensing the dangers of a gene-centered approach to evolution, beginning early in the 1980s Eldredge, Gould and the other proponents of PE shifted their attention to the inconsistencies of what they called "ultra-Darwinism." It would be a mistake to overlook the multiplicity of evolutionary levels that are woven into evolutionary dynamics: large-scale biological systems must be considered as important as genes, and genetic and environmental causes are closely intertwined in defining the properties of organisms and populations (Gould and Eldredge 1977).

Habitat fragmentation is proportional to the number of species that can arise, and to the rate at which they arise. If a territory is divided into thousands of tiny environmental niches, into thousands of communicating or separate enclaves, effective local populations are similarly reduced in size. There will therefore be a proliferation of rapid genetic drifting events within small populations, carrying changes that are not necessarily adaptive but are rather due to drift phenomena such as founder effects. That is why, for example, the habitat-fragmenting succession of Ice Ages produced adaptive radiation in many African mammal groups, maybe including our own.

As paleontologist Elisabeth Vrba discovered in the 1980s, thanks to the PE perspective, we can satisfactorily interpret the intricate events that marked the Pleistocene in Africa and that endowed the hominin bush with a veritable explosion of forms (Vrba 1984, 1993). Rapid environmental change overwhelmed the capacity of hominin forms to adapt, forcing them to migrate to other habitats (even out of Africa), to go extinct, or to survive through branching specializations (as in the genus *Paranthropus*) and re-adaptations. Fragmentation of environmental niches then multiplied both speciations and the colonization of new habitats, triggering rapid processes of interspecies replacement (Vrba's turnover pulse hypothesis).

Ultimately, the MS underestimated (both empirically and theoretically) the importance of phenomena acting at the macroevolutionary level: speciation from small populations, punctuated equilibria, species sorting, turnover pulses, adaptive radiation, interspecies competition, and large-scale extinctions. Beyond better explaining the patterns typically observed in the fossil record, punctuationism thus had a positive effect on the theory of evolution itself by increasing its explanatory capacity, as even John Maynard Smith admitted in 1984.

7.4 PE and Stasis

Fifty years on, the notion of punctuated equilibria has taken on multiple meanings. In the version proposed by Eldredge and Gould in 1993, it represents a well-defined and verifiable theory of the origin of species and their geological unfolding, as well as the recognition that standard events on the ecological or biological scale (allopatric speciation) are sudden on the geological scale, and the idea that macroevolutionary phenomena can be addressed within known evolutionary mechanisms and causes (albeit requiring a hierarchical expansion to understand which causes are operating at higher levels). Punctuated equilibria is thus a theory that embraces macroevolution and has been extended to a pluralist conception of evolutionary units.

Summing up, PE consist of three patterns: (1) stasis at the paleontological level (which does not necessarily imply stasis at the molecular level, as shown for example by genomic analysis of morphologically very stable and ancient species such as the coelacanths); (2) rapid speciation that condenses most morphological change in time; and (3) macroevolutionary trends that are dependent on sorting among species.

Historically, what most shocked the scientific community about PE was not so much the assumption that speciation occurred quickly (which could be interpreted in terms of an acceleration of normal neodarwinian processes, and therefore was a question merely of the tempo, and not the mode, of speciation), but that it was the counterpart of a substantial stability that would last for much of the rest of a species' life. Stasis, even more than sudden speciation, was viewed by gradualists as a scandal. Normalizing selection is not enough to explain all the stasis evident in evolutionary histories, and neither is optimal adaptation to a niche, since species were often noted to remain stable even as the environment changed. Yet, as one of us has argued (e.g., Tattersall 2021), there are powerful reasons for considering classical natural selection as a more effective guardian of the status quo than of change, morphological or otherwise.

What is more, according to PE species, like individuals, are homeostatic systems (Ghiselin 1974; Hull 1976). Allopatric speciation is effective (and rapid) by virtue of a sudden increase in selective pressures due to unfamiliar environments and the isolation of small populations: the coincidence of these two forces would be expected to break the regulatory effect of homeostasis, resulting in a new equilibrium. There is a clear connection here with the theory of nonlinear complex systems, which exhibit not only pronounced resilience and stability (i.e., the absorption of external perturbations) but also the possibility of rapid systemic change if a perturbation affects internal positive feedback mechanisms that generate an overall phase-transition in the system. The recent Italian Nobel Laureate in Physics, Giorgio Parisi, said he was inspired by PE in his early studies of complex systems in the 1970s (an interesting case of fruitful cross-fertilization from biology to physics; see: www.youtube.com/watch?v=I_ewXxpukjY).

We can explain species stability by the cohesive homeostatic and regulatory forces of the genetic pool or by the ecological mechanism of geographic movement from one habitat to another (habitat tracking). Species, rather than adapting to a changing ecological niche, chase habitats that are more congenial to them, thus avoiding any need for changes in their morphology. It is also true that some species show only superficial stasis, beneath which gradual genetic change accumulates without expressing phenotypic and morphological consequences. Developmental constraints also have a stabilizing effect on species; and, finally, it is likely that phenotypic and developmental plasticity also have a stabilizing role, epigenetic in origin, which promotes adaptability to unstable ecological contexts and makes directional morphological change unnecessary.

Many findings have corroborated the existence of the patterns predicted by PE, despite the practical difficulties of demonstrating such events in stratigraphic contexts. Moreover, more recent integrated phylogenies (combining paleontological, molecular, and paleo-ecological data) confirm that punctuations may occur at a high frequency that is complementary to that of more gradual changes (Pagel et al. 2006). But stasis is even better documented: many cases are known today of highly stable species (referred to as "fossil species" by Darwin) that are morphologically the same at the beginning of their lives as they are at the end, millions of years later, or as they are today. Morphological stasis seems to be a general phenomenon, also present in the "bushy tree" of hominin species.

Since evolution occurs at multiple levels, morphological stasis obviously does not exclude the possibility of underlying changes and oscillations at the molecular level (Eldredge et al. 2005). Only the complex geographic structure of species can explain how taxa commonly exhibit both short-term evolutionary dynamics and long-term stasis. Furthermore, the multiplicity of factors and dynamics that characterize evolutionary stasis confirm the "Brocchian" individuality of species as discrete entities, as spatiotemporally bounded units with births, histories, and deaths.

7.5 PE and Evolutionary Pluralism

After five decades of debate, the general consensus around the mechanisms of speciation is that we need to envisage a multiplicity of processes and modes of birth of new species (punctuated in many ecological circumstances, and gradual in others), and to accept that we need to consider a multiplicity of possible rates of speciation and a multiplicity of levels of change from both ecological and genealogical points of view (Coyne and Orr 2004). By "epistemological pluralism," we mean here the idea that these modes, processes, and rates are not to be considered alternatives, but coexisting (like a toolkit of possible patterns) (Pievani 2016). Depending on the ecological circumstances, a speciation can occur in many ways. Likewise, in each case, an evolutionary process can be due more to natural selection or more to random genetic drift. These patterns are not mutually exclusive.

Hence, the recent interest in calculating the relative frequencies of one general pattern (punctuationism) with respect to another (gradualism). A leading example is the contribution by Pagel et al. (2006), the title of which made its theme clear: "Large Punctuational Contribution of Speciation to Evolutionary Divergence at the

Molecular Level." We do not need to choose between radically different and incompatible patterns.

Following this epistemological pluralism, we see a balance between a need to reject the past methodological stances of the MS (such as phyletic gradualism) and a need for theoretical continuity. The underpinnings of rejection are: (1) punctuations are not due to imperfections of the geological record; (2) speciation is not only anagenetic, but frequently cladogenetic; (3) speciation is connected to major episodic evolutionary changes; and (4) the high rate of apparent stasis in natural histories. The points of continuity are: (a) the evolutionary mechanisms that act during speciation are still Darwinian; (b) that gradual trends should not be excluded; and (c) that punctuation and stasis exist on the geological scale, so do not clash with standard mechanisms of change at the population level (Gould 2002).

Clearly, then, the postulates of PE do not at all imply a literal return to Goldschmidt's anti-Darwinian "hopeful monsters." Rather, they seek to broaden the spectrum of the possible changes and the magnitude of mutations that intervene during processes of adaptation and speciation: accumulations of small mutations can integrate with systemic mutations that profoundly alter entire morphologies, as in the case of sticklebacks (Chouard 2010). After 50 years, then, we can conclude not only that the saltationist interpretations of PE, as promoted by Richard Dawkins and Daniel Dennett among others, were wholly unfounded and improper, but that these authors largely misunderstood the multi-level complexity of evolutionary processes. As even Ernst Mayr (1989) belatedly admitted, PE is not only compatible with Darwinian theory but also defies the gene-centered version of the MS that forced the Darwinian legacy in just one direction.

The best evidence for the verifiability of PE is the cladogenetic pattern of species origin: ancestral forms often outlive their descendants, while evolution proceeds by irregular branching and not by substitution across whole lineages. This highly branching evolution is precisely what the fossil record often reveals, and it is what is investigated by cladistics, i.e., phylogenetic or historical systematics. The ideas behind PE are also compatible with the phylogenetic notion of species that cladists have placed alongside the biological one. Here, the focus is not so much on the means by which species are generated, but on the pattern of phylogenetic kinship: by species, we are referring to a particular segment within the phylogenetic tree, that is, the stretch of lineage lying between two speciation events (or between speciation and extinction).

Finally, PE had at least three significant impacts on evolutionary thinking. The first was on the history of biological thought: the critique of strict phyletic gradualism in paleontology and paleoanthropology as *pars destruens*. The second was the theoretical impact of arguing for an extension and revision of the neo-Darwinian structure of the theory of evolution. The third impact is empirical: the substantial alignment with expectation of the emerging data on rates of evolution and modes of speciation that emerges because PE concerns both the patterns and the processes of evolution: not only the timing, but also the mode, of speciation.

In sum, PE is an articulated set of explanations that brings together heterogeneous facts, absorbs new information, and updates its theoretical toolkit when required. As such, it should be described not simply as "theory" or a "paradigm," but as a "scientific research program" as defined by the epistemologist Imre Lakatos (1978). This active definition means that our evolutionary explanations are subject to continuous change, sometimes on a large scale. They thus appear like an open and busy workyard rather than like an old, fixed, traditional building. As a result of PE, the theory of evolution is evolving; or rather, the evolutionary scientific program is evolving. The process of criticism and growth of evolutionary knowledge after Darwin seems to be a continuous one—made by extensions, revisions, and even by reversions to originally Darwinian insights—rather than a discontinuous, dramatic, and paradigmatic one (Pievani 2012). We persist in returning to Darwin, sometimes to rethink him, while always appreciating his original intuitions. In this process, PE continues to have a fertile and provocative role as part of a more general movement.

7.6 PE and Multilevel Theories

Other fruitful fields of research threaten the inclusive capacity of gene-centered neo-Darwinism because they touch fundamental components of its theoretical architecture, or in some cases add entire new domains of experimental evidence that argue for a powerful theoretical updating. Among them are neutralism, symbio-genesis, Evo-Devo, epigenetics, enlarging the range of the sources of variation and inheritance (Jablonka and Lamb 2005); phenotypic and developmental plasticity which modifies the relationships between genomes, phenotypes and ecological niches (West-Eberhard 2003); "niche construction" (Odling-Smee et al. 2003) which emphasizes the active role of organisms in evolution and the reciprocal modifications of organisms and niches; exaptation; and functional shifts; and on and on.

The upshot has been wide agreement that we need a new kind of neo-Darwinism, revised and extended: a transition from the traditional MS to the so-called "Evolutionary Extended Synthesis" (EES: Pigliucci and Müller 2010). Let us specify that an update of evolutionary theory does not require moving away from Darwin, because its original formulation was more flexible than the gradualist and adaptationist neo-Darwinism of the Modern Synthesis (Gould 2002). However, the EES has yet to be embraced by the community of evolutionary biologists and has so far had little real impact. This is because of two inherent weaknesses. The first is that this research program lacks a clear internal structure (for a proposed pluralistic structure, based on multiple sources of variation, multiple selective processes, neutral factors, and macroevolutionary factors, see Pievani 2016). The second is that it is entirely centered on the organism (as opposed to the genes or the species) as the essential actor in the evolutionary play, thereby failing to consider the multiple levels at which evolution occurs. Paradoxically, the EES, like the MS, underestimates macroevolutionary and ecological factors.

An alternative proposal is the "hierarchy theory of evolution" (Eldredge et al. 2016), with two interacting nested hierarchies: a genealogical hierarchy of nested levels of transmission of the genetic material (organisms, populations like "demes," species, monophyletic taxa); and an ecological hierarchy of nested levels of transfers of matter and energy (organisms, populations, local ecosystems, regional ecosystems) (Eldredge 1985, 1999). Ecological hierarchy describes processes of energy and matter transfer, while genealogical hierarchy describes patterns of genetic heritability at many scales.

The ecological hierarchy affects the genealogical one to varying extents, while the genealogical one accumulates changes and produces diversity. This gives us an integrated framework in which genealogical levels and ecological levels coexist and interact. The Darwinian notions of the organism (at the same time genetic replicator and ecological interactor) and natural selection (differential survival and reproduction) both remain at the fundamental junction of the two hierarchies (see the "sloshing bucket" model of evolution: Eldredge 2008; Spiridonov and Eldredge 2024). The result is a better articulated and more pluralistic structure for the theory of evolution: one that is more realistic in its assumptions than either the MS or the EES. In this way, PE remains an important stimulus for the continuing larger debate over evolutionary theory.

Recently, Spiridonov and Eldredge described the so-called "Bretskyan hierarchy" (named after paleo-ecologist Peter W. Bretsky): a hybrid eco-genealogical hierarchy that consists of nested sets of different-sized, usually polyphyletic communities of interacting individuals separated from other such communities in space and time at multiple scales. It is a connection between economic-ecological hierarchies and genealogical hierarchy of evolutionary units, stemming from the growing understanding of symbiogenesis. The Bretskyan hierarchy at lower tiers is populated by holobionts, individuals composed of multiple polyphyletic lineages. At larger spatial tiers and longer time scales, the units of the Bretskyan hierarchy are of a more diffuse nature, longer-lasting structures, such as geo-biomes, intended as truly evolutionary individuals. The global biota, Gaia, might be understood as the largest eco-genealogical individual compartmentalized by the outer space of the Earth, and integrated at long time scales by biotic interactions and plate tectonic mixing of biota. The existence of a hierarchy of barriers and multilevel allopatry suggests that geographic isolation contributes not only in individuating species lineages, but also in producing coherent complexes of separate lineages forming "bioprovinces" at multiple space and time scales (Spiridonov and Eldredge 2024).

7.7 Punctuated Equilibria in Paleoanthropology: A Lesson Yet to Be Absorbed

In contrast to the paleontologists in other sub-disciplines who grumbled initially but eventually incorporated the new perspective into their own thinking, students of human evolution staunchly resisted PE, and continue to do so. It is not easy to fully grasp the epistemological and perhaps paradigmatic reasons (i.e., linked to the tacit influence of a powerful general vision of evolution as a linear process culminating in *Homo sapiens*) for this recalcitrance (Gould 1990).

And while this recalcitrance has been severely damaging to our understanding of the human past, in a way this is hardly surprising because, in contrast to its cousin vertebrate paleontology which has its roots in biostratigraphy and comparative anatomy (both fields that in one way or another depend on biotic diversity), the insular field of paleoanthropology developed independently out of the extremely inward-looking specialty of human anatomy. This occurred as antiquarians, and eventually archeologists, acquired the habit of sending the ancient human remains they excavated to physicians and anatomists for analysis. In itself, this was entirely understandable because the anatomists had a uniquely intimate acquaintance both with the morphology of the only living hominin species and of the structural variation within it. But it meant entrusting the analysis of the human fossil record to clinically trained specialists who neither knew nor cared very much about the riotous biodiversity of which *Homo sapiens* formed merely an infinitesimal part, and who hence perceived little need to pay attention to the processes of evolution that had produced both that biodiversity and their specific objects of study. They acknowledged that human beings were the products of evolution, but they paid little more than lip service to the processes involved.

As a result, if they thought about it at all, right up until the midpoint of the twentieth century paleoanthropologists conceived the business of reconstructing human phylogeny as a matter of projecting the one extant hominin species *Homo sapiens* back into the past in an ever-receding chain—basically, the *scala naturae* that had given rise to the notion of the "missing link." And, as contrary to the signals of morphology as this way of thinking might have been, in an era of authoritative pronouncement, it was relatively easily maintained. More remarkably, perhaps, even experienced vertebrate paleontologists tended to be infected by this exceptionalist mindset when they ventured into the paleoanthropological arena. For example, the fossil elephant specialist (and advocate of "aristogenesis") Henry Fairfield Osborn proposed (Osborn 1915) that the lineage culminating in today's *Homo sapiens* had split from the other primates in the Eocene, the better to distance our own splendid species from the appalling apes.

7.8 Mayr: The Continuous Transformation of a Single Lineage

Into this largely theory-free world strode Ernst Mayr, one of the principal architects of the then-dominant MS whose tenets PE would eventually seek to overthrow. In 1950, at a hugely influential conference hosted by the Cold Spring Harbor Biological Laboratory and entitled "Origin and Evolution of Man," the ornithologist presented a peroration on "Taxonomic Categories in Fossil Hominids" in which he energetically excoriated the entire paleoanthropological profession for committing just about every procedural and theoretical sin possible (Mayr 1950). And admittedly, he had a very soft target. With their lackadaisical approach to diversity, evolution, and taxonomy, paleoanthropologists had indeed—and purely for reasons of convenience—allowed the number of species and genus names applied to a very modest quantity of fossils to balloon uncontrollably (see Tattersall 2018). This malfeasance, thundered Mayr, had bestowed on the human fossil record an entirely illusory and misleading appearance of diversity. Whereas, he insisted, because "Man occupies more different ecological niches than any known animal," indeed "all … that are open for a man-like creature, it is obvious that he *cannot* speciate" (1950: 116; emphasis added). Which meant, in turn, that the human fossil record must necessarily bear witness to the continuous transformation of a single lineage, an idea previously propounded by Mayr's MS colleague Theodosius Dobzhansky (1944; with some rethinking and refinements in Dobzhansky 1962). What is more, according to Mayr that lineage had only ever contained three species, all three of which had belonged to the single genus *Homo*. Specifically, *Homo transvaalensis* (broadly, the australopiths) had insensibly evolved into *Homo erectus*, which had in turn gradually transformed into *Homo sapiens* (including the Neanderthals). For good and ironic measure, the rising expert on allopatric speciation insisted that no discontinuities in the sequence should ever be expected, and that any that were perceived were necessarily due to deficiencies in the material record.

The distinguished conference attendees were stunned, as were the vast majority of anglophone paleoanthropologists who were not in attendance, but who anxiously absorbed the conference proceedings. And because they were smart enough to have known that they had been skating on thin theoretical ice all along, almost everybody capitulated instantly to Mayr's onslaught. Only a single contrary voice was raised. From faraway South Africa, John Robinson (1953) complained that there was indeed variety not only among the australopiths, but even among the Java hominins. Mayr himself (1953) rapidly if grudgingly conceded this specific point, but English-speaking paleoanthropology nonetheless went into shock. Virtually unanimously, anglophone paleoanthropologists rejected past practice, in an epic repudiation that was spurred along by Sherwood Washburn's (1951) energetic promotion of the New Physical Anthropology, a reorientation of biological anthropology that emphasized the dynamic nature of organisms and elevated functional anatomical and behavioral studies high above fusty old taxonomy. So effective was this concatenation of influences that, for well over a decade, traumatized paleoanthropologists eschewed phylogenetic diagrams. Instead, where it was not strictly unilinear (e.g. Campbell 1962, Fig. 1), the typical hominin family tree of the immediate post-Mayr period consisted of billowing blobs enclosing the identifiers of individual fossils (e.g. Howell 1957, Fig. 4; see Tattersall 2018). Anglophone paleoanthropology, at least, had fallen both instantly and firmly under the thrall of the MS in its most "hardened" (Gould 1983) form.

7.9 Mayrian Notion of Linearity Persisted Despite Evidence to the Contrary

The next paleoanthropological shock arrived in 1964, when Louis Leakey, Phillip Tobias, and John Napier presented to the world the new species *Homo habilis* from (the then) Tanganyika's Olduvai Gorge. This addition to the human genus of a fossil hominin that was morphologically vastly more akin to the gracile South African australopiths than to the type species of *Homo*, and that was a million years older than any other putative member of the genus (Leakey et al. 1961), owed nothing to any rational rethinking of evolutionary or taxonomic process. Instead, it owed everything to Leakey's attachment to the widely accepted notion of "Man the Toolmaker," in the context both of his single-minded search for the hominin that had made the Mode 1 tools found in the lowest stratigraphic levels of the Gorge, and of his willingness to create new taxa with all the abandon of an old-time anatomist.

Small wonder, then, that Leakey and colleagues' unusual taxonomic move initially ran into significant resistance. And in the event, the new *Homo* species did not become widely accepted until the mid-1970s, when Richard Leakey began finding a variety of hominin fossils in sediments of broadly similar age in the Turkana Basin of northern Kenya (see Howell 1978). In a typical display of paleoanthropological dissonance, both the most thoughtful attack on *Homo habilis* (Pilbeam and Simons 1965), and the ultimate general acceptance of Leakey's new species, had their roots deep in the extreme Mayrian notion of continuity: a seductively reductionist idea that was most visibly expressed in the tenacious and ongoing search for the "earliest *Homo*" that was launched by the acceptance of the ancient and tiny-brained new *Homo* species, and that both preceded and followed the temporary florescence of the "single-species hypothesis" (Brace and Montagu 1965) and its derivative the "multiregional hypothesis" (Wolpoff et al. 1984). In direct obeisance to Mayr, advocates of both hypotheses continued to insist that only one hominin species could ever have existed on the planet at any one time, even as an increasing appreciation of the role in evolution of external factors such as competition and climate change (e.g., Vrba 1988) was making the unilineal idea appear naïve even in principle.

As the hominin fossil record grew apace during the last quarter of the twentieth century, it became impossible to ignore the fact that substantially more taxonomic categories than the parsimonious Mayr had admitted were needed to accommodate the emerging morphological, temporal, and geographic variety of fossil hominins known. But the Mayrian spirit still stubbornly persisted in paleoanthropology, most notably in the form of a resolute taxonomic minimalism that was usually justified not only by invoking the undisputable fact that all species are morphologically variable, but by alleging that any suggestion that such variation might not be limitless was simply to yield to old-fashioned and "nonbiological" typology. Any attempt to apply the established standards of mammalian taxonomy to hominins was likely to be derided as mere "quibbling about names." The anatomists' active disdain for systematics and nomenclature was resurfacing with a vengeance, and it found the perfect vehicle for expression in the grotesque bloating of the species *Homo erectus*—which, ironically, happened mostly in the service of a blind adherence to the MS.

Originally named on the basis of fossils that are now known to date from around 800 Ma or possibly earlier (Hilgen et al. 2023), and that were found in Java in the 1890s when there was little in the fossil record to compare them to, *Homo erectus* automatically became the "hominin in the middle" after the South African australopiths had become generally accepted as early hominins in the late 1940s (just before Mayr delivered his broadside: see Tattersall 2009). When Richard Leakey's team began finding fossils in Kenya that dated in the 1.7–1.5 Ma range, and that were reasonably assigned to the genus *Homo*, a blanket ban was initially instituted on identifying them to species. This was perhaps the most dramatic example ever of the insularity of paleoanthropology: it is unimaginable that any such maneuver would ever have been adopted elsewhere in paleontology, where the very first order of business is to identify fossils to species. But when this bizarre policy inevitably proved untenable, the reflex response was to assign the Kenyan fossils to *Homo erectus* (Leakey and Walker 1978)—even though that species was documented only on the far side of the world and a million years later in time, and despite the fact that the Javan type material was notably dissimilar to the Kenyan fossils in anything other than brain volume. Since then, the ranks of *H. erectus* have continued to swell, as a motley assortment of fossils not just from eastern Asia and east Africa, but from Europe as well, was crammed into the species regardless of morphology. Of course, with each addition, the morphological limits of the species bulged out yet farther, making it easier to shoehorn others in. The latest ploy to rationalize this, usually trotted out in the context of the EES (Laland et al. 2015), is to invoke phenotypic plasticity (the propensity for the same genome to produce different phenotypes under different environmental circumstances) as the excuse for stuffing a wildly implausible variety of morphologies into *H. erectus* (Antón et al. 2016; see also Tattersall 2022a).

At the same time, though, none of this criticism should be taken to imply that *Homo erectus* does not itself furnish us with one of the best examples in the hominin fossil record of stasis—the lack of change, over significant stretches of time, that was so strongly emphasized in the original formulation of PE (Eldredge and Gould 1972). Indeed, while the species *Homo erectus*, initially recognized in the Indonesian island of Java, and subsequently found in China as well, is geographically limited to eastern Asia, its temporal range is a very considerable one. Early *Homo erectus* found in Java's Sangiran Dome date to as early as 1.3 Ma (Matsu'ura et al. 2020), while the latest Javan representatives of the species from Ngandong date from less than 120 Ka (Rizal et al. 2020). Well over a million years is an unusually long time in the history of the genus *Homo*, although there are plenty of other examples of significant species longevity in the hominin fossil record. Among them is the eastern African australopith species *Australopithecus afarensis* (2.9–2.0 Ma), which until recently also gave us the most famous claimed example in that record of documented anagenesis—the insensible transition, over time, of one species into another. In this case, it was alleged that *A. afarensis* had gradually emerged out of the earlier *A. anamensis* (Kimbel et al. 2006), a proposition that was recently falsified by Haile-Selassie et al. (2019), who discovered temporal overlap between the two species in Ethiopia.

Perhaps the most egregious abuse of the concept of *Homo erectus* has involved the hominin fossils found at the 1.8 Ma (Ferring et al. 2011) site of Dmanisi, in the Republic of Georgia. Beginning in the mid-1990s, five hominin crania and some matching postcranial elements were recovered at Dmanisi. The crania share small brain volumes (smaller than anything else attributed to *Homo erectus*), but otherwise they are not reminiscent of anything known from Java and additionally show a great morphological variety among themselves (Schwartz et al. 2014; Rightmire et al 2019). The mandible initially found was compared to east African *Homo habilis* (Gabunia and Vekua 1995). But as the Dmanisi hominin crania accumulated, the describers settled on the designation of *Homo erectus* (e.g., Rightmire et al. 2006). This assignment was maintained even after the discovery of a very small-brained and large-faced skull ("Skull 5") that is very different from anything else at Dmanisi and, indeed, from anything else known in the entire hominin fossil record. The only possible rationale for placing this specimen in the genus *Homo* is supplied by the undeniable observation that it is not *Australopithecus*. And even though that exclusion must have been the reason why its describers (Lordkipanidze et al. 2013) classified it as *Homo* (which was, after all, the genus already—for better or worse—established for the less distinctive hominin fossils coming out of Dmanisi), the systematic naïveté involved is clearly evident in the fossil's extraordinary designation as *Homo erectus ergaster georgicus* (Lordkipanidze et al. 2013). Only paleoanthropologists would ever see fit to invent the category of sub-subspecies for their fossils, and although this action (non-Code-compliant as it is) gives the appearance of paying due attention to systematics, it actually reflects a slavish adherence to the linear strictures of the Synthesis by maintaining the "hominin in the middle" status of *H. erectus*.

Exceptions to the implicit taboo on the creation of new hominin genera and species to accommodate new morphologies were eventually made, as a rather heterogenous assortment of very ancient (7–4 Ma) and putatively hominin fossils was named in the years around the turn of the millennium. Thus, *Sahelanthropus*, *Orrorin*, and *Ardipithecus* were rapidly accepted into the paleoanthropological canon, and even Meave Leakey et al.' (2001) proposal of the new genus *Kenyanthropus* for a somewhat younger 3.5 Ma cranium from West Turkana was met with only one shrill response (White 2003). But all were from a remote period in which no intellectual positions had previously been staked out, and to which no stifling received wisdoms yet applied. In an implicit nod to Robinson, the genus *Paranthropus* was grandfathered in as well, for the highly distinctive southern and eastern African "robust" australopiths. But for the rest, and particularly in the critical period following 2 Ma, no proposal of a new genus would have been tolerated, even as fossil morphologies multiplied dramatically. The most destructive result of this reluctance to contemplate that branching had been a common feature of hominin phylogeny, and that those divergences required some nomenclatural expression, was that paleoanthropologists working with non-*Paranthropus* hominin fossils in the all-important period of the Pleistocene were faced with the stark choice we have already seen illustrated at Dmanisi: either those fossils belonged to *Australopithecus*, or they were *Homo*. Which meant, in practice, that diagnosis to the genus *Homo* became not one of inclusion (i.e., justified by the monophyly indicated by the possession of common derived morphologies), but one of exclusion: "It isn't *Australopithecus*, so it must be *Homo*" (Tattersall 2022a, b). The same reasoning, clearly born out of the hardline linear thinking of the Synthesis, applies equally to the alternative diagnosis to *Australopithecus*, and the fact that it provides a completely irrational basis for alpha taxonomy appears to have bothered paleoanthropologists hardly at all, even as the hominin fossil record continued to expand dramatically in the early twenty-first century.

7.10 The Cases of *H. floresiensis* and *H. naledi*

This is very clearly seen in the case of two very late, very small-statured, and very small-brained hominins that have been classified in the genus *Homo* (which is, after all, defined by the huge-brained type species *Homo sapiens*) in the absence of any substantial morphological justification. Neither the ~100 Ka *H. floresiensis* from Flores (Brown et al. 2004), nor the ~300 Ka *H. naledi* from South Africa (Berger et al. 2015), shares significant synapomorphies with *H. sapiens*; and neither species self-evidently belongs to a monophyletic genus *Homo*. It is admittedly also the case that neither belongs to *Australopithecus*, but that is hardly adequate justification for cramming them into *Homo*. And until we free ourselves from the minimalist inheritance of the MS and admit that stark reality, it will be impossible for us to properly perceive the phylogenetic structure that unquestionably lies both within our subfamily and within our own particular genus.

Before concluding, we should admit that paleoanthropologists have to bear a burden that most paleontologists do not. That is because students of human evolution are very tightly focused on a single closely-knit and recently evolved group, within which differentiation is often subtle and species recognition is often difficult (with apparent reticulation among lineages further complicating the picture late in the story: Villanea and Schraiber 2018). The resulting complications may, at least in part, explain why the majority of contemporary vertebrate paleontological analyses begin with a tree establishing relationships among the organisms of interest, while paleoanthropological ones typically do not. However, it is equally likely that this omission is an inheritance from the early days of PE, a time when paleoanthropologists were wont to confound the new characterization of evolutionary process and pattern with the unrelated but then equally novel practice of cladistics. This was extremely unfortunate. For although cladistics assumed dichotomies, it was purely an approach to determining the relationships among species and higher taxa in terms of recency of ancestry. It was entirely unconcerned with process. Still, whichever of the two disciplines paleoanthropologists chose to spurn, the baby inevitably went out with the bathwater, reinforcing the rejection, already so deeply inculcated by the linear thinking of the MS, of the value of systematics.

Equally unfortunately, in tandem with this erosion, general awareness declined of the importance of the intimate linkages between systematics and theory of process, between speciation and diversification, between smaller and larger patterns in nature. Over the last few decades, and despite the intellectual obstacles we have noted, the sheer weight of paleoanthropological discovery has imposed a significant expansion in the number of generally recognized hominin species. This diversity has clearly demonstrated that, far from being linear, hominin evolutionary history has involved the continual testing of the subfamily's evolutionary potential through speciation and diversification and a wide variety of other influences, exactly as the punctuational approach to evolution predicts. Instead of the burnishing of a single variable lineage envisaged by Ernst Mayr and his fellow advocates of the hardened synthesis, human phylogeny was a matter of energetic evolutionary experimentation. Hominin species were repeatedly sent out on to the evolutionary stage to succeed or to fail, producing a complex set of evolutionary dichotomies, new lineages, and dead ends. The result was that, typically, multiple hominin species coexisted at any one point in time. In other words, we modern human beings are not, as we often like to think, the apogee of a process of constant improvement over the eons.

7.11 Conclusion

In the half-century since it was first proposed, PE has had a profound effect on the shape of evolutionary theory, replacing a reductionist view of process with a far more nuanced one, and bringing a science dominated by the relentless gradualism of the MS much more directly into line with what paleontologists have actually been observing all along. It has also deeply affected the practical exercise of systematics and most areas of paleontology, in which the necessity of unraveling phylogenetic structure is now axiomatic. Duran-Nebreda et al. recently proposed to call "punctuated evolution" the more general theory that aggregates the evidence for punctuated equilibria in geology, molecular biology, genetics, anthropology, and socio-technology, expanding punctuational evolution into somatic and cultural evolutionary realms (2024). To understand how a revised form of PE theory can be fruitful in explaining biocultural evolution today, see O'Brien et al. (2024).

Paleoanthropology, however, has been slower to absorb the pluralist vision of evolution, due perhaps to its short and unusual history as a discipline, and maybe to additional resistance related to the fact that it deals retrospectively with the evolution of *Homo sapiens*. One reason for hope may lie in the accelerating growth of the empirical basis of paleoanthropology, as the addition of new areas of investigation—from molecular anthropology to archeology—reduces its dependence on the old linear and progressionist biases. However, both the history and the philosophy of science teach us that hardened paradigms can be very resistant to change, and that the transition to a new research program can be long and troubled. Sadly, we will fully recognize the field's full potential only when paleoanthropology has properly absorbed the lessons of PE: a development that does not appear to be on the immediate horizon.

References

Anderson SAS, Weir JT (2022) The role of divergent ecological adaptation during allopatric speciation in vertebrates. Science 378:1214–1218

Antón SC, Taboada HG, Middleton ER et al (2016) Morphological variation in *Homo erectus* and the origins of developmental plasticity. Philos Trans R Soc Lond B Biol Sci 371:20150236. https://doi.org/10.1098/rstb.2015.0236

Barrett PH, Gautrey PJ, Herbert S, Kohn D, Smith S (eds) (1989) Charles Darwin's notebooks. 1836–1844. Cornell University Press, Ithaca

Berger LR, Hawks J, De Ruiter DJ et al (2015) Homo naledi, a new species of the genus Homo from the Dinaledi Chamber, South Africa. elife 4:e09560. https://doi.org/10.7554/eLife.09560

Brace CL, Montagu MFA (1965) Man's evolution. Macmillan, New York

Brown P, Sutikna T, Morwood MJ et al (2004) A new small-bodied hominin from the Late Pleistocene of Flores, Indonesia. Nature 431:1055–1061. https://doi.org/10.1038/nature02999

Campbell B (1962) The systematics of man. Nature 194:225–232. https://doi.org/10.1038/194225a0

Chouard T (2010) Evolution: revenge of the hopeful monster. Nature 463:864–867. https://doi.org/10.1038/463864a

Coyne JA, Orr HA (2004) Speciation. Oxford University Press, Oxford

D'Orbigny A (1842) Paléontologie française. Vol. 1, Terrains oolitiques ou Jurassiques. Arthus Bertrand, Masson, Paris

Dobzhansky T (1937) Genetics and the origin of species. Columbia University Press, New York

Dobzhansky T (1944) On species and races of living and fossil man. Am J Phys Anthropol 2:251–265

Dobzhansky T (1962) Mankind evolving. Yale University, New Haven/London

Duran-Nebreda S, Bentley RA, Vidiella B et al (2024) On the multiscale dynamics of punctuated evolution. Trends Ecol Evol 39:734–744. https://doi.org/10.1016/j.tree.2024.05.003

Eldredge N (1985) Unfinished synthesis. Biological hierarchies and modern evolutionary thought. Oxford University Press, New York

Eldredge N (1995) Reinventing Darwin. The great debate at the high table of evolutionary theory. Wiley, New York

Eldredge N (1999) The pattern of evolution. W.H. Freeman and Company, New York

Eldredge N (2008) Hierarchies and the sloshing bucket: toward the unification of evolutionary biology. Evol Educ Outreach 1:10–15. https://doi.org/10.1007/s12052-007-0007-6

Eldredge N (2015) Eternal ephemera. Columbia University Press, New York

Eldredge N, Gould SJ (1972) Punctuated equilibria: an alternative to phyletic gradualism. In: Schopf TJ (ed) Models in paleobiology. Freeman, San Francisco, pp 82–115

Eldredge N, Tattersall I (1982) The myths of human evolution. Columbia University Press, New York

Eldredge N, Thompson JN, Brakefield PM et al (2005) The dynamics of evolutionary stasis. Paleobiology 31:133–145. https://doi.org/10.1666/0094-8373(2005)031[0133:TDOES]2.0.CO;2

Eldredge N, Pievani T, Serrelli E, Tëmkin I (eds) (2016) Evolutionary theory: a hierarchical perspective. The University of Chicago Press, Chicago

Ferring R, Oms O, Agustí J et al (2011) Earliest human occupations at Dmanisi (Georgian Caucasus) dated to 1.85–1.78 Ma. Proc Natl Acad Sci 108:10432–10436. https://doi.org/10.1073/pnas.1106638108

Gabunia L, Vekua A (1995) A Plio-Pleistocene hominid from Dmanisi, East Georgia, Caucasus. Nature 373:509–512. https://doi.org/10.1038/373509a0

Gabunia L, Abesalom V (1995) A Plio-Pleistocene hominid from Dmanisi, East Georgia, Caucasus. Nature 373(6514):509–512

Ghiselin MT (1974) A radical solution to the species problem. Syst Biol 23:536–544. https://doi.org/10.1093/sysbio/23.4.536

Gould SJ (1983) The hardening of the modern synthesis. In: Grene M (ed) Dimensions of Darwinism. Cambridge University Press, Cambridge, pp 71–93

Gould SJ (1990) Wonderful life: the burgess shale and the nature of history. Norton & Co, New York

Gould SJ (2002) The structure of evolutionary theory. Belknap Press of Harvard University Press, Cambridge, MA
Gould SJ, Eldredge N (1977) Punctuated equilibria: the tempo and mode of evolution reconsidered. Paleobiology 3:115–151. https://doi.org/10.1017/S0094837300005224
Gould SJ, Eldredge N (1993) Punctuated equilibrium comes of age. Nature 366:223–227. https://doi.org/10.1038/366223a0
Haile-Selassie Y, Melillo SM, Vazzana A et al (2019) A 3.8-million-year-old hominin cranium from Woranso-Mille, Ethiopia. Nature 573:214–219. https://doi.org/10.1038/s41586-019-1513-8
Hilgen SL, Pop E, Adhityatama S et al (2023) Revised age and stratigraphy of the classic Homo erectus-bearing succession at Trinil (Java, Indonesia). Quat Sci Rev 301:107908. https://doi.org/10.1016/j.quascirev.2022.107908
Howell FC (1957) The evolutionary significance of variation and varieties of "Neanderthal" man. Q Rev Biol 32:330–347. https://doi.org/10.1086/401978
Howell FC (1978) Hominidae. In: Maglio VJ, Cooke HBS (eds) Evolution of African mammals. Harvard University Press, Cambridge, MA, pp 154–248
Hull DL (1976) Are species really individuals? Syst Zool 25:174. https://doi.org/10.2307/2412744
Jablonka E, Lamb MJ (2005) Evolution in four dimensions. The MIT Press, Cambridge, MA
Kimbel W, Lockwood C, Ward C et al (2006) Was Australopithecus anamensis ancestral to A. afarensis? A case of anagenesis in the hominin fossil record. J Hum Evol 51:134–152. https://doi.org/10.1016/j.jhevol.2006.02.003
Lakatos I (1978) The methodology of scientific research programmes, Philosophical papers, vol 1. Cambridge University Press, Cambridge
Laland KN, Uller T, Feldman MW et al (2015) The extended evolutionary synthesis: its structure, assumptions and predictions. Proc R Soc Lond B Biol Sci 282:20151019. https://doi.org/10.1098/rspb.2015.1019
Leakey LSB, Evernden JF, Curtis GH (1961) Age of Bed I, Olduvai Gorge, Tanganyika. Nature 191:478–479. https://doi.org/10.1038/191478a0
Leakey LSB, Tobias PV, Napier JR (1964) A new species of the genus homo from Olduvai Gorge. Nature 202:7–9. https://doi.org/10.1038/202007a0
Leakey MG, Spoor F, Brown FH et al (2001) New hominin genus from eastern Africa shows diverse middle Pliocene lineages. Nature 410:433–440. https://doi.org/10.1038/35068500
Lordkipanidze D, Ponce De León MS, Margvelashvili A et al (2013) A complete skull from Dmanisi, Georgia, and the evolutionary biology of early *Homo*. Science 342:326–331. https://doi.org/10.1126/science.1238484
Matsu'ura S, Kondo M, Danhara T et al (2020) Age control of the first appearance datum for Javanese *Homo erectus* in the Sangiran area. Science 367:210–214. https://doi.org/10.1126/science.aau8556
Mayr E (1942) Systematics and the origin of species from the viewpoint of a zoologist. Columbia University Press, New York
Mayr E (1950) Taxonomic categories in fossil hominids. Cold Spring Harb Symp Quant Biol 15:109–118. https://doi.org/10.1101/SQB.1950.015.01.013
Mayr E (1953) Comments on evolutionary literature. Evolution 7:273–281. https://doi.org/10.2307/2405739
Mayr E (1959) Isolation as an evolutionary factor. Proc Am Philos Soc 103(2):221–230
Mayr E (1970) Populations, species and evolution. Harvard University Press, Cambridge, MA
Mayr E (1989) Speciational evolution or punctuated equilibria. In: Somit A, Peterson SA (eds) The dynamics of evolution. Cornell University Press, Ithaca, pp 21–53
O'Brien MJ, Valverde S, Duran-Nebreda S et al (2024) Punctuated equilibrium at 50: anything there for evolutionary anthropology? Yes; definitely. Evol Anthropol 33:e22009. https://doi.org/10.1002/evan.22009
Odling-Smee J, Laland K, Feldman MW (2003) Niche construction. Princeton University Press, Princeton

Osborn HF (1915) Men of the old stone age: their environment, life, and art. Charles Scribner, New York

Pagel M, Venditti C, Meade A (2006) Large punctuational contribution of speciation to evolutionary divergence at the molecular level. Science 314:119–121. https://doi.org/10.1126/science.1129647

Pievani T (2012) An evolving research programme: the structure of evolutionary theory from a Lakatosian perspective. In: Fasolo A (ed) The theory of evolution and its impact. Springer, New York, pp 211–228

Pievani T (2016) How to rethink evolutionary theory: a plurality of evolutionary patterns. Evol Biol 43:446–455. https://doi.org/10.1007/s11692-015-9338-3

Pigliucci M, Müller GB (eds) (2010) Evolution: the extended synthesis. MIT Press, Cambridge/London

Pilbeam DR, Simons EL (1965) Some problems of hominid classification. Am Sci 53:237–259

Rightmire GP, Lordkipanidze D, Vekua A (2006) Anatomical descriptions, comparative studies and evolutionary significance of the hominin skulls from Dmanisi, Republic of Georgia. J Hum Evol 50:115–141. https://doi.org/10.1016/j.jhevol.2005.07.009

Rightmire GP, Margvelashvili A, Lordkipanidze D (2019) Variation among the Dmanisi hominins: multiple taxa or one species? Am J Phys Anthropol 168:481–495. https://doi.org/10.1002/ajpa.23759

Rizal Y, Westaway KE, Zaim Y et al (2020) Last appearance of Homo erectus at Ngandong, Java, 117,000–108,000 years ago. Nature 577:381–385. https://doi.org/10.1038/s41586-019-1863-2

Robinson JT (1953) Meganthropus, australopithecines and hominids. Am J Phys Anthropol 11:1–38. https://doi.org/10.1002/ajpa.1330110112

Schwartz JH, Tattersall I, Chi Z (2014) Comment on "A complete skull from Dmanisi, Georgia, and the evolutionary biology of early *Homo*". Science 344:360–360. https://doi.org/10.1126/science.1250056

Sepkoski D (2012) Rereading the fossil record. University of Chicago Press, Chicago

Spiridonov A, Eldredge N (2024) The Bretskyan hierarchy, multiscale allopatry, and geobiomes—on the nature of evolutionary things. Paleobiology 50:194–213. https://doi.org/10.1017/pab.2023.37

Sterelny K, Griffiths PE (1999) Sex and death. An introduction to philosophy of biology. University of Chicago Press, Chicago

Tattersall I (2009) The fossil trail: how we know what we think we know about human evolution, 2nd edn. Oxford University Press, New York

Tattersall I (2018) History and reality of the genus "Homo": what is it and why do we think so? Metode Sci Stud J 8:79–89. https://doi.org/10.7203/metode.8.9111

Tattersall I (2021) Natural selection as agent of evolutionary change: a view from paleoanthropology. In: Delisle RG (ed) Natural selection. Springer International Publishing, Cham, pp 419–439

Tattersall I (2022a) Evolutionary theory, systematics, and the study of human origins. J Anthropol Sci 100:19–43

Tattersall I (2022b) New pigeons in old holes: anthropology, algorithms and alpha taxonomy. Anthropologie 126:103055. https://doi.org/10.1016/j.anthro.2022.103055

Tattersall I (2023) Endocranial volumes and human evolution. F1000Res 12:565. https://doi.org/10.12688/f1000research.131636.1

Villanea FA, Schraiber JG (2018) Multiple episodes of interbreeding between Neanderthal and modern humans. Nat Ecol Evol 3:39–44. https://doi.org/10.1038/s41559-018-0735-8

Vrba ES (1984) Patterns in the fossil record and evolutionary processes. In: Ho M-W, Saunders PT (eds) Beyond neo-Darwinism: an introduction to the new evolutionary paradigm. Academic Press, London/Orlando, pp 115–142

Vrba ES (1985) Environment and evolution: alternative causes of the temporal distribution of evolutionary events. S Afr J Sci 81:229–236

Vrba ES (1988) Late Pliocene climatic events and hominid evolution. In: Grine FE (ed) Evolutionary history of the robust australopithecines. Aldine de Gruyter, New York, pp 405–426

Vrba ES (1993) Turnover-pulses, the red queen and related topics. Am J Sci 293-A:418–452

Washburn SL (1951) Section of anthropology: the new physical anthropology. Trans N Y Acad Sci 13:298–304. https://doi.org/10.1111/j.2164-0947.1951.tb01033.x

West-Eberhard MJ (2003) Developmental plasticity and evolution. Oxford University Press, Oxford

White T (2003) Early hominids—diversity or distortion? Science 299:1994–1997. https://doi.org/10.1126/science.1078294

Wolpoff MH, Wu XZ, Thorne AG (1984) Modern Homo sapiens origins: a general theory of hominid evolution involving the fossil evidence from East Asia. In: Smith FH, Spencer F (eds) The origins of modern humans. A world survey of the fossil evidence. A. R. Liss, New York, pp 411–483

Part III

From Data to Evidence: Concepts, Models, and Representations

8 Images of Phylogeny and Visualizations of Evolution: Scales, Trees, Networks, and Maps

Marianne Sommer

Abstract

The dominant image to communicate hominin phylogeny is the 'tree'. In the aftermath of Ernst Haeckel's work, an image that had been used in family genealogy became the predominant way of arguing for a certain understanding of the evolution of 'races', species, and higher taxa. However, tree building was strongly related to other forms of visualization, such as the scale and the map. Thus, old concepts could be maintained when new icons were introduced, and aspects of the imagery of scales, trees, maps, and even networks might be combined. There is no neat historical development through these forms, all of which were and are being used to depict human evolution and kinship to express diverse views. In this contribution, I look at the history of images of human phylogeny and ask what roles they play: Do such images support or contradict the theories they stand for? Is the tree image the dominant one throughout history? Are there rivaling images, and if so, what difference do they make? Besides the epistemics of visualizations of human phylogeny, is there something like a politics of diagrams of human kinship and evolution? While there is no linear, progressive history such as from scale to tree, I will show how the emergence of an evolutionary framework in the nineteenth century, the integration of the human origins sciences into the modern synthesis, and the 'molecularization of anthropology' were associated with shifts in the ways in which hominin and human phylogenies were represented.

Keywords

Visualization · Diagram · Scale/chain of being · Tree · Net · Map · 'Race'

M. Sommer (✉)
Department of Cultural and Sciences Studies, University of Lucerne, Lucerne, Switzerland
e-mail: marianne.sommer@unilu.ch

M. Lequin, J. M. Rodríguez Caso (eds.), *The Field of Human Evolution: Critical Perspectives from History and Epistemology*, Evolutionary Biology – New Perspectives on Its Development 11, https://doi.org/10.1007/978-3-032-13925-2_8

8.1 Introduction

In this chapter, I throw a necessarily limited number of spotlights on the history of visualizing hominin and human phylogeny in the field of human evolution. For a full account, the reader is referred to the book-length *The Diagrammatics of 'Race': Visualizing Human Relatedness in the History of Physical, Evolutionary, and Genetic Anthropology,* ca. *1770–2020* (Sommer 2024). While limiting myself strictly to iconography that incorporates or is restricted to aspects of *human* evolution and descent, I focus on three major periods in which different areas of knowledge were integrated—processes that influenced human classification and phylogeny: the emergence of an evolutionary framework after Charles Darwin's *On the Origin of Species* (1859); the integration of the human origins sciences into the modern synthesis; and the 'molecularization of anthropology'. Understandings of the nature of progress were at the core of many of the ideas under discussion here and underlay the ways in which different models of human evolution and phylogeny conceptualized relationships between linearity and diversification, local evolution and migratory movements. Also in this regard, the impact of evolutionary theory on the scientific understanding and depiction of human kinship and diversity must be considered against the background of pre-evolutionary concepts. So this is where I begin.

8.2 Pre-evolutionary Images

In the European Middle Ages, the genealogy of humans was derived from Genesis. In the course of early-modern encounters with non-Christian West Africa and 'the heathens of the New World', the tribes of Noah's sons Japheth, Shem, and Ham were identified with the inhabitants of the continents of Europe, Asia, and Africa. The notion that humans stemmed from a single pair at one locality, from where they eventually spread across the known world, was closely associated with imagining human kinship and diversity in the form of maps and in tree-like shapes. Indeed, since the Middle Ages, biblical history could be represented in tree-like forms in which Noah constituted the root, while his three sons were the founding fathers of the branches. Joachim of Flora, in a twelfth-century diagram, had the branches of Japheth (Christians) and Shem (Jews) intertwine, while the branch of Cham was only of short life. In the course of the Early Modern period, more clearly arboreal shapes came to represent the descent of nations from Noah. And such a tree could be linked also visually (not only conceptually) to the map, as in the case of a Swedish scholar who planted Noah's tree in the old world, which was drawn on a kind of globe (Hellström 2019, 165–72).

The tree image—that has become so overwhelmingly dominant in representations of human phylogeny and evolution—presents several complications in connection with the tree of life, the tree of Jesse, and the *arbor consanguinitatis*. From the eleventh century, the tree of Jesse could have vegetal or tree-like iconography but not appear in a truly bifurcating manner, and the *arbores*, though also called thus,

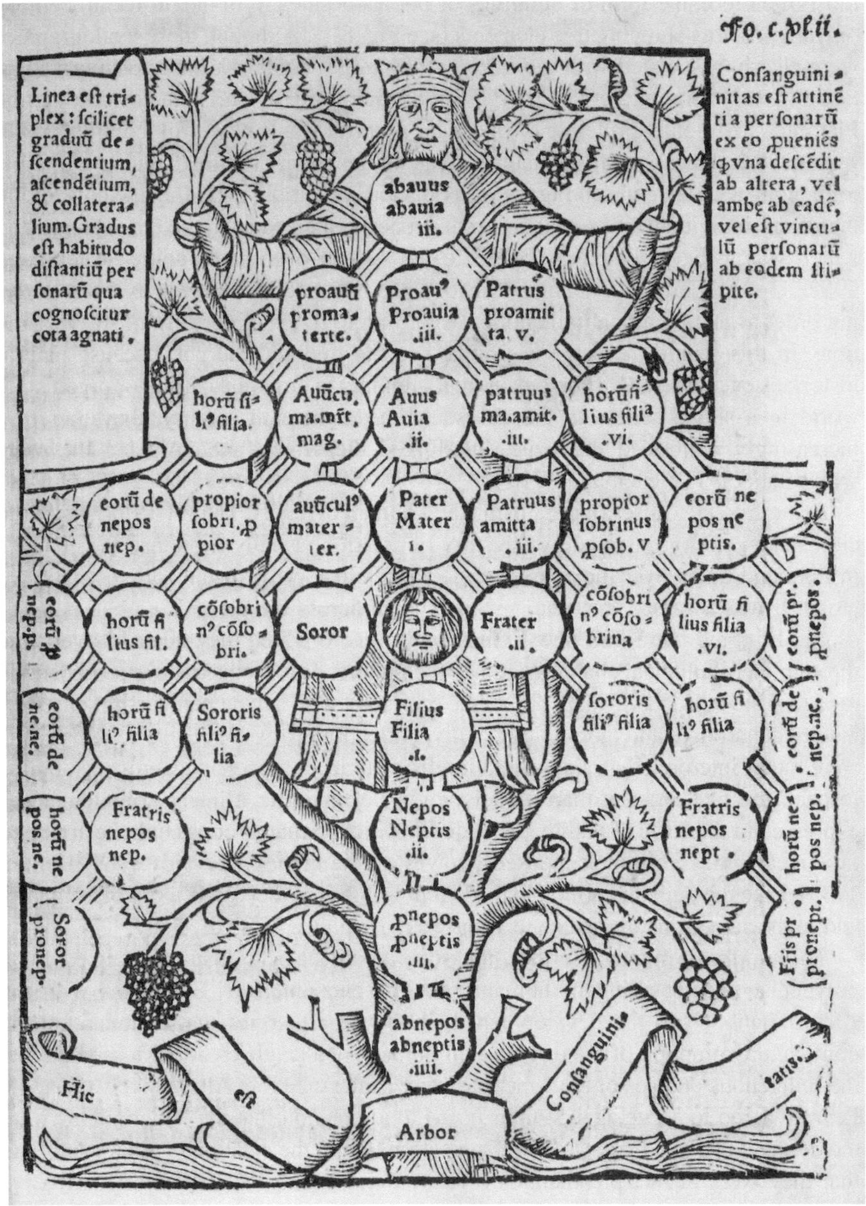

Fig. 8.1 "A tree of consanguinity in Jean Boutillier's LA GRANT SOMME RURAL (Paris, 1539), call no. France 34 B66 1539." Yale Law Library, 13 Nov. 2006, Wikimedia Commons: https://commons.wikimedia.org/wiki/File:Flickr_-_Yale_Law_Library_-_Arbor_consanguinitatis,_1539.jpg. Accessed 19 May 2025

most often took the form of buildings or bodies besides from the thirteenth century onward possibly showing tree elements (see Fig. 8.1). Although these iconographies were clearly associated with genealogy, they might contain a linear or lineal logic (see Fig. 8.2) (Sommer et al. 2024, 5–7). Furthermore, in Christian cosmology, the Platonic-Aristotelian principle of the *scala naturae* was a powerful iconography. It was in the eighteenth century that the image of the chain or scale of nature was most influential, seen as representing the order in God's creation, including the conception that every individual and social group had a designated place in the order of society, which was to be accepted and filled to the best of everyone's ability (Diekmann 1992, 53–81). Certainly, in natural history, the linear scale was seen as the underlying order arranging all natural entities according to rungs of perfection. At the same time, in this realm, the scale was linked to the tree and the net, because natural historians encountered difficulties in their attempts to order the natural or organismic world in a single series. In the course of the eighteenth century, they therefore increasingly multiplied the scale, interlinked the scales, and provided the scale with offshoots (Thienemann 1910).

Regarding the causes of organismic and human diversity, natural historians of the eighteenth century, inspired by the story of creation, usually explained the changes in flora and fauna over the course of the earth's history in spatial and geographical terms. It was assumed that humans had spread across the globe from one point of origin, bringing them into new living conditions to which they adapted over time through morphological changes and new behaviors. This seems to have provided the model of the tree and the map, or the tree on a map, with a dynamism in the sense of a historical narrative and morphological diversification. However, these developments were rather interpreted as a degeneration from an ideal type than as progress; in this respect, the tree and the map models could be linked to a linear arrangement as expressed in the scale of nature – a ladder in which human groups could be arranged in a linear series from the supposedly lowest to the highest (Lovejoy 1964; Bynum 1975). Like the tree, this principle made it possible to preserve the unity of humanity and at the same time do justice to what was interpreted as its diversity.

To complicate matters further, the scale could also be instrumentalized for a polygenist understanding of humankind. The encounters of Europeans with the Amerindians led to the development of polygenist theories that contradicted the dominant assumption of the origin of all humans in a single creation. It seemed that the Indigenous peoples of the Americas were a later creation. Alternatively, Isaac La Peyrère dared to argue against their post-Flood creation that the Amerindians were the descendants of a first act of God's creation before the creation of Adam and Eve, that they were in fact preadamites (*Praeadamitae* 1655). A particularly offensive pre-evolutionary image of such a divisive, non-genealogical, scale of nature was introduced by the English physician Charles White. In his *An Account of the Regular Gradation in Man* (1799), he used the 'law of gradation' to present the human 'races' as aboriginally different, as deriving from different origins (Bynum 1975, 12; Bowler 2021, 71). Figure 8.3 makes evident, even without reference to the text, that this was an appropriation of Petrus Camper's famous diagrams demonstrating the facial angle (even though Camper was not a proponent of the scale of being). White

Fig. 8.2 "Patrilineage in Matthew was traditionally illustrated by a Tree of Jesse showing the descent of Jesus from Jesse, father of King David". Illumination on parchment, Bible of the Capuchins, ca. 1180, French National Library. SimonP, 17 May 2005, Wikimedia Commons: https://commons.wikimedia.org/wiki/File:JesseTree.jpg. Accessed 19 May 2025

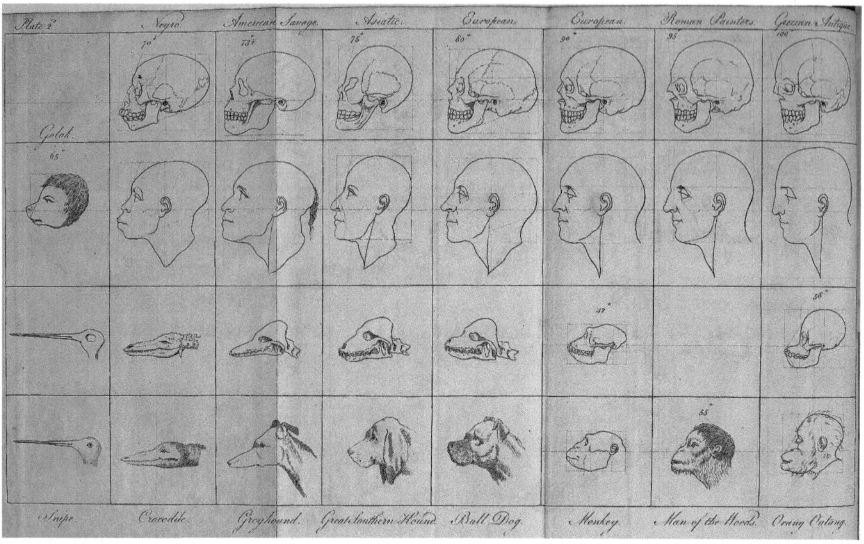

Fig. 8.3 White C (1799) An account of the regular gradation in man, and in different animals and vegetables; and from the former to the latter. Printed for C. Dilly, in the Poultry, London, n.p

was further inspired by the gradation of skulls presented by the Scottish surgeon John Hunt to the Royal Society in lectures on midwifery. In the portrait shown as Fig. 8.4, we can see such an image from 'monkey to man'. The scale in the shape of a 'racial' skull series—or a series of other kinds of bones, later including fossils—would advance to one of the most ubiquitous images in physical-anthropological treatises, with visual superimpositions being one particular expression of the series (Sommer 2024, Chs. 6 and 7).

When transformist theories were introduced, the scale was still intact. Jean-Baptiste Lamarck and Étienne Geoffroy Saint-Hilaire at the Muséum national d'Histoire naturelle in Paris advocated transformist theories and derived humans from an ape-like precursor. In his *Philosophie zoologique* (1809), Lamarck described the evolutionary development from infusoria to man—still following the logic of the scale of perfection. In this model, bifurcation was only due to diversions from the straight path due to adaptation. The transformist ideas found little favor with the scientific establishment. Nevertheless, Lamarck's theory of evolution lived on in nineteenth-century anthropology. In particular, the aspects (already present in natural philosophy) of the inheritance of acquired characteristics, an inner drive for perfection that propels organisms up the steps of the natural ladder, and the analogy between individual development and 'racial' history formed the cornerstones of the new evolutionary anthropology (the term *evolution* first appeared in embryology) (e.g., Bowler 1986, Ch. 2). Adrian Desmond and James Moore (2009, 92–3, 111) have discussed how Darwin engaged with human evolution at a time when such notions were applied to human biological diversity. The explanations for 'racial' differences had developed from the environmental framework of the eighteenth

Fig. 8.4 "The pioneering Scottish surgeon and anatomist John Hunter (1728–1793) in old age. Oil painting with tinting by an unknown artist after the painting by Reynolds 1786". Paul D. Stewart / Science Foto Library, https://www.sciencephoto.com/media/150765/view/1788-john-hunter-portrait-surgeon

century to a stronger emphasis on hereditary influence. And the first half of the nineteenth century was also a heyday of polygenism.

However, rather than providing a separate evolutionary ladder or tree for each human variety—and thus a series of trees—as might have been suggested by Lamarck's understanding of evolution, the French naturalist, military man, politician, and radical transmutationist, Jean-Baptiste Bory de Saint-Vincent, for example, communicated his understanding of human diversity through a map (Fig. 8.5). In *L'Homme (Homo)* (1827), he refuted the notion that 'the Red and Black races' were consanguineous with 'the White'. Bory asserted that his expeditions had provided evidence for the existence of 11 distinct human species, a claim that he later revised to include 15. Bory's work followed in the footsteps of the first physical anthropologists, such as Camper and Johann Gottfried Blumenbach, in describing *Homo Japeticus* (after Japhet) as the most beautiful in proportions, with the largest facial angle, an oval face, and the greatest number of geniuses (Bory 1827, Vol. I, 102–162). *Homo Aethiopicus* was also described regarding the form of the skull, its volume (smaller than that of *H. Japeticus*, Vol. II, 30), the protruding face, etc. (Vol. II, 29–86). The Khoekhoe or Khoisan, Bory's species number 15, were presented as a transitional group between humans and apes (Vol. II, 113–34). At the same time, Bory vehemently denounced the cruelty perpetrated by 'Whites' against Black Africans, including in the context of the slave trade and slavery. He also (paternalistically) observed that Haiti's Black population had taken revenge and demonstrated

Fig. 8.5 "Distribution primitive du genre humain a [sic] la surface du Globe, Par M. le Colonel Bory de St. Vincent" (1827). "Grave [sic] par Ambroise Tardieu. (A Paris, Chez Mme. veuve Agasse, Imprimeur-Libraire, rue des Poitevins, no. 6. M.DCCCXXVII)". David Rumsey Map Collection, David Rumsey Map Center, Stanford Libraries: see https://www.davidrumsey.com/luna/servlet/detail/RUMSEY~8~1~28545~1120434:Distribution-primitive-du-genre-hum?sort=Pub_List_No_InitialSort&qvq=q:Distribution%20primitive%20du%20genre%20humain%20a%20la%20surface%20du%20Globe%2C%20Par%20M.%20le%20Colonel%20Bory%20de%20St.%20Vincent;sort:Pub_List_No_InitialSort;lc:RUMSEY~8~1&mi=0&trs=1. Accessed 19 May 2025

their capacity for freedom. Bory's overt scientific racism serves to illustrate the multidimensionality of the relation between science and politics. He was a staunch opponent of slavery and an antimonarchist, liberal thinker who espoused the tenet of equality before the law (Sommer 2024, 47–8, Part I in general).

With color codes for human species such as *Homo Japétique*, Bory's polygenist map also referred back to the T and O 'maps' that first appeared in the Middle Ages. In these diagrams, the T, or the cross, cuts the space within the O, or the orb, into three parts. These three parts were associated with Noah's sons Shem, Ham, and Japheth, who, as we have seen at the beginning of this section, were posited as the fathers of the main human lines. Possibly from the fifteenth century, the map's three parts and Noah's three sons became more stably associated with the continents Asia, Africa, and Europe (Sommer 2024, Ch. 6). Furthermore, with Bory's distinction of the two major stems of 'Leiotrique Espèces' ('straight-haired species') and 'Ulotriques' ('wooly-haired species'), we have a bridge to the great map- and tree-builder Ernst Haeckel, who maintained the belief in an original distinction of humankind according to hair structure. With this in mind, I turn to those efforts that, in the wake of Darwin's *On the Origin of Species by Means of Natural Selection, or the Preservation of Favoured Races in the Struggle for Life* (1859), sought to integrate aspects of biology and anthropology, as well as 'sociology' and psychology, within the framework of an evolutionary theoretical approach.

8.3 The Beginnings of Human-Evolutionary Diagrams

When Darwin integrated humankind into his evolutionary understanding of the organismic world in *The Descent of Man and Selection in Relation to Sex* (1871a, b), he did not provide a diagram for human history, kinship, and diversity. Darwin had experimented with tree-like structures. And he did draw a family tree of the primates by hand in the context of his work on *Descent* that included 'Man'. But he did not incorporate it into the publication. At the same time, in *Descent* he did make use of aspects of the tree metaphor to help the reader envision this tree through words. Darwin was inspired by the view of ontogeny promoted by the Prussian-Estonian Carl Ernst von Baer as a process of differentiation and individuation (von Baer 1828, 1837). Finally, there was the strong analogy to philologists' depiction of the history and evolution of languages as tree-shaped. However, although Darwin drew his tree of the primates in words such as *stock*, *stem*, and *branch*, his descriptions of human evolution and differentiation also relied on the vocabulary of the scale of nature (Sommer 2021). Darwin was in agreement with his contemporaries that the 'White race' represented the highest stage in the gradations formed by physical and mental traits. In *The Expression of the Emotions in Man and Animals* (1998 [1872]) as well as in *Descent*, he introduced 'savage peoples' as a kind of missing link to establish the continuum from 'primitive' to 'civilized'. In certain respects, women, children, and 'idiots' could also represent earlier stages and thus provide insights into the general course of human evolution (e.g., Darwin 1998 [1872], 241; 1871a, 121–2).

Even before the publication of Darwin's *Origin* (1859), there was the influential thought of the English philosopher and social theorist Herbert Spencer. From the 1850s onward, Spencer advocated universal evolutionism, which united biological, psychological, and social progress in a synthetic philosophy. Like Darwin, Spencer referred to von Baer's theory of development in his *Principles of Biology* (1864; 1867). He thought of the tree structure, revealed by comparative embryology, as parallel to the classification system and the phylogenetic kinship system (e.g., Spencer 1864, Part II, Ch. 2, and Part III, Ch. 5). For Spencer, this analogy was the integrating factor between biological and social evolution, both of which he understood as processes of increasing differentiation, division of labor, and increase in complexity with simultaneous integration. Spencer thought that the mechanism of progress in biology, of the psyche, and of society was the inheritance of acquired characteristics. New habits led to changes in the anatomy and mind, which were passed on to the next generation. A more highly developed mind and a more efficient body would in turn lead to further progress in human behavior in the social and cultural sphere.

If we look for visual renditions of this kind of fusion of differentiation and progress or hierarchy with regard to hominid and human evolution and kinship, we have to move to the German context. There, the two most important exponents of 'Darwinism', Carl Vogt and Haeckel, also applied the theory of evolution to humans. Vogt's integration of humans into Darwin's theory of evolution earned him the name *Affenvogt*. However, while promoting scientific materialism and denouncing those scientists (such as the Göttingen physiologist Rudolf Wagner) who adhered to the biblical account of creation, Vogt, too, seems to have been inspired by the scale of nature. In the first volume of his *Vorlesungen über den Menschen, seine Stellung in der Schöpfung und in der Geschichte der Erde* (1863a), Vogt compared measurements of the skull, brain, and skeleton of contemporary human types. In doing so, he gave a number of lists of measurements ordering human groups (with a wild mix of labels, including racial ones) in hierarchical scales, such as a table containing means of cranial capacity for which he drew on the renowned polygenist craniologists Samuel George Morton in the United States and Paul Broca in France (see Fig. 8.6) (Sommer 2015a, 18–9, 46–74; on Morton and Broca, see also Gould 1996 [1981], Chs. 2 and 3; Sommer 2024, Part I).

As integral to the image of the scale, Vogt also searched for connecting links in the chain of being and described the differences and similarities between humans and apes. He arrived at the conclusion that Black Africans constituted such a link between apes and 'higher human races'. According to Vogt, women also narrowed this gap, as they were close to the 'lower races' and higher animals in certain anatomical characteristics. It was Vogt on whom Darwin based his treatment of the 'microcephalic' condition as another missing link, because for Vogt, humans who exhibited an inhibition of brain development represented in this respect a past life form that had emerged from the ape and developed into humans. Thus, according to Vogt, 'microcephalic' individuals remained at an earlier stage of evolutionary history in their phylogenetic as well as individual development. In the second volume of *Lectures on Man* (1863b), Vogt dealt with the prehistoric human finds at the same time as Charles Lyell and Thomas Henry Huxley. Also, the sparse fossil finds and younger bones served him as intermediate links in the anatomical series from what he considered 'the lowest human races' to the great apes. The hierarchy of 'human races' established in the first volume and the assertion that these differed sufficiently to be considered different species prepared the ground for the theory introduced in the second volume that the main human varieties had evolved from different apes.

Haeckel, too, developed a kind of evolutionary polygenism, and it was him who introduced a plethora of tree diagrams into anthropology. At the same time, Haeckel did not adopt all aspects of Darwin's theory, because for him evolution remained a directed development that was predestined to culminate in humans. German natural philosophy regarded the purposeful and directed evolution of organisms as a consequence of embryonic development, which passed through its predetermined stages. And for Haeckel, ontogenesis was a compressed and accelerated repetition of the evolution of a species (biogenetic law). Phylogeny was the mechanical cause of ontogeny, which is why investigations into phylogeny were a kind of search for

Tabelle des Schädelinhaltes bei verschiedenen Rassen.

Nr.	Völkerschaft.	Zahl der gemessenen Schädel.	Volumen in Cubikcentimetern.	Beobachter.	Bemerkungen.
1.	Australier.	8.	1228,27	Aitken Meigs.	
2.	Polynesier.		1230	Morton.	
3.	Hottentotten.		1230	"	
4.	"	3.	1233,78	Aitken Meigs.	
5.	Peruaner.	152.	1233,78	"	
6.	"		1246	Morton.	
7.	Oceanische Neger.	2.	1253,45	Aitken Meigs.	
8.	Mexikaner.		1296	Morton.	
9.	Amerikaner im Allgemeinen.	341.	1315,71	Aitken Meigs.	
10.	In Amerika geborene Neger.	12.	1323,90	"	
11.	Malayen.		1328	Morton.	
12.	Mexikaner.	25.	1338,65	Aitken Meigs.	
13.	Grönländer.	1.	1340	Welcker.	
14.	Chinesen.		1345	Morton.	
15.	Neger im Allgemeinen.	76.	1347,66	Aitken Meigs.	
16.	Neger im Allgemeinen.		1361	Morton.	
17.	Altperuaner		1361	"	
18.	In Afrika geborene Neger.	64.	1371,42	Aitken Meigs.	
19.	Wilde Indianer.	164.	1376,71	"	
20.	Pariser aus der gemeinschaftlichen Grube.	35.	1403,14	Broca.	Schädel aus dem 19. Jhdt.
21.	Pariser vom Kirchhofe des Innocents.	117.	1409,31	"	Aus dem 12. bis 18. Jhdt.
22.	Eskimos.		1410	Morton.	
23.	Pariser des 12. Jahrhunderts.	115.	1425,98	Broca.	Aus einem Gewölbe.
24.	Kaukasier im Allgemeinen.		1427	Aitken Meigs.	
25.	Malayen.	1.	1430	Welcker.	
26.	Deutsche.	30.	1448	"	
27.	Pariser des 19. Jahrhunderts.	125.	1461,53	Broca.	
28.	Angloamerikaner.	7.	1474,65	Aitken Meigs.	
29.	Pariser aus Privatgräbern.	90.	1484,23	Broca.	Schädel aus dem 19. Jhdt.
30.	Pariser aus der Morgue.	17.	1517	"	Desgl.
31.	Germanen im Allgemeinen.	88.	1534,127	Aitken Meigs.	
32.	Brachycephale von Meudon.	1.	1540	Broca.	Aus einem Dolmen.
33.	Engländer.	5.	1572,95	Aitken Meigs.	

Fig. 8.6 Vogt C (1863) Vorlesungen über den Menschen, seine Stellung in der Schöpfung und in der Geschichte der Erde, vol 1. De Gruyter, Berlin, p. 104–105

ontogenetic laws (e.g., Haeckel 1874, 55–72). It was Haeckel himself who coined the terms *Phylogenese* and *Ontogenese*. In his works *Generelle Morphologie der Organismen* (1866) and *Natürliche Schöpfungsgeschichte* (1868), he applied the theory of evolution to the entire biological classification system, combining Lamarck's and Darwin's evolutionary theories into a progressive model. Despite this strong reliance on scales, Haeckel thought the findings from different scientific fields made it clear that there was a threefold parallelism between embryology, phylogeny, and systematics, and that that parallelism could only be adequately explained by a theory of divergent descent. If only the individual development or evolution of a species were taken into consideration, the result would be a straight line of organic forms. If, on the other hand, one compared the embryonic and phylogenetic development of different species, one would arrive at a tree structure congruent with the classification system (Haeckel 1868, 227–58). This allowed the maintenance of a ladder of beings alongside a diversifying evolution.

Haeckel not only developed a synthetic theory of evolution and kinship systems, but he also painted these into real phylogenies, which had a great influence on later anthropologists. As he considered the Asian apes (gibbon and orangutan) to be the most similar to humans, Haeckel thought that the first hominids (today hominins)[1] must have originated in southern Asia, possibly on a continent now submerged in the Indian Ocean. Later, new geological evidence made this idea seem unlikely, and Haeckel instead postulated India as the birthplace of mankind, particularly the Himalayan region. In line with this hypothesis, he claimed that the Indigenous people of Ceylon (Veddas), now Sri Lanka, were the most primitive living humans and saw in them the link to the highest ape species. They could therefore serve as a model for the earliest human forms. For his scenario of human evolution, Haeckel relied heavily on philology (which he called the comparative anatomy of language and the ontogeny and phylogeny of language). Based on the analogy to language evolution, Haeckel designed a hypothetical link between Asian apes and 'human savages', which he referred to as *Pithecanthropus* or *Homo primigenius*. These prehistoric humans, also called *Alali*, were speechless. Further on the basis of the language analogy, Haeckel went so far as to claim that the living human types had evolved from various *Pithecanthropi* precursors. Comparative philological studies, according to which the main languages could not be traced back to a common origin, were regarded by Haeckel as proof of polygenism. The prehistoric humans or ape-men thus split into the stems of the *Lissotrichi* ('straight-haired') and the *Ulotrichi* ('woolly-haired'), which spread across the earth and forked into different branches; Haeckel considered the Indo-European branch to be an offshoot of the *Lissotrichi*. Although this scenario assumed a common origin of humanity far back in prehistory, Haeckel in fact regarded the living human types as different species

[1] *Hominids* then referred to the group today called *hominins*, i.e., it was used in a more restrictive sense: Living humans, extinct species of *Homo*, and all immediate ancestors of humankind (excluding the great apes and their immediate ancestors). To reflect changes in taxonomy, I follow this usage where appropriate.

that could even be divided into different genera. The human species had developed the characteristics that distinguish them from animals at different times in evolutionary history, and they had also perfected them to varying degrees up to the present day (e.g., Haeckel 1868, 486–520).

These views found expression in a number of phylogenetic trees that separated the two main human stems from the beginning and provided the different recent 'human species' with parallel lines of descent that ended in a hierarchy expressed through distribution in the image space. Furthermore, Haeckel also immediately provided a map of the world on which such a tree, such a story of a barely common origin with subsequent spread and differentiation, was projected. The hierarchy expressed in Haeckel's treemaps stood for Haeckel's idiosyncratic interpretation of Darwin's selection between tribes (which for Darwin, as for Spencer, was limited to prehistoric times). Haeckel justified contemporary imperialism and genocide as natural processes that had driven human evolution since time immemorial. From this perspective, the progress of human anatomy and culture depended on the displacement of 'lower' by 'higher human species'—from prehistoric times to the present. Haeckel claimed that 'the woolly-haired species' were incapable of developing a high culture. It was therefore the 'Midlandic species' (especially the 'Indo-European race') that presided his tree and who were envisioned to extend their dominion over the entire globe by means of their high intelligence and civilization. Haeckel predicted that the species of the temperate zones would displace the less developed species of humans everywhere, except in the inhospitable tropical and polar regions. He saw this process already underway with regard to Indigenous Americans and Australians, the Khoekhoe and Papuans, as well as other Indigenous groups (see also Haeckel 1898 [1868], 729–65).

Both Vogt and Haeckel still looked to substitutes for (premodern) human ancestors—their missing links—to arrive at progressive series, and throughout the nineteenth century, bones from contemporary and historical 'races' were collected for anthropological measurements, and surveys were carried out on living populations. The discovery of the Stone Age led to the investigation and (re-) excavation of caves and open sites and drew the attention of scientists and amateurs, and not least of workers in quarries or road construction, to human bones and stones worked by humans. With Paleolithic artifacts, awareness of the diversity of prehistoric European cultures grew, and a classification of them was sought. In the 1860s, organic evolution and the development of material culture were thought to be analogous. In Britain, researchers such as George Busk, John Evans, Thomas Henry Huxley, John Lubbock, Pitt Rivers, Edward Tylor, and Alfred Russel Wallace made the Ethnological Society a platform for discussions about Darwinism and the application of an evolutionary view to humans. But also the French archeologist and anthropologist Gabriel de Mortillet, or the German anthropologist Hermann Schaaffhausen, for example, spoke of a law according to which 'die körperliche Bildung des Menschen dieselbe Entwicklung zurückgelegt hat, wie die menschliche Civilisation und mit dieser in der Vervollkommnung gleichen Schritt hält!' ('the physical formation of man has undergone the same development as human civilization and is keeping pace with it in its perfection!' Schaaffhausen 1872 [1869], 127).

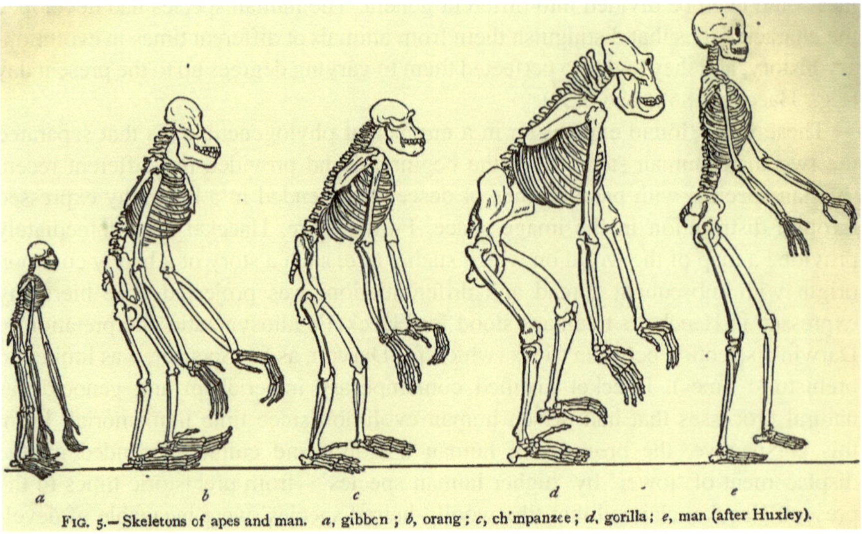

Fig. 8.7 Tylor EB (1881) Anthropology: an introduction to the study of man and civilization. Macmillan, London, p. 39

While the comparative method was used in anatomy to determine hierarchies between 'human races', it was used in ethnology to reconstruct the evolutionary series of cultures (or cultural elements). Since 'savages' were regarded as links between 'civilized races' and apes, they could now shed light on the Paleolithic (Sommer 2024, Chs. 8–10).

Tylor's thinking was particularly characteristic of such an evolutionary view of the history of cultures, and he gave it expression, also visually, in *Anthropology: An Introduction to the Study of Man and Civilization* (1881). Tylor provided an introduction to the various subfields: comparative anatomy of apes and humans, physical anthropology, historical and comparative linguistics, ethnology, and Paleolithic archeology. In an attempt to show the integrative potential of anthropology for education, Tylor traced the development of human culture from 'savage' to 'barbaric' to 'civilized'. He showed readers the marvelous progress from the most primitive weapons and tools, which were nothing but stones picked up from the ground, to the achievement of the steam engine, jumping back and forth with his illustrative examples, not only between the Paleolithic and Neolithic cultures of Europe and the numerous cultures of 'savage and barbarian peoples' around the world, but also between ancient Eastern and modern Western civilizations. The story was not told according to chronological or geographical patterns, but along the global gradual improvement of material culture (Tylor 1881, esp. Chs. 8 and 16).

This approach was mirrored, in terms of actual visualization, in diverse kinds of 'progressive' serial arrangements of body parts, bones, and artifacts. Tylor, for example, included Thomas Henry Huxley's famous series from apes to man in his treatise (see Fig. 8.7). And again, continuing a visual tradition that by now extended

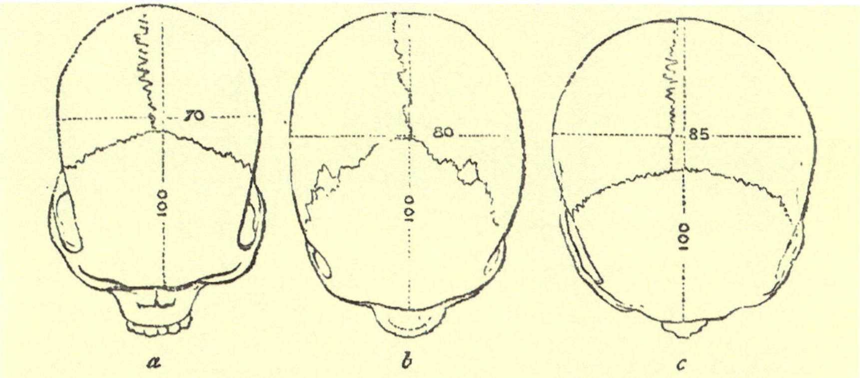

FIG. 9—Top view of skulls. *a*, Negro, index 70, dolichokephalic; *b*, European, index 80, mesokephalic; *c*, Samoyed, index 85, brachykephalic.

Fig. 8.8 Tylor EB (1881) Anthropology: an introduction to the study of man and civilization. Macmillan, London, p. 61

from Blumenbach's eighteenth-century row of skulls of the human varieties, to polygenist 'racial' skull arrangements as, for example, found in Josiah Nott's and George Gliddon's *Types of Mankind* (1854), to images by staunch evolutionary monogenists like Tylor himself, there was the obligatory skull series—this time signifying gradual evolutionary change (see Fig. 8.8). Skull types could evolve into each other: 'A model skull of a flexible material like gutta-percha, if of the middle shape, like that of an ordinary Englishman, might, by pressure at the sides, be made long like a [Black African's], or by pressure at back and front be brought to the broad Tatar form' (Tylor 1881, 61). Like Blumenbach some hundred years earlier, Tylor diagrammatically experimented with skulls to morph them into each other (Sommer 2024, Part I). The same could be done with the crania viewed from the side with Camper's facial angle, with the correlated characteristics of a protruding jaw and a receding forehead, in 'the two lower races and our own' (Tylor 1881, 62), as shown in Fig. 8.9. This time, be it due to the limited space of the page or for the communication of a certain meaning, the arrangement of the skulls seems to suggest a branching diagram underlying the connection of the 'ape-like slope' to 'the more nearly upright European face' (62).

At the same time, George Stocking in particular has demonstrated the broad understanding in nineteenth-century anthropology of human varieties ascending the same line, but in parallel, and at unequal tempi and to different levels. He argued that the Victorian evolutionists modified the eighteenth-century notion of a chain of being that could be climbed by all 'human races' to a parallel evolutionary interpretation of the results of comparative ethnology. Stocking suggested that rather than a unilinear image of evolution, we are confronted with a pyramid-shaped diagram that stands for the understanding that 'human races' followed the same evolutionary sequence but independently from each other, progressing, stagnating, regressing, or dying out on their way. In addition, the belief that 'the savage races' may not be able

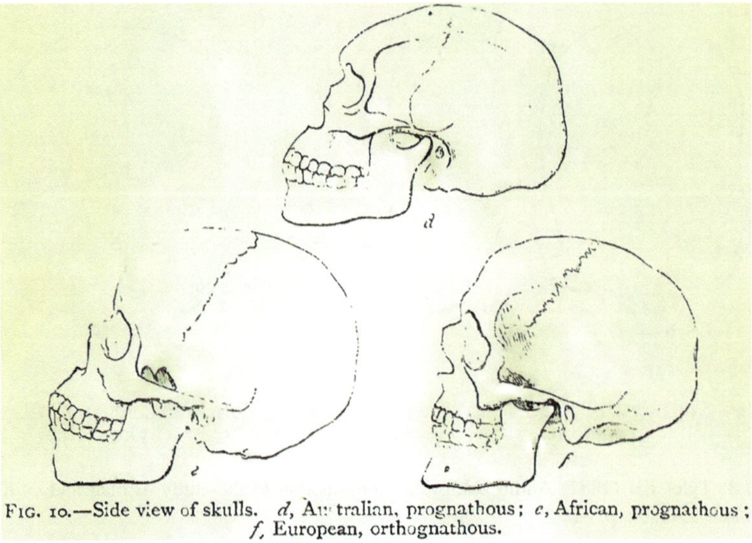

Fig. 8.9 Tylor EB (1881) Anthropology: an introduction to the study of man and civilization. Macmillan, London, p. 62

to reach the cultural level of the European civilizations on their own, and the observable changes brought about by imperialism and colonialism, seemed to suggest (the necessity for) a certain diffusion of cultural progress from the civilized to 'the savage races'. Thus, we might add horizontal connections to the parallel vertical lines that make up the pyramid to illustrate diffusion and overall progress (e.g., Stocking 1968, Ch. 6, also Chs. 4 and 5).

It is important to bear in mind that these nineteenth-century debates were limited to Europe and only referred to finds that were considered to be evidence of Paleolithic, Neolithic, and contemporary human *races*. Haeckel's *Pithecanthropus* and de Mortillet's *Anthropopithecus* (1883) remained empty taxa until the end of the century. Right at that time, in 1895, however, the Dutch physician Eugène Dubois brought the femur, the skullcap, and some teeth of an organism to Europe, which he named *Pithecanthropus erectus* (today *Homo erectus*). He had discovered these pieces, which seemed to fulfill Haeckel's prophecy of an Asian ape-man, during his service as a military doctor on Java. The German anatomist Gustav Schwalbe integrated the new missing link into a linear series of hominid evolution that now led from *Pithecanthropus erectus* via Neanderthals to modern humans. Schwalbe defined the Neanderthals as a separate fossil species (*Homo primigenius*) and no longer as a fossil 'race' within the human species, as Thomas Henry Huxley and others had done. This step enabled him to describe human evolution at the genus and species level (Schwalbe 1906). However, Schwalbe's linear series was soon to be reinterpreted as a branching phylogeny. This is indicated by Schwalbe's figure in a treatise on the influential French anthropologist Marcellin Boule who was

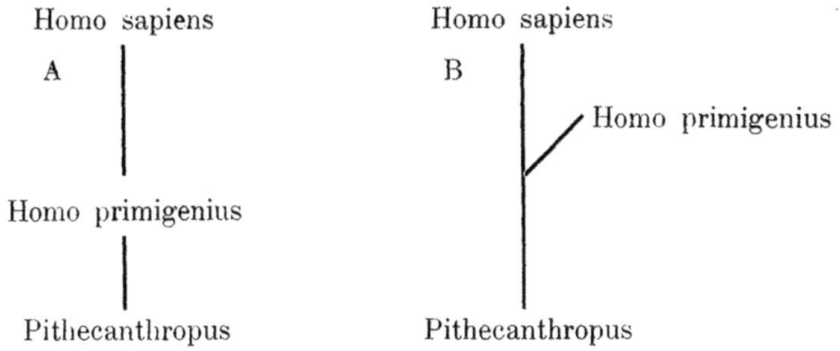

Fig. 8.10 Schwalbe G (1914) Kritische Besprechung von Boule's Werk: 'L'homme fossile de La Chapelle-aux-Saints' mit eigenen Untersuchungen. Zeitschrift für Morphologie und Anthropologie 16(3):527–610, p. 601

responsible for the general move from the unilinear (model A) to the branching model (B) (see Fig. 8.10). We will see in the course of the next section that the phylogenetic trees came to not only exclude the Neanderthals but also *Pithecanthropus* and other 'fossils' from the main human line (Sommer 2024, 145–50).

8.4 The Treeing Frenzy

Shortly after Schwalbe's inclusion of *Pithecanthropus* (now *Homo erectus*) in the line leading to modern humans, the anthropological community shifted to various tree diagrams to represent human phylogeny. In the early 1900s, Boule proposed that the Neanderthals and *Pithecanthropus* should be regarded as side branches of the human family tree. This interpretation was subsequently endorsed by other prominent paleoanthropologists, including the British triumvirate of Arthur Keith, Arthur Smith Woodward, and Grafton Elliot Smith, as well as Henry Fairfield Osborn and Earnest Hooton in the United States. The fact that known and newly discovered hominid finds were often interpreted as the remains of later survivors of older forms had not least to do with *Eoanthropus dawsonii* ('Dawson's Dawn Man' or 'Piltdown Man'), evidence of which began to emerge in Sussex in 1911. This 'fossil', which was not definitively exposed as a forgery until the 1950s, was made up of a large skull with an ape-like lower jaw. The 'find' therefore suggested, on the one hand, that brain development must have been the driving force in the evolution to modern humans and, on the other hand, that forms such as *Pithecanthropus*, which were anatomically more 'primitive' but not of a sufficiently older age, could not be placed on the direct ancestral line of modern humans. Modern human anatomy was thus dated far back. This was further supported by supposed stone tools ('eoliths') found in England, but also in continental Europe, that were attributed to the Tertiary period. Smith may function as an example to illustrate certain aspects of this 'hyper-diversity' approach, most importantly the tendency to create family trees in which

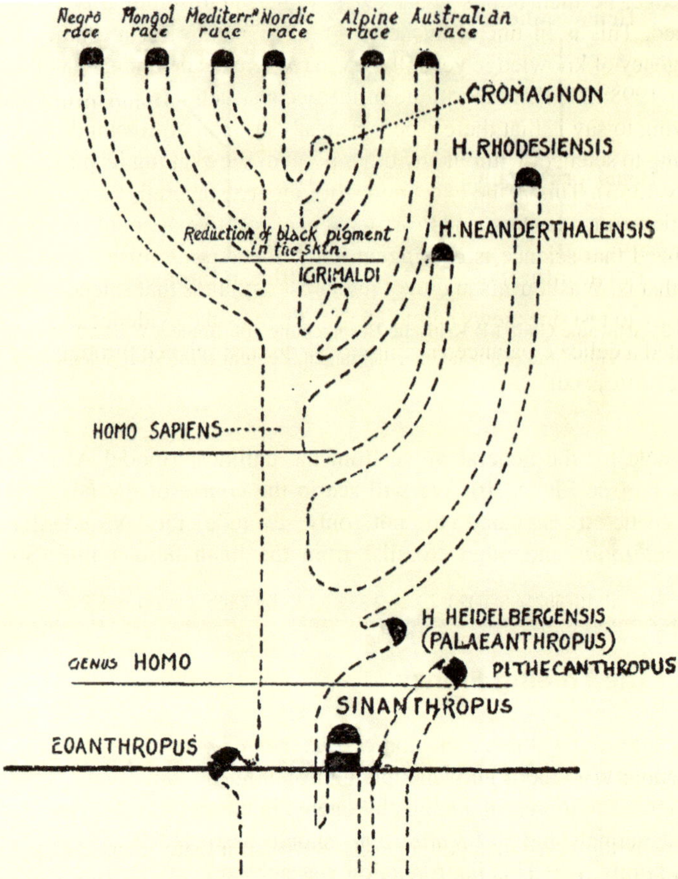

FIG. 16. – **Scheme to suggest the possible relationships in time and affinities of the extinct genera and species, and the living races of the Human Family. The places assigned to the Cro-Magnon and Grimaldi Races are wholly tentative.**

Fig. 8.11 Smith GE (1929) Human history. W. W. Norton, New York, Fig. 16, p. 54

all hominid forms were accommodated on dead-ending branches (see Fig. 8.11) (Sommer 2007, Part II; also, Bowler 1986, Ch. 1).

As we have seen for Haeckel, such phylogenetic trees tended to be abstractions of narratives of evolution that had spatial implications. Smith initially followed Darwin in the assumption of Africa as the continent of human origins. However, with more hominid remains at hand, their distribution led him to move the center of hominid evolution from Africa to Asia. There were: *Pithecanthropus erectus* from Java and a similar hominid, *Sinanthropus pekinensis* (today also *Homo erectus*), from near Beijing in China (1927–9); *Eoanthropus dawsonii* from Great Britain, i.e., in the far west, from around the same time; and the slightly younger (Neanderthal-like)

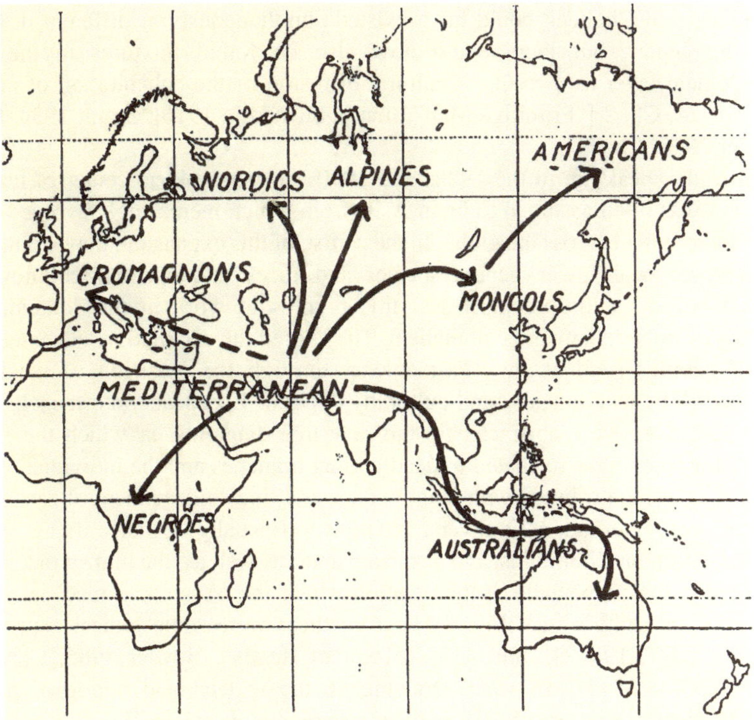

FIG. 13. – Map to suggest the directions of the wanderings of the various races into which the species *Homo sapiens* became subdivided to reach the areas in which they were originally found.

Fig. 8.12 Smith GE (1929) Human history. W. W. Norton, New York, Fig. 13, p. 49

Homo heidelbergensis (1907) from today's Germany; finally, a similar form (*Homo rhodesiensis*, 1921) from Zambia (formerly Northern Rhodesia). From a biogeographical perspective, Smith assumed a starting point of evolution and distribution somewhere in the middle between these sites, most likely in southwest Asia (Smith 1929, 46–8 and Ch. 2).

Smith's phylogenetic tree, shown as Fig. 8.11, is of course also striking because the 'human races' were given long parallel branches without indication of intermixture, with the 'Nordic race' as the apex and as continuation of the main stem. This might be supported by the 'tree-on-a-map' diagram given as Fig. 8.12 that shows 'racial' dispersal. However, Smith criticized the Tylor school's lineal (mental) imagery that assumed a necessary cultural developmental series from the simpler to the more complex archeological cultures due to the unity of the human psyche. Instead of assuming parallel linear evolution in situ (multilinear, multiregional model), Smith explained the archeological findings with his notion of a center of evolution from which peoples in possession of successively more developed cultures had spread, so that the series of Magdalenian, Solutrean, Aurignacian, Mousterian,

Acheulian, and Chellean could have existed simultaneously at different distances from the center, but there could now also be found mixtures in the same archeological layer as a result of cultural diffusion or the cohabitation of peoples (Smith 1924, Ch. 2 ["Primitive Man", first published in 1916]; Smith 1933, Ch. 4; Sommer 2007, Ch. 11).

Generally speaking, in the 1920s and 1930s, a geographical center of hominid evolution was often assumed to be in Asia, from which increasingly evolved forms would have spread across the globe. In the course of this expansion, they would have displaced the local 'lower' genera, species, and 'races' and thus their cultures. Like Haeckel before, anthropologists saw this process at work in the contemporary practices of imperialism and colonialism. They noted the decimation and extermination of peoples such as the 'Tasmanians' through the intrusion of what they presumed to be anatomically and culturally superior Europeans into their habitats. These narratives were abstracted in the new tree diagrams, in which the known 'fossil' hominids were relegated to dead-ending branches and the individual human 'races', too, were attributed with lengthy ancestral lines independent of each other.

There were internationally diverse attempts to (visually) distance living humans from fossil humans, hominids, and also from apes, as well as 'the human races' from each other—look especially at the position where the 'Australian race' leaves the main stem in Smith's tree (Fig. 8.11). And there were some radical expressions (Sommer 2007, 197–212 and 337; 2016, Part II; also, Bowler 1986, Chs. 4–6; Delisle 2007, 135–41, 161–74). Drawing on the widely held notion of parallel morphological evolution, the British anthropologist Frederic Wood Jones (e.g., 1919, 1929), for example, went as far as to claim that 'man' and the anthropoid apes had arisen and evolved independently from a basal primate stock. Another extreme example would be the imagery of the Argentinian Florentino Ameghino. Ameghino derived *Homo pampaeus* (skeletal remains from Argentina, supposedly of Pliocene Man) directly from a small bipedal *Homunculus* from the early Tertiary of Patagonia through the stages (of the apparently South American Miocene fossils) *Tetraprothomo*, *Triprothomo*, *Diprothomo*, and *Prothomo*. *Homo pampaeus* was seen as ancestral to the species *Homo sapiens*, comprising today's 'American and Caucasian races', while the species *Homo ater* consisted of the contemporary 'African and Australian types'. Thus, inspired by Haeckel, Ameghino espoused a form of evolutionary polygenism. The old-world hominid fossils were conjectured as side branches of the main stem leading to *Homo pampaeus*. The recent old world apes (*Anthropomorphidae*) were hypothesized to have evolved from primitive humans. Like in other theories of extreme parallelism, the apes were degenerated humans (Podgorny 2015). As shown in Fig. 8.13, Ameghino's view of 1909 concerning South America as the place of origin and center of dispersion in human evolution was visualized, and criticized, by Boule in his *Les hommes fossiles* (1923 [1921]).

Ameghino was not alone with his evolutionary polygenism. The independent evolutionary lines of 'the human races' could be conceptualized as drastically long. Maurus Horst (1913), to give a particularly radical example, separated 'the human races' into different phyla that originated as far down in the primate order as the

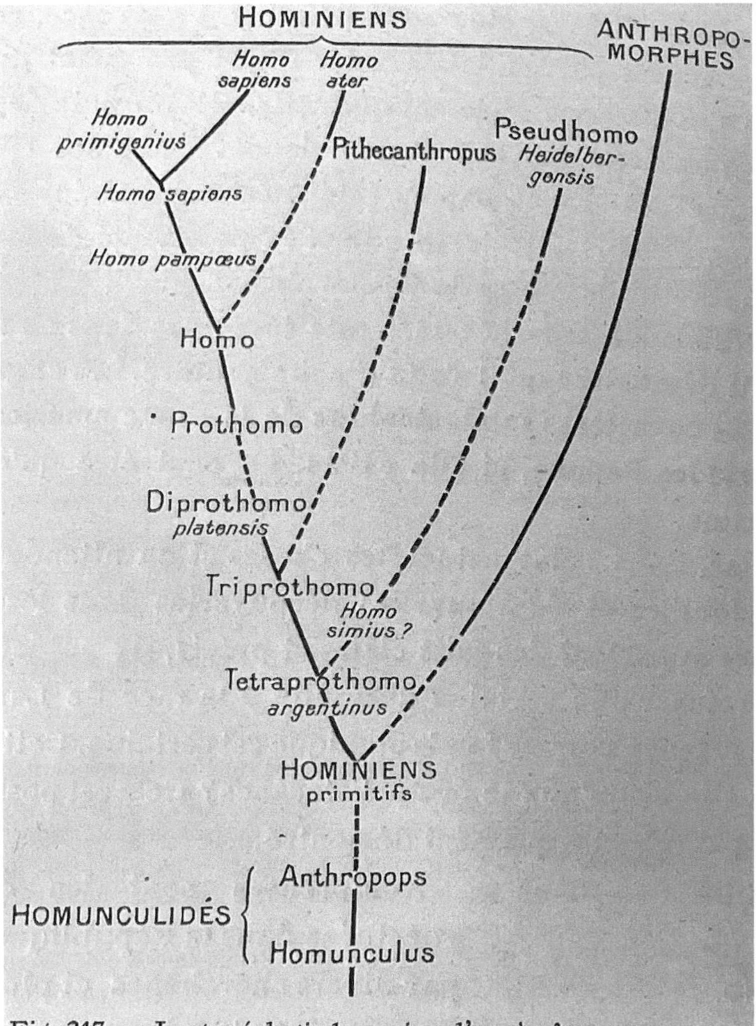

Fig. 247. — La genéalogie humaine d'après AMEGHINO.

Fig. 8.13 Florentino Ameghino in Boule M (1923 [1921]) Les hommes fossiles: Éléments de paléontologie humaine, second ed. Masson, Paris, Fig. 247, p. 440

lemurs. With his posthumously published *Der Werdegang der Menschheit und die Entstehung der Kultur* (1920) and a diagram therein (Fig. 273, p. 330), the German anthropologist Hermann Klaatsch had been a proponent of a kind of parallel evolution that aligned different recent 'races' with different ape genera. The fact that the Neanderthals were associated with an African fauna, while the Cro-Magnons seemed to belong to Asian animals, made Klaatsch speculate on human migrations. His subsequent comparative studies of the anthropoid apes of the respective continents led him to infer a particularly close relationship between the Neanderthals

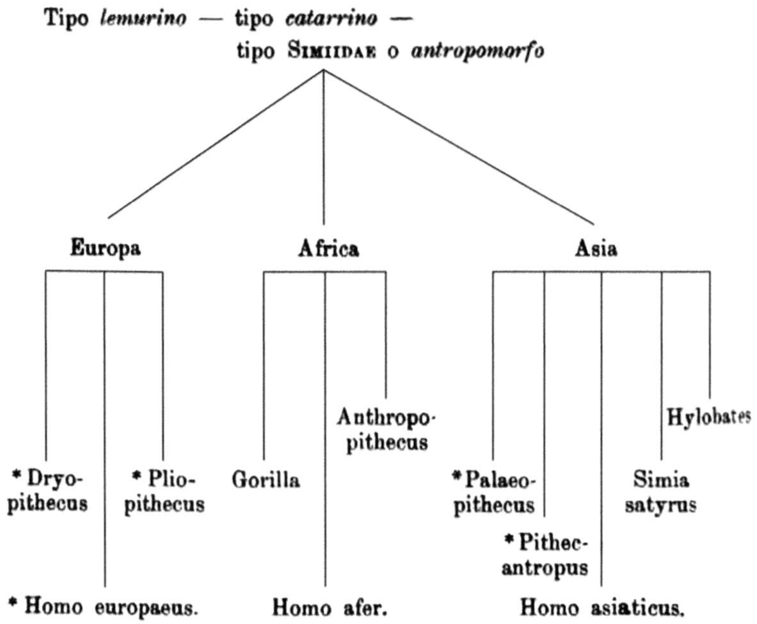

Fig. 8.14 Sergi G (1908) Europa: L'origine dei popoli Europei e loro relazioni coi popoli d'Africa, d'Asia e d'Oceania. Bocca, Milano, p. 530

and gorillas on the one hand, and the Cro-Magnons and orangutans on the other. Generally speaking, he believed in, and diagrammatically visualized, an early separation of the hominid-anthropoid group into a western and an eastern branch from which had evolved the anthropoids and 'human races' still found in these regions. Similar to Haeckel, Klaatsch conjectured that the common origin of both groups might have been a now-submerged continent in the Indian Ocean. Like Ameghino, Klaatsch considered the common hominid-anthropoid ancestors to have been more humanoid than anthropoid; the apes had degenerated from that original state. Klaatsch even speculated that the Indigenous Australians represented an isolated survival of this original stock of which *Pithecanthropus erectus* gave testimony (e.g., 89–92, 255–386).

There existed other polyphyletic theories and diagrams of this extreme form that linked different human groups to different apes or even monkeys. In these cases, human groups were seen as more closely related to some nonhuman primates than to the remainder of humankind and usually correspondingly as constituting at least different species, but rather different genera. Criteria for grouping particular humans with particular nonhuman primates were, for example, head shape (dolicho- vs. brachycephaly) or facial characteristics. Figure 8.14 is taken from the Italian physical anthropologist Guiseppe Sergi's *Europa: L'origine dei popoli Europei e loro relazioni coi popoli d'Africa, d'Asia e d'Oceania* (1908), which besides series and maps contains strange polygenist phylogenetic 'trees' (see also, Sera 1917;

Sommer 2024, 183–4; Chs. 11 and 13 in general). As indicated by Fig. 8.14, even if researchers referred to such diagrams as trees, they may rather constitute assemblages of scales.

Although, as we will see, family trees in which the various 'modern human races' were derived from different fossil hominid taxa were proposed until after World War II, the 'hyper-diversity' approaches and diagrams came to be challenged by several developments. The discovery of new hominids was decisive. The australopithecines not only strengthened the arguments in favor of Africa over Asia as the place of origin. They also ultimately called into question the thesis supported by Piltdown that human evolution began with brain enlargement (and not with upright walking). This made it possible to reintegrate *Pithecanthropus erectus* into the direct lineage of modern humans. In 1924, fossils were found in a limestone quarry near Taung in South Africa of a young individual that was ape-like, but also had surprisingly human features. The fossils were scientifically named *Australopithecus africanus* and introduced as a 'missing link', but the paleoanthropological community in general rejected this interpretation. However, the American comparative anatomist William King Gregory, who worked under Osborn at the American Museum of Natural History, used the australopithecine with its advanced anthropoid brain and primitive human dentition for 'A Critique of Professor Osborn's Theory of Human Origins' (Gregory 1930). The *Australopithecus* morphology contradicted Osborn's assertion that the hominids had already split off from the anthropoids in the Oligocene. After all, the australopithecines were no older than the late Tertiary. Overall, *Australopithecus* provided important evidence for the close relationship between apes and humans and against Osborn's orthogenetic theory of the evolution of 'man' past the apes. Despite continued skepticism, the australopithecine story came to a good end with Wilfrid Le Gros Clark's *The Fossil Evidence for Human Evolution* (1955), which emphasized the importance of the australopithecines in human evolution (Le Gros Clark, Ch. 4; for the history of australopithecine interpretation, see Gundling 2005).

Le Gros Clark's book, which became required reading for a new generation of physical anthropologists, was also indicative of the changes that had taken place in paleoanthropology in other respects. Le Gros Clark not only retracted from the importance of evolutionary parallelism, but he also made a strong reference to George Gaylord Simpson and the new systematics. Le Gros Clark emphasized the relevance of the concepts of the modern synthesis and genetics as well as those of anatomical functional studies. Accordingly, he followed the trend of simplifying nomenclature and reducing taxonomic diversity, a trend which went along with other kinds of visualizations of human phylogeny than we have seen in this section. The 'diagram' shown as Fig. 8.15 does not look very tree-like, but for the *Hominidae* rather consists of something like tongues representing genera that grow out of each other and may encompass lower taxa. It was meant to communicate that genera could give rise to new ones while specimens of the parent form continued to exist, rendering parent and offspring genera contemporaries for a certain time. This is what the tree shape can achieve, as Schwalbe and others had realized (see Fig. 8.10): Taxa put on branches can still be imagined as stages on the main-stem lineage leading to

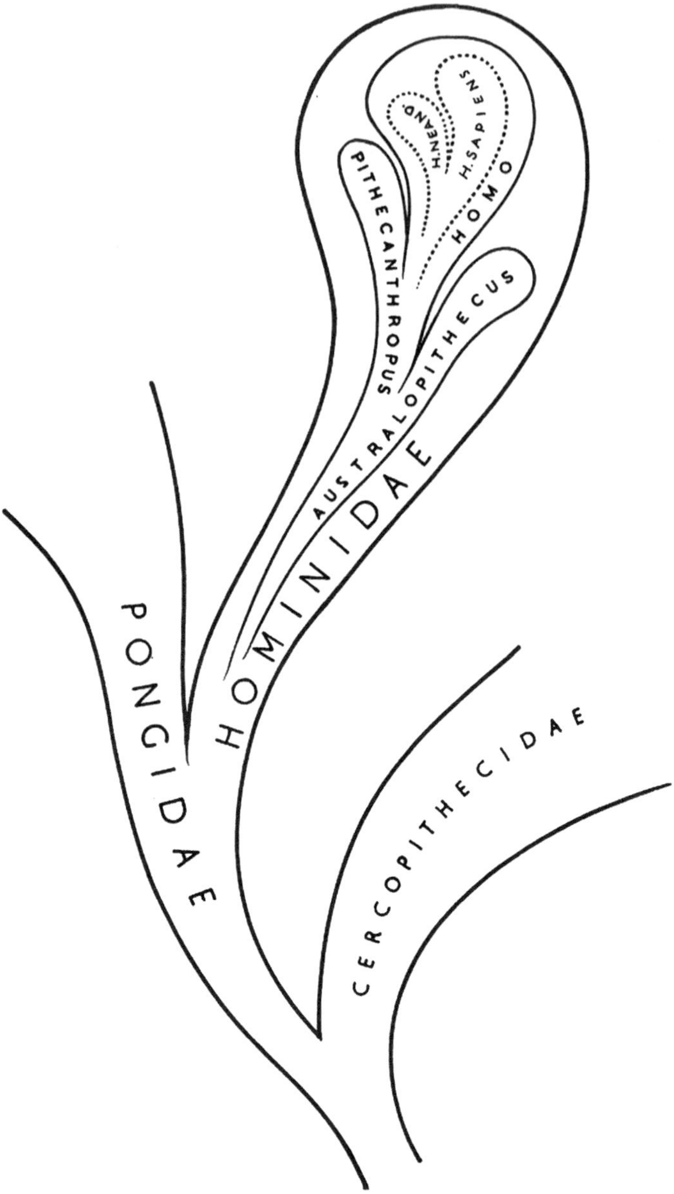

FIG. 1.—Diagram illustrating the probable phylogenetic relationships of the known genera of the Hominidae, and the relationship of the latter to the Pongidae. This diagram is also intended to emphasize that, as paleontological records in general clearly establish, representatives of an earlier genus may survive and persist long after other representatives of the *same* genus have given rise to a later genus. In other words, an ancestral genus may for some time actually be contemporary with its derivative genus.

Fig. 8.15 Reprinted by permission of the University of Chicago Press from THE FOSSIL EVIDENCE FOR HUMAN EVOLUTION: AN INTRODUCTION TO THE STUDY OF PALEO-ANTHROPOLOGY (Fig. 1) by W. E. Le Gros Clark. © 1955 by The University of Chicago. All rights reserved modern humans, if parts of them persisted in the older form after other representatives had already 'moved on to higher levels' (Delisle 2007, 128–34; Sommer 2007, Part II and Appendix B). This is why Fig. 8.15 retains elements of the tree while simultaneously being a 'vertical' rendering of the genera of the *Hominidae*, 'in the sense that they are indicative of separate main and subsidiary evolutionary sequences [...]' (Le Gros Clark 1955, 9).

In fact, as we will see in the next section, the modern synthesis increasingly influenced the interpretation of fossils. The understanding of species as polymorphic populations was sometimes accompanied by an extreme simplification of hominid taxonomy. The number of fossil genera and species was drastically reduced, and a linear concept of hominid evolution regained ground. The so-called single-species concept stated that, because all hominids occupied the same ecological niche, only one species could be described for a geological horizon. This made it possible to reconcile multiregional evolution with a unilinear model.

8.5 Synthetic and Cladistic Models

The modern synthesis brought a novel interest in systematics, and with it debates about the nature or existence of subspecies variation in living humans and in the course of human evolution. For some anthropologists, populations replaced 'races' as the groups under concern. This meant that intraspecific evolutionary processes received more attention. The populational approach focused interest on evolutionary processes like natural selection, gene flow, and genetic drift that worked on variation within and between populations. Although there were many anthropologists who simply replaced the term *race* with *population* and continued to conceive of such groups as isolated breeding communities, others, to the contrary, promoted the study of individual traits and their distribution, in combination with the causes underlying their variation (e.g., Caspari 2010). The integration of anthropology into the theoretical framework of the modern synthesis thus led to the reassessment of known fossil forms and ultimately to the reintroduction of some of them into the lineage leading to modern humans, such as *Pithecanthropus erectus*, now called *Homo erectus*. Many experts reclassified Neanderthals as part of the human species, a development that was influenced by the work of the German-born anatomist Franz Weidenreich, among others, and that inspired alternative ways of visualizing phylogeny. Right at the beginning of his book *Apes, Giants, and Man* (1946, 1–3), which contains five lectures he had given as Hitchcock Professor at the University of California in 1945, Weidenreich positioned himself against polygenism—against such models as discussed in the previous section. He deconstructed the conventional typological concept of race and propagated a population-oriented understanding of human diversity:

> As the matter stands now, the only thing that can be done about the definition of races is [...] to trace these features through the whole of mankind without regard to any previous racial definition or classification. Then the frequency of each combination and its geographical

> distribution should be noted and a framework built of the final subgrouping of the populations of the earth. (Weidenreich 1946, 90; also Ch. 4)

Instead of starting from preconceived categories, types, or 'races', Weidenreich demanded that the distribution of trait combinations throughout humanity and their geographical distribution be determined. Only on this basis would it be possible to determine populations. For Weidenreich, humanity was a single species with continuous variation. He even argued that not only modern but also all fossil hominids should be attributed to the same species. However, for reasons of communication, he maintained the prevailing nomenclature. According to Weidenreich, the Australian, Asian, African, and Eurasian populations had developed locally along the series *Archanthropinae* (*Meganthropus/Pithecanthropus/Sinanthropus*)—*Paleoanthropinae* (Neandertaloids)—*Neanthropinae* (anatomically modern humans), without speciation taking place. He recognized local specializations in the fossil record but assumed that the populations maintained a certain unity through the continuous exchange of genes. Weidenreich therefore rejected phylogenetic trees and monocentric models such as Osborn's Central Asian hypothesis, according to which modern humans evolved from an ape-like precursor to their modern form and only then spread across the globe, where they displaced the 'less advanced' species. The australopithecines, however, were not represented in Weidenreich's system of hominids, as he classified them as apes (as well as the Piltdown jaw fragment) (1946, Chs. 1–4). Weidenreich's human evolution and phylogeny was visualized as a squared network diagram structured by space and time, with vertical lines signifying local lineal evolution, horizontal lines standing for migrations, and diagonal lines for gene exchange between all local lines throughout hominid history (1946, Fig. 30, 30).

The synthesist Theodosius Dobzhansky saw in the assumption that at no time more than one hominid species existed and that hominid evolution had taken place at the level of 'races' the possibility of a synthesis of the classical evolutionary model or images, with a center from which successively more highly evolved forms spread, with Weidenreich's multiregional and unilinear evolution. According to this synthesis, Weidenreich could be right that several local fossil forms were ancestral to living humans, but it seemed possible that not from each local type emerged a different modern 'race'. The classical view could therefore be correct in that some of the historical 'races' contributed more than others to the genetic make-up of modern humans. Local adaptations could have spread throughout the species (Dobzhansky 1944). The systematist Ernst Mayr also participated in the efforts to merge taxa under the assumption of the single-species hypothesis. He included the australopithecines in the spectrum of *Homo* variation and regarded all hominids known up to that point as representatives of a single lineage (*Homo transvaalensis*—*erectus*—*sapiens*) (Mayr 1950).

While criticizing the tree diagrams for hominid and human evolution, there seems to have been a certain suspicion toward concrete visual expressions of such phylogenies among the synthesists. One rather encounters metaphors, such as Dobzhansky's cable consisting of many strands, or the merging and diverging streams of Julian Huxley, who took issue with the tree diagram for human evolution early on (Sommer 2024, Chs. 12 and 14). In fact, the human family tree has had its critics at least from the early twentieth century onward for political as well as scientific reasons. It depicts human, also intra-human, evolution analogous to the phylogenies of different species and higher taxa (Sommer, e.g., 2014). At the time of Weidenreich's network diagram, even the great tree-builder Keith underwent a change of opinion and in visualization. Keith, too, increasingly moved away from the original 'hyper-diversity' model expressed in his 'family trees', in which each fossil usually sat on its own dead-ending branch. Indeed, this change has often been linked to Weidenreich's multiregional and unilineal model. However, Keith's new model differed from Weidenreich's in that Keith thought of the local evolutionary lineages as having evolved in isolation from each other. Such alternatives as Keith's are therefore not unilinear, but rather multilinear (one could also say polyphyletic, as opposed to monophyletic). While the unilinear model assumed constant gene exchange and subsumed the known variation under a few taxa, Keith and others argued for parallel evolution of the human 'races' independently of each other through different fossil forms.

In *A New Theory of Human Evolution* (1948), Keith thus proposed regional fossil hominid genera as the ancestors of different modern 'races' and he offered a diagram to naturalize his evolutionary polygenism (158–9). In the new kind of diagram, local varieties of ground-dwelling anthropoids and australopithecines gave rise to parallel branches according to which: Indigenous Australians evolved from *Pithecanthropus* via intermediate stages; Kanam Man and *H. rhodesiensis* were the ancestors of Africans; Asians arose from *Sinanthropus*; and 'Caucasians' were descended from Neanderthals and Cro-Magnons. Keith's work on Neanderthals from the Carmel Mountains (Israel) with Berkeley paleoanthropologist Theodore D. McCown led him to believe that the evolution of the 'Caucasian type' had passed through a pre-Neanderthal stage after all (Keith 1950, 599). Keith's diagram had therefore more in common with visualizations of multilinear and polygenist theories such as discussed in the previous section than with Weidenreich's multiregional but unilinear one.

This also becomes evident from Harvard professor Carleton Coon's idea of 'racial history'. Like Keith, Coon referred to Weidenreich, and even dedicated *The Origin of Races* (1962) to him—an unfortunate fact that is almost certainly the reason why Weidenreich's theory has been associated with racism in historiography on various occasions. However, in Coon's scenario that, too, was given visual expression, a polymorphic species of *Australopithecus* had evolved into *Homo* in the Old World tropics. The evolution of polymorphic *Homo* locally passed through the stages of *Homo erectus*, Neanderthal-like forms, and modern humans. *Pithecanthropus* was the ancestor of the 'Australoid' type, *Sinanthropus* that of the 'Mongoloid' type, and the 'Caucasian' type evolved from Heidelberg Man. The five human subspecies (Coon divided the 'Negroid' type into 'Capoid' and 'Congoid') had developed in parallel and at different speeds in the various geographical regions of the world. Those who were the first to complete the process of becoming human, according to Coon the 'Caucasoid' and the 'Mongoloid', were the most advanced in the present.

However, and on this point more in line with Weidenreich than with Keith, Coon explained that the different lineages had remained within the framework of a common species during their evolution through genetic exchange (especially 305–309, 332–7, for quotes for example Fig. 44, 335). Coon's chosen diagram was nonetheless a table emphasizing the separate and parallel columns of evolution up to the current 'races' (Sommer 2024, Ch. 13).

Although multilinear models persisted into the years after the Second World War, it can generally be said that the influence of the modern synthesis and the acceptance of australopithecines as hominids with predominantly human but also anthropoid features led to a decline in explanations of morphological similarities on the basis of parallel evolution. The 'ape-men' turned the tide in favor of Africa as the place of origin and bipedality as an early adaptation in hominid evolution. The anatomy of the australopithecines showed similarities to *Pithecanthropus/Sinanthropus*, allowing the reintroduction of *Homo erectus* into the lineage of modern humans. At the same time, the small brain and human-like teeth of the australopithecines made the Piltdown chimera of human-like skull and ape-like jaw look stranger than ever (e.g., Bowler 1986, 239–50). This forgery was finally banished from the phylogenetic diagrams thanks to the revolutionary advances in dating technology, which were largely driven by the geologist Kenneth Oakley at the British Museum of Natural History. He began working on a chemical method for relative dating in 1947. The systematic dating of the fossil record culminated with regard to Piltdown in the results published by the physical anthropologist Joseph Weiner in *The Piltdown Forgery* (1955). At this time, however, radiocarbon dating had not yet been carried out, as extracting the required amount of carbon would have meant destroying 'the find'. It was not until 1959, when the technique had been refined and smaller samples were sufficient, that a radiocarbon analysis revealed a Holocene origin for both the skull and the lower jaw. Other controversial finds such as Louis Leakey's Kanam Man were also reclassified (no older than Middle Pleistocene) (Oakley 1963, 1980).

But the Leakeys, including Louis' wife Mary Leakey and their children Richard and Meave Leakey, unearthed numerous other spectacular finds in East Africa—the most famous of which is probably *Homo habilis* or Handy Man (found in the Olduvai Gorge in 1960), the estimated age of which was in some cases supported by new dating technologies. The Leakeys not only continued the old tradition of introducing a new species or even genus for each fossil, they also continued to interpret them in the scheme of the old age of *Homo* to the exclusion of the australopithecines from human evolution. The claim that the skull KNM-ER 1470 (i.e., Kenya National Museum-East Rudolf Fossil No. 1470) belonged to a nearly three-million-year-old *Homo* eventually led to a public debate with Donald Johanson, whose team had found the famous australopithecine Lucy (*Australopithecus afarensis*) in Hadar (Afar Triangle, Ethiopia) in 1974, which appeared to be older than three million years. Johanson and his colleague Timothy White promoted Lucy as the ancestor of modern humans, while Richard Leakey ascribed that place to his *Homo*. Although the Leakeys attracted attention for their findings and these often became the subject of controversy, the Leakey era marked a decisive turning point for evolutionary anthropology. Not only was the fossil record

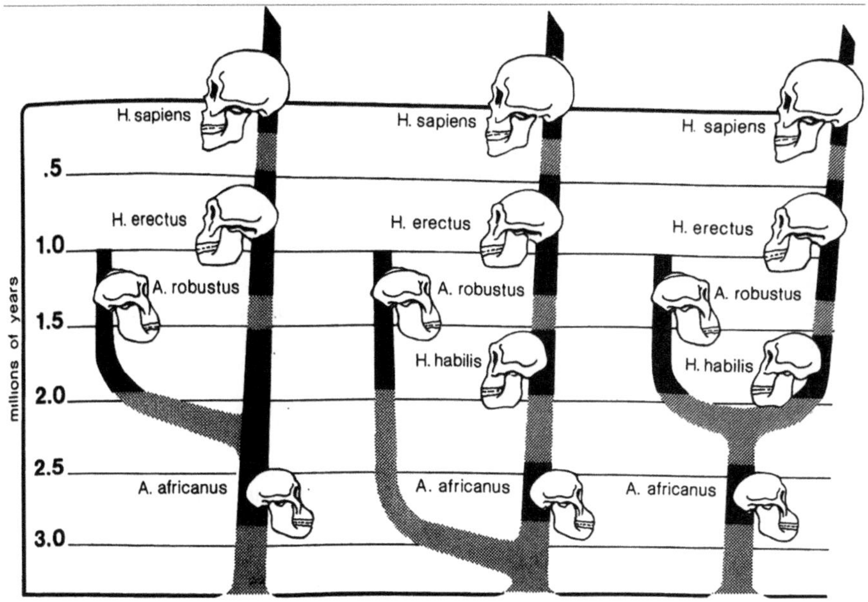

Fig. 8.16 Johanson D, Edey M (1981) Lucy: the beginnings of humankind: the dramatic discovery of our oldest human ancestor—and the controversial change it makes in our view of human origins. Simon and Schuster, New York, p. 281

greatly enriched, but the methodology was also refined. Mary Leakey transferred the more rigorous excavation practices typical of Neolithic sites to Paleolithic sites, and increasingly professionalized interdisciplinary teams were put together.

While, in line with the developments discussed above, some paleoanthropologists opposed the Leakeys' multilinear trees and assumed the ancestral lineage *A. africanus—H. habilis—H. erectus—H. sapiens*, the fossil record soon became too complex for this. The third model in the series of Fig. 8.16 was common in the 1960s and 1970s. However, most scientists eventually followed Johanson's view that *Australopithecus afarensis* represented the common ancestor of two lineages, *Homo* and *Australopithecus* (Johanson and Edey 1981, Chs. 10, 13–14; Willis 1989, 1992). Johanson and Maitland Edey attributed the first model in Fig. 8.16 to the single-species theory proponents, as it 'lumped' *H. habilis* with *A. africanus*, and more specifically to the views of the American anthropologist C. Loring Brace. Brace talked of *The Stages of Human Evolution* (1967) and also provided suiting visualizations. I have included the cover of this book as Fig. 8.17 below, because I read the artistic diagram as communicating a single evolutionary lineage from at least the australopithecines to modern humans and a rather uniform fossil record (without great taxonomic diversity). It is a kind of picture puzzle showing vases somewhat matryoschka-like contained in vases, or vases that grow out of each other, as well as (a series of) heads and skulls. However, as the first model in Fig. 8.16 suggests, as an increasing number of anatomically diverse fossil hominids were

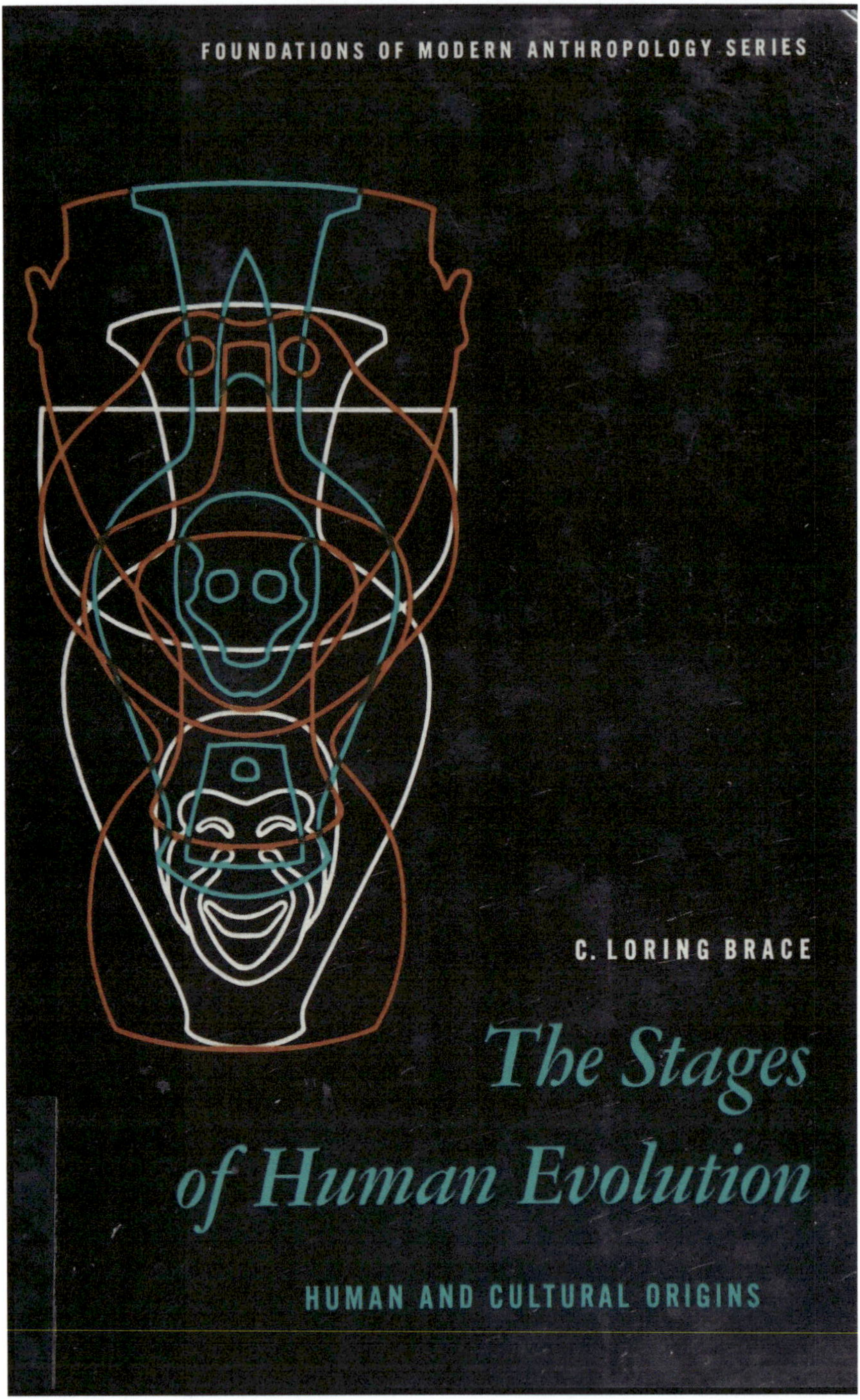

Fig. 8.17 Brace CL (1967) The stages of human evolution: human and cultural origins. Prentice-Hall, Englewood Cliffs, New Jersey, cover designed by Harry Rinehart

found to have lived contemporarily, even Brace could no longer accommodate the known fossils in a single line of descent but had to exclude the robust australopithecines on a side branch (see also Delisle 2007, 336–7).

Besides the increasingly bewildering fossil record, such notions were also questioned from a theoretical and methodological point of view. The American paleoanthropologists Niles Eldredge and Ian Tattersall from the American Museum of Natural History applied the punctuated equilibrium theory developed by Eldredge and Stephen J. Gould to human remains. The model states that large populations in particular can remain relatively stable over long periods of time due to factors such as the absence of selection pressure. In contrast, geographical isolation and niche changes in smaller populations can lead to speciation comparatively quickly. Species formation may therefore not only be gradual, but also episodic. The gaps in the fossil record are not—as Darwin had thought—primarily gaps in finds, but often represent evolutionary leaps. This theory shows parallels with Simpson's *Tempo and Mode in Evolution* (1944) and could therefore be seen as a further development of the modern synthesis. However, instead of assuming evolution within one or a few human species, Eldredge and Tattersall suggested that human evolution was characterized by multiple events of speciation and extinction, thus contradicting Mayr's single-species hypothesis.

Eldredge and Tattersall applied cladistics (developed in the 1950s by German entomologist Willi Henning) to study evolutionary relationships and patterns in the hominid record. While traditional comparative morphology is based on the general similarity between organisms, a cladogram is based on the distribution of individual characteristics. In cladistic analysis, the distribution of derived (advanced) traits in related groups of organisms is determined and, on this basis, common ancestors (which may be hypothetical) are inferred. In contrast to derived traits, primitive traits are widespread in a group or in related groups, appear relatively early in embryonic development and are geologically older. The cladistic method has been widely accepted since the 1980s, and it has produced human phylogenies with great taxonomic diversity and numerous extinct branches (e.g., Tattersall and Eldredge 1977; Tattersall 1995, Ch. 12). Finally, the phylogenies and scenarios of human evolution developed within the framework of the synthetic theory were also called into question by findings from biochemical and molecular methods, which seemed to prove in particular a close relationship and recent history of differentiation between apes and humans. But it is also the genetic studies that seem to have re-centered an interest in intra-*Homo-sapiens* diversity.

8.6 Molecular Maps, Trees, and Networks

The history of molecular evolution, the application of molecular technologies to questions of human history and diversity, can be traced back to the beginning of the twentieth century. The German immunologist Karl Landsteiner—the discoverer of the ABO blood group system—had suggested that people could be divided into groups based on their immunological reactions to treatment with another person's blood. Anthropologists had also noticed the work of the British bacteriologist George H. F. Nuttall, who studied the reciprocal immunological reactions of blood sera from different species of monkeys and humans. His experiments showed that humans are very closely related to apes, and he found a particularly close resemblance to the African apes (Nuttall 1904). In *Ursprung der Menschheit* (1932), for example, the German anthropologist Hans Weinert presented blood serum tests as one of the pillars supporting his claim of a close relationship between humans and African apes, especially between humans and chimpanzees, which he also visualized in a new family tree (Fig. 122, 351). Immunological analyses, especially from the early 1960s onward, were to further support Weinert's criticism of systematics that regarded hominids and pongids as two separate families. Weinert (1932, 316–7) separated humans from apes at the genus level.

However, an application of the knowledge about blood groups had also developed, which consisted of studying their distribution in different groups of people. During the First World War, Ludwik and Hanka Hirszfeld used the opportunity offered by the coming together of members of different nations in the troops of the Entente for such investigations. They believed they could show that the four ABO blood group types occurred in different proportions and that the distribution of frequencies was related to the geographical distribution of the groups studied (Hirschfeld and Hirschfeld 1919). These studies are often described as the origin of the genetic analysis of human populations (Schneider 1996, 282). Subsequently, the American immunologist William C. Boyd and other researchers saw the blood group distribution studies as a robust approach to arriving at a classification of 'races'. The preferred image here was the distribution map, as it had also functioned for anthropometric and physical characteristics (Gannett and Griesemer 2004). Fig. 8.18 by Boyd constructs 'six genetically defined races'.

In the era of such blood-group studies, a center emerged that was able to connect a network of researchers. Scientists turned to Arthur Mourant's serology laboratories in Great Britain when they needed to identify populations for sampling and the best way to collect samples. They asked for experts in specific areas, informed about their projects and expeditions, sent in blood samples for analysis, and received sera and blood-group data. Mourant's team compiled huge amounts of data on blood-group frequencies in human populations, partly from publications and partly directly from research. Unlike Boyd, Mourant saw in blood-group research the possibility of a 'scientific anthropology' (Mourant 1954, 1) that would do away with the old racial approach. He was a central figure in the endeavor to obtain as much data as possible on the distribution of the so-called classical or conventional genetic markers, i.e., on blood groups, enzymes, serum proteins, the immunoglobulin, and human leukocyte antigen systems. He and his allies made this data available to the scientific community in books such as *The Distribution of the Human Blood Groups and Other Polymorphisms* (1954), *The ABO Blood Groups* (1958), and *Blood Relations: Blood Groups and Anthropology* (1983). The latter contained an appendix with maps indicating the distribution of blood-system genes among Indigenous peoples of the world. In these, gene frequencies were turned into clearly demarcated intervals (e.g., percentages of 0–10 or 90–100) and assigned their own code of shading

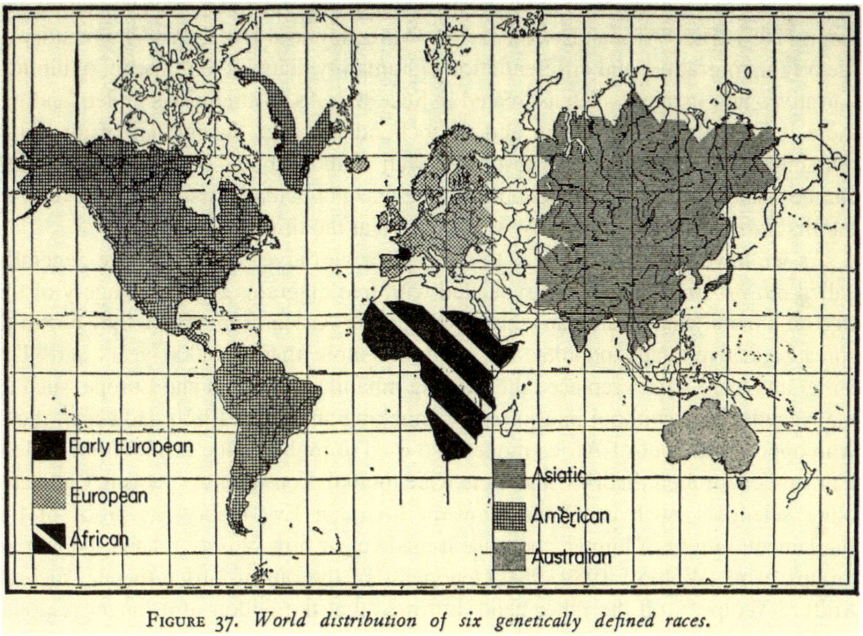

Fig. 8.18 Boyd WC (1950) Genetics and the races of man: an introduction to modern physical anthropology. Little, Brown and Company, Boston, Fig. 37, p. 271

(on Mourant, see Sommer 2016, 259–61; more generally on blood groups, Bangham 2020).

By that time, in fact as early as the 1960s, the relationships between migration patterns and the distribution of blood groups in humans had attracted the attention of the Italian population geneticist Luigi Luca Cavalli-Sforza, and he wanted to develop statistical and computational methods for the production of family trees from such data (e.g., Cavalli-Sforza and Edwards 1965). It was also Cavalli-Sforza's team at Stanford University that published the first molecular family tree of living humans based on sequence analyses of mitochondrial DNA (mtDNA) in the early 1980s. Allan Wilson's laboratory at the University of California, Berkeley, however, overtook Cavalli-Sforza in mtDNA research, which should make it possible to reconstruct the differentiation history of modern humans. Rebecca Cann, Mark Stoneking, and Wilson used restriction fragment length polymorphisms (differences in DNA sequences in sites recognized by restriction enzymes) to compare the mtDNA of about 150 people taken to represent African, Asian, Australian, 'Caucasian', and New Guinean populations. Based on the differences they discovered, they created a phylogenetic tree (of mtDNA sequences) that was first published in *Nature* (Cann et al. 1987). Beyond the phylogenetic tree, the analysis also seemed to suggest a scenario for the evolution of modern humans. Assuming a molecular clock for the evolution of mtDNA, it was estimated that all mtDNA of humans living today can be traced back to a woman who lived in Africa about 200,000 years ago (Stoneking and Cann 1989). This 'African Eve theory' or 'Mitochondrial Eve theory' was a story of the origin, migration, and differentiation of humanity that was reminiscent of biblical narratives and imagery such as treated in Sect. 8.2. As in a religious understanding, there is a single human origin and (mitochondrial) Eve, the mother of all living humanity. The story therefore resonated well in the public—only that the naming can be misleading, since mitochondrial Eve was not actually assumed to have been the first woman or even the only woman alive at the time.

As we have seen in Sect. 8.2, this model or narrative, which is more generally called *out-of-Africa model*, is associated with tree diagrams and the imagery of the tree on a map: Modern humans thus originated in Africa, from where they began to spread and eventually populate the globe no more than 140,000 years ago. The emigrants from Africa replaced the existing inhabitants of Asia and Europe, such as the Neanderthals, without intermixing (Stoneking and Cann 1989). However, those who opposed the out-of-Africa model also tried to mobilize the results from molecular studies for their theories, and there were mutual accusations of racism. Cann and other scientists, such as Gould, saw the 'African Eve theory' as proof of the fundamental unity of humans because it made clear how closely related the human groups were (Wolpoff 1989, 98). Opponents of the out-of-Africa theory, such as Milford Wolpoff, on the other hand, affirmed that the notion of a longer regional evolution of the respective human groups was not more racist. On the contrary, in their view, the idea of constant (genetic) exchange between these populations provided a stronger foundation for the unity of humanity. According to Wolpoff, the model of an early differentiation into diverse geographical groups that lived side by side for a long time and had contact and cooperated with each other was a more satisfying prehistory than a scientifically refreshed story of Cain and Abel, which essentially consisted of a single population spreading rapidly and probably violently across the globe and displacing long-established populations in the process. Wolpoff recognized in this a narrative along the lines of 'making war and not love' (Wolpoff 1989, 98), which, if true, would have unpleasant implications (Sommer 2008, 510–15).

The multiregional evolutionary hypothesis, of which Wolpoff is one of the proponents, followed in the tradition of Weidenreich's multiregional but unilinear model of the evolution of human populations. This becomes particularly evident in a diagram that Wolpoff and Alan G. Thorne provided in 1992 and updated in 2003, and that is reproduced as my Fig. 8.19 (Thorne and Wolpoff 1992, 82; 2003 [1992], 52). It is clearly an adaptation of Weidenreich's network diagram of human evolution and phylogeny of 1946. Although the diagram indicates continuities in the evolution of local lineages, these local lineages are suggested to have been in constant exchange of genes so that, throughout the Pleistocene (or since the '*Homo-erectus* stage'), the differentiated local lines were kept within a single species.

The out-of-Africa scenario also had its 'predecessors' (beyond pre-evolutionary imagery), such as the monocentric theories of the first decades of the twentieth century, according to which the stratigraphic sequence of cultural layers and their

		EUROPE AND LEVANT	AFRICA	EAST ASIA	AUSTRALASIA
UPPER PLEISTOCENE	LATE	Lagar Velho Předmostí Mladeč	Afalou Lukenya	Shandingdong Ziyang Liujiang	Kow Swamp Keilor Willandra Lakes 50
	MIDDLE	Vindija Kebara La Ferrassie La Chapelle	Dar es Soltan	Maba	Lake Mungo 1, 3
	EARLY	Qafzeh Krapina	Klasies Omo Kibish	Dingcun Xujiayao	
MIDDLE PLEISTOCENE	LATE	Ehringsdorf Biache Zuttiyeh	Ngaloba Florisbad	Dali Jinniushan	Sambungmachan 1, 3 Ngandong
	MIDDLE	Sima de los Huesos Petralona Arago Steinheim	Kabwe Ndutu	Zhoukoudian H Hexian Nanjing	
	EARLY	Gran Dolina	Bodo Ternifine Olduvai 12	Zhoukoudian D, E, L Chenjiawo Yunxian	Sangiran 2, 10, 12, 17 Trinil
LOWER PLEISTOCENE	LATE		Buia, Bouri Olduvai 9	Gongwangling Yuanmou	
	MIDDLE		Konso Gardula Lake Turkana (east) 992		Sangiran 4, 27, 31
	EARLY	Dmanisi	Lake Turkana (east) 730, 3883, 3733 (west) 15000		Mojokerto

WELL-DATED FOSSILS point to the continuous, linked evolution of modern humans at sites around the world. Modern human groups in different regions developed distinct anatomical identities. Nevertheless, gene flow between the groups through interbreeding spread important changes throughout and was sufficient to maintain humans as a single species.

Fig. 8.19 Thorne AG, Wolpoff MH (2003 [1992]) The multiregional evolution of humans. Scientific American. Special Edition 'New Look at Human Evolution' 13(2):46–53, p. 52. Reproduced with permission.

associated hominid fossils did not point to local evolution, but to a series of immigration and displacement processes. Current proponents of the theory usually refer to William W. Howells, who succeeded his former teacher Hooton as Professor of Anthropology at Harvard University in 1954. Howells had already identified evolutionists (model of multiregional continuity) and migrationists among contemporary anthropologists (while Smith spoke even earlier of evolutionary versus historical approaches [Smith 1924, Ch. 2 ("Primitive Man", first published in 1916), 111]). Although Howells favored the model of a single origin of *Homo sapiens* and subsequent migratory movements, he did not rule out genetic exchange between the modern and local archaic forms—or their absorption. As such, his theory only partially fit the extreme versions of the migration model, which Howells himself dubbed 'Noah's Arc hypothesis' (Howells 1976, 480)—thus already recognizing the reference to the biblical story and more precisely to the genealogy of Noah's sons that, as we have seen in Sect. 8.2, were rendered in trees and/or maps in the early modern period.

The arguments for both—out-of-Africa as well as multiregional histories and phylogenies—strongly relied on diagrams. The first was associated with population trees that were thought of in terms of histories of common origin and dispersal, also figured as migration or distribution maps. The second worked with network-like images that, too, had map character, as is the case for Fig. 8.19. However, it was the out-of-Africa model that gained most ground and integrated different approaches to human evolution. That this process was propelled by human population genetics is paradigmatically shown in the abundantly visualized and codifying *The History and Geography of Human Genes* (Cavalli-Sforza et al. 1994). It was the time of the grand Human Genome Diversity Project that had been called out in 1991. Yet, the work that led to the book had begun almost 15 years earlier. Cavalli-Sforza and his Italian colleagues Paolo Menozzi and Alberto Piazza reconstructed the history of differentiation and migration of entire humankind over the course of the species' existence. To do this, they integrated knowledge from different disciplines. The focus was on the comparative analysis of gene frequencies in populations whose gene pools were assumed to have changed the least since the Stone Age: Indigenous or so-called isolated populations. It was assumed that their members largely reproduced among themselves. Cavalli-Sforza took the view that such populations could best be identified and delimited on the basis of a common language. Drawing on the work on the global classification of languages carried out by Cavalli-Sforza's colleagues Joseph Greenberg and Merritt Ruhlen at Stanford University, the authors of *History and Geography* had an inventory and a worldwide atlas of populations understood in this way. Furthermore, it was assumed that all the world's languages had a single origin and that they had developed in parallel with genetic evolution, so that the population-genetic and linguistic family trees would coincide to some extent.

Far beyond this recourse to linguistics, the analysis of genetic data was combined with knowledge from archeology and (paleo)anthropology as well as with insights into climate, environmental, and human history. In order to date splitting and migration events, the genetic distances between populations were compared with archeological data on the first settlement of the various continents, and archeological excavation sites provided further important information on the dating and localization of movements. Thus, Cavalli-Sforza and his colleagues worked on global genetic mapping. They reconstructed the great migrations, which they understood to have been set in motion by population growth and technological innovations. At the beginning of the journey out of Africa was the first language in the true sense of the word, and at the end of the Paleolithic period, the emergence of agriculture triggered the expansion of Indo-European speakers into Europe.

From the painstaking and complicated processing of various types of data, the picture of a journey emerged that represented a refinement and extension of the theory of the African Eve and should be described and visualized in updated and increasingly complex fashion again and again: Modern humans left Africa about 100,000 years ago and began to spread across the globe via the Middle East. One branch moved along the southern coast of Asia to Australia (more than 40,000 years ago); another migrated through Central Asia (possibly about 60,000 years ago) and from there to South Asia, Europe (about 35,000 years ago) and the New World (between 35,000 and 15,000 years ago). Following this overview, Cavalli-Sforza, Menozzi, and Piazza described in *History and Geography* the more detailed multidisciplinary historical reconstructions for Africa, Asia, Europe, and the Americas, as well as for Australia, New Guinea, and the Pacific Islands. The diagrams—the maps, trees, tables, and graphs—to present data, models, and results from statistical analyses were integral to this goal. In fact, *The History and Geography of Human Genes* contains so many maps that it may be described as an atlas. It has an appendix with maps mostly giving frequency distributions of alleles. This appendix makes up by far the larger part of the book, also providing tables of allele frequencies for the populations analyzed and a reference list for the alleles studied (Sommer 2015b, 2024, Ch. 16).

Cavalli-Sforza's population genetics has been criticized, particularly by cultural anthropologists, for identifying populations in the biological sense on the basis of languages and other cultural characteristics, but also for conceiving of Indigenous societies, whose identity was in many cases shaped by colonial practices, as monolithic and ahistorical entities (from which it was believed to be possible to infer humanity before 1492). Other objections concerned the relationship between disciplines (e.g., MacEachern 2000). Cavalli-Sforza emphasized the need to take different approaches into account, partly because historical knowledge involves a greater degree of uncertainty than non-historical methods in mathematics, physics, chemistry, and biology. The mutual compatibility of the research results was therefore particularly relevant. This did not necessarily involve cooperation between experts. It was usually considered sufficient to incorporate the results from other (sub)disciplines. Cavalli-Sforza's scientific biographers even suggested that he considered the 'two cultures' to be incommensurable (Sommer 2016, Ch. 11):

> In terms of theory, cultural anthropology is and always has been in flux. Unlike the hard science of genetics (and also less like the subdisciplines of archeology and physical

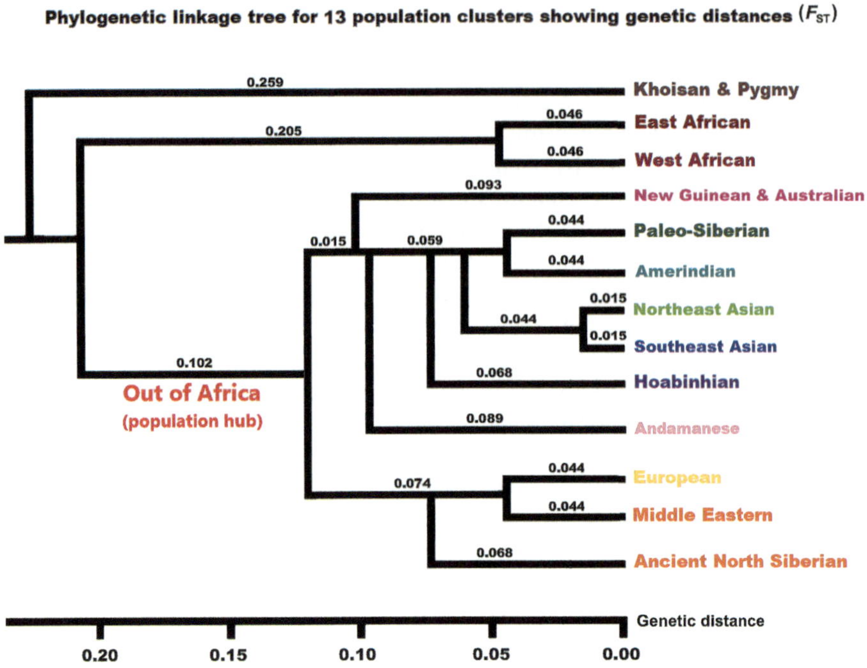

Fig. 8.20 "Phylogenetic clustering tree and genetic distance; based on Ananyo Choudhury, Shaun Aron, Dhriti Sengupta, Scott Hazelhurst, Michèle Ramsay, African genetic diversity provides novel insights into evolutionary history and local adaptations, Human Molecular Genetics, Volume 27, Issue R2, 1 August 2018, Pages R209–R218, https://doi.org/10.1093/hmg/ddy161 and Leonardo Vallini, Giulia Marciani, Serena Aneli, Eugenio Bortolini, Stefano Benazzi, Telmo Pievani, Luca Pagani, Genetics and Material Culture Support Repeated Expansions into Paleolithic Eurasia from a Population Hub Out of Africa, Genome Biology and Evolution, Volume 14, Issue 4, April 2022, evac045, https://doi.org/10.1093/gbe/evac045". By Lightningstrike3 (own work), 8 October 2022, Wikimedia Commons: https://commons.wikimedia.org/w/index.php?curid=124436516

> anthropology), it does not appear to make cumulative advances in the collection of data or the understanding of its subject—culture—but rather reflects intellectual trends and fashions in the broader humanities. (Stone and Lurquin 2005, 18)[2]

What has for a long time eluded critical reflection are the types of representation associated with this approach: Contrary to what seemed plausible to geneticists, the phylogenetic tree or the tree on a map (signifying common origin and subsequent diversification through migration) suggests that human populations at one point started to migrate and successively split and differentiated in isolation, without there being intermixture. As exemplified in Fig. 8.20, modern human populations

[2]This contribution strongly relies on Sommer (2015a) and has partly been (pre)translated from the German with DeepL.com (free version).

are thus not only shown as having isolated biological histories, they also appear as clearly demarcated, seemingly homogenous entities at the end of branches, an impression that in Fig. 8.20 is intensified through the coloring of their names (also mark the kinds of populations that are being constructed in this tree of 2022). Such a tree-like rendering of modern human evolution did reflect older views going as far back as Haeckel, according to which there existed typological races or according to which contemporary human groups were even as different from each other as to deserve the status of different species or genera. Human population genetics, however, while partly maintaining that modern humans did not mix with local archaic lines, had a completely different understanding of the nature of living humanity and its more recent history. Thus, even Cavalli-Sforza, Menozzi, and Piazza grappled with processes of admixture in their *The History and Geography of Human Genes*:

> In conclusion, there has probably been enough intermingling of the clusters that a network representation (i.e., a tree with interconnections between branches) would be highly desirable. But the tree in figure 2.3.3 is probably the best result that can be obtained using present methods, that is, a phylogenetic tree without interconnections. (Cavalli-Sforza et al. 1994, 81, quoted in Sommer 2024, 227, see also Ch. 17, where the issue of admixture is discussed in detail)

Actually, developments in the desired direction were not that long in the waiting. By the beginning of the third millennium, new statistical and computational approaches were being brought to bear on the analysis of an unprecedented amount of human genomic data (collected by the Human Genome Diversity Project/Panel and similar efforts). Extending the power of clustering methods such as principal component analysis, which had been extensively used by Cavalli-Sforza and colleagues, statistical software such as STRUCTURE (Pritchard et al. 2000), FRAPPE (Tang et al. 2005), and ADMIXTURE (Alexander et al. 2009) made it possible to group genetic samples into clusters and to analyze the extent to which present-day individuals and populations are the result of genetic mixtures. With the introduction of programs for the graphical display of population structure, such as DISTRUCT (Rosenberg 2004), the visual black box of these seemingly discrete and homogeneous entities—human populations—was opened. Individuals and populations began to be represented as colored bar plots indicating their histories of admixture. Accordingly, individual human genomes as well as human genetic diversity as such were now widely conceived of as a mosaic (see Fig. 8.21) (Sommer and Amstutz 2024). This is not to say that this kind of visualization does not incorporate some long-held as well as long-criticized notions, such the existence of originally pure populations (indicated in the image by different colors), the contributions of which to the now admixed individual genomes and populations can be reconstructed (Sommer, e.g., 2015b, 134–6).

Furthermore, and as I have discussed in length elsewhere (Sommer 2024, Chs. 18–20), the neatness of the human family tree has increasingly been deconstructed, in particular on the basis of the analysis of 'fossil' or ancient DNA (aDNA), beyond the deepening understanding of the genetic admixture among modern human

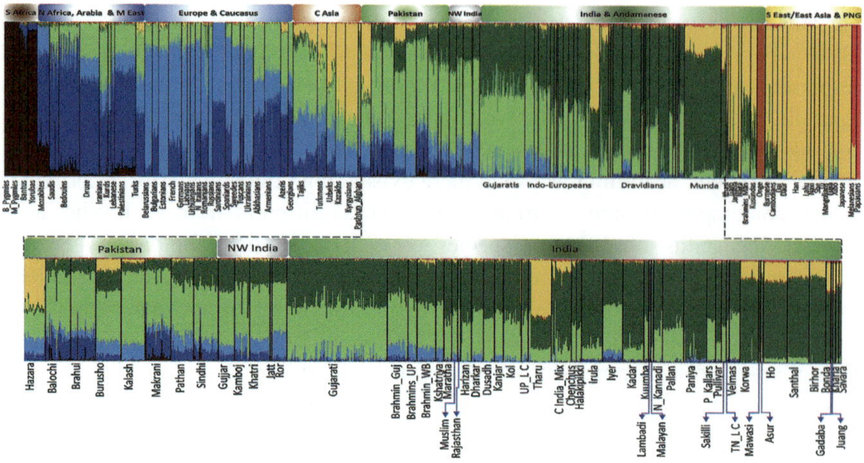

Fig. 8.21 "ADMIXTURE analysis at K8 ancestral components with global populations. The populations are ordered geographically in a bar plot." By Ajai K. Pathak, Anurag Kadian, Alena Kushniarevich, Francesco Montinaro, Mayukh Mondal, Linda Ongaro, Manvendra Singh, Pramod Kumar, Niraj Rai, Jüri Parik, Ene Metspalu, Siiri Rootsi, Luca Pagani, Toomas Kivisild, Mait Metspalu, Gyaneshwer Chaubey, and Richard Villems, https://www.cell.com/ajhg/fulltext/S0002-9297(18)30398-7, 22 November 2018, Wikimedia Commons: https://commons.wikimedia.org/w/index.php?curid=133940748. Accessed 19 May 2025

populations. With advanced technologies and greater availability of data, including aDNA data, the prevailing tree (on a map) view of hominin[3] and human evolution as consisting of a common origin with subsequent population splits without visible admixture at any level has become more problematic than ever. The pioneering aDNA research of Svante Pääbo's group finally suggested that Neanderthals and Eurasian modern humans had interbred after all (Green et al. 2010). In the ensuing years, technological developments such as next-generation sequencing and the increasing accessibility of aDNA in quality, quantity, and temporal depth have been said to have exploded the potential of aDNA research from the limited ancient genomes of individuals to the scale of population genomics (e.g., Lan and Lindqvist 2019, 21). And some of the most groundbreaking and surprising findings in aDNA research relate to the genetic contribution of extinct lineages to lineages leading to modern human populations, as well as evidence for ancient genetic exchange between different archaic lineages (Resendez et al. 2019, 379).

Not only was the fact that Neanderthals had contributed to contemporary 'non-African genomes' more firmly established, but Denisovans—an archaic hominin introduced on the basis of DNA from a little finger bone found in a cave in the Altai

[3]Reflecting changes in taxonomy, over the last decades, the term *hominins* has come to replace *hominids* to refer to the same group: Living humans, extinct species of *Homo*, and all immediate ancestors of humankind (excluding the great apes and their immediate ancestors).

Mountains of southern Siberia—appeared to have contributed to the genomes of modern Papuans, Melanesians, Indigenous Australians, and other Southeast Asian Islanders, as well as, to a lesser extent, mainland East and South Asians (Green et al. 2010; Reich et al. 2010; for a review, see Leonardi et al. 2017). Further research has suggested possible deep-rooted gene flow from an ancient 'ghost population' and a modern 'ghost population' (genetic traces of unknown ancestors) into West African populations, as well as large-scale Neanderthal inbreeding into African populations (reviewed in Vicente and Schlebusch 2020, 13). Thus, aDNA studies have increasingly undermined the pure tree model of human evolution, and at lower and lower levels. Nonetheless, among other things due to the long history of visualizing human phylogeny as trees (which, as we have seen, for many researchers matched their theories), to the history of mathematical and technological innovations and their present limitations, as well as due to unsatisfactory (aDNA) data, the above findings regarding archaic and modern genetic exchange between populations and 'species' mainly only resulted in connections between the branches of the tree. Tree or graph topologies are influenced 'by prevailing ideas and prior expectations' (Scally 2021, 181). But even if scientists in their studies have often assumed that gene flow took place over long periods of time and was most likely reciprocal, their models might only test for 'introgression events' visualized by arrows suggesting a singular and unilateral genetic input of short duration, and they may start from the assumption of 'treeness' (Sommer 2024, Ch. 18).

At the same time, the perspective is gaining momentum that '[f]or most time periods and in most regions it is no longer reasonable to make the default assumption of no gene flow between human populations [...]' (Scally 2021, 181). Accordingly, scientists are looking for new models that might more closely approach the complexity that seems to characterize hominin and human history and diversity. They come up with more and more reticulate metaphors for and diagrams of hominin and human evolution and kinship. While we have seen that throughout the history of evolutionary anthropology, hominin and human evolution and phylogeny have predominantly been depicted in diagrams that were referred to or at least remotely resembled trees, this icon of evolution is increasingly in jeopardy. At present, with the assumption of intermixture between Neanderthals and modern humans as well as between modern humans and other 'archaic' lines, concepts such as 'archaic' versus 'modern' and the species status of groups like the Neanderthals are open for discussion (Sommer 2024, Ch. 20).

Indeed, the final Fig. 8.22 seems to transcend such currently widespread trees with connecting arrows to indicate pulses of gene flow. It includes *Homo habilis*, *erectus*, and *heidelbergensis* in the line leading to modern humans, and it shows 'A winding past' and a modern 'Patchwork planet' (Stringer 2012, 34). However, let us end with a note of caution: While these metaphors and Fig. 8.22 seem closer to the semantic field of flower power than to that of processes of othering, such as racism, alternative interpretations of the same knowledge are possible. The newer insights suggest that some people living today carry a larger degree of 'archaic' ancestry than others, or from different 'archaic' forms. Speaking on the level of populations, they suggest that some groups might be 'more Neanderthal' than others—understood in

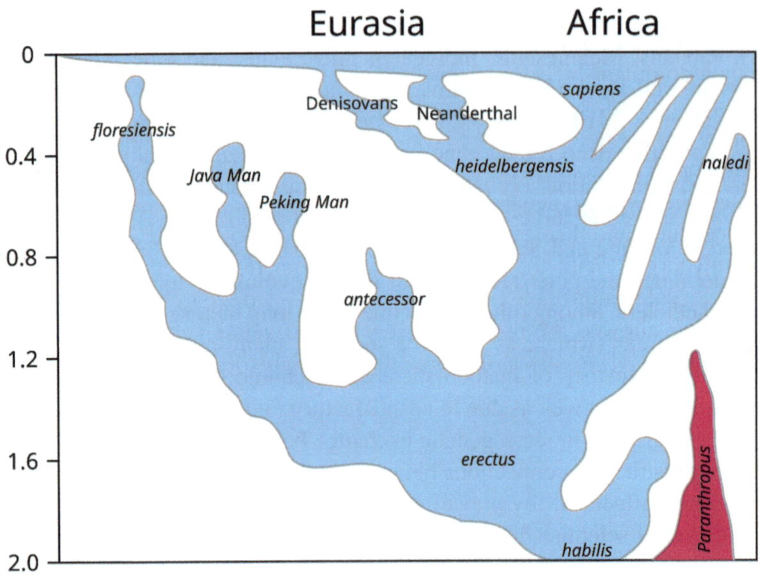

Fig. 8.22 "Phylogeny of the *Homo* genus, based on Stringer, C. (2012). "What makes a modern human". Nature 485 (7396): 33–35. doi:10.1038/485033a, with modifications." By user: Conquistador, user:Dbachmann—updated version of File:Homo-Stammbaum, Version Stringer-en.svg, 19 December 2017, Wikimedia Commons: https://commons.wikimedia.org/w/index.php?curid=64917473. Accessed 19 May 2025

either a good or a bad sense. In fact, already in 2012, Chris Stringer of the Natural History Museum in London perceived the novel possibilities that this kind of understanding posed for social misuse:

> These [archaic] genes might be expressed in the phenotype, and may require a rethink about how and when regional variation developed in *H. sapiens*. Those with alternative agendas may also try to use these new data to rank modern human populations in terms of supposedly different degrees of modernity. Already I'm reading blogs that speculate about whether some groups are less 'modern' than others, and I fear that such discussions endanger the considerable progress promised by paleogenetic research. (Stringer 2012, 34)

8.7 Concluding Remarks

The above condensed and spotty history of diagrams of human evolution, kinship, and phylogeny has not revealed a clear linear history, such as from chain to tree to web. There have always been different kinds of images in competition, and combinations of elements of different types. Iconographies of scales, trees, and maps in the context of human genealogy long predated evolutionary views of the organic world, and they seem to have influenced the way human phylogeny was conceptualized in evolutionary anthropology. In the latter field, trees could still convey the meaning of progressive evolution along a main line, while in natural history scales had already multiplied to form 'trees' and networks or other three-dimensional structures. Trees and maps have often been combined, and tree-like structures may contain network aspects. We have seen that a diagrammatic form, once established, can survive new ways of conceptualizing human kinship. In particular, the human family tree that included current human diversity has been used for almost diametrically opposed purposes. The tree flourished in the evolutionary frameworks to express diversification, but it could express approaches as radically different as evolutionary polygenism and human population genetics, even findings from aDNA admixture research. It is not surprising, then, that 'trees' have indeed come in a wide range of forms, and diagrams have sometimes been called trees that do not look much like them.

In spite of the above remarks, it all the same seems that certain imagery is more strongly associated with particular theoretical and methodological approaches. The tree was introduced into anthropology as *the* icon to communicate evolutionary outlooks, and at the beginning of the twentieth century, its capacity to show diversity in a typological sense was radicalized by certain anthropologists to the purpose of distancing 'the recent human races (species or genera)' from each other and to hierarchize them, as well as in order to strongly demarcate the living 'human types' from anthropoids and fossil 'human types'. At the same time, at least from the early decades of the twentieth century, the rendering of current human diversity in the shape of trees came under scrutiny, and with the modern synthesis impacting anthropology, verbal and visual images of hominin evolution and phylogeny tended to convey less diversity and more linearity. Up to this day, a particular school of anthropologists promotes multiregional unilineal ideas that take the visual shape of networks. Some aspects of cutting-edge genetic research, especially when taking aDNA data into consideration, too, seem to point in this direction. At the very least, the population-genetic trees have acquired ever more interconnections. And it seems the critics of the tree structure for human phylogeny are increasing.

And what about the politics of human relatedness? Are meandering rivers, cables, entangled skeins, networks, or trellises inherently 'better' relational diagrams than 'trees' in an ethical sense? Some of these diagrams have been brought into position (verbally and/or visually) against those in use in racist and polygenist anthropologies. The fact that some current scientists deconstruct trees that communicate the notion of self-contained races, populations, and even species in hierarchical relations seems to carry subversive potential. But as Stringer reminded us, we should be wary of making any inferences from knowledge in the biological sciences about social relations. They have a tendency to lead us astray.

References

Alexander DH, Novembre J, Lange K (2009) Fast model-based estimation of ancestry in unrelated individuals. Genome Res 19(9):1655–1656

Bangham J (2020) Blood relations: transfusion and the making of human genetics. The University of Chicago Press, Chicago

Boule M (1923 [1921]) Les hommes fossiles: Éléments de paléontologie humaine, 2nd ed. Masson, Paris

Bowler PJ (1986) Theories of human evolution: a century of debate, 1844–1944. Blackwell, Oxford

Bowler PJ (2021) Progress unchained: ideas of evolution, human history and the future. Cambridge University Press, Cambridge

Boyd WC (1950) Genetics and the races of man: an introduction to modern physical anthropology. Little, Brown and Company, Boston

Brace CL (1967) The stages of human evolution: human and cultural origins. Prentice-Hall, Englewood Cliffs

Bynum WF (1975) The great chain of being after forty years: an appraisal. Hist Sci 13(1):1–28

Cann RL, Stoneking M, Wilson AC (1987) Mitochondrial DNA and human evolution. Nature 325(6099):31–36

Caspari R (2010) Deconstructing race: race, racial thinking, geographic variation, and implications for biological anthropology. In: Larsen C (ed) A companion to biological anthropology. Wiley-Blackwell, Chichester, pp 104–122

Cavalli-Sforza LL, Edwards AWF (1965) Analysis of human evolution. In: Geerts SJ (ed) Genetics today. Proceedings of the XI. International congress of genetics, The Hague, The Netherlands, September 1963, vol. 3, Oxford, pp 923–933

Cavalli-Sforza LL, Menozzi P, Piazza A (1994) The history and geography of human genes. Princeton University Press, Princeton

Coon CS (1962) The origin of races. Alfred A. Knopf, New York

Darwin C (1859) On the origin of species by means of natural selection, or the preservation of favoured races in the struggle for life. John Murray, London

Darwin C (1871a) The descent of man, and selection in relation to sex, vol 1, 1st edn. John Murray, London

Darwin C (1871b) The descent of man, and selection in relation to sex, vol 2, 1st edn. John Murray, London

Darwin C (1998 [1872]) The expression of the emotions in man and animals. With an introduction, afterword and commentaries by Paul Elkman, 3rd ed. Oxford University Press, New York

Bory de Saint-Vincent JBGM (1827) L'Homme (Homo): Essai zoologique sur le genre humain, 2 vols, 2nd edn. Rey et Gravier, Paris

de Mortillet G (1883) Le Préhistorique: Antiquité de l'homme. C. Reinwald, Paris

Delisle RG (2007) Debating humankind's place in nature 1860–2000: the nature of paleoanthropology. Pearson, Upper Saddle River

Desmond AJ, Moore J (2009) Darwin's sacred cause: how a hatred of slavery shaped Darwin's views on human evolution. Houghton Mifflin Harcourt, Boston

Diekmann A (1992) Klassifikation, System, 'scala naturae': Das Ordnen der Objekte in Naturwissenschaft und Pharmazie zwischen 1700 und 1850. Wissenschaftliche Verlagsgesellschaft, Stuttgart

Dobzhansky T (1944) On species and races of living and fossil man. Am J Phys Anthropol 2(3): 251–265

Gannett L, Griesemer JR (2004) The ABO blood groups: mapping the history and geography of genes in *Homo sapiens*. In: Rheinberger H-J, Gaudillière J-P (eds) Classical genetic research and its legacy: the mapping cultures of twentieth century genetics. Routledge, London, pp 119–172

Gould, Stephen J. (1996 [1981]) The Mismeasure of man, 2nd ed. W. W. Norton, New York

Green RE, Krause J, Briggs AW et al (2010) A draft sequence of the Neandertal genome. Science 328(5979):710–722

Gregory WK (1930) A critique of Professor Osborn's theory of human origins. Am J Phys Anthropol 14(2):133–163

Gundling T (2005) First in line: tracing our ape ancestry. Yale University Press, New Haven

Stocking GW Jr (1968) Race, culture, and evolution: essays in the history of anthropology. Free Press, New York

Haeckel E (1866) Generelle Morphologie der Organismen, vol 2. Walter de Gruyter, Berlin

Haeckel E (1868) Natürliche Schöpfungs-Geschichte: Gemeinverständliche wissenschaftliche Vorträge über die Entwickelungs-Lehre im Allgemeinen und diejenige von Darwin, Goethe und Lamarck im Besonderen, über die Anwendung derselben auf den Ursprung des Menschen und andere damit zusammenhängende Grundfragen der Naturwissenschaft. Georg Reimer, Berlin

Haeckel E (1874) Anthropogenie oder Entwickelungsgeschichte des Menschen: Gemeinverständliche wissenschaftliche Vorträge über die Grundzüge der menschlichen Keimes- und Stammes-Geschichte. Engelmann, Leipzig

Haeckel E (1898 [1868]) Natürliche Schöpfungs-Geschichte: Gemeinverständliche wissenschaftliche Vorträge über die Entwickelungs-Lehre im Allgemeinen und diejenige von Darwin, Goethe und Lamarck im Besonderen [...] 9th ed. Georg Reimer, Berlin

Hellström P (2019) Trees of knowledge: science and the shape of genealogy. Uppsala Universitet, Uppsala

Hirschfeld L, Hirschfeld H (1919) Serological differences between the blood of different races: the results of researches on the Macedonian front. Lancet 194(5016):675–679

Horst M (1913) Die 'natürlichen' Grundstämme der Menschheit, Hildburghausen

Howells WW (1976) Explaining modern man: evolutionists versus migrationists. J Hum Evol 5(5): 477–496

Johanson D, Edey M (1981) Lucy: the beginnings of humankind: the dramatic discovery of our oldest human ancestor—and the controversial change it makes in our view of human origins. Simon and Schuster, New York

Keith A (1948) A new theory of human evolution. Watts, London

Keith A (1950) An autobiography. Watts, London

Klaatsch H (1920) Der Werdegang der Menschheit und die Entstehung der Kultur. Bong, Berlin

La Peyrère I de (1655) Prae-Adamitae: sive exercitatio super versibus duodecimo: decimotertio et decimoquarto capitis quinti Epistolae D. Pauli ad Romanos: quibus inducuntur primi homines ante Adamum conditi. Louis & Daniel Elzevier

Lamarck JB (1809) Philosophie zoologique [...], vol 2. Duminil-Lesueur, Paris

Lan T, Lindqvist C (2019) Technical advances and challenges in genome-scale analysis of ancient DNA. In: Lindqvist C, Rajora OP (eds) Paleogenomics: genome-scale analysis of ancient DNA. Springer, Cham, pp 3–29

Le Gros Clark W (1955) The fossil evidence for human evolution: an introduction to the study of paleoanthropology. The University of Chicago Press, Chicago

Leonardi M, Librado P, Sarkissian CD et al (2017) Evolutionary patterns and processes: lessons from ancient DNA. Syst Biol 66(1):e1–e29

Lovejoy AO (1964) The great chain of being: a study of the history of an idea. Harvard University Press, Cambridge, MA

MacEachern S (2000) Genes, tribes, and African history. Curr Anthropol 41(3):357–384

Mayr E (1950) Taxonomic categories in fossil hominids. Cold Spring Harb Symp Quant Biol 15(1): 109–118

Mourant AE (1954) The distribution of the human blood groups. Blackwell, Oxford

Mourant AE (1958) The ABO blood groups: comprehensive tables and maps of world distribution. Blackwell, Oxford

Mourant AE (1983) Blood relations: blood groups and anthropology. Oxford University Press, Oxford

Nott JC, Gliddon GR (1854) Types of mankind: or ethnological researches, based upon the ancient monuments, paintings, sculptures, and crania of races, and upon their natural, geographical, philological and biblical history, illustrated by selections from the inedited papers of Samuel George Morton and by additional contributions from L. Agassiz, W. Usher, and H. S. Patterson. J. B. Lippincott, Grambo and Co, Philadelphia

Nuttall GHF (1904) Blood immunity and blood relationship: a demonstration of certain blood-relationships amongst animals by means of the precipitin test for blood. Cambridge University Press, Cambridge

Oakley KP (1963) Dating skeletal material. Science 140(3566):488

Oakley KP (1980) Relative dating of the fossil hominids of Europe. Bull Brit Mus Nat Hist (Geology) 34(1):1–63

Podgorny I (2015) Human origins in the New World? Florentino Ameghino and the emergence of prehistoric archaeology in the Americas (1875–1912). PaleoAmerica 1(1):68–80. https://doi.org/10.1179/2055556314Z.0000000008

Pritchard JK, Stephens M, Donnelly P (2000) Inference of population structure using multilocus genotype data. Genetics 155(2):945–959

Reich D, Green RE, Kircher M et al (2010) Genetic history of an archaic hominin group from Denisova Cave in Siberia. Nature 468(7327):1053–1060

Resendez SD, Bradley JR, Xu D et al (2019) Structural variants in ancient genomes. In: Lindqvist C, Rajora OP (eds) Paleogenomics: genome-scale analysis of ancient DNA. Springer, Cham, pp 375–391

Rosenberg NA (2004) Distruct: a program for the graphical display of population structure. Mol Ecol Notes 4(1):137–138

Scally A (2021) Complexity in human ancestral demography. J Anthropol Sci 99:179–182. https://doi.org/10.4436/JASS.99009

Schaaffhausen H (1872 [1869]) Ueber die Methode der vorgeschichtlichen Forschung. Arch Anthropol 5: 113–128

Schneider WH (1996) The history of research on blood group genetics: initial discovery and diffusion. Hist Philos Life Sci 18(3):277–303

Schwalbe G (1906) Studien zur Vorgeschichte des Menschen. E. Schweizerbartsche Verlagsbuchhandlung, Stuttgart

Schwalbe G (1914) Kritische Besprechung von Boule's Werk: 'L'homme fossile de La Chapelle-aux-Saints' mit eigenen Untersuchungen. Z Morphol Anthropol 16(3):527–610

Sera GL (1917) La testimonianza dei fossili di antropomorfi per la questione dell'origine dell'uomo. Atti Soc Ital Sci Nat Mus Civ Stor Nat Milano 56:25–156

Sergi G (1908) Europa: L'origine dei popoli Europei e loro relazioni coi popoli d'Africa, d'Asia e d'Oceania. Bocca, Milano

Simpson GG (1944) Tempo and mode in evolution. Columbia University Press, New York

Smith GE (1924) Evolution of man: essays. Humphrey Milford/Oxford University Press, London

Smith GE (1929) Human history. W. W. Norton, New York

Smith GE (1933) The diffusion of culture. Watts, London

Sommer M (2007) Bones and Ochre: the curious afterlife of the red lady of Paviland. Harvard University Press, Cambridge, MA

Sommer M (2008) History in the gene: negotiations between molecular and organismal anthropology. J Hist Biol 42(3):473–528

Sommer M (2014) Biology as a technology of social justice in interwar Britain: arguments from evolutionary history, heredity, and human diversity. Sci Technol Hum Values 39(4):560–585. https://doi.org/10.1177/0162243913520330

Sommer M (2015a) Evolutionäre Anhtropologie zur Einführung. Junius, Hamburg

Sommer M (2015b) Population-genetic trees, maps and narratives of the great human diasporas. Hist Hum Sci 28(5):108–145. https://doi.org/10.1177/0952695115573032

Sommer M (2016) History within: the science, culture, and politics of bones, organisms, and molecules. The University of Chicago Press, Chicago

Sommer M (2021) The meaning of absence: the primate tree that did not make it into Darwin's *The Descent of Man*. Br J Hist Sci Themes 6(1):45–61

Sommer M (2024) The diagrammatics of 'race': visualizing human relatedness in the history of physical, evolutionary, and genetic anthropology, ca. 1770–2020. Open Book Publishers, Cambridge. https://doi.org/10.11647/OBP.0396

Sommer M, Amstutz R (2024) Diagrams of human genetic kinship and diversity: from the tree to the mosaic and the network? In: Strand D, Källen A, Mulcare C (eds) Critical perspectives on ancient DNA. MIT Press, Cambridge MA, pp 25–52

Sommer M, Arni C, Müller-Wille S, Teuscher S (2024) In the shadow of the tree: the diagrammatics of relatedness in genealogy, anthropology, and genetics as epistemic, cultural, and political practice. Hist Hum Sci 37(3–4):3–15. https://doi.org/10.1177/09526951241261837

Spencer H (1864) The principles of biology, vol 1. Williams and Norgate, London

Spencer H (1867) The principles of biology, vol 2. Williams and Norgate, London

Stone L, Lurquin PF (2005) A genetic and cultural odyssey: the life and work of L. Luca Cavalli-Sforza. Columbia University Press, New York

Stoneking M, Cann RL (1989) African origin of human mitochondrial DNA. In: Stringer C, Mellars P (eds) The human revolution: behavioural and biological perspectives on the origins of modern humans. Edinburgh University Press, Edinburgh, pp 17–30

Stringer C (2012) What makes a modern human. Nature 485(7396):33–35. https://doi.org/10.1038/485033a

Tang H, Peng J, Wang P et al (2005) Estimation of individual admixture: analytical and study design considerations. Genet Epidemiol 28(4):289–301

Tattersall I (1995) The fossil trail: how we know what we think we know about human evolution. Oxford University Press, New York

Tattersall I, Eldredge N (1977) Fact, theory, and fantasy in human paleontology. Am Sci 65(2): 204–211

Thienemann A (1910) Die Stufenfolge der Dinge, der Versuch eines natürlichen Systems der Naturkörper aus dem achtzehnten Jahrhundert. Zool Ann 3:185–274

Thorne AG, Wolpoff MH (1992) The multiregional evolution of humans. Sci Am 266(4):76–83

Thorne AG, Wolpoff MH (2003 [1992]) The multiregional evolution of humans. Sci Am. Special Edition 'New look at human evolution' 13(2):46–53

Tylor EB (1881) Anthropology: an introduction to the study of man and civilization. Macmillan, London

Vicente M, Schlebusch CM (2020) African population history: an ancient DNA perspective. Curr Opin Genet Dev 62(1):8–15

Vogt C (1863a) Vorlesungen über den Menschen, seine Stellung in der Schöpfung und in der Geschichte der Erde, vol 1. De Gruyter, Berlin

Vogt C (1863b) Vorlesungen über den Menschen, seine Stellung in der Schöpfung und in der Geschichte der Erde, vol 2. J. Ricker'sche Buchhandlung, Giessen

von Baer KE (1828) Über die Entwicklungsgeschichte der Thiere: Beobachtungen und Reflexion, vol 1. Bei den Gebrüder Bornträger, Königsberg

von Baer KE (1837) Über die Entwicklungsgeschichte der Thiere: Beobachtungen und Reflexion, vol 2. Bei den Gebrüder Bornträger, Königsberg

Weidenreich F (1946) Apes, giants, and man. The University of Chicago Press, Chicago

Weiner JS (1955) The Piltdown forgery. Oxford University Press, London

Weinert H (1932) Urspung der Menschheit: Über den engeren Anschluss des Menschengeschlechts an die Menschenaffen. Ferdinand Enke, Stuttgart

White C (1799) An account of the regular gradation in man, and in different animals and vegetables; and from the former to the latter. Printed for C. Dilly, in the Poultry, London

Willis D (1989) The hominid gang: behind the scenes in the search for human origins. Viking, New York

Willis D (1992) The Leakey family: leaders in the search for human origins. Facts on File, New York

Wolpoff MH (1989) Multiregional evolution: the fossil alternative to Eden. In: Mellars P, Stringer C (eds) The human revolution: behavioural and biological perspectives on the origins of modern humans. Edinburgh University Press, Edinburgh, pp 62–108
Wood Jones F (1919) The origin of man. In: Dendy A (ed) Animal life and human progress. Constable, London, pp 99–132
Wood Jones F (1929) Man's place among the mammals. E. Arnold, London

Proximate–Ultimate Causation in Human Evolution? Implications from Archeology

9

Ivan Colagè and Francesco d'Errico

Abstract

The Modern Synthesis in evolutionary biology cemented a gene-centric view of variation production, transmission and selection; and Ernst Mayr's proximate–ultimate causation distinction reinforced the claim that behaviour is evolutionarily secondary. Recent developments, encapsulated in the Extended Evolutionary Synthesis (EES), challenge this view, integrating behavioural innovation, social learning, niche construction and non-genetic inheritance as evolutionary forces in their own right. In this chapter, we argue that culture should be regarded as a primary cause of human evolution and not merely as an epiphenomenon of genetic evolution. Drawing on critiques of the proximate–ultimate divide, and on archeological reconstructions of long-term cultural trajectories, we propose a "culture-driven evolutionary" framework. This model implies that cultural innovations can originate at the behavioural level, spread and stabilize through non-genetic transmission channels and alter both cognitive capacities and neural substrates—via mechanisms such as cultural exaptation and cultural neural reuse—without requiring prior, causally related genetic change. Archeological evidence demonstrates that such processes have shaped human life forms across different *Homo* populations, independently of speciation events or key genetic mutations. Cultural innovations emerge primarily through recombination, repurposing and cumulative integration of previously devised cultural traits at

I. Colagè (✉)
Pontifical University of the Holy Cross, Rome, Italy
e-mail: i.colage@pusc.it

F. d'Errico
PACEA, University of Bordeaux/CNRS/MCC, Pessac, France

Centre for Early Sapiens Behaviour, Department of Archaeology, History, Cultural Studies and Religion, University of Bergen, Bergen, Norway

M. Lequin, J. M. Rodríguez Caso (eds.), *The Field of Human Evolution: Critical Perspectives from History and Epistemology*, Evolutionary Biology – New Perspectives on Its Development 11, https://doi.org/10.1007/978-3-032-13925-2_9

the genuinely behavioural-cultural level. We thus advance a circular, rather than linear, conception of causation in human evolution where genetic, behavioural and environmental factors mutually influence each other. By foregrounding the autonomy of cultural dynamics in evolutionary explanations, our approach has implications for the epistemology of human evolution, challenges reductionist assumptions and advocates for integrative frameworks bridging biology, archeology and cognitive science.

Keywords

Behavioural plasticity · Cognitive archeology · Cultural evolution · Extended evolutionary synthesis · Innovation · Non-genetic inheritance · Niche construction · Proximate–ultimate causation · Social transmission

9.1 Introduction

In this chapter, we argue that culture is a primary cause of human evolution, not only a factor relevant solely in its interaction with genetic dynamics. Beyond its significance for understanding the human past, this thesis has profound epistemological implications for the field of human evolution. We approach this argument from two perspectives. First, we consider the expanding framework of evolutionary theory, particularly the Extended Evolutionary Synthesis (EES) (Laland et al. 2015; Müller 2017; see also Kissel and Fuentes 2021). The EES implies that behaviour, as a key component of an organism's phenotype, is a fundamental drive of evolution. Since culture is a core behavioural trait of both past and present humans—being also relevant, as we will briefly see, in other animals—it plays a central role in evolutionary processes (see, e.g. overviews in: Read 2012; Tomasello 2014; Henrich 2016; Laland 2017; Fuentes 2017). Second, we adopt a prehistoric archeological perspective. By analysing past material culture, prehistoric archeology provides the most direct insights into the cultural behaviour of our ancestors and how it evolved.

Evolutionary biology begun in the nineteenth century with a core question: how does life *change*? The task soon became that of ascertaining the *mechanisms* of that "change", that is, the processes underlying the emergence, spreading and stabilization of biological *novelty*. This brought forth the query for the *causes* of evolution.

At the beginning of the nineteenth century, Jean-Baptiste de Lamarck elaborated the first accomplished theory of biological evolution (de Lamarck 1809). This is not the place to review and comment his theory at large. Interestingly, however, Lamarck stressed that biological novelty ultimately comes from what he considered to be the third fundamental law to understand life, known as the "law of use and disuse" (1809, vol. I, Ch. 7). The law states that the *use* an organism makes of an organ determines its shaping and development. Lamarck was also convinced that this same law is responsible for the appearance of new organs and, more generally, of the variety observed in the animal kingdom. It is often stressed that, according to Lamarck, the differential use of organs is due to the varied and changing environments organisms encounter during their life. This is, of course, a key tenet in Lamarck's thought, one that will shape all ensuing developments in evolutionary biology—Darwin's work included. What we would like to stress here is that, with his third law, Lamarck de facto considered animal *behaviour* (the use or disuse of organs) as the main source of biological variability and novelty. Though the expression "use or disuse of organs" may at first sight appear not to capture the variety and complexity of animal behaviour—and much more so after more than two centuries of ethological research—Lamarck's statement was intended to be quite general in scope, specifically concerning animal behaviour at large ("the habits of animals and their mode of life") and even the "faculties" behind it.[1]

With Charles Darwin's work, 50 years later (Darwin 1859), the theory of natural selection was established in evolutionary biology. The success of the theory was rapid and impressive. As it is well known, it states that the interaction between an organism with its traits and the characteristics of the environment in which it lives determines the organism's level of fit. Organisms with higher fitness will cope more effectively with the environment, will live better and longer and will reproduce more than organisms with lower fitness. Hence, because of characters' heredity, the traits conferring higher fitness—relative to a given environment—will spread within a population in the ensuing generations. The relevance given to the environment in the process is common to Lamarck's and Darwin's positions. The most stressed difference between the two evolutionary biologists is about the inheritance of acquired characters, which was admitted by Lamarck and excluded by Darwin. However, there is a second important difference—linked with the one just mentioned—that we would like to emphasize. Both Lamarck and Darwin considered variation a key point in understanding life's evolution; both documented it to a great deal in their works. Darwin also gathered an impressive amount of observations about variation in domestic species, not to mention what he learnt journeying on the *Beagle*. However, Darwin thought that, though being such an evident fact that could not be denied, variation was produced by as yet unknown laws, and never proposed developed hypotheses for the very origin of variation—focusing much more on the laws of inheritance and, of course, on the mechanism of variation selection.[2] Lamarck,

[1] "it is not the shape either of the body or its parts which gives rise to *the habits of animals and their mode of life*; but [...] it is, on the contrary, the habits, mode of life and all the other influences of the environment which have in course of time built up the shape of the body and of the parts of animals. With new shapes, new *faculties* have been acquired, and little by little nature has succeeded in fashioning animals such as we actually see them." (de Lamarck 1809, end of Ch. VII, emphases added)

[2] Darwin did not entirely overlook the relevance of inherited habits, as it is clear from the Preface to the second edition of *The Descent of Man*: "I may take this opportunity of remarking that my critics frequently assume that I attribute all changes of corporeal structure and mental power exclusively to natural selection of such variation as are often called spontaneous; whereas, even in the first edition of the Origin of Species, I distinctly stated that great weight must be attributed to the inherited effects of use and disuse, with respect both to the body and the mind" (Darwin 1874: 3–4). However, he mainly considered it about humans and the relationships between animal and human evolution (see, e.g. Richerson and Boyd 2010). Moreover, Darwin's emphasis essentially

instead, as we have seen, regarded variation as ultimately stemming from organisms' own behaviour in different and changing environments.

Darwin's theory of natural selection, admirably explaining how organisms differentiate in time and across environments, could not explain the *origin* of biological novelty. This limit was pointed out already in Darwin's time, for example by St. George Mivart (1871). The famous sentence "natural selection may explain the survival of the fittest, but it cannot explain the arrival of the fittest" is found at the very end of Hugo de Vries' *Species and Varieties: Their Origin by Mutation* (de Vries 1904: 825–26), where he develops one of the first genetic theories for the evolution of species and, indeed, for the origin of biological novelty by mutations (as the work's subtitle emphasizes). The same year, William Bateson (1904: 528) stated that "selection is a true phenomenon; but its function is to select, not to create".

By 1904, Gregor Mendel's work on the hybridization of plants—which he presented in 1865 at the Brünn (now Brno) Natural History Society and published the following year in the proceedings of the Society—was known in the biological community, especially because William Bateson translated Mendel's report into English and published it on the *Journal of the Royal Horticultural Society* in 1901 (Mendel 1901). At this point, the process bringing together Darwin's natural selection and Mendel's genetics began. The so-called Modern Synthesis in evolutionary biology stems from these developments; it was articulated in all its fundamental tenets by the 1940s (Huxley 1942), and the discovery of the DNA structure (Watson and Crick 1953) established it as an accomplished scientific theory a few years later. It was only with the development of molecular biology and genetics, by the mid-twentieth century, that a clear mechanistic cause for variation and the origin of biological novelty was ascertained: genetic mutation and recombination.

With the Modern Synthesis, biological evolution was still seen, quite like Darwin's original insights, as a three-step process of variation production, variation transmission and variation selection. Now, however, the three steps are given a fairly genetic basis. Production of biological novelty comes from genetic mutations, which provide "material" for natural selection to work on. Variation transmission is essentially genetic inheritance, whereby (only) genetic information is transmitted from parents to offspring. Variation selection is Darwinian natural selection. Of course, natural selection does not act directly on genetic variation; it acts on real organisms, with their traits, setting the level of fitness relative to the environment where they live. However, as for the Modern Synthesis organisms' traits depend on their genomes, which are thought to store a developmental programme on the basis of which organisms come to have the traits they exhibit, natural selection ultimately selects for organisms endowed with genomes bringing about the most adaptive traits in a given environment, and selects against organisms with genomes providing less adaptive or maladaptive traits. This is, in a nutshell, a grasp of the gene-centrism

remained on the processes of variation inheritance and selection, whereas here and in the remainder of this section we are mainly dealing with the origin of variation.

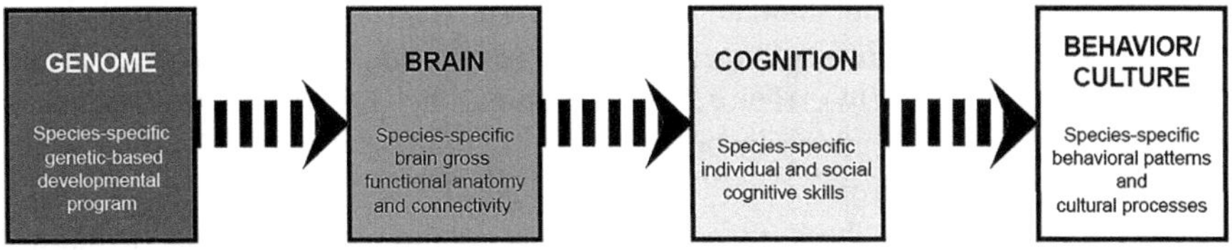

Fig. 9.1 From genes to behaviour. Schematic representation of the gene-centrism characterizing the Modern Synthesis in evolutionary biology. (Modified after Figure 1 in Colagè and d'Errico 2020)

characterizing the Modern Synthesis, which is so fundamental as to imply that biological evolution as such is defined as *change in gene frequency* within populations of organisms.

On this background, Ernest Mayr (1961) stated the distinction between proximate and ultimate causation in biology and made of it a cornerstone of evolutionary thinking. Proximate causes address "how-questions" and concern organisms' actual functioning: their physiology and *behaviour*. According to the distinction, proximate causes tell nothing about organisms' evolution, which is the domain of ultimate causes. Ultimate causes answer "why-questions": they focus on the historical reasons why organisms come to have certain traits. In the Modern Synthesis framework, these reasons amount essentially to the history of natural selection acting on genetic variability within populations of organisms for many generations. As a matter of fact, such a view excludes from the range of evolutionary (ultimate) causes all what pertains to organism actual development, phenotype and behaviour, not because the latter are irrelevant for natural selection (they are indeed what natural selection primarily works on) but because they are regarded as direct consequences of the genetic programme stored in the DNA constituting organisms' genome (Fig. 9.1). Only ultimate causes are considered relevant for evolution as they are "responsible for the shaping of the genetic program" (Mayr 1993: 94).

Such an overall understanding of biological evolution has been applied to human evolution as well (see, e.g. Shea 2011). As a consequence, the behaviour of human ancestors has been sometimes disregarded as a relevant cause in human evolution; specifically, it was not considered as an ultimate evolutionary cause. Coherently, this was applied to culture as well. As above, this does not amount to say the human ancestors' behaviour and culture are irrelevant to human evolution, but it does imply that they have been considered just as a consequence of the evolving "genetic programme" along the human lineage. And this, to the point that the emergence of modern cognition, behaviour and culture (i.e. the life form typical of modern humans, characterized by complex material and symbolic culture, fully developed articulate language, ornaments, art) was believed to have emerged either as a consequence of the speciation event leading to our species *H. sapiens* ca. 250,000 years ago in Africa (e.g. McBrearty and Brooks 2000; Mithen 2005; Mellars and Stringer 1989; Neubauer et al. 2018), or to key mutations within our species some 50–100,000 years ago (e.g. Coolidge and Wynn 2007, 2017; Wynn et al. 2016; Klein 1989, 2000; Tattersall 1995). Arguments in the same line have been proposed also about human brain evolution (e.g. Herculano-Houzel 2017; Geschwind and Rakic 2013; Somel et al. 2013) and language origin (e.g. Fisher 2017; Bolhuis et al. 2014). Hence, modern human behaviour was regarded as a consequence of modifications in our genetic programme, which all the novelties would *ultimately* stem from.

In the last decades, both the overall understanding of biological evolution sketched above and the just mentioned view on human evolution have been challenged from many standpoints. As we will discuss in the next section, the raise of the so-called Extended Evolutionary Synthesis (Laland et al. 2015, 2024; Müller 2017) overcomes the Modern Synthesis assumption that gene change is the only source of biological novelty. Mayr's distinction between ultimate and proximate causation has been directly criticized on the basis of many empirical and conceptual developments in the life sciences (Laland et al. 2011a, 2013; Mesoudi et al. 2013). Scenarios tracing back human behavioural and cultural modernity to speciation or key mutation has been challenged not only as a consequence of the just mentioned developments in the life sciences at large, but also thanks to advancements in archeology ascertaining the pattern of emergence of cultural novelties along our lineage (e.g. Colagè and d'Errico 2020; Scerri and Will 2023).

9.2 The Extended Evolutionary Synthesis and Behaviour

As we have mentioned, evolutionary biology has been developed, since its beginnings in the nineteenth century, as a three-step process of variation production, variation transmission and variation selection. We have seen that the Modern Synthesis regarded all these three steps as essentially grounded in genetic processes. One way to capture the novelty of the Extended Evolutionary Synthesis (EES) in a nutshell is to say that it overcomes the gene-centrism of the Modern Synthesis. Somewhat more specifically, in the EES framework, none of the three steps are regarded as hinged solely on genes.

As we have seen at the end of the previous section, the EES acknowledges a number of sources of biological novelty beyond genetic mutation and change. Novelty can emerge at the developmental (Müller 2007; Peterson and Müller 2016) and phenotypic (West-Eberhard 2003, 2005; Moczek et al. 2011) level, as a consequence of niche construction (Odling-Smee et al. 2003; Odling-Smee 2024), in the form of behavioural innovation (Bateson 2004, 2014), and as an outcome of social learning (Hoppitt and Laland 2013; Whiten 2017) or of varied social learning strategies (Aoki and Mesoudi 2015; Mesoudi et al. 2016; Kendal et al. 2018).

The EES also acknowledges a number of intergenerational transmission channels beyond genetic inheritance (Jablonka and Lamb 2006). Transgenerational epigenetic inheritance (whereby molecular *non*-genetic signals influencing DNA expression can pass from parents to offspring) is nowadays acknowledged, in both plants and animals, as a relevant channel of inheritance not requiring or implying any genetic mutation (e.g. Burggren 2014; Fitz-James and Cavalli 2022). In animals, social learning constitutes a powerful means of inheritance (Galef and Whiten 2017;

Whiten 2019, 2021). Importantly, though young animals often learn from their parents, social learning works also among organisms not necessarily bearing kinship relationships within the same group or population. As an aspect of niche construction, ecological inheritance (sometimes also referred to as "parental effect") can play a key role in transgenerational transmission of information both in plants and animals (Odling-Smee 1988; Badyaev and Uller 2009; Uller 2012). Environmental modifications brought about by organisms of one generation in a group or population sometimes last longer than the lifespan of their makers; such modifications may affect the development of future generations as well as their learning processes. Hence, relevant biological information is not transmitted only through genes.

Niche construction is considered as one of EES's main pillars. Its conception can be traced back to the work of Richard Lewontin (1983). It is generally defined as the ability of organisms to actively modify environmental states and features so as to affect the selection pressures that they, their descendants and members of other species in the same environment experience (Odling-Smee et al. 1996). Through niche construction, "the organism influences its own evolution, by being both the object of natural selection and the creator of the conditions of that selection" (Levins and Lewontin 1985: 106). Niche construction being a core tenet of EES also implies, therefore, that natural selection is no longer considered as the sole process directing biological evolution and establishing organisms' adaptation. Organisms' own activities are part and parcel of the "selective environment". It could be said that not only the environment selects organisms, but organisms themselves select the environment, either modifying it (perturbational niche construction) or displacing themselves into new environments (relocational niche construction), thus giving raise to new selective dynamics (inceptive niche construction) and/or buffering existing ones (counteractive niche construction). Though virtually all aspects of organisms' activity can have niche-constructing outcomes—including the mere presence of an organism with its physiology and metabolic activity—the behavioural component is overwhelmingly important (see, e.g. Odling-Smee et al. 2003, Ch. 2).

Comprehensively accounting for all the novel data, conceptual developments, theoretical implications and empirical predictions of the EES is beyond the scope of this chapter. However, it is worth expanding on topics whose consequences are momentous in human evolution, such as: behavioural innovation, social learning, environmental induction of (phenotypic and) behavioural change, ecological inheritance and niche construction.

Animals invent and innovate. Above all, they invent new behaviours. Behavioural innovation can be prompted by environmental change, but it can also seemingly be an endogenous process in some individual organism (Duckworth 2009). There are striking and well-documented cases. Blue tits (*Cyanistes caeruleus*) around the year 1950, in Scotland, begun to open milk bottles (by beak-puncturing the bottles' foil caps) left by milkmen on rural Scottish houses' thresholds and drink the milk (Fisher and Hinde 1949; Hinde and Fisher 1951). In the year 2000, Japanese ravens (*Corvus corone*) developed the skill of making nuts cracked by automobiles, and of using traffic lights to recover the edible parts safely (Nihei and Higochi 2002). Importantly, these two behaviours spread via social learning in the local populations in Scotland and Japan, respectively. A macaque monkey called Imo (of the species *Macaca fuscata*) excogitated the so-called sweet-potato-washing behaviour in 1953 on the Japanese islet of Koshima (Kawai 1965). The behaviour consists in washing sweet potatoes to remove sand from them. The behaviour spread consistently in Imo's group, especially among the females, and evolved in time: at first, fresh water was used to wash potatoes, but then monkeys preferred salty sea water, probably because it makes potatoes tastier (Hirata et al. 2001). Some years later, Imo also invented a new, related behaviour: the wheat-placer-mining behaviour (Hirata et al. 2001). It consists in gathering several wheat grains from the sand and throwing them into the water to recover the floating grains without the sand, which sinks in the water. Though these new behaviours may also depend on "peaceful" interactions with humans on Koshima (Kawai 1965), this does not detract from the point that pioneer individuals in animal groups can introduce new strategies at the behavioural level. Innovation in macaques' foraging strategies is, indeed, increasingly documented (e.g. Pal et al. 2018).

A very interesting case is that of the terrestrial hermit crabs (*Coenobita compressus*), which invented a "market of shells" (Laidre 2012, 2019). This deserves a brief description. These crabs need a shell to survive and reproduce, but they are unable to produce one; thus, they use gastropod shells found in the environment. Importantly, these shells are extensively modified by the crabs to be lighter, thinner and more sizeable. Individuals have to change shells as they grow, so that the abandoned shells, already modified, can be reused by younger crabs. Field observation and experimental research showed that terrestrial hermit crabs developed a sort of "market of shells" in which specimens of different ages and sizes gather and line up from the largest to the smallest. As soon as the first finds a new suitable shell and abandons the one it is "wearing", a cascade of shell exchanges is triggered such that each crab in the line gets the shell of the specimen ahead of it.

Now, all these behaviours have been initiated spontaneously by some pioneer at the genuinely behavioural level, and have subsequently spread and stabilized via social learning. These inventions de facto changed the life form of the populations which acquired them. In the case of the hermit crabs, the new behaviour profoundly affected the life form of these organisms, actually turning a solitary species into a social one. Moreover, probably, none of these behaviours has brought about changes in gene frequency within the populations devising them; and, very likely, no gene change triggered the new behaviour. From the standpoint of the Modern Synthesis, these processes would be considered as proximate causes, not relevant for biological evolution. The EES, on the contrary, consider them fully relevant for evolution, even if no genetic effect has been produced (as yet).

A key point for considering behavioural innovation as part and parcel of biological evolution is that the emerging behaviours are transmitted via social learning so as to have transgenerational consequences. In other words, once a behaviour has been invented and has spread throughout a population and across generations via social learning, it has actually changed the life form of that population independently of its possible effects on allele distribution at the genetic level (see also Laland 2015).

Social learning in animals is another key issue for the EES, also contributing to overcome the distinction between proximate and ultimate causation in evolution. Its scope in the animal kingdom is wide, encompassing primates, cetaceans, fish, songbirds and insects; it has been documented in several contexts, such as foraging and hunting strategies, prey selection, predator recognition, breeding or egg-laying locations, migratory routes, tool use, vocal dialects and social customs (Whiten 2019, 2021). Interestingly, the distribution of social learning across animal species does not seem to reflect strict phylogenetic links (e.g. Laland et al. 2011b). It shows, for example, relevant differences in phylogenetically related species (e.g. Webster et al. 2019), as well as convergent evolution in distant taxa (Shettleworth 2001; Emery and Clayton 2004). Besides social learning overall capabilities, also individual and species variation in social learning strategies (i.e. the biases and "heuristics" influencing an organism's choices about what, when and whom to copy) unveil patterns not easily reducible to phylogeny (e.g. Mesoudi et al. 2016; Kendal et al. 2018). Moreover, there is evidence suggesting that social learning itself may be socially learnt, so that variation in social learning and social learning strategies (SLSs) may stem from the behavioural level both across and within populations (Mesoudi et al. 2016). Hence, social learning is responsible for transgenerational transmission of novel behaviours to represent a potential ultimate evolutionary cause, and can in its turn evolve behaviourally.

We have seen that animals frequently develop new behaviours. A subset of such innovations can be induced by the environment (Badyaev 2005, 2009; Lea et al. 2020). Some of the abovementioned examples of behavioural innovations (mainly blue tits' bottle-opening and ravens' use of traffic lights to feed on nuts) may have indeed been induced by the specific environments created by human communities in those areas. Now, organisms' behaviour can greatly affect the environment in which they live (niche construction) and such modifications can affect several subsequent generations (ecological inheritance). The case of the hermit crabs described above is highly significant here, as their own niche-constructing behaviour of modifying shells has induced a novel *social* behaviour. The point is that such modified environments can, in their turn, prompt further behavioural innovations, thus giving rise to a cascade of behavioural innovation events. One could even speak, in these cases, of "cumulative *behavioural* evolution" (Fig. 9.2). Incidentally, this is a clear instance of *reciprocal causation* in biological evolution, which constitutes another argument against the sharp separation between ultimate and proximate causes in biology (Laland et al. 2011a). We will see that human cultural evolution provides further arguments in favour of this view.

9.3 From Gene-Culture Co-evolution to Culture-Driven Evolution

The previous section actually presents the first standpoint according to which we intend to argue for our thesis (i.e. that culture is a primary cause of human evolution per se and not only in its interactions with genetic dynamics). Indeed, though

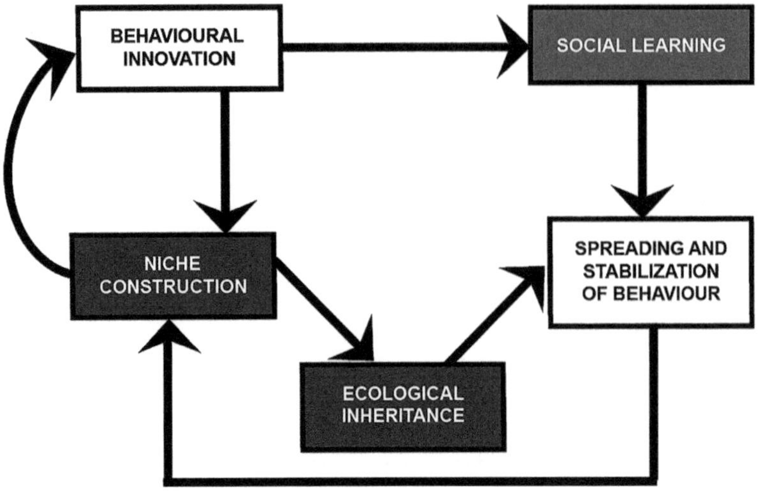

Fig. 9.2 Cumulative behavioural evolution. Schematic representation of cumulative *behavioural* evolution, highlighting the roles of niche construction, social learning and ecological inheritance

understanding culture just as behaviour might be undue, it clearly unfolds through a series of complex behaviours put in practice by members of communities. Hence, maintaining that behaviour is a leading force of evolution per se, and that it affects all the steps in the evolutionary process (variation production, transmission and selection), dovetails with and strengthens the idea that culture can play the same role in human evolution.

Before entering the second standpoint—the archeological one—some broader considerations about the relationships between cultural, biological and genetic evolution are needed. To begin with, even before the EES started to be explicitly elaborated (Boyd and Richerson 2024), pioneering works by Luigi Cavalli-Sforza, Marcus Feldman, Peter Richerson and Robert Boyd (Cavalli-Sforza and Feldman 1973; Feldman and Cavalli-Sforza 1975, 1976; Richerson and Boyd 1978; Boyd and Richerson 1985) brought forth the idea that culture is a relevant evolutionary force able to affect genetic evolution in a population. This paved the way for the development of so-called gene-culture co-evolution theory (Feldman and Laland 1996; Richerson and Boyd 2005; Laland et al. 2010; Gintis 2011). Since the first proposals and the early developments, both empirical and theoretical research on the topic has expanded (e.g. Kasser et al. 2024), and gene-culture co-evolution theory is currently applied also to non-human animals (Whitehead et al. 2019).

The best known case of gene-culture co-evolution is human adult lactose tolerance, the lifelong capability of digesting lactose (the sugar contained in milk). After weaning, human beings should stop producing lactase, the enzyme digesting lactose, like most other mammals. Given the usual genetic variability in populations, mutations delaying the period in which lactase is no longer produced may occur, and has occurred. If this happens in populations that do not have abundant milk supply, the mutation is neutral (or even slightly costly metabolically) and nothing

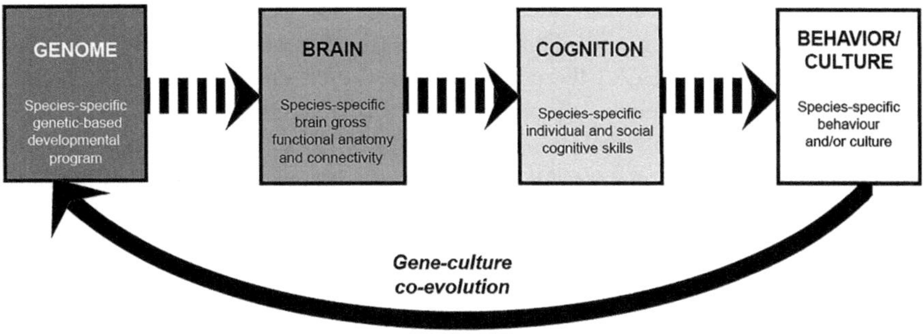

Fig. 9.3 Gene-culture co-evolution. Schematic representation of gene-culture evolution, highlighting that behavioural and cultural dynamics may affect genetic evolution

favours its fixation within the population. However, if this happens in populations that have developed dairy farming, so to have a diet rich in milk and dairy products, that mutation becomes advantageous and its fixation favoured. This is, succinctly, how a consistent part of humanity came to easily digest milk in adulthood—and this is what the early work by Cavalli-Sforza and Feldman demonstrated. Now, dairy farming is a cultural practice. So, a cultural practice is shown to affect the genetic evolution of a population: culture becomes a factor in evolution (Fig. 9.3). Moreover, in this as in other cases, it acts as a *primary driver*: without the advent of farming the mutations allowing adult lactose tolerance in human would have likely never become fixed. It is worth noting that, although significant progress has been made in gene-culture coevolution theory, substantial advancements are still expected particularly in relation to the integration of biological, human and social sciences (Boyd and Richerson 2024).

Notwithstanding the great historical, conceptual and empirical merits of gene-culture co-evolution theory, we think a step further could and should be taken towards a "culture-driven evolution" theory (e.g. Heyes 2012; Dor and Jablonka 2014; see also D'Ambrosio and Colagè 2017). Indeed, gene-culture coevolution actually follows "classical" rules and mechanisms in population genetics, whereby genetic variability within a population is under selection pressures driving the distribution of alleles and determining the spreading of adaptive mutations as well as the progressive elimination of maladaptive ones. The novelty is that cultural traits count among the relevant selective factors. However, gene-culture co-evolution: (1a) does not say anything about the emergence or origin of the relevant cultural traits and behaviours, which must be already developed to a sufficient extent in order to play the role of selective factors, and (2a) requires genetic evolution (the fixation of mutations favoured by the concerned cultural practice) to count as an evolutionary process.

Per se, these are not limits of gene-culture co-evolution theory, which: (1b) does not aim at explaining the origin of cultural novelties and (2b) is explicitly concerned with genetic evolution (as constrained and canalized by existing cultural factors). However, in the field of human evolution at large: (1c) understanding how culture evolves and how cultural novelties *emerge* (and not only stabilize) has indeed become a priority (being one of the key issues in the whole field of cumulative cultural evolution), and (2c) there are cases in which cultural practices emerge, spread, stabilize and deeply affect the life form of entire (human) populations without implying genetic evolution. It is exactly on these points that the study of human (cultural) evolution interestingly converges with the development of the EES. Indeed, two of the main problems with the Modern Synthesis in evolutionary biology triggering the elaboration of the EES were, as previously discussed: (1d) its limits in explaining the *emergence* of evolutionary novelties besides random genetic mutations, and (2d) its sharp distinction between proximate and ultimate causes in biology, with the consequent dismissal of the former as truly evolutionary causes because not involving gene change, and the consideration of the latter as the sole relevant evolutionary causes in their affecting allele distributions within populations in time. On the contrary, current advancements in cultural evolution studies unveil clear cases in which: (1e) the emergence of cultural innovations happens at a genuine behavioural/cultural level without necessarily requiring concomitant genetic changes as a cause, and (2e) such innovations spread and stabilize (entailing significant changes in the life form of the populations devising them) without necessarily implying fixation at the genetic level.

Emergence of cultural novelty is not an easy research task as far as ancient innovations are concerned, since sufficient documentation of the invention processes is obviously lacking: we are just left with the remains of the accomplished innovations in the archeological record. However, a number of possible mechanisms have been proposed in the literature. Cultural innovations may emerge by exploiting a *new* natural resource or phenomenon (Derex 2021). The resource or phenomenon may be new in three fundamental senses. It can be introduced as, or by, an environmental change independent of the population's activity in that environment. It can be the consequence of the population modifying or changing its environment (also through niche construction). It can be new in the sense that an environmental aspect already present in the population's environment begins to be perceived as an *affordance* to be exploited (e.g. Colagè et al. 2024). Cultural innovations may also emerge by building upon already existing cultural traits. Specifically, existing cultural traits can be combined together in novel ways to obtain new cultural traits (*recombination*); a cultural trait previously devised in view of a certain function or purpose can be put at the service of a different task, even only loosely related with the original one (*repurposing*); cultural traits may be re-combined and the resulting combination repurposed to obtain a cultural innovation (Colagè and d'Errico 2023). Note that these are all processes happening mainly at the behavioural level.

Especially in the context of recombination and repurposing involved in the production of cultural novelties, *transmission strategies* acquire a crucial relevance (Colagè and d'Errico 2023, 2025). Indeed, only adequate transmission strategies—those facilitating social learning, thus enabling the successful acquisition of cultural elements—can ensure both the faithful and effective (horizontal) spread and the (vertical) perpetuation of existing cultural traits to be repurposed and recombined. Importantly, the fact that cultural transmission can occur within the same generation and among non-relatives constitutes a key difference between genetic and cultural inheritance (Creanza et al. 2017). Not by chance, virtually all scholars engaged in cultural evolution studies maintain that social learning is a cornerstone of cultural evolution, and that teaching might be one of the keys for human exalted cumulative culture (Csibra and Gergely 2011; Dean et al. 2012; Sterelny 2012; d'Errico and Banks 2015; Gärdenfors and Högberg 2017).

There are a number of interesting examples of cultural innovations exemplifying the recombination and repurposing dynamics mentioned above, and emphasizing the role of social learning and teaching. A particularly eloquent one is the process culminating with literacy and the creation of number symbols (d'Errico et al. 2017; d'Errico and Colagè 2018). Human ancestors left cutmarks on animal bones at least as far back as 2.6 million years ago, as a consequence of butchery. These were by-products of subsistence strategies. At a later time, however, those marks came to be perceived in a new way, as a potential affordance to improve tool efficiency (e.g. increasing tool-hand grip) or to convey some sort of meaning. For example, incisions on shells or bones may constitute (already 540,000 years ago) abstract decorative patterns or marks of ownership (e.g. Mania and Mania 1988; Joordens et al. 2015). Incisions were also combined with ornaments, use of pigments on hard surfaces, and depictional representations such as rock art. Starting from 44,000 years ago, incised objects also took the form of so-called "artificial memory systems" (AMS)—objects whose marks store discrete bits of information outside the body (d'Errico 1995, 1998, 2002). The first known proper writing system, the Mesopotamian cuneiform invented 5.400 years ago, still consisted of marks on surfaces.

Literacy has had, and still has, an enormous impact on the human life form. Without it, theoretical culture, complex institutions, modern science and high-tech would probably be impossible. Though literacy is not, as yet, present in each and every human group (there are traditional peoples currently lacking a writing system), its consequences affect these populations too, and changed humanity's history. It is also worth mentioning that writing has been invented at least three times independently—the first being the Mesopotamian cuneiform, the last occurring in Mesoamerica probably between 600 and 300 BCE, at least; writing in China is attested before 1000 BCE though the process is thought to have begun possibly more than 1000 years earlier. Literacy is widespread across the globe nowadays. However, as a cultural innovation, it is quite recent on the evolutionary timescale (thousands of years younger than farming), and it has been restricted to a minority of members in many societies up to a few centuries ago. Hence, it is very unlikely that it emerged because of some key genetic mutations, and it is possibly even less likely that it has spread, fixed and stabilized because of genetic evolution. That notwithstanding, there is compelling evidence that learning to read impacts cognitive faculties and changes brain functional and anatomical connectivity at the ontogenetic scale, during the lifetime of individuals (Dehaene et al. 2010). At the level of cognitive faculties, data show that learning to read improves phonological awareness (Perre et al. 2009) and affects the ability to visually recognize/categorize specific classes of objects such as shoes, faces or symbols (Cantlon et al. 2011; Dehaene et al. 2015). At the neural level, clear data show that learning to read has a measurable effect, in the left hemisphere, on co-activation patterns of ventral visual areas, on one hand, and core language areas (such as inferior parietal areas in the so-called "Geshwind territory", and frontal regions in the Broca's area), on the other (Caspers et al. 2013, 2014). Still more interestingly, learning to read has been shown to induce changes in the anatomical connectivity between the ventral temporal cortex (specifically, the inferior fusiform gyrus) and inferior parietal areas through the white-matter fibre tract known as the arcuate fasciculus (Klingberg et al. 2000; Thiebaut de Schotten et al. 2012).

Accordingly, the case of literacy suggests that cultural practices arising at the genuinely behavioural level without involving concomitant gene change for their emergence can have measurable effects on both the cognitive and the neural level. Moreover, such practices may also spread and stabilize within and across populations without requiring genetic fixation. This prompts the proposal of a general model for cultural evolution that takes into account the possibility that cultural dynamics may drive evolutionary processes via definite mechanisms (Colagè and d'Errico 2020). This model emphasizes that cultural innovations, once they emerged at the proper cultural level—once excogitated behaviourally—can influence cognition through "cultural exaptation". They can spread behaviourally via social learning and teaching, eventually reshaping neural substrates through "cultural neural reuse" (d'Errico and Colagè 2018; Fig. 9.4). Cultural exaptation refers to the co-option of an existing cultural trait for a new purpose, either as such or in combination with other traits. Starting to use cutmarks to count or to store information on AMSs is a cultural exaptation as the function of the marks changes significantly. Starting to employ ochre pigments on ornaments is another cultural exaptation—as pigments are repurposed with respect to their use on the body and combined with another cultural trait, the ornament. When this happens, the exapted trait (repurposed and/or recombined) impinges on cognitive processes and faculties (like symbolic capacities or social cognition in the just-mentioned cases). If the new trait obtained via cultural exaptation meets "social acceptance", is considered of some interest or worthiness by the relevant group and, hence, is socially transmitted, it can begin to affect the neural substrates involved in its performance. In this way, cultural neural reuse may be triggered. Cultural neural reuse (also labelled "neural recycling") consists in putting a brain region at a service of a different function with respect to the one it usually supports (Dehaene and Cohen 2007; also Colagè and D'Ambrosio 2014; D'Ambrosio and Colagè 2017). The reused brain region must have processing capabilities—or "workings"[3] (Anderson 2010, 2014)—suitable to serve the new, incipient function, but has not

[3]The notion of "working" of a brain region indicates the basic information-processing capabilities of that region. It is not a cognitive function, which is usually supported by many brain regions with their respective "workings". A single brain region with its working can be part of more than one neural network and concur to more than one cognitive function. Cultural neural reuse emphasizes the possibility that a brain region evolved with a specific working to be part of a neural network supporting one function may become part of another network and concur to fulfil a different function without changing its local and specific working.

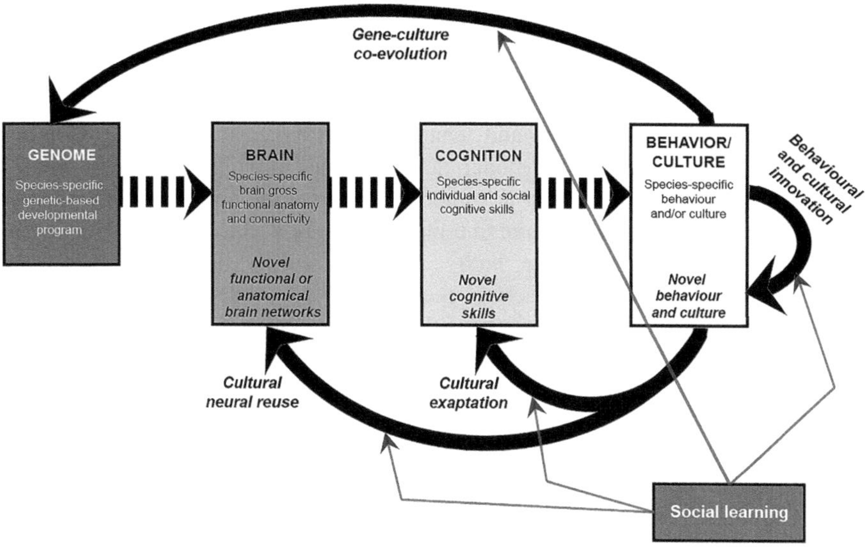

Fig. 9.4 Culture-driven evolution. Schematic representation of the proposed model of cultural evolution, highlighting the role of cultural exaptation in affecting cognitive skills, and cultural neural reuse in affecting brain substrates. As discussed in the text, social learning plays a key role in the emergence of cultural innovations (ensuring elements to be combined and/or repurposed), in gene-culture co-evolution (stabilizing the cultural practices affecting genetic evolution), as well as in cultural exaptation and cultural neural reuse. (Modified and expanded after Figure 2 in Colagè and d'Errico 2020)

evolved "in view of" it. In some cases (such as literacy, as we have seen), cultural neural reuse also involves formation of a new functional or anatomical network in which the reused region begins to work with other areas or circuits to better support the new function. Cultural neural reuse is grounded in the exceptional plasticity of the (human) brain (e.g. Pascual-Leone et al. 2005), but is triggered by the emergence (via cultural exaptation) and spreading (via social learning) of a new cultural trait, and does not involve gene change: cultural neural reuse happens at the ontogenetic scale, within the lifespan of single organisms.

It is worth mentioning that there is evidence suggesting that other cultural practices may involve such kind of processes. Arithmetic (the ability to make symbolic calculations with exact quantities) is a very recently invented (centuries after reading) cultural practice affecting both cognitive faculties and neural substrates about which there is evidence of cultural exaptation and cultural neural reuse (Dehaene and Cohen 2007; Dehaene 2011; d'Errico et al. 2017). Preliminary data suggesting reuse of already existing cognitive faculties and neural substrates are also available for the human capacity to infer social status from face decoration like paintings and ornaments (Salagnon et al. 2024). Tool making is another ability possibly involving neural reuse (Hecht et al. 2014; Stout et al. 2015). Even tool-use learning in monkeys seems to trigger measurable changes in anatomical connectivity in the parietal cortex (Ishibashi et al. 2002; Hihara et al. 2006).

The understanding of cultural evolution implied by the just-sketched model does *not* claim or imply that genetic evolution is irrelevant. Gene-culture co-evolution has certainly played a crucial role in human evolution. Key events on our lineage (such as acquisition of erect posture, encephalization, neoteny) has very likely been triggered by genetic mutations and then stabilized by ensuing genetic evolution (and directional selection). The model, however, does highlight the role of cultural behaviour in *triggering* cultural novelties and in ensuring spreading, stabilization and progressive refinement of cultural traits. It shows how cultural behaviour has the potential to affect and change both cognitive faculties and neural substrates mobilized by new cultural practices themselves. Consequently, emergence and stabilization of novel cultural practices do not necessarily require to be triggered by genetic mutations causing changes at the neural and/or cognitive level. The model also underlines the key role of cultural transmission strategies in cultural evolution and, specifically, in the stabilization, spreading and refinement of cultural traits. Hence, the model does imply that genetic evolution might be *inessential* for novel cultural traits to emerge and stabilize, at least in *some* relevant cases. When this happens for so-called key cultural innovations—innovations that are so momentous that they spawn an array of secondary products or behaviours (Reader and Laland 2003), and that consequently deeply affect the form of life of the population devising them—then culture should be considered a proper, *ultimate* evolutionary cause in itself. In such cases at least, culture can bring about evolutionary change even without (or prior to) gene change.[4]

9.4 Cultural Evolution: An Archeological Standpoint

Paleolithic archeology is a field in constant development. In recent decades, the advancements have been impressive, both quantitatively (findings and discoveries accumulate at a sustained pace) and qualitatively (novel analytic techniques and research methods emerge and improve). Moreover, the interactions between archeology and other research fields have increased. Cognitive and neural sciences interact with archeology in trying to ascertain the implications of past material culture for the evolution of cognition and the brain (e.g. Renfrew and Zubrow 1994; Wynn et al. 2024; Stout and Hecht 2015, 2017). Disciplines like paleogenetics, archeogenetics and paleogenomics have indicated repeated genetic admixture among late Pleistocene *H. sapiens*, Neanderthals and Denisovans in Eurasia (Koller et al. 2022; Mafessoni 2019; Villanea and Schraiber 2019; Harris et al. 2023; Iasi et al. 2024; Li et al. 2024) and widespread recent Neanderthal ancestry among Initial Upper Paleolithic humans in Europe (Hajdinjak et al. 2021; Stringer and Crété 2022; Zilhão et al. 2024). Among other implications, these

[4] As we have already mentioned, this does not imply that gene changes did not prompt relevant cultural changes (it has certainly happened in the evolution of our lineage and, of course, even beyond), neither that gene-culture co-evolution did not play an important role in human evolution.

findings suggest that genetic difference and belonging to different fossil "species" did not constitute an unbridgeable gap between different past hominin cultures. Ethnography—the study of present-day traditional peoples across the globe—also interacts with archeology (e.g. London 2000; David and Kramer 2001), and provides valuable insights about the use and significance of prehistoric cultural products, as well as information on the transmission strategies likely involved in the perpetuation of those products and subsistence strategies (e.g. Lew-Levy et al. 2020, 2023; Milks et al. 2021; Hewlett et al. 2024; Blackwell and d'Errico 2021).

The mentioned interdisciplinary endeavours, along with archeological discoveries and reappraisal of known findings, allow to build up comprehensive scenarios for the cultural evolution of definite classes of cultural traits and innovations, as well as still larger pictures of Paleolithic cultural evolution. Attempts in this direction are already available regarding the evolution of, for example, bone technology (Ma and Doyon 2021), body decoration and modification (d'Errico et al. 2023; d'Errico 2024), clothing and sewing technology (d'Errico et al. 2018; Gilligan et al. 2024), production and use of artificial memory systems (d'Errico et al. 2017), and ochre pigment use (Rosso et al. 2017; Dapschauskas et al. 2022). Still larger pictures, trying to capture the main trends in Paleolithic cultural evolution on the basis of available archeological record, have also been proposed (e.g. Colagè and d'Errico 2020; d'Errico and Colagè 2022; d'Errico et al. 2025).

Figure 9.5 displays the ascertained presence of *Homo* species in Africa, Europe and Asia in the last 1.6 My, along with the timings of the main archeological periods within the Paleolithic.[5] For the last about 600 ka, the figure also synthesizes the principal cultural adaptations and innovations in the three continents. Overall, the gradual complexification of lithic technology and the continuing accumulation of novel cultural practices, including more symbolic and less utilitarian ones, is documented in the three regions, in spite of the different species present in each of the three continents at different times. This already suggests, at a general level, that cultural evolution does not strictly parallel genetic evolution or speciation events. Moreover, the patterns shown in Fig. 9.5 also suggest how cultural practices can be *recombined* in different ways among each other, and how some of them can be *repurposed* when integrated in emerging cultural complexes.

For example, ochre use has a complex evolution, progressively acquiring novel functions and being integrated in different cultural practices across Africa, Europe and Asia. Ochre may have both utilitarian and symbolic functions: it can protect the skin from UV radiation (Rifkin et al. 2015) and from parasites or insects, but it can also be used for body decoration, on ornaments, as well as in both portable and parietal art. In Africa, ochre use is attested from 500 ka, possibly with mainly utilitarian functions up to about 160 ka (Dapschauskas et al. 2022). After this

[5]Figure 5 does not include *Homo naledi*'s cultural behaviours—such as deliberate burial and symbolic marking—as they are currently subject to active scientific debate (e.g. Pettitt 2022; Foecke et al. 2024; Martinón-Torres et al. 2024). Pending broader consensus, we adopt a conservative stance in this summary.

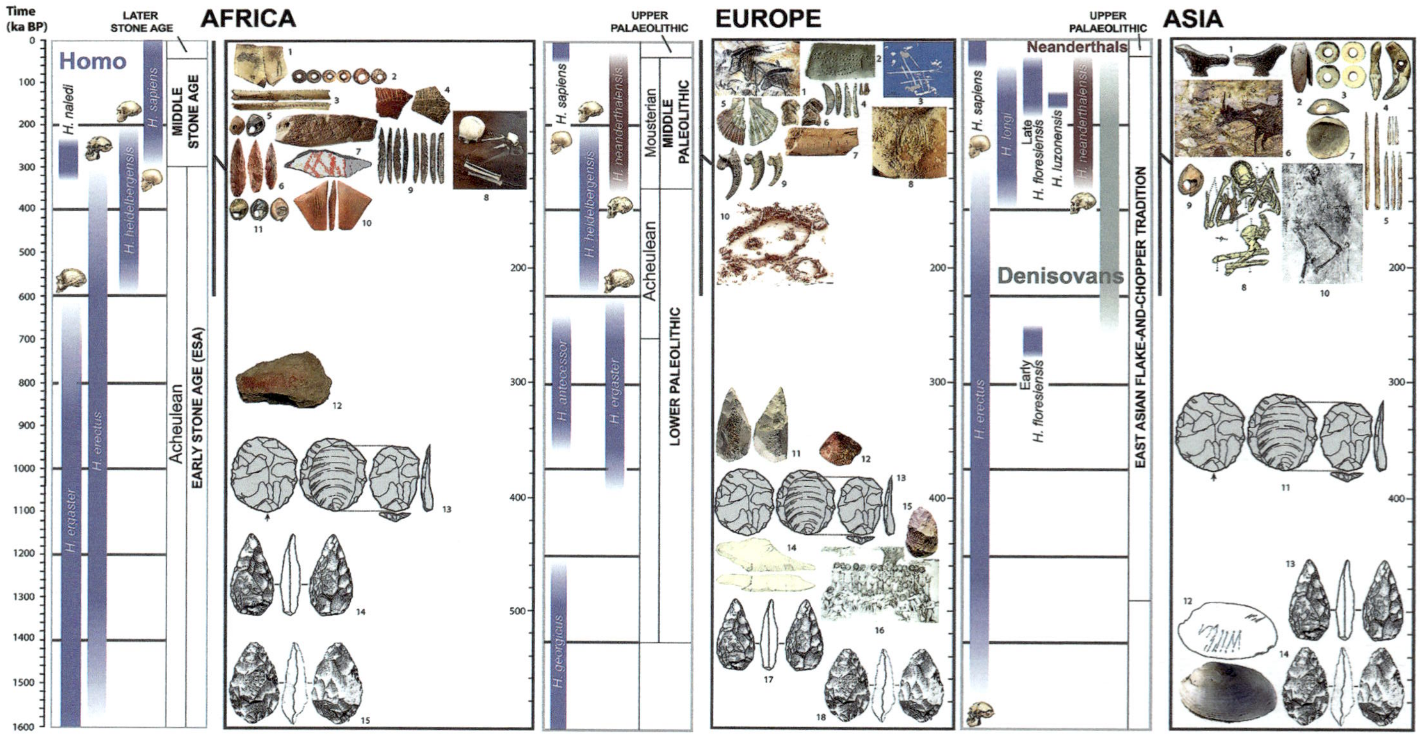

Fig. 9.5 Timeline of cultural innovations (previous page). Timelines for the fossil record of principal human ancestors over the last 1.6 My, the periodization of main cultural adaptations in Africa, Europe and Asia, and discoveries documenting increasing complexity in material culture. **Africa**: 1, slab with painted figurative art, Apollo 11, southern Namibia, c. 30 ka; 2, naturally perforated *Conus* spire cups used as beads and ostrich egg shell beads, Panga Ya Saidi, Kenya, c. 60–33 ka; 3, notched baboon fibula carrying notational notches, Border Cave, KwaZulu-Natal, South Africa, c. 43 ka; 4, fragments of ostrich egg containers engraved with abstract patterns, Diepkloof, c. 64 ka; 5, perforated *Nassarius kraussianus* shell used as bead, Blombos Cave, c. 73–70 ka; 6, Still Bay bifacial points, Blombos Cave, South Africa, c. 73–70 ka; 7, engraved ochre piece and silcrete ochre piece, Blombos Cave, South Africa, c. 73 ka; 8, child burial from

Fig. 9.5 (continued) Panga Ya Saidi, Kenya, c. 78 ka; 9, bone barbed points from Katanda, Democratic Republic of Congo, c. 90 ka; 10, ochre crayon, Klasies River, South Africa, c. 110 ka; 11, perforated *Trivia gibbosula* from Morocco and Tunisia, c. 140–80 ka; 12, modified ochre, Olorgesailie, Kenya, c. 300 ka; 13, preferential flake Levallois knapping technology, Middle Stone Age, c. 400–40 ka; 14, Acheulean handaxe knapped with soft hammer, c. 800–400 ka; 15, Acheulean handaxe knapped with hard hammer, c. 1.6–0.5 My. **Europe**: 1, painted horse heads, Chauvet Cave, Drôme, France, c. 37 ka; 2, Aurignacian incised ivory spatula, Blanchard rock shelter, Dordogne, France, c. 38 ka; 3, engraved criss-cross pattern on the cave bedrock, Gorham's Cave, Gibraltar, c. 45–40 ka; 4, personal ornaments from the Châtelperronian layers of Grotte du Renne, Arcy-sur-Cure, Yonne, France, c. 45–42 ka; 5, *Pecten maximus* stained with ochre, Cueva Antón, Murcia, Spain, c. 48 ka; 6, incised giant deer phalange, Einhornhöhle Cave, Germany, c. 51 ka; 7, incised shaft fragment of a hyena femur, Les Pradelles, Charente, France, c. 57 ka; 8, finger fluting on cave wall, La Roche-Cotard, Indre-et-Loire, France, c. 57 ka; 9, cut-marked raptor claws from European Middle Palaeolithic sites interpreted as personal ornaments, Combe Grenal, Rio Secco, Pech-de-l'Azé, 80–45 ka; 10, subcircular structures made of broken stalagmites, Bruniquel Cave, Tarn-et-Garonne, France, c. 176 ka; 11, engraved elephant bone shaft fragment, Bilzingsleben, Germany, c. 350 ka; 12, modified red ochre piece, Terra Amata, France, c. 380 ka; 13, preferential flake Levallois knapping technology, Mousterian, c. 400–40 ka; 14, Acheulean handaxe made of elephant bone, Castel di Guido, Italy, c. 400 ka; 15–16, Sima de los Huesos deposition of 28 individuals with an associated Acheulean handaxe ('Excalibur') c. 430 ka; 17, Acheulean handaxe knapped with soft hammer, c. 500–50 ka; 18, Acheulean handaxe knapped with hard hammer, c. 1 My-500 ka. **Asia**: Lingjing miniature bird sculpture, Henan, China, c. 12 ka; 2, perforated *Oliva* sp. shell bead, Asitau Kuru, Jerimalai, Timor, c. 30 ka; 3, ostrich eggshells beads, Shuidonggou, China, c. 31 ka; 4, perforated and ochred badger canine, Zhoukoudian Upper Cave, China, 38–35 ka; 5, eyed bone needles from the early Upper Palaeolithic sites, China, 38 ka; 6, painted scene on cave wall with small human-like figures and wild bovid, Sulawesi, Leang Bulu' Sipong 4, Sulawesi, c. 44 ka; 7, naturally perforated and ochred *Glycymeris insubrica* shell, Qafzeh Cave, Israel, c. 95 ka; 8, modern human burial with wild boar mandibule, Skhul Cave, Israel, 110 ka; 9, perforated *Tritia gibbosula*, Skhul Cave, Israel, c. 120 ka; 10, Neanderthal or modern human burial, Tabun Cave, Israel, 120 ka; 11, preferential flake Levallois knapping technology, Middle Stone Age, c. 350 ka; 12, engraved freshwater shell from Trinil, Indonesia, c. 540 ka; 13, Acheulean handaxe knapped with soft hammer, c. 1.2 My; 14, Acheulean handaxe knapped with hard hammer, 1.5 My. Graphism by Lila Geis time, symbolic and ritual uses may have taken over (though possibly never completely replacing the utilitarian functions). There is evidence suggesting symbolic use of ochre and other mineral pigments in MSA sites: preference for specific colours and high-quality pigments from distant sources (Dayet et al. 2013; Bouillot et al. 2017; Salomon et al. 2012), deliberate heating of iron-rich rocks to change their colour (d'Errico et al. 2010), creating abstract engravings on ochre pieces (Henshilwood et al. 2009), drawing with ochre pencils (Henshilwood et al. 2018), producing small quantities of powder useless for utilitarian purposes (Rosso et al. 2016), and applying ochre to personal ornaments (d'Errico et al. 2009; Sehasseh et al. 2021). Figure 9.5 shows, for example, an ochre crayon from Klasies River, South Africa, dated ca. 110 ka (Fig. 9.5, **Africa**, n. 10) and an engraved ochre piece from Blombos Cave, South Africa, at ca. 73 ka (Fig. 9.5, **Africa**, n. 7). These data and literature are consistent with the idea of cultural exaptation discussed above (Fig. 9.4) in that ochre has been reused and repurposed in several different cultural artefacts, practices and contexts.

Interestingly, archeological evidence on ochre use is also available outside Africa. 77 pieces of ochre were found in archeological layers dated to around 380 ka in Terra Amata, Nice, France (Fig. 9.5, **Europe**, n. 12). Ochre use is also attested in caves, likely with symbolic and ritual implications, like in Cueva de Ardales (Andalucía, Spain) where, between ca. 45 and 65 ka, the pigment was brought into the cave and applied on the walls of a stalagmitic dome (Pitarch Martí et al. 2021). Cave visits with symbolic roles are much older, as suggested by the subcircular structures made of broken stalagmites recently discovered at Bruniquel Cave (Tarn-et-Garonne, France) and dated ca. 176 ka (Fig 9.5, **Europe**, n. 10). Ochre was also used on ornaments (or on clothes in association with ornaments) likely as early as 115 ka in Cueva de Los Aviones, Spain (Hoffmann et al. 2018). Hence, symbolic use of ochre could have evolved in Europe with similar mechanisms as in Africa, via cultural exaptation, reuse and repurposing. This dovetails not only with the idea that emergence of symbolic culture is not bound to our species, but also with the idea that the *mechanisms* of cultural evolution, as discussed in the previous section, apply to several "species" in different geographical areas.

Ornaments are another possible and interesting case of behavioural evolution of symbolic culture. The archeological record presents a variety of findings where many techniques are combined: naturally or artificially perforated shells; teeth or claws; ornaments coloured with pigments or by heating them; finely fully-shaped beads done out of ostrich eggshell. These findings are spread across Africa, Europe and Asia and cover a period of more than 100,000 years. Shells used as ornaments are known in north Africa and Israel before 100 ka. They are naturally perforated and coloured. For example, perforated *Tritia gibbosula* shells are present in Morocco and Tunisia back to ca. 140–80 ka (Fig. 9.5, **Africa**, n. 11), as well as in Skhul Cave, Israel at ca. 120 ka (Fig. 9.5, **Asia**, n. 9). In Israel, Qafzeh Cave, naturally perforated and ochred *Glycymeris insubrica* shells date 95 ka (Fig. 9.5, **Asia**, n. 7). Cut-marked raptor claws interpreted as personal ornaments are present in European Middle Paleolithic sites dated 130–45 ka, such as Combe Grenal and Pech-de-l'Azé in

France, or Rio Secco in Italy (Fig. 9.5, **Europe**, n. 9). In South Africa, at Blombos Cave, perforated *Nassarius kraussianus* shells used as beads are dated ca. 73–70 ka (Fig. 9.5, **Africa**, n. 5), and naturally perforated *Conus* spire cups are used as beads at Panga ya Saidi, Kenya, ca. 60–40 ka. A *Pecten maximus* shell stained with ochre is from Cueva Antón, Murcia, Spain, ca. 48 ka (Fig. 9.5, **Europe**, n. 5). Several personal ornaments including perforated and grooved mammal teeth, bones and fossils are known in Europe, such as those from the Châtelperronian layers of Grotte du Renne, Arcy-sur-Cure, France, dated at 45–42 ka (Fig. 9.5, **Europe**, n. 4). Then, after about 42 ka, personal ornaments are widespread in Asia as well, at a growing number of sites (Fig. 9.5, **Asia**, n. 4); ostrich eggshells beads have been found in Shuidonggou, China and dated ca. 31 ka (Fig. 9.5, **Asia**, n. 3); perforated *Oliva* sp. shell beads are found at Asitau Kuru, Jerimalai, Timor, ca. 30 ka, and beads become widespread after that date in the region (Fig. 9.5, **Asia**, n. 3). At around 50 ka, fully shaped ostrich eggshell beads appear in East Africa and become gradually common in eastern and southern Africa (Fig. 9.5, **Africa**, 2).

The gradual complexification of personal ornaments across Africa, Europe and Asia nicely shows the capacity of our ancestors to combine different skills (e.g. selecting and/or perforating beads, colouring them with pigments or by heat-treatment, choosing specific objects as personal ornaments and modifying them). Some such skills emerged earlier and for other purposes: ochre pigments, as we have seen above, but also the ability to produce cut-marks and incisions on a variety of supports (d'Errico et al. 2017). Cut-marks likely emerged as by-products of butchery activities on bones already 2.6 My. This ability was transferred to shells more than half a million years ago, as documented by the freshwater shell from Trinil, Indonesia, engraved with a zigzag pattern and dated ca. 540 ka, likely done by *H. erectus* (Joordens et al. 2015; Fig. 9.5, **Asia**, n. 12). Much later, this ability was reused and repurposed to produce engravings with possible symbolic or even notational import. In Europe, for example, we have an incised shaft fragment of a hyena femur from Les Pradelles, France, dated 57 ka (Fig. 9.5, **Europe**, n. 7), and an incised giant deer phalange in Einhornhöhle Cave, Germany, at 51 ka (Fig. 9.5, **Europe**, n. 6). 15–20 thousand years later, again in Europe, it is worth mentioning the engraved criss-cross pattern in Gibraltar, on the Gorham's Cave bedrock, dated 45–40 ka (Fig. 9.5, **Europe**, n. 3), and Aurignacian incised ivory objects and figurines such as the spatula from Blanchard rock shelter (Dordogne, France) at 40–38 ka (Fig. 9.5, **Europe**, n. 2). Roughly of the same period (ca. 43 ka), on the African continent, the notched baboon fibula from Border Cave, South Africa, witnesses the earliest known example of a system of notation based on the accumulation of information over time (Fig. 9.5, **Africa**, n. 3).

Finally, pigments—which have a very long history, as we have seen above with ochre, and have been reused to modify the appearance of ornaments—have also been crucial for the development of cave paintings across Africa, Europe and Asia, as exemplified by the scene with small human-like figures and wild bovid painted on the cave wall of Leang Bulu' Sipong 4, Sulawesi, Indonesia, 44 ka (Fig. 9.5, **Asia**, n. 6), the painted horse heads in Chauvet Cave, Drôme, France, 37 ka (Fig. 9.5,

Europe, n. 1) and the slab with painted figurative art at Apollo 11, southern Namibia, dated to about 30 ka (Fig. 9.5, **Africa**, n. 1).

Hence, evidence clearly suggests that ornaments, engravings and parietal art interacted in the past in terms of the techniques employed for their production. Cultural exaptation appears to be at work in the evolution of Paleolithic cultures along an extended period of time, across distant geographical areas, and in different hominin populations. This is consistent with the notion of culture-driven evolution discussed above. Interestingly, the interaction of material-culture techniques may have scaffolded similar interactions among socio-cognitive contexts and related brain substrates as suggested by the understanding of cultural evolution presented above and synthesized in Fig. 9.4. Moreover, it is worth stressing that the cultural developments discussed in this section from an archeological standpoint crucially contributed to shape human cultural niches and, consequently, to what we have labelled "cumulative behavioural evolution" (Fig. 9.2). This contributes to unveil the links between our understanding of cultural evolution and the EES as discussed in Sect. 9.2, especially as far as niche construction theory and non-genetic channels of inheritance constitute central pillars of EES (see also, e.g. Fuentes 2016; Prentiss 2021; Laland and O'Brien 2023; O'Brien and Laland 2023).

9.5 Conclusion: Evolutionary Causes on the Human Lineage

In this contribution, we have first set the stage summarizing the most basic tenets of the Modern Synthesis in evolutionary biology (Sect. 9.1; Fig. 9.1), highlighting its gene-centrism and the sharp distinction between proximate and ultimate causation. We focused on those aspects with major implications for human evolution. In particular, the development of the Modern Synthesis in the mid-twentieth century solved the issue of how biological novelty emerges—left open by Darwin's theory—in genetic terms. Moreover, the Modern Synthesis interpreted the three fundamental steps of biological evolution (variation production, transmission and selection) in a gene-centric fashion.

In Sect. 9.2, we have seen how the above tenets characterizing the Modern Synthesis are overcome by the emerging Extended Evolutionary Synthesis. None of the three fundamental steps of biological evolution (variation production, transmission and selection) is any longer conceived in strict genetic terms. The sharp distinction between proximate and ultimate causation in evolution is blurred, and the "linear" understanding of biological causation is replaced by a fairly circular one. The main point, here, is in acknowledging the key role that actual organisms' behaviour and behavioural innovation have on the evolution of life forms (Fig. 9.2).

Hence, in Sect. 9.3, we have focused on that peculiar and highly complex form of behaviour represented by human culture and cumulative cultural evolution. Here, we have discussed the relevance of gene-culture co-evolution theory and emphasized its historical role in the acknowledgement of culture as an evolutionary force (Fig. 9.3). We also noticed how it is, in a sense, gene-centred, as the focus is on cases where cultural traits influence allele distribution within populations. This is not a limit of gene-culture co-evolution—but its primary aim. However, the development of the Extended Evolutionary Synthesis suggests that behaviour may also be an evolutionary force "irrespective" of its genetic consequences, also prompting the emergence of relevant biological novelty. Specifically, gene-culture co-evolution can hardly explain the origin of the concerned cultural traits—a point that we have seen is explained in fundamentally genetic terms by the Modern Synthesis. Therefore, we discussed the notion of "culture-driven evolution" (Fig. 9.4) to emphasize how behavioural/cultural novelty can emerge at the genuinely behavioural/cultural level, and its perpetuation can be ensured even before (and besides) genetic fixation, mainly through social learning and cultural transmission, and their implications in terms of niche construction and ecological inheritance. Accordingly, the strict bottom-up, gene-centric and linear understanding of causation in the Modern Synthesis is changed into a more *circular* understanding of causation—as it should be evident from Figs. 9.2 and 9.4.

With Sect. 9.4, we entered the archeological level. Fig. 9.5 summarizes the main cultural innovations emerged in the last 1.6 My—with a focus on the last 600 ky—in Africa, Europe and Asia. The archeological record suggests that cumulative cultural evolution is supported by similar mechanisms across the three continents in part independently of the *Homo* species that implemented it. It also suggests how cultural innovations may well have been built at the truly behavioural/cultural level, possibly resorting to the mechanisms of behavioural innovation, social learning, cultural exaptation and cultural neural reuse whereby cultural traits are transmitted, reused, combined and repurposed.

The implications of our contribution to the critical history and epistemology of human evolution can, in conclusion, be thus summarized: (1) in inquiring into human evolution, the truly behavioural/cultural dimension should never be overlooked, and not even considered as secondary or epiphenomenal to the genetic one; (2) the causal mechanisms beneath human evolution likely are fundamentally circular, consistently with the developments in the life sciences at broad; (3) as culture—though not being unique to the human life-form and evolution—is peculiarly developed on the human lineage, the field of human evolution needs theories able to recognize the actual role of cultural evolution in human evolution as a whole. We wish that the notion of culture-driven evolution as discussed above may be a helpful contribution to the future development of our understanding of human evolution.

Acknowledgements We warmly thank Mathilde Lequin and Juan Manuel Rodríguez Caso for inviting us to contribute to this insightful and needed volume, as well as for their discrete but effective support in all phases of our chapter preparation. Many thanks to Lila Geis for her help with Figure 5. We also thank two anonymous Reviewers as their comments helped us improving our chapter. This study was supported by the Grand Programme de Recherche "Human Past" of the Initiative d'Excellence (IdEx) of the University of Bordeaux. IC acknowledges a research grant by the Pontifical University of the Holy Cross on "Creativity" (grant PRI-HCR). FD research is supported by the European Research Council (Synergy Grant for the project Evolution of Cognitive Tools for Quantification [QUANTA], no. 951388); the Research Council of Norway through its Centres of Excellence funding scheme, SFF Centre for Early Sapiens Behaviour (SapienCE), project number 262618 and the Talents Programme of the University of Bordeaux (grant number: 191022_001).

References

Anderson ML (2010) Neural reuse: a fundamental organizational principle of the brain. Behav Brain Sci 33:245–266

Anderson ML (2014) After phrenology: neural reuse and the interactive brain. MIT Press, Cambridge, MA

Aoki K, Mesoudi A (2015) Introduction. In: Mesoudi A, Aoki K (eds) Learning strategies and cultural evolution during the Palaeolithic. Springer, Tokyo, pp 1–8

Badyaev AV, Uller T (2009) Parental effects in ecology and evolution: mechanisms, processes and implications. Philos Trans R Soc B Biol Sci 364:1169–1177

Badyaev AR (2005) Stress-induced variation in evolution: from behavioural plasticity to genetic assimilation. Proc. R. Soc. B 272:877–886

Badyaev AR (2009) Evolutionary significance of phenotypic accommodation in novel environments: an empirical test of the Baldwin effect. Phil. Trans. R. Soc. B 364:1125–1141

Bateson W (1904) Heredity and evolution. Pop Sci Mon 65:522–531

Bateson P (2004) The active role of behaviour in evolution. Biol Philos 19:283–298

Bateson P (2014) New thinking about biological evolution. Biol J Linn Soc 112:268–275

Blackwell L, d'Errico F (2021) The San Elders Speak: ancestral knowledge of the Kalahari San. Witts University Press, Johannesburg

Bolhuis JJ, Tattersall I, Chomsky N, Berwick RC (2014) How could language have evolved? PLoS Biol 12:e1001934

Bouillot LD, Wurz S, Daniel F (2017) Ochre resources, behavioural complexity and regional patterns in the Howiesons Poort: new insights from Klasies River main site, South Africa. J Afr Archaeol 15(1):20–41

Boyd R, Richerson PJ (1985) Culture and the evolutionary process. University of Chicago Press, Chicago

Boyd R, Richerson PJ (2024) Cultural evolution: where we have been and where we are going (maybe). Proc Natl Acad Sci U S A 121(48):e2322879121

Burggren WW (2014) Epigenetics as a source of variation in comparative animal physiology – or – Lamarck is lookin' pretty good these days. The Journal of Experimental Biology 217:682-689

Cantlon JF, Pinel P, Dehaene S, Pelphrey KA (2011) Cortical representations of symbols, objects, and faces are pruned back during early childhood. Cereb Cortex 21:191–199

Caspers J, Zilles K, Eickhoff SB, Schleicher A, Mohlberg H, Amunts K (2013) Cytoarchitectonical analysis and probabilistic mapping of two extrastriate areas of the human posterior fusiform gyrus. Brain Struct Funct 218:511–526

Caspers J, Zilles K, Amunts K, Laird AR, Fox PT, Eickhoff SB (2014) Functional characterization and differential coactivation patterns of two cytoarchitectonic visual areas on the human posterior fusiform gyrus. Hum Brain Mapp 35:2754–2767

Cavalli-Sforza LL, Feldman MW (1973) Cultural versus biological inheritance: phenotypic transmission from parents to children. (A theory of the effect of parental phenotypes on children's phenotypes). Am J Hum Genet 25:618–637

Colagè I, D'Ambrosio P (2014) Exaptation and neural reuse: a research perspective into the human specificity. Antonianum 89:333–358

Colagè I, d'Errico F (2020) Culture: the driving force of human cognition. Top Cognit Sci 12:654–672

Colagè I, d'Errico F (2023) The roots of creativity: investing in cultural transmission. Acta Philosophica 31(1):95–116

Colagè I, d'Errico F (2025) An empirically-based scenario for the evolution of cultural transmission in the human lineage during the last 3.3 million years. PLoS One 20(5):e0325059

Colagè I, Oliva S, d'Errico F (2024) Rewriting the environment, remaking humanity: niche construction, creativity, and cultural evolution. Versus 138(1):25–38
Coolidge FL, Wynn T (2007) The working memory account of Neandertal cognition – how phonological storage capacity may be related to recursion and the pragmatics of modern speech. J Hum Evol 52(6):707–710
Coolidge FL, Wynn T (2017) The rise of Homo sapiens: the evolution of modern thinking. Oxford University Press, Oxford
Creanza N, Kolodny O, Feldman MW (2017) Cultural evolutionary theory: how culture evolves and why it matters. Proc Natl Acad Sci U S A 114(30):7782–7789
Csibra G, Gergely G (2011) Natural pedagogy as evolutionary adaptation. Philos Trans R Soc B Biol Sci 366:1149–1157
D'Ambrosio P, Colagè I (2017) Extending epigenesis: from phenotypic plasticity to the bio-cultural feedback. Biol Philos 32:705–728
d'Errico F (1995) A new model and its implications for the origins of writing: the La Marche antler revisited. Camb Archaeol J 5:163–206
d'Errico F (1998) Palaeolithic origins of artificial memory systems: an evolutionary perspective. In: Renfrew C, Scarre CC (eds) Cognition and material culture: the archaeology of symbolic storage. The McDonald Institute Monographs, Cambridge, pp 19–50
d'Errico F (2002) Memories out of mind: the archaeology of the oldest artificial memory systems. In: Nowell A (ed) In the mind's eye. International Monographs in Prehistory Archaeological Series, Ann Arbor, pp 33–49
d'Errico F (2024) An integrated evolutionary scenario for the culturalisation of the human body. In: Manni F, d'Errico F (eds) The Oxford handbook of the archaeology and anthropology of body modification. Oxford University Press, Oxford. https://doi.org/10.1093/oxfordhb/9780197572528.001.0001
d'Errico F, Banks WE (2015) The archaeology of teaching: a conceptual framework. Camb Archaeol J 25:859–866
d'Errico F, Colagè I (2018) Cultural exaptation and cultural neural reuse: a mechanism for the emergence of modern culture and behaviour. Biol Theory 13:213–227
d'Errico F, Colagè I (2022) The emergence of symbolic cognition. In: Skov M, Nadal M (eds) The Routledge international handbook of neuroaesthetics. Routledge, pp 539–554
d'Errico F, Vanhaeren M, Barton N, Bouzouggar A, Mienis H, Richter D, Hublin J-J, McPherron S, Lozouet P (2009) Additional evidence on the use of personal ornaments in the Middle Paleolithic of North Africa. Proc Natl Acad Sci U S A 106(38):16051–16056
d'Errico F, Salomon H, Vignaud C, Stringer C (2010) Pigments from the Middle Palaeolithic levels of Es-Skhul (Mount Carmel, Israel). J Archaeol Sci 37(12):3099–3110
d'Errico F, Doyon L, Colagè I, Queffelec A, Le Vraux E, Giacobini G, Vandermeersch B, Maureille B (2017) From number sense to number symbols. An archaeological perspective. Philos Trans R Soc B Biol Sci 373:20160518
d'Errico F, Doyon L, Zhang S, Baumann M, Lázničková-Galetová M, Gao X, Chen F, Zhang Y (2018) The origin and evolution of sewing technologies in Eurasia and North America. J Hum Evol 125:71–86
d'Errico F, van Niekerk KL, Geis L, Henshilwood CS (2023) New Blombos Cave evidence supports a multistep evolutionary scenario for the culturalization of the human body. J Hum Evol 184:103438
d'Errico F, Colagè I, Straffon LM, Bender A (2025) The emergence and complexification of symbolic practices in the *Homo* lineage: an archaeological and cognitive perspective. Submitted
Dapschauskas R, Göden MB, Sommer S, Kandel AW (2022) The emergence of habitual ochre Use in Africa and its significance for the development of ritual behavior during The Middle Stone Age. J World Prehist 35:233–319
Darwin C (1859) On the origin of species by means of natural selection, or preservation of favoured races in the struggle for life. John Murray, London

Darwin C (1874) The descent of man and selection in relation to sex, 2nd edn. John Murray, London
David N, Kramer C (2001) Ethnoarchaeology in action. Cambridge University Press
Dayet L, Texier PJ, Daniel F, Porraz G (2013) Ochre resources from the Middle Stone Age sequence of Diepkloof Rock Shelter, Western Cape, South Africa. J Archaeol Sci 40(9): 3492–3505
de Lamarck JB (1809) Philosophie zoologique. Dentu, Paris
de Vries H (1904) Species and varieties: their origin by mutation. The Open Court Publishing Company, Chicago
Dean LG, Kendal RL, Schapiro SJ, Thierry B, Laland KN (2012) Identification of the social and cognitive processes underlying human cumulative culture. Science 335:114–118
Dehaene S (2011) The number sense: how the mind creates mathematics. Oxford University Press, Oxford
Dehaene S, Cohen L (2007) Cultural recycling of cortical maps. Neuron 56:384–398
Dehaene S, Pegado F, Braga LW, Ventura P, Nunes Filho G, Jobert A, Dehaene-Lambertz G, Kolinsky R, Morais J, Cohen L (2010) How learning to read changes the cortical networks for vision and language. Science 330:1359–1364
Dehaene S, Cohen L, Morais J, Kolinsky R (2015) Illiterate to literate: behavioural and cerebral changes induced by reading acquisition. Nat Rev Neurosci 16:234–244
Derex M (2021) Human cumulative culture and the exploitation of natural phenomena. Philos Trans R Soc B Biol Sci 377:20200311
Dor D, Jablonka E (2014) Why we need to move from gene-culture co-evolution to culturally driven co-evolution. In: Dor D, Knight C, Lewis J (eds) The social origins of language. Oxford University Press, Oxford, pp 15–30
Duckworth RA (2009) The role of behavior in evolution: a search for mechanism. Evol Ecol 23: 513–531
Emery NJ, Clayton NS (2004) The mentality of crows: convergent evolution of intelligence in corvids and apes. Science 306:1903–1907
Feldman MW, Cavalli-Sforza LL (1975) Models for cultural inheritance: a general linear model. Ann Hum Biol 2:215–226
Feldman MW, Cavalli-Sforza LL (1976) Cultural and biological evolutionary processes, selection for a trait under complex transmission. Theor Popul Biol 9:238–259
Feldman MW, Laland KN (1996) Gene-culture coevolutionary theory. Trends Ecol Evol 11:453–457
Fisher SE (2017) Evolution of language: lessons from the genome. Psychon Bull Rev 24:34–40
Fisher J, Hinde RA (1949) The opening of milk-bottles by birds. Br Birds 42:347–357
Fitz-James MH, Cavalli G (2022) Molecular mechanisms of transgenerational epigenetic inheritance. Nat Rev Genet 23:325–341
Foecke KK, Queffelec A, Pickering R, Harvati K (2024) No sedimentological evidence for deliberate burial by Homo naledi – a case study highlighting the need for best practices in geochemical studies within archaeology and paleoanthropology. PaleoAnthropology
Fuentes A (2016) The extended evolutionary synthesis, ethnography, and the human niche toward an integrated anthropology. Curr Anthropol 57:S13–S26
Fuentes A (2017) The creative spark: how imagination made humans exceptional. Penguin
Galef BG, Whiten A (2017) The comparative psychology of social learning. In: APA handbook of comparative psychology: perception, learning, and cognition, vol 2. American Psychological Association, Washington, DC, pp 411–439
Gärdenfors P, Högberg A (2017) The archaeology of teaching and the evolution of Homo docens. Curr Anthropol 58:188–208
Geschwind DH, Rakic P (2013) Cortical evolution: judge the brain by its cover. Neuron 80:633–647
Gilligan I, d'Errico F, Doyon L, Wang W, Kuzmin YV (2024) Paleolithic eyed needles and the evolution of dress. Sci Adv 10(26):eadp2887

Gintis H (2011) Gene-culture coevolution and the nature of human sociality. Philos Trans R Soc B Biol Sci 366:878–888

Hajdinjak M, Maffessoni F, Skov L, Vernot B, Hübner A, Fu Q, Essel E, Nagel S, Nickel B, Richter J, Moldovan OT (2021) Initial Upper Palaeolithic humans in Europe had recent Neanderthal ancestry. Nature 592(7853):253–257

Harris DN, Platt A, Hansen ME, Fan S, Mcquillan MA, Nyambo T, Mpoloka SW, Mokone GC, Belay G, Fokunang C, Njamnshi AK (2023) Diverse African genomes reveal selection on ancient modern human introgressions in Neanderthals. Curr Biol 33(22):4905–4916

Hecht EE, Gutman DA, Khreisheh N, Taylor SV, Kilner J, Faisal AA, Bradley BA, Chaminade T, Stout D (2014) Acquisition of paleolithic toolmaking abilities involves structural remodeling to inferior frontoparietal regions. Brain Struct Funct 220:2315–2331

Henrich J (2016) The secret of our success: how culture is driving human evolution, domesticating our species, and making us smarter. Princeton University Press, Princeton

Henshilwood C, d'Errico F, Watts I (2009) Engraved ochres from the middle stone age levels at Blombos Cave, South Africa. J Hum Evol 57(1):27–47

Henshilwood C, d'Errico F, van Niekerk KL, Dayet L, Queffelec A, Pollarolo L (2018) An abstract drawing from the 73,000-year-old levels at Blombos Cave, South Africa. Nature 562(7725): 115–118

Herculano-Houzel S (2017) Numbers of neurons as biological correlates of cognitive capability. Curr Opin Behav Sci 16:1–7

Hewlett BS, Boyette AH, Lew-Levy S, Gallois S, Dira SJ (2024) Cultural transmission among hunter-gatherers. Proc Natl Acad Sci 121(48):e2322883121

Heyes C (2012) Grist and mills: on the cultural origins of cultural learning. Philos Trans R Soc B Biol Sci 367:2181–2191

Hihara S, Notoya T, Tanaka M, Ichinose S, Ojima H, Obayashi S, Fujii N, Iriki A (2006) Extension of corticocortical afferents into the anterior bank of the intraparietal sulcus by tooluse training in adult monkeys. Neuropsychologia 44:2636–2646

Hinde RA, Fisher J (1951) Further observations of the opening of milk bottles by birds. Br Birds 44: 393–396

Hirata S, Watanabe K, Kawai M (2001) "Sweet-potato washing" revisited. In: Matsuzawa T (ed) Primate origins of human cognition and behaviour. Springer, Berlin/Heidelberg, pp 487–508

Hoffmann DL, Standish CD, García-Diaz M, Pettitt PB, Milton JA, Zilhão J (2018) U-Th dating of carbonate crusts reveals Neandertal origin of Iberian cave art. Science 23(359):912–915

Hoppitt W, Laland KN (2013) Social learning: an introduction to mechanisms, methods, and models. Princeton University Press, Princeton

Huxley J (1942) Evolution: the modern synthesis. George Allen & Unwin Ltd., London

Iasi LN, Chintalapati M, Skov L, Mesa AB, Hajdinjak M, Peter BM, Moorjani P (2024) Neandertal ancestry through time: insights from genomes of ancient and present-day humans. bioRxiv

Ishibashi H, Hihara S, Takahashi M, Heikeb T, Yokota T, Iriki A (2002) Tool-use learning selectively induces expression of brain-derived neurotrophic factor, its receptor trkB, and neurotrophin 3 in the intraparietal multisensorycortex of monkeys. Cognit Brain Res 14:3–9

Jablonka E, Lamb M (2006) Evolution in four dimensions. The MIT Press, Cambridge

Joordens JCA, d'Errico F, Wesselingh FP, Munro S, de Vos J, Wallinga J, Ankjærgaard C, Reimann T, Wijbrans JR, Kuiper KF, Mücher HJ, Coqueugniot H, Prié V, Joosten I, van Os B, Schulp AS, Panuel M, van der Haas V, Lustenhouwer W, Reijmer JJG, Roebroeks W (2015) *Homo erectus* at Trinil on Java used shells for tool production and engraving. Nature 518:228–231

Kasser SM, Fortunato L, Lala KN (2024) Gene-culture coevolution: a broader evolutionary perspective. EcoEvoRxiv. https://doi.org/10.32942/X2862X

Kawai M (1965) Newly-acquired pre-cultural behavior of the natural troop of Japanese monkeys on Koshima islet. Primates 6(1):1–30

Kendal RL, BoogerNJt RL, Laland KN, Webster M, Jones PL (2018) Social learning strategies: bridge-building between fields. Trends Cognit Sci 22:651–665
Kissel M, Fuentes A (2021) The ripples of modernity: how we can extend paleoanthropology with the extended evolutionary synthesis. Evol Anthropol Issues News Rev 30(1):84–98
Klein RG (1989) The human career. University of Chicago Press, Chicago
Klein RG (2000) Archeology and the evolution of human behavior. Evol Anthropol 9:17–36
Klingberg T, Hedehus M, Temple E, Salz T, Gabrieli JDE, Moseley ME, Poldrack RA (2000) Microstructure of temporo-parietal white matter as a basis for reading ability: evidence from diffusion tensor magnetic resonance imaging. Neuron 25:493–500
Koller D, Wendt FR, Pathak GA, De Lillo A, De Angelis F, Cabrera-Mendoza B, Tucci S, Polimanti S (2022) Denisovan and Neanderthal archaic introgression differentially impacted the genetics of complex traits in modern populations. BMC Biol 20(1):1–15
Laidre ME (2012) Niche construction drives social dependence in hermit crabs. Curr Biol 22(20): R861–R863
Laidre ME (2019) Architectural modification of shells by terrestrial hermit crabs alters social dynamics in later generations. Ecology 100(9):e02767
Laland KN (2015) On evolutionary causes and evolutionary processes. Behav Process 117:97–104
Laland KN (2017) Darwin's unfinished symphony: how culture made the human mind. Princeton University Press, Princeton/Oxford
Laland KN, O'Brien MJ (2023) The cultural contribution to evolvability. PaleoAnthropology 2: 164–180
Laland KN, Odling-Smee J, Myles S (2010) How culture shaped the human genome: bringing genetics and the human sciences together. Nat Rev Genet 11:137–148
Laland KN, Sterelny K, Odling-Smee FJ, Hoppitt W, Uller T (2011a) Cause and effect in biology revisited: is Mayr's proximate-ultimate dichotomy still useful? Science 334:1512–1516
Laland KN, Atton N, Webster MM (2011b) From fish to fashion: experimental and theoretical insights into the evolution of culture. Philos Trans R Soc B Biol Sci 366:958–968
Laland KN, Odling-Smee J, Hoppitt W, Uller T (2013) More on how and why: a response to commentaries. Biol Philos 28:793–810
Laland KN, Uller T, Feldman MW, Sterelny K, Muller GB, Moczek A, Jablonka E, Odling-Smee J (2015) The extended evolutionary synthesis: its structure, assumptions and predictions. Proc R Soc B Biol Sci 282:20151019
Laland KN, Uller T, Feiner N, Feldman M, Gilbert SF (2024) Evolution evolving: the developmental origins of adaptation and biodiversity. Princeton University Press, Princeton
Lea SEG, Chow PKY, Leaver LA, McLaren IPL (2020) Behavioral flexibility: A review, a model, and some exploratory tests. Learning & Behavior 48:173–187
Levins R, Lewontin RC (1985) The dialectical biologist. Harvard University Press, Cambridge, MA
Lew-Levy S, Milks A, Lavi N, Pope SM, Friesem DE (2020) Where innovations flourish: An ethnographic and archaeological overview of hunter-gatherer learning contexts. Evol Hum Sci 2 (e31):1–23
Lew-Levy S, van den Bos W, Corriveau K, Dutra N, Flynn E, O'Sullivan E, Pope-Caldwell S, Rawlings B, Wood L (2023) Peer learning and cultural evolution. Child Dev Perspect 17:97–105
Lewontin RC (1983) Gene, organism and environment. In: Bendall DE (ed) Evolution from molecules to men. Cambridge University Press, Cambridge, pp 273–285
Li L, Comi TJ, Bierman RF, Akey JM (2024) Recurrent gene flow between Neanderthals and modern humans over the past 200,000 years. Science 385(6705):eadi1768
London G (2000) Ethnoarchaeology and interpretation. Near Eastern Archaeol 63:2–8
Ma S, Doyon L (2021) Animals for tools: the origin and development of bone technologies in China. Front Earth Sci 9:784313
Mafessoni F (2019) Encounters with archaic hominins. Nat Ecol Evol 3(1):14–15
Mania D, Mania U (1988) Deliberate engravings on bone artefacts of *Homo erectus*. Rock Art Res 5:91–95

Martinón-Torres M, Garate D, Herries AI, Petraglia MD (2024) No scientific evidence that Homo naledi buried their dead and produced rock art. J Hum Evol 195:103464
Mayr E (1961) Cause and effect in biology: kinds of causes, predictability, and teleology are viewed by a practicing biologist. Science 134(3489):1501–1506
Mayr E (1993) Proximate and ultimate causations. Biol Philos 8:93–94
McBrearty S, Brooks AS (2000) The revolution that wasn't: a new interpretation of the origin of modern human behavior. J Hum Evol 39:453–563
Mellars PA, Stringer CB (eds) (1989) The human revolution: behavioral and biological perspectives on the origins of modern humans. Edinburgh University Press, Edinburgh
Mendel GJ (1901) Experiments on plant hybridization (transl. by W. Bateson). J R Hort Soc 26:1–32
Mesoudi A, Blanchet S, Charmantier A, Danchin É, Fogarty L, Jablonka E, Laland KN, Morgan TJH, Müller GB, Odling-Smee FJ, Pujol B (2013) Is non-genetic inheritance just a proximate mechanism? A corroboration of the extended evolutionary synthesis. Biol Theory 7:189–195
Mesoudi A, Chang L, Dall SRX, Thornton A (2016) The evolution of individual and cultural variation in social learning. Trends Ecol Evol 31:215–225
Milks A, Lew-Levy S, Lavi N, Friesem DE, Reckin R (2021) Hunter-gatherer children in the past: an archaeological review. J Anthropol Archaeol 64:101369
Mithen SJ (2005) The singing Neanderthals: the origins of music, language, mind and body. Weidenfeld & Nicolson, London
Mivart St GJ (1871) On the genesis of species. D. Appleton and Company, New York
Moczek AP, Sultan S, Foster S, Ledon-Rettig C, Dworkin I, Nijhout HF, Abouheif E, Pfennig DW (2011) The role of developmental plasticity in evolutionary innovation. Proc R Soc B Biol Sci 278:2705–2713
Müller GB (2007) Evo-devo: extending the evolutionary synthesis. Nat Rev Genet 8:943–949
Müller GB (2017) Why an extended evolutionary synthesis is necessary. Interface Focus 7: 20170015
Neubauer S, Hublin J-J, Gunz P (2018) The evolution of modern human brain shape. Sci. Adv. 4: eaao5961
Nihei Y, Higochi H (2002) When and where did crows learn to use automobiles as nutcrackers. Tohoku Psyhcologica Folia 60:93–97
O'Brien MJ, Laland KN (2023) Culture and evolvability: a brief archaeological perspective. J Archaeol Method Theory 30(4):1079–1108
Odling-Smee FJ (1988) Niche constructing phenotypes. In: Plotkin HC (ed) The role of behavior in evolution. The MIT Press, Cambridge, MA, pp 73–132
Odling-Smee J (2024) Niche construction. How life contributes to its own evolution. The MIT Press, Cambridge, MA
Odling-Smee FJ, Laland KN, Feldman MW (1996) Niche construction. Am Nat 147:641–648
Odling-Smee FJ, Laland KN, Feldman MW (2003) Niche construction. Princeton University Press, Princeton
Pal A, Kumara HN, Mishra PS, Velankar AD, Singh M (2018) Extractive foraging and tool-aided behaviors in the wild Nicobar long-tailed macaque (Macaca fascicularis umbrosus). Primates 59:173–183
Pascual-Leone A, Amedi A, Fregni F, Merabet LB (2005) The plastic human brain cortex. Annu Rev Neuroscie 28:377–401
Perre L, Pattamadilok C, Montant M, Ziegler JC (2009) Orthographic effects in spoken language: on-line activation or phonological restructuring? Brain Res 1275:73–80
Peterson T, Müller GB (2016) Phenotypic novelty in evo-devo: the distinction between continuous and discontinuous variation and its importance in evolutionary theory. Evol Biol 43:314–335
Pettitt P (2022) Did Homo naledi dispose of their dead in the Rising Star Cave system? S Afr J Sci 118(11–12)

Pitarch Martí A, Zilhão J, d'Errico F, Cantalejo-Duarte P, Domínguez-Bella S, Fullola JM, Weniger GC, Ramos-Munoz J (2021) The symbolic role of the underground world among Middle Paleolithic Neanderthals. Proc Natl Acad Sci U S A 118(33):e2021495118
Prentiss AM (2021) Theoretical plurality, the extended evolutionary synthesis, and archaeology. Proc Natl Acad Sci U S A 118(2):e2006564118
Read D (2012) How culture makes us human: primate evolution and the formation of human societies. Left Coast Press, Walnut Creek
Reader S, Laland KN (2003) Animal innovation: an introduction. In: Reader S, Laland KN (eds) Animal innovation. Oxford University Press, Oxford, pp 1–9
Renfrew C, Zubrow EBW (1994) The ancient mind. Cambridge University Press, Cambridge
Richerson PJ, Boyd R (1978) A dual inheritance model of the human evolutionary process I: basic postulates and a simple model. J Soc Biol Struct 1(2):127–154
Richerson PJ, Boyd R (2005) Not by genes alone: how culture transformed human evolution. University of Chicago Press, Chicago
Richerson P, Boyd RT (2010) The Darwinian theory of human cultural evolution and gene-culture coevolution. In: Bell A, Futuyma DJ, Eanes WF, Levinton JS (eds) Evolution since Darwin: the first 150 years. Gruter Institute Squaw Valley Conference
Rifkin RF, Dayet L, Queffelec A, Summers B, Lategan M, d'Errico F (2015) Evaluating the photoprotective effects of ochre on human skin by in vivo SPF assessment: implications for human evolution, adaptation and dispersal. PLoS One 10:e0136090
Rosso DE, Pitarch Martí A, d'Errico F (2016) Middle Stone Age ochre processing and behavioural complexity in the horn of Africa: evidence from Porc-Epic Cave, Dire Dawa, Ethiopia. PLoS One 11(11):e0164793
Rosso DE, d'Errico F, Queffelec A (2017) Patterns of change and continuity in ochre use during the late Middle Stone Age of the Horn of Africa: the Porc-Epic Cave record. PLoS One 12: e0177298
Salagnon M, d'Errico F, Rigaud S, Mellet E (2024) Assigning a social status from face adornments: an fMRI study. Brain Struct Funct 229(5):1103–1120
Salomon H, Vignaud C, Coquinot Y, Beck L, Stringer C, Strivay D, d'Errico F (2012) Selection and heating of colouring materials in the Mousterian level of Es-Skhul (c. 100 000 years BP, Mount Carmel, Israel). Archaeometry 54(4):698–722
Scerri EML, Will M (2023) The revolution that still isn't: the origins of behavioral complexity in Homo sapiens. J Hum Evol 179:103358
Sehasseh EM, Fernandez P, Kuhn S, Stiner M, Mentzer S, Colarossi D, Clark A, Lanoe F, Pailes M, Hoffmann D, Benson A (2021) Early middle stone age personal ornaments from Bizmoune Cave, Essaouira, Morocco. Sci Adv 7(39):eabi8620
Shea J (2011) Homo sapiens is as homo sapiens was: behavioral variability versus "Behavioral modernity" in Paleolithic archaeology. Curr Anthropol 52(1):1–35
Shettleworth SJ (2001) Animal cognition and animal behaviour. Anim Behav 61:277–286
Somel M, Liu X, Khaitovich P (2013) Human brain evolution: transcripts, metabolites and their regulators. Nat Rev Neurosci 14:112–127
Sterelny K (2012) The evolved apprentice: how evolution made humans unique. MIT Press, Cambridge, MA
Stout D, Hecht E (2015) Neuroarchaeology. In: Bruner E (ed) Human paleoneurology. Springer, Cham, pp 145–175
Stout D, Hecht EE (2017) Evolutionary neuroscience of cumulative culture. Proc Natl Acad Sci U S A 114:7861–7868. https://doi.org/10.1073/pnas.1620738114
Stout D, Hecht E, Khreisheh N, Bradley B, Chaminade T (2015) Cognitive demands of Lower Paleolithic toolmaking. PLoS One 10:e0121804
Stringer C, Crété L (2022) Mapping Interactions of H. neanderthalensis and Homo sapiens from the Fossil and Genetic Records. PaleoAnthropology 2022(2). https://doi.org/10.48738/2022.ISS2.130

Tattersall I (1995) The fossil trail: how we know what we think we know about human evolution. Oxford University Press, New York

Thiebaut de Schotten M, Cohen L, Amemiya E, Braga LW, Dehaene S (2012) Learning to read improves the structure of the arcuate fasciculus. Cereb Cortex 24:989–995

Tomasello M (2014) A natural history of human thinking. Harvard University Press, Cambridge, MA

Uller T (2012) Parental effects in development and evolution. In: Royle NJ, Smiseth PT, Kölliker M (eds) Evolution of parental care. Oxford University Press, Oxford, pp 247–266

Villanea FA, Schraiber JG (2019) Multiple episodes of interbreeding between Neanderthal and modern humans. Nat Ecol Evol 3(1):39–44

Watson JD, Crick FHC (1953) Molecular structure of nucleic acids: a structure for deoxyribose nuclei acid. Nature 171(4356):737–738

Webster MM, Chouinard-Thuly L, Herczeg G, Kitano J, Riley R, Rogers S, Shapiro MD, Shikano T, Laland KN (2019) A four-questions perspective on public information use in sticklebacks (Gasterosteidae). R Soc Open Sci 6:181735

West-Eberhard MJ (2003) Developmental plasticity and evolution. Oxford University Press, Oxford

West-Eberhard MJ (2005) Phenotypic accommodation: adaptive innovation due to developmental plasticity. J Exp Zool Part B Mol Dev Evol 304B:610–618

Whitehead H, Laland KN, Rendell L, Thorogood R, Whiten A (2019) The reach of gene–culture coevolution in animals. Nat Commun 10:2405

Whiten A (2017) A second inheritance system: the extension of biology through culture. Interface Focus 7:20160142

Whiten A (2019) Cultural evolution in animals. Ann Rev Ecol Evol Syst 50:27–48

Whiten A (2021) The burgeoning reach of animal culture. Science 372:eabe6514

Wynn T, Overmann KA, Coolidge FL (2016) The false dichotomy: a refutation of the Neandertal indistinguishability claim. J Anthropol Sci 94:1–22

Wynn T, Overmann K, Coolidge F (2024) Oxford handbook of cognitive archaeology. Oxford University Press, Oxford

Zilhão J, d'Errico F, Banks WE, Teyssandier N (2024) A data-driven paradigm shift for the Middle-to-Upper Palaeolithic transition and the Neandertal Debate. Q Environ Hum 2(6):100037

On Genetic Data, Models, Storytelling and Consensus: (Re)Constructing the Demographic History of *Homo sapiens* and Its Putative Admixture with Neanderthals

10

Lounès Chikhi, Jonathan Racine, Rémi Tournebize, and Arnaud Pocheville

Abstract

Genetic data have been increasingly used in the last decades to reconstruct (and tell) what is presented as the "evolutionary history of humans". The central role of genetic data has become particularly clear in the last three decades, with the increasing role played by genomic and paleogenomic data in dominant narratives. In this chapter, we argue that genetic data are powerful and defend their use to improve our understanding of human evolutionary history. For instance, geneticists have shown that the human species is surprisingly less diverse genetically than chimpanzees and orangutans, despite a much larger population size and distribution. They have also shown that humans exhibit very little genetic differences, even at long geographic distances compared to chimpanzees. However, we also question a number of statements made by some geneticists in the last decades. Genetic data have indeed been used to make strong and possibly misleading claims even when well-supported alternative interpretations of the genetic data existed. Here, we try to clarify some important results that genetic data brought, but we also discuss the links between the assumptions of the models used by geneticists and the statements they make. We do this by using two example cases: the inference of ancient bottlenecks and hybridization events. The inferences made by geneticists can be problematic when the model

L. Chikhi (✉) · A. Pocheville
Centre de Recherche sur la Biodiversité et l'Environnement (CRBE), Université de Toulouse, CNRS, IRD, Toulouse INP, Université Toulouse 3 – Paul Sabatier (UT3), Toulouse, France
e-mail: lounas.chikhi@cnrs.fr

J. Racine
Erraphis, Université de Toulouse Jean Jaurès, Toulouse, France

R. Tournebize
UMR DIADE, Université de Montpellier, IRD, CIRAD, Montpellier, France

M. Lequin, J. M. Rodríguez Caso (eds.), *The Field of Human Evolution: Critical Perspectives from History and Epistemology*, Evolutionary Biology – New Perspectives on Its Development 11, https://doi.org/10.1007/978-3-032-13925-2_10

assumptions are unlikely to hold and when departures from such assumptions significantly change the conclusions and inferences. This may lead a whole scientific community to characterize events that may never have taken place. We discuss whether genetic data unambiguously support the different bottlenecks that have been invoked by geneticists, and the theory that some ancient *Homo sapiens* populations admixed with Neanderthals. We also consider these issues from a wider point of view in which we discuss the notions of pluralism, monism, and indeterminacy or identifiability in science and in population genetics. Given the increasing importance of genomic and archeogenomic data in the construction of narratives that are themselves increasingly used to define national or ethnic identities, it seems urgent to develop a more critical attitude towards their interpretation. We also call for a research programme where sociologists of science could help clarify why and how a community moved from the rejection of Neanderthal admixture, when genetic data did not allow us to do so, to its near-universal acceptance today, when genetic data still do not seem to allow us to make such strong claims.

Keywords

Archeogenomics · Storytelling · Bottlenecks · Admixture · Hybridization · Pluralism · Identifiability · Realism

10.1 Introduction

Genetic data have been increasingly used in the last decades to reconstruct the evolutionary history of humans on the basis of patterns observed in present-day populations (Alves et al. 2016; Auton et al. 2015; Beaumont 2004; Goldstein and Chikhi 2002; Li and Durbin 2011) as well as on the basis of genetic data obtained from ancient samples (ancient DNA or aDNA studies: Currat and Excoffier 2004; Green et al. 2006; Hajdinjak et al. 2021; Krings et al. 1997; Kuhlwilm et al. 2016; Prüfer et al. 2014; Tournebize and Chikhi 2025; Villanea and Schraiber 2019). In addition, comparisons with living great apes have also contributed to the interpretation of human genetic diversity and have thus improved our understanding of human genetic diversity (Brand et al. 2022; Lester et al. 2021; Prado-Martinez et al. 2013; Steux et al. 2025).

Altogether, genetic data have established important results that no other type of data could have established as clearly. While the objective of this chapter is to discuss some areas of human population genetics that are potentially problematic, this cannot be done without insisting on the clarifying role that genetic data have had over several decades. In this chapter, we will thus start with the positive and constructive role of genetics in establishing a fundamental or central model of human evolutionary history. We will then quickly identify two types of events (bottlenecks and admixture/hybridization), which have been inferred by geneticists, and will ask whether they are as clearly identified as geneticists claim. We will then try and explain how geneticists reach their conclusions and show that these conclusions can change dramatically depending on whether one accounts for or ignores population structure. To do that, we will try to explain some important results from the coalescent theory, the statistical theory that is central to population genetics inference. We argue that human populations are, and have probably always been, spatially organized and hence spatially structured. We will thus argue that signals of population size change (such as bottlenecks) and signals of admixture with Neanderthals that geneticists claim to see in the genetic data are actually expected, because most inferential methods ignore population structure. In other words, we will argue that it is *currently* difficult to separate spurious from real signals. We will then try to understand the current situation of human population genetics in relation to classical questions of epistemology, namely indeterminacy in science and pluralism.

10.2 On the Importance (and Some Traps) of Genetic Data

10.2.1 The Importance and Clarifying Role of Genetic Data in the Last Decades

Currently, most human evolutionary biologists (paleoanthropologists, geneticists, archeologists, etc.) agree that humans most likely expanded recently from some regions of the African continent (Bräuer and Stringer 1997; Scerri et al. 2018, 2019; Stringer 2014), following a scenario which we will call the RAO for Recent African Origin model[1] in the rest of the chapter. While researchers may not all agree on the details of that model, we can safely state that the following four results coming from genetic data have been central in establishing the RAO and that these four results are difficult to question (in the present state of knowledge).[2] Firstly, the genetic diversity of humans is lower than that of chimpanzees and orangutans, despite a much wider geographic distribution and a much larger population size in humans compared to the distribution of these and all other apes (Lester et al. 2021; Prado-Martinez et al. 2013; Steux et al. 2025). Given that genetic diversity is expected to be positively correlated with population size, the lower genetic diversity of humans compared to other apes is difficult to reconcile with models where present-day humans derive from a species that was present in all continents (mainly Asia and Africa) for very long periods of time, hence the "recent" origin.

Secondly, genetic differentiation between human populations is also more limited than between chimpanzee populations at comparable spatial scales (Lester et al.

[1] The name RAO appears to us preferable to previous names such as "Noah's Ark" (Howell 1976; Wolpoff 1989) or "the Garden of Eden" model (Ambrose 1998; Harpending et al. 1993), or Out of Africa (OOA) or ROA (Recent Out of Africa) because of the religious or colonial flavours of these previous designations. To make the latter clear: no, humans did not "leave Africa". Last time we heard, humans still live in Africa.

[2] But who knows what the future will bring?

2021; Ramachandran et al. 2005). This means that when we compute the amount of genetic differences between humans living on different continents, we find values that are smaller than between chimpanzees living hundreds of kilometres away from each other. This suggests that human populations have been genetically more connected to each other than chimpanzee populations among themselves. It is also likely due to the fact that humans arrived where they are (i.e. in all continents) recently, and have been sufficiently connected to impede the accumulation of genetic differences between them. This also explains why human populations harbour a large proportion of total human genetic diversity (Barbujani et al. 1997; Barbujani and Chikhi 2007; Fujimura et al. 2014; Lewontin 1972).

Thirdly, genetic diversity is higher across the African continent and in African populations compared to human populations from other continents, and that genetic diversity decreases with geographic distance from the African continent (Prugnolle et al. 2005; Ramachandran et al. 2005).

Fourthly, genetic differences between human individuals and populations tend to increase with geographic distance. This pattern, known as isolation by distance (IBD) (Wright 1943), is indicative of spatial population structure in humans, and expected in models where gene flow is mainly (mainly, but not only) taking place between neighbouring populations. This pattern, together with the previous point, is consistent with a colonization process starting in Africa with a stepwise or continuous loss of genetic diversity as human groups colonized new lands across Africa and elsewhere (e.g. Liu et al. 2006). This global expansion process can be seen as a series of founder events that took place whenever groups of individuals expanded into new regions, and only took with them a sample of the genetic diversity from their area of origin.

Altogether, these genetic patterns (low diversity, low differentiation, higher diversity in African populations, with a decrease of diversity as distance from the African continent increases, and IBD pattern) have been confirmed across genetic studies and thus are likely here to stay. Genetic data have thus helped establish a general model that states that modern humans, in their current distribution, result from a relatively recent expansion from African populations that colonized the planet in the last 100,000 years (100 ky) to give a very rough time scale (Nielsen et al. 2017). Yet, some details of the RAO model have changed through time, as is expected for a model or set of models that were proposed more than 40 years ago, when genetic data were much more limited. We would like now to mention two types of "demographic" events that have been put forward within the RAO model, namely bottlenecks and hybridization (or admixture) events.

10.2.2 The RAO With or Without Bottlenecks and Admixture?

The idea that bottlenecks are important and have impacted our genome appears to trace back to a seminal article by Haigh and Maynard Smith (1972). The authors suggested that the low diversity observed in some human proteins was incompatible with a human population that had always been large, but could be explained if humans had gone through an ancient bottleneck.[3] After this article, and with the accumulation of genetic data, the idea that humans might have survived as a small population at a crucial moment of their evolutionary history gained momentum and started to be mentioned in an increasing number of studies, with variations regarding whether humans had been subject to one or several bottlenecks. Also, some studies suggested that bottlenecks had influenced all human populations, while others stated that bottlenecks only occurred in non-African populations (Ambrose 1998; Jones and Rouhani 1986).

A number of details, including the dating of the main events, were debated then, and are still debated today. For instance, Wolpoff (1989)[4] noted that "the main elements of the Garden of Eden hypothesis are that modern humans *arose recently* from a small source population, approximately 200,000 years ago in Africa, and that this population was *a new species*. It is contended that this new species split into two main branches in Africa and that one of these passed through a bottleneck of small population size in the process of emigration from Africa".[5] Ambrose (1998), reviewing the literature of the time, defended the idea that *all humans* suffered a bottleneck, possibly in relation to the Toba eruption around 75 kya, after an expansion from a small African region (hence implicitly assuming that humans located elsewhere were replaced). Despite the differences between the narratives, a recurring idea, however, has been that modern humans expanded recently from a small African region and that populations were or became small as a result of one or several bottlenecks.

The idea that humans may have nearly gone extinct or were surviving in small numbers seems to invoke images of a fragile species that fought for survival in a dangerous environment, for a possibly long period, risking its survival every day. This idea or scenario seems to be a popular alternative to another opposite scenario where humans are the dominant species that was meant to rule the planet from the very beginning. It would be interesting to analyse why, when and how these two opposite narratives are invoked and by whom. Coming back to bottlenecks, the vision of a species on the brink of extinction, whose success is attributable to random factors, and which could have gone extinct any time, appears to have caught the attention and interest of the scientists, journalists and wider public. So much so that recently, Hu et al. (2023) claimed that the human lineage nearly went extinct around 900 kya (i.e. earlier than the Neanderthal-Sapiens split). This statement that has been criticized by two recent studies claiming that the data do not support such a bottleneck (Cousins and Durvasula 2025; Deng et al. 2025), and the original authors

[3]These authors were very careful in showing that the observed results (low diversity) could be explained by very different types of bottlenecks, such as long and mild bottlenecks or shorter but stronger bottlenecks.

[4]Wolpoff did not defend the idea of bottlenecks but rather a Multi-Regional (MR) model of human origins.

[5]Italics are the author's.

are currently answering, claiming that the bottleneck did happen (Haipeng Li, personal communication). Interestingly, a decade ago, Mazet and collaborators (Mazet et al. 2016) analysed genomic data from humans and suggested that an "apparent bottleneck" around 900 kya could be explained by changes in connectivity if we assumed that humans were structured. We will come back to these questions in the next section, but the point is that genetic data have been repeatedly used to claim that bottlenecks of different types may explain some patterns of genetic diversity observed today in humans, even when alternative interpretations existed.

With the advent of ancient genetic data (i.e. obtained from samples dating hundreds, thousands or tens of thousands of years), the other issue that has generated a major interest and fascination is the putative hybridization of *Homo sapiens* with so-called "archaic" *Homo* species, such as Neanderthals (Green et al. 2006; Krings et al. 1997),[6] that they likely encountered during their spatial expansion across the world (but see Amos 2019, 2020; Eriksson and Manica 2012; or Tournebize and Chikhi 2025) for different interpretations of the so-called signals of "archaic" introgression).[7]

The current consensus on that question among geneticists and paleoanthropologists is that admixture with Neanderthals is well established (e.g. Hajdinjak et al. 2021). It has now replaced the opposite consensus, which dominated the field 20 years ago, and which rejected such hybridization (Krings et al. 1997), despite the existence of studies questioning this narrative on the basis of genetic or non-genetic data (Goldstein and Chikhi 2002; Nordborg 1998; Zilhão 2006, 2010). The last 10–15 years have thus seen a major shift in the human population genetics community. Most of (if not all) the data that have led to this change of "paradigm" have been produced by studies interpreting paleogenomic data. Ancient DNA data have thus become central in the construction of a new narrative where what was deemed negligible or impossible in the past, had become indisputable.

One common argument to justify the change in consensus is that the genetic data and methods changed and improved, and this is how science works: new evidence, new conclusions. What we will defend here is that genetic data are difficult to interpret, and this has been well known for at least three decades. People who rejected admixture vigorously did it with arguments that they would have difficulty

[6]Like all simplifications, the vision presented here misses the diversity of studies and opinions that has always existed. A number of authors defending Multi-Regional (MR) models of human evolution have also claimed that there was no clear data favouring a bottleneck in the recent past. These authors defend the idea that people currently living in Asia are the descendants of *Homo erectus* ancestors that colonized Eurasia more than a million years ago. These theories have been discredited even though they regularly come back under different guises (Chen and Huang 2020; Wolpoff 2020).

[7]This question is indirectly related to the MR models mentioned in the previous note. MR models sustain the idea that the expanding *sapiens* African populations did not replace the Eurasian "archaic" populations that they encountered. Instead, they contributed marginally to their genetic make-up, thus maintaining some form of geographical continuity, with little "African" input in Eurasia. See Bräuer and Stringer (1997).

to defend today. And those who now defend admixture tend to ignore the many inconsistencies of admixture models. In short, statements against admixture were inconsistent with existing knowledge at the time. The argument of "new data, new conclusions" does not seem to actually hold.

10.2.3 The Pitfall of Model Reification

The common thread between the detection of bottlenecks or admixture events using genetic data is that to make such inferences, geneticists need to use so-called demographic models. Such models need to make a number of assumptions, and one central assumption is that human populations have been panmictic (i.e. mating between humans is random) over very large areas (typically across continents) and for very long periods (typically tens of thousands of years). In other words, the models detecting admixture or population size change generally assume that human populations are not structured or that this structure is negligible. The problem with that assumption is that population geneticists have known for at least two and a half decades that this assumption will lead to the inference of spurious events (Beaumont 2004; Chikhi et al. 2010; Eriksson and Manica 2012; Mazet et al. 2016; Tournebize and Chikhi 2025; Wakeley 1999). They also know that humans are spatially structured, as noted in the first section.

This chapter defends the idea that genetic data are extremely important and supports a model where humans expanded from the African continent in the recent past. We claim that the details of what happened are unknown to us but that the best model or framework is a still-to-be-clarified version of the RAO that must incorporate population structure. We defend the idea that population structure is and was likely important throughout human evolution (e.g. Arredondo et al. 2021; Goldstein and Chikhi 2002; Ragsdale et al. 2023; Scerri et al. 2018, 2019).

At the same time, we stress that population structure should not be reified. The models of population structure that we are referring to have nothing to do with multiregional models as rightly said by Chris Stringer more than a decade ago (Scerri et al. 2019; Stringer 2014). The use of structured models does not lend support to a division of humans into homogeneous (racial) groups, or any ancient separation of humans into homogeneous (racial) units. The structured models discussed here are not *the* reality. They are just that, models, and as such are here to help us analyse and compare alternative evolutionary scenarios in a controlled manner (Chikhi 2023), though they cannot capture the full diversity of present-day humans. The demes in these models do not correspond to real populations. *Homo sapiens*, as we know it, is a species where genetic diversity is distributed continuously. The idea that it is difficult if not impossible to identify neatly and clearly defined groups in humans was clear to Darwin more than 150 years ago (Darwin 1872),[8] and has been demonstrated over and over again since then. If anything, it is a failure of the genetics and anthropological community towards society that we still need, in 2025, to write such words. Population geneticists should try to stop falling into Whitehead's fallacy (i.e. believing in theoretical objects as if they were real).

If, as we argue, humans are and were spatially structured throughout their evolution (like most known primates and mammals), and if, as we argue, incorporating population structure is *sufficient* to explain genetic signatures that are interpreted as bottlenecks or admixture events, this raises serious questions regarding the reality of demographic events that are discussed by the scientific community. In 2020, anthropologist Jonathan Marks discussed the concept of apophenia, which he described as "the perception of a pattern where there is no pattern" (Marks 2019).[9] He argued that the history of biological anthropology and human population genetics exhibited such apophenia, for instance when discussing race or ancient species.[10] We wonder whether our community has not again constructed apophenias. We discuss these problems by asking questions about narratives, models and paradigms, the epistemic and pragmatic values of models, monism and pragmatic pluralism, and ask if we should not tend towards a pragmatic pluralism and a prudent agnosticism.

10.2.4 Population Genetics and the Lab Coat Effect

In the next section, we try to clarify the process by which population geneticists infer demographic events such as bottlenecks, expansions or admixture with genetic data. To do that, we introduce some concepts from the coalescent theory because it is the theory that has been at the heart of demographic inference for at least three decades. Readers not interested in these technical issues may want to ignore this section. It is however important to realize that one reason why it is easy to make up narratives

[8] In Chapter VII of The Descent of Man, Darwin wrote: "Man has been studied more carefully than any other animal, and yet there is the greatest possible diversity amongst capable judges whether he should be classed as a single species or race, or as two (Virey), as three (Jacquinot), as four (Kant), five (Blumenbach), six (Buffon), seven (Hunter), eight (Agassiz), eleven (Pickering), fifteen (Bory de St-Vincent), sixteen (Desmoulins), twenty-two (Morton), sixty (Crawford), or as sixty-three, according to Burke. This diversity of judgment does not prove that the races ought not to be ranked as species, but it shows that they graduate into each other, and that it is hardly possible to discover clear distinctive characters between them".

[9] Statisticians Efron and Tibshirani wrote: "Left to our own devices we are not very good at picking out patterns from a sea of noisy data. To put it another way, we all are too good at picking out non-existent patterns that happen to suit our purposes. Statistical theory attacks the problem from both ends. It provides optimal methods for finding a real signal in a noisy background, and also provides strict checks against the overinterpretation of random patterns" (Tibshirani and Efron 1993).

[10] J. Zilhão has also argued that the separation of Neanderthals from the rest of *Homo sapiens* can be seen as such an apophenia (pers. comm.).

with genetic data is because too many people do not (wish to) understand (or care about) the specific properties of genetic data (Chikhi 2009, 2023). It is much easier to write sentences describing the spread of mutations or haplogroups, based on visual associations the way people interpreted morphological data in the nineteenth century, than to test alternative hypotheses using the simulation or theoretical tools that have been available for several decades. It is easier to make sweeping statements mixing terminology from anthropology, archeology, linguistics and genetics without clarifying concepts that may have little to do with each other, and without clarifying that the objects analysed have very different properties. Genetic data have specific properties that people called population geneticists try to understand. It is thus potentially problematic when researchers with little or no training in population genetics make statements that go beyond what most population geneticists would consider reasonable. One potential consequence is that genetic data can be used to reify poorly defined concepts such as that of ancient "people". If it is difficult to define words like "French" or "American" today, then how likely is it that the words like "Dorians", "Myceneans" or "ancient Greeks" (and the people behind them) are properly understood today, hundreds or thousands of years later.[11] The use of genetics to "validate" such groups is worrying. This is however not new.

From the mid-1990s to the end of the 2000s, a number of geneticists (together with archeologists) used what are called haplotypes or haplogroups to tell stories of ancient migrations without (to our knowledge) testing them against alternative scenarios. These stories were based on mitochondrial DNA and Y chromosome data (which behave as a single genetic locus, each) and have had a huge success in some scientific journals such as the *American Journal of Human Genetics* and in the public, with several books published. In parallel to these publications, genetic ancestry companies started using genetic data to tell people who they "really were" (Kennett et al. 2018). This happened despite the fact that these genetic studies were criticized from the very beginning for their lack of connection with what population geneticists understood of genetic data (Barbujani et al. 1998; Chikhi et al. 1998, 2002). For instance, some studies linked the "age" of haplogroups with demographic events, and mutations with migrations of ancient people. Several studies seem to read genetic data like one reads palms or tells fortune.[12] This led one of us to use the terms "astrologenetics" and "mytho-chondrial" (Chikhi 2009). Genetic data and geneticists benefit from a high social status within science, making many of their claims look like serious science. This is a bit like these adverts where a

[11]Middleton, GD. 2022. Mycenaeans, Greeks, Archaeology and Myth: Identity and the Uses of Evidence in the Archaeology of Late Bronze Age Greece. Eras journal. Available at:
https://www.monash.edu/arts/philosophical-historical-indigenous-studies/eras/past-editions/edition-three-2002-june/mycenaeans-greeks-archaeology-and-myth-identity-and-the-uses-of-evidence-in-the-archaeology-of-late-bronze-age-greece

[12]Mark TG. 2013. To Claim Someone Has "Viking Ancestors" Is No Better than Astrology. The Guardian. February 25. Available at:
http://www.theguardian.com/science/blog/2013/feb/25/viking-ancestors-astrology

person with a lab coat sells you a toothpaste or washing powder. And some toothpastes and washing powders are very good, indeed.

Genetic data are absolutely unique in the way they allow us to improve our understanding of human evolution and disease but, using the words of human population geneticist Mark Thomas (and co-authors), there is a difference between storytelling and story testing (Gerbault et al. 2014).

Readers who are still not interested in the technical issues which we identify around the interpretation of genetic data, but wanting to impress colleagues at the pub or in a bar, can ignore most of Sect. 10.3, and start reading again from the last paragraph of Sect. 10.3.7, or from Sect. 10.3.8.

10.3 Demographic Inference: Panmictic and Structured Models Under the Coalescent

10.3.1 Demographic Inference and Genetic Data

Population geneticists use genetic data from present-day or ancient samples (DNA from fossils or museum specimens) to reconstruct or infer the recent evolutionary history of species. They try to make sense of the *genetic patterns* observed in the data, by inductively trying to reconstruct the putative historical processes that contributed to the observed patterns. Typically, population geneticists will write that they aim at reconstructing the "demographic history" of a species. This phrase is often used to describe a history of population size changes (e.g. Beaumont 1999; Boitard et al. 2016; Fu 1997; Hu et al. 2023; Li and Durbin 2011; Rogers and Harpending 1992; Slatkin and Hudson 1991) even if the demographic event of interest can vary across studies and could correspond to other types of events. For instance, these genetic data can also be used to detect and date ancient admixture or gene flow events (Beerli and Felsenstein 1999; Chikhi et al. 2001; Hey and Nielsen 2004), or selective processes in the genome. More recently, the geneticists' focus has shifted towards the detection of hybridization events with so-called archaic humans (Abbott et al. 2016), and this has even led some authors to try and infer their putative consequences for human health (Dannemann 2021; Gregory et al. 2021; Pauly et al. 2024; Simonti et al. 2016; Zeberg et al. 2020; Zeberg and Pääbo 2020). These statements play a major role in how we see ourselves as a species and how we describe humans in relation to their geographical origin.

To make quantitative statements about such putative ancient events, geneticists need to determine if the observed genetic data are more or less likely under alternative or competing demographic models. For example, they will ask whether the patterns of genetic diversity observed in humans are more likely the result of a stationary, expanding or decreasing human population. Similarly, geneticists may ask whether a model of recent hybridization between *Homo sapiens* and Neanderthals is more likely than a model of shared ancestry but without hybridization, to explain the presence of DNA fragments in the human genome that resemble Neanderthals DNA fragments (Eriksson and Manica 2012; Tournebize and Chikhi 2025).[13] Calculating the relative likelihood of parameters (or models) can be very complicated even under the simplest models of evolution, and has been at the heart of population genetics inference for several decades (Beaumont 1999; Beerli and Felsenstein 1999; Chikhi et al. 2001; Hey and Nielsen 2004; Kuhner et al. 1995; Nielsen 2001). Most of that research has been based on the coalescent theory (Donnelly and Tavaré 1995; Griffiths and Tavaré 1994; Kingman 1982) that allowed geneticists to make sense of increasingly complex data.

Raw genetic data can be indeed highly complex, even when researchers analyse small samples genotyped for a limited number of genetic markers. The data are thus typically summarized through the computation of numbers (summary statistics) or distributions (histograms).

Let us assume that we sample 20 individuals and look at 1000 different DNA fragments from their genomes (called loci, singular: locus), and let us assume that each one of these loci is of length 500 base pairs (bp)[14]. Since each individual has two parents and inherits one chromosome from each parent, each locus will be found in two copies within the individual (these copies are not necessarily identical, as the two parents may have different copies). Under this hypothetical example, population geneticists would typically summarize the data using a few meaningful statistics, like the number of distinct alleles at a given locus, or the heterozygosity (the probability that two alleles in a random individual at the same random locus are different). By making such a choice, geneticists reduce the amount of genetic information, but this is necessary for computational or theoretical reasons. Different genetic studies will thus use different types of genetic data and different types of summaries. Some of these summary statistics have been studied for decades, and their statistical properties are increasingly well understood (Lapierre et al. 2017; Nei et al. 1975; Slatkin and Hudson 1991; Tajima 1989a, b). Others are presented or introduced with little testing and are perhaps interpreted too rapidly (Durand et al. 2011), and the published interpretation may only be valid under specific assumptions (as we will see below).

In brief, geneticists summarize possibly large genetic data sets with statistics and histograms. They then need a theoretical framework to predict how these statistics and histograms would behave under alternative scenarios. The coalescent theory represents such a theoretical framework (Kingman 1982; Tajima 1983). For more detailed descriptions, see the following references: Hudson 1990; Nordborg 2019.

---

[13] See also Amos (2019, 2020) and Amos and Elhaik (2025) for another interpretation of the signals of admixture.

[14] DNA is split into chromosomes, each representing a continuous linear sequence of nucleotide bases.

10.3.2 The Coalescent Theory as a Way to Simulate Genetic Data

The standard coalescent theory focuses on a DNA fragment (a locus) that is long enough to accumulate mutations (and thus diversity) and short enough to not recombine.[15] These assumptions mean that when we sample this locus in different individuals today, we can consider that there is one ancestral sequence from which all observed (sampled) sequences descend. In other words, these sequences are connected by a genealogical tree (a gene tree, in fact) that leads backward in time to their most recent common ancestor (MRCA).[16] As we move backward in time, every time the lineages of the sampled sequences share a common ancestor, we say that there is a *coalescent* event, or that the lineages *coalesce*. Forward in time, we would say that the lineages *split* or *diverge*.

Let us now take the simplest and smallest genetic sample possible, one individual for which we genotype one non-recombining locus. This individual has two copies of the locus, one transmitted by its mother and the other by its father. Since we assumed that this locus did not recombine, there are two lineages that go through the two parents of the sampled individual. These copies were themselves inherited from one of the two parents of the individual's parents. This goes on backward in time, generation after generation, through the different ancestors until the two lineages *coalesce* in the unique ancestral sequence. If we now looked at another locus, we would have the same process, but the ancestral copy would be in a different individual at a different time. There is thus a statistical distribution of coalescence times across the genome.[17] The coalescent theory tries to understand the statistical properties of this distribution of coalescent times across the genome under different evolutionary models.

The process that we have just described for two sequences of the same locus sampled today in one individual can be extended to more sequences of the same locus sampled across several individuals. The problem is more complex for several reasons. If we start with 20 DNA sequences today (sampled from 10 individuals, each having two copies), there will be 20–1 = 19 coalescence events until we reach the MRCA of the sampled DNA sequences (assuming that exactly two lineages coalesce at each coalescence event). After the first coalescence, there will be 20–1 = 19 lineages, which will become 18 lineages after the second coalescence, and so on as lineages coalesce. These 19 coalescence events are represented by 19 coalescence times, which can also be seen as random variables. The coalescent theory is thus the probabilistic theory that studies the statistical properties of these

[15] During the production of the germ cells ("meiosis"), the maternal and paternal copies of each pair of chromosomes pair up and can exchange segments at random points. This process is known as "recombination". Here, we assume for simplicity that the locus has not been affected by such recombination events in the recent past.

[16] The MRCA of another locus will likely have a distinct origin and age. Mutations, if they do not have an effect on the reproductive success of the individuals, do not change genealogical relationships.

[17] And across space.

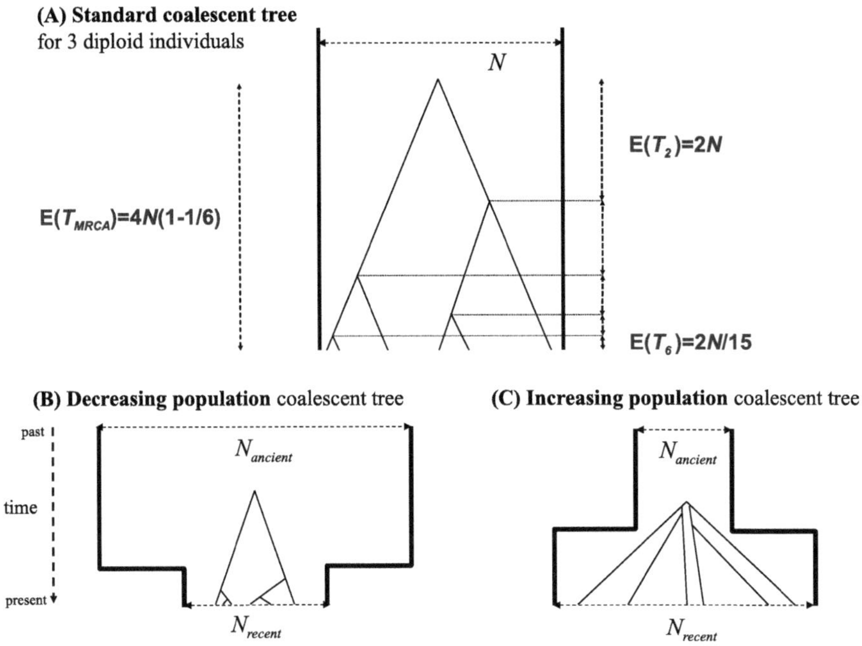

Fig. 10.1 Expected coalescent trees under (**A**) stationary, (**B**) decreasing and (**C**) increasing populations. For the standard tree expected under stationarity, the coalescence times are provided for a sample size of three diploid individuals—this corresponds to six copies of the same locus, which are represented as six leaves or tips in the tree. $E(T_{\mathrm{MRCA}})$ is the expected time to the most recent common ancestor for all sampled individuals. $E(T_k)$ is the duration of the period where the coalescent tree has k lineages. For the cases where the population size is decreasing (**B**) or increasing (**C**), we represent schematic topologies of the typical trees expected under the coalescent. In all cases, these topologies are expected only under the assumption of panmixia and isolation. Note that the tree in (**B**) (decreasing population size) has a shape that is similar to that of the stationary structured population represented in Fig. 10.2

coalescence times across individuals, and along the genome (as we saw above). The coalescent theory thus tries to understand the properties of gene trees along the genomes of sampled individuals. Panel A of Fig. 10.1 shows a standard coalescent tree for three sampled individuals (thus six sequences of one locus). When the first coalescence event takes place, the number of lineages decreases from six to five, and so on until one reaches the MRCA (the "unique ancestral DNA sequence") of that locus for that sample.

10.3.3 Some Simple Consequences of the Standard Coalescent Theory

It may be important to give here a few results from the coalescent theory because some of these results are not intuitive and are a good reminder of the fact that we humans do not always have a very good intuition for numbers. For example, one could ask whether the coalescence time of the two sequences sampled in the individual above should be counted in tens, hundreds or thousands of years. The standard coalescent theory shows that if we assume that the constant-size population where our individual was sampled is panmictic (well-mixed)[18] and isolated, this time is expected to be $2N$ generations on average, where N is the size of the population. For instance, if the population size is $N = 1{,}000$, it will take 2,000 generations, on average, to reach the coalescence time of two DNA sequences. Assuming a generation time of 20 years for humans, the average coalescence time of the two alleles is about 40,000 years. If $N = 10{,}000$, this average time becomes 400,000 years. This time is surprisingly long, and much older than the oldest fossils attributed to *Homo sapiens*. This is a very important result because it suggests that finding very divergent alleles in humans does not necessarily require the intervention of archaic populations (whatever this term may mean[19]) in *Homo sapiens* history[20].

Thus, one central result of the standard coalescent theory is that coalescence times are functions of N, the population size. This means that genetic data contain information about population size, and this is exactly why population geneticists have tried to detect changes in population size over the last decades, as we discussed above.

Another interesting and non-trivial result is that if we now sample a larger number of individuals, the coalescent theory shows that the MRCA of all the sampled copies should be found around $4N$ generations ago.[21] Thus, the time to the MRCA of a large sample is only twice as large as the age of the MRCA of two randomly sampled genes. Again, we need to repeat some of the main assumptions: constant size N, panmixia, isolation, no selection, no recombination within the locus of interest.

10.3.4 Population Sizes: Tricky Concepts

Population geneticists claim to reconstruct past population size trajectories. It is important to realize that the concept of population size is complex when addressed from a genetics point of view. There is large literature on this subject and we can only mention a few issues here.

---

[18] Panmixia refers to random mating, the standard, by default assumption of population genetics models. Populations are said to be panmictic when individuals from every new generation are the result of the random mating between individuals of the previous generation. Mating thus happens irrespective of putative environmental, spatial or social constraints. Even structured models assume the existence of subpopulations, or demes that are themselves panmictic. This means that one assumes that mating is more or less random at a local scale but not random at a continental scale.

[19] Who was not "archaic" at the time, anyway? See Scerri et al. (2019).

[20] Coalescence times follow exponential distributions, which are skewed towards small values but have large variances.

[21] For a sample of size $2k$, where k is the number of diploid individuals, the expected time to the MRCA is $T_{\text{MRCA}} = 4\,N(2k - 1)/(2k)$. For $k = 1$ individual, we obtain $2N$ generations with this formula, as noted above.

Real populations are very complex, with individuals belonging to different age classes (infants, juveniles, adults, etc.) living at the same time. This means that the number of individuals actually reproducing is not the same as the total (or census) population size. Population geneticists use simple population models to represent the reality, such as the Wright-Fisher model (WF model) which assumes just one type of individual in any generation (i.e. all individuals can and do reproduce). In addition, geneticists often put themselves at the scale of the *sequences* rather than that of the human individuals (which have *two* copies of each sequence), considering what they call *haploid* models rather than *diploid* models.[22] In the haploid WF model, the population size is therefore unambiguously defined as the number of sequences at each locus. In order to connect the WF model to real-life populations, geneticists use the concept of *effective population size*, N_e. N_e can be seen as the size of a constant-size and isolated population that would have the same *properties* as the real population of interest. Population geneticists rapidly realized that depending on the property of interest (genetic diversity, genetic "differentiation" through time, inbreeding…), different concepts and values of N_e were necessary (Charlesworth 2009; Crow and Kimura 1970).

To avoid discussing technical details of this rather confusing issue, we will assume here that the N_e concept is acceptable as a first approximation. We will thus assume here (just in this section) that estimating N_e makes sense.[23] For *Homo sapiens*, population geneticists tend to consider that N_e is roughly ten thousand (conditional on a largely accepted mutation rate), because this is the size of a constant-size population that would have a level of genetic diversity similar to that observed in humans.[24] As a comparison, chimpanzees have more genetic diversity than humans and have thus been attributed[25] a larger N_e. If we take the N_e of *Homo sapiens* as a first approximation, it means that when we sample several individuals, the time to the MRCA for a non-recombining locus should be on average ~ 800,000 years. This time is older than the estimated split between Neanderthals and *Homo sapiens*.

We must stress that these numbers are averages and that many loci will have MRCAs that are much more recent[26] and quite a few will have MRCA times that are

[22] The results we present here are also based on haploid models, as those are much more common.

[23] When species are spatially distributed or organized in complex social groups with different mating systems, they will exhibit properties that differ from those of panmictic populations as shown by many studies (Chikhi et al. 2018; Parreira and Chikhi 2015; Parreira et al. 2025; Sjödin et al. 2005), making the notion of N_e disputable under some circumstances.

[24] Some studies suggest higher values, between 10,000 and 20,000. Times would then be even larger.

[25] Generously, we must admit.

[26] Around 63% of the values of an exponential distribution of mean one are smaller than one, for example. Thus, if Ne is 10,000, that means that for a large sample 37% of the genome has coalescence times older than 800,000 years. This also means that there has been a long period during which coalescences could have taken place in the species that was ancestral to the Sapiens and Neanderthal lineages. Note though that in real data, the situation is more complex as population size and structure must have changed, and this could shift the distribution towards recent times.

older, possibly much older than one million years, when *Homo erectus* roamed the earth.[27] Our genome is thus a mosaic of loci, some of which accumulated diversity over thousands of years and others over hundreds of thousands of years, and a tiny proportion may have accumulated mutations for more than one million years without being lost by genetic drift (the stochastic process representing variation of allele frequencies as a consequence of the finite size of populations).

These computations are approximations which assume that there is such a number as N_e and that we can approximate reality by a simple WF model, and that mutations are neutral (that they have no effect on the reproductive success of the individuals which carry them). One consequence of such assumptions is that the rate of coalescence is inversely proportional to N_e: the smaller the population, the higher the coalescence rate (and the smaller the coalescence times). Another consequence of assuming that populations are panmictic and isolated is that if we cannot explain the distribution of coalescence times and genetic diversity observed in real data, by assuming one single N_e value, we will need to assume that N_e changed through time, because this is the only parameter that can be varied in a WF model, if population structure is ignored. Alternatively, we may invoke hybridization events (or selection) to improve the fit to the data.[28] We will thus need to "invent" population size changes, selection or hybridization events to improve the fit between our data and those produced by the model because the data cannot be explained by a stationary WF model. This does not mean that hybridization, population size change events or selection do not exist or did not influence our populations. This just means that when we make some assumptions (panmixia and isolation) this will likely generate some effects that are well known on the inference (the detection of spurious size changes and hybridization events).

10.3.5 From Coalescence Times to Genetic Data and Patterns in a Complex World

If one takes the two sequences of our sampled *Homo sapiens* individual, the total number of mutations between the two copies should be equal to the number of mutations arising on each of the two branches connecting the two sequences to their MRCA.[29] Thus, the coalescent times and the genealogical tree expected under a particular demographic model help predict the number of mutations between the sampled sequences and thus the amount of genetic diversity. The coalescent theory can therefore help us compare the genetic data expected (or simulated) under

[27] Eery music to accompany this sentence, thrills in the public.

[28] And this hybridization event can originate from a source population for which there is potentially no evidence that it ever existed.

[29] Thus equal to $2 \times T_{\mathrm{MRCA}} \times \mu$, where μ is the locus mutation rate and T_{MRCA} the time to the MRCA. But note that this makes other assumptions regarding the mutational process, that we will not develop here (e.g. infinite allele model).

different evolutionary models, with the genetic data observed from real individuals. Again, the calculations or computer simulations carried out by geneticists must be taken with a grain (and sometimes a shovel) of salt. The predicted coalescent times depend on a number of strong assumptions. In real life, populations and species have complex demographic histories with bottlenecks which may eliminate many alleles, and thus reduce the age of the MRCAs across the genome. Populations can expand spatially and then spread some alleles more than others just by chance (a process called "surfing" that can be mistaken for selection, and first described by Edmonds et al. 2004). Some regions of the genome are under selection and will drive the dynamics of the genome around them, even if the latter behaves neutrally, a process called "hitchhiking" or "background selection" (Charlesworth 1994; Smith and Haigh 1974). Some individuals may have significantly more surviving children as a result of stochastic processes (no selection being involved). This is referred to as variance in reproductive success, a process that will change the shape of coalescent trees, at least in the recent past. Social structure may also generate departures from the standard coalescent and possibly from the structured coalescent (Parreira and Chikhi 2015; Parreira et al. 2025).

We must thus be very modest in the way we interpret genetic or genomic data under specific models. The coalescent theory also assumes that the sample size is much smaller than the population size. For human evolution, this is generally not seen as a problem, but it should be kept in mind when bottlenecks or structured populations are invoked. For instance, many researchers think that Neanderthals likely lived in small populations, at least for those individuals for whom we have genomic data.[30] There is ongoing research that tries to understand the robustness of the coalescent framework to many of these departures that are known to occur in real species. This may explain why some of us are surprised by research invoking events for which we have zero or close to zero observational data when so many factors that are known to occur are so easily brushed away if not under the proverbial carpet.

10.3.6 From the Standard WF Model to Models with Population Size Variation and Population Structure

Theoreticians have extended the standard coalescent theory to the case where population sizes change through time (Griffiths and Tavaré 1994) and to the case where populations are structured in space (Herbots 1994; Notohara 1990; Wilkinson-Herbots 1998). These and other studies allowed the development, in the 1990s, of methods and statistics to detect, date and quantify population size changes from genetic data (Slatkin and Hudson 1991; Tajima 1989a, b). Population geneticists showed that some statistics were sensitive to selection, expansions and bottlenecks, making them interesting indicators of selection or population size

[30] Some of us also work on endangered or domesticated species where changes in population size are known to have taken place, and makes us perhaps more aware of this issue.

changes (Beaumont 1999; Fu 1997; Rogers and Harpending 1992). They also quickly understood that many statistics were not the silver bullets some had expected them to be. The reason for this is that these statistics were sensitive to several processes (Nielsen and Beaumont 2009; Tajima 1989a, b).

To understand why and how population structure and population size changes (or admixture) are connected, it is necessary to understand how gene trees are influenced by these factors. To do that, we will describe how coalescence rates are influenced by small and large population sizes and by gene flow, and will focus on two extreme examples. We must remember that, for the same sample size, the coalescence rate is higher when N is smaller. If we now take a population that has changed size T generations in the past, from a large N in the past to a low N in the present (Panel B in Fig. 10.1), we can predict that coalescences will happen frequently in the recent past because N is small and less frequently in the ancient past when N was large. We thus expect short branches close to the tips of the tree and long branches close to the MRCA, as can be seen in the gene tree Fig. 10.1B. If we now imagine the opposite scenario with a large N in the recent past and a small N in the ancient past (Fig. 10.1C), we expect, by the same reasoning, to see longer branches in the recent past and an acceleration of coalescent events in the ancient past, close to the MRCA, when N was small.

These two extreme scenarios are just toy examples, and the full mathematical treatment of the distribution of coalescence times under more complex histories of population size change was done by several theoreticians in the 1990s (Donnelly and Tavaré 1995; Griffiths and Tavaré 1994), under the assumption that the population studied is panmictic and isolated. This is why and how population geneticists have been able to infer changes in population size, to explain the data when the latter exhibited departures from the standard constant-size coalescent model (Fig. 10.1A).

We can now turn to structured populations to understand how they will generate genealogies that look like those of a panmictic population changing size. Let us now assume that a species is organized as a set of n panmictic populations connected to each other by gene flow, and let us assume that they all have the same size, N, that is constant through time. We also assume that connectivity (gene flow between populations) is constant through time. This model is one of the simplest imaginable because there is population structure (we have n connected populations), but there is no space (since every population is connected to all the others in the same way: every population is at the same "distance" from all other populations) and there is no change in size, connectivity and number of populations. This model (the n-island model) was introduced by Wright in the 1930s (Wright 1931). Altogether, the total population size of the species is $n \times N$.

If we now sample individuals in *one* of the populations, we expect from standard coalescent theory that the coalescence rate will depend on N, the size of the sampled population. However, since that population is connected by gene flow to other populations, there is a certain probability that some of the lineages sampled today in the local population arrived by gene flow in the past (as in Fig. 10.2). This probability depends on the rate of gene flow (migration rate, m) between populations. If a lineage arrived by gene flow at some point in the past, it means

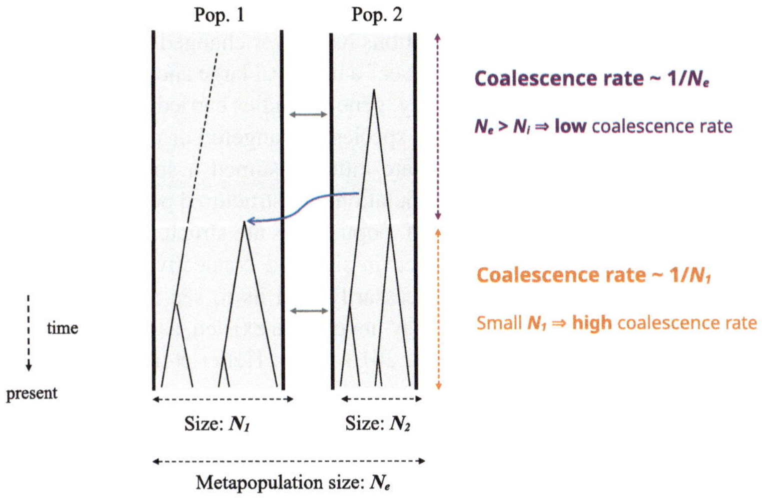

Fig. 10.2 Expected coalescent tree under a structured population model. This figure represents two populations that are connected by gene flow (grey solid arrows). A locus was sequenced for three diploid individuals sampled in one population ("Pop. 1"). As we move from the present towards the past, the gene genealogy of that locus is first influenced by N_1, the size of Population 1. However, one of the ancestral sequences in Population 1 migrated from Population 2 during a gene flow event (blue arrow). This means that the coalescence times will now depend not just on the population sizes but also on the migration rates between populations. If we assume that the species is structured in an n-island model with a migration rate m and island size N, the coalescence rate will be a function of these three parameters. Altogether, population structure leads to a reduction in coalescence rates in the past, which is interpreted under a panmictic model as the result of a large ancient population, and a small recent population: a population size decrease, or bottleneck towards the present. The gene tree will be similar to the tree (**B**) in Fig. 10.1

that the ancestor of the sampled lineages was at some point in a different population. This means that, looking now backward in time, the older coalescence events between the remaining lineages can only happen when they "meet" again, in the same population. This will require at least one additional migration event. The coalescence rate of these remaining lineages will thus decrease compared to the coalescence rate in the sampled population. If we now analyse our sampled sequences by assuming that they come from a panmictic and isolated population, the only possible interpretation for this change in coalescence rate is that the population was large in the past because the coalescence rate decreased (see Fig. 10.2). Thus, if a gene flow event has influenced the genealogy of the locus of interest, this locus will exhibit a gene genealogy influenced by N in the recent past and influenced by $n \times N$ and m in the ancient past. This is exactly the kind of genealogies expected when a panmictic and isolated population was small in the recent past and large in the ancient past (Fig. 10.1B).

These examples suggest that if populations are spatially structured and stationary, we will necessarily see a decrease in population size (towards the present) if we analyse the data under the assumption that the sampled population is panmictic and isolated, even if the size of the populations had never changed for millions of years (as in Fig. 10.2). Geneticists will thus "see" a mythical large ancestral population that may never have existed. Unsurprisingly, genetic studies carried out in the 1990s and 2000s uncovered bottlenecks in many species, endangered or not.

The simple examples discussed here either assumed a single population size change in a panmictic and isolated population or a structured population of constant size and connectivity. In the real world, populations are structured in more complex ways, and they have probably changed in size and connectivity. This means that ignoring population structure will necessarily lead us to see not one but possibly several population size changes that may never have existed, as discussed by several authors (Beaumont 2004; Chikhi et al. 2010, 2018; Heller et al. 2013; Mazet et al. 2016).

We mainly focused on the *qualitative* equivalence between the two families of models (panmictic with size changes versus structured population with neither size nor connectivity changes). The objective of population geneticists is to determine whether they can reconstruct important aspects of that complex history. This is one of the greatest challenges of population geneticists because this history likely involved complex population structure with changes in connectivity and population size (Arredondo et al. 2021; Rodríguez et al. 2018; Steinrücken et al. 2019; Wang et al. 2020). There are good reasons to think that we can separate to some extent some of these scenarios (panmictic vs. structured), but we also have results that show that under some conditions, it is mathematically impossible to do so.

For instance, Mazet et al. (2016) showed that if we have the genome of a single individual, even if this individual came from a population with a highly complex evolutionary history—e.g. extinctions, changes in population size, changes in migration rates, etc.—, one can always find a history of population size change that will perfectly fit and explain the distribution of coalescence times of this individual genome (see Box and Fig. 10.3). In other words, there is a problem of identifiability (i.e. the underlying models cannot be separated solely based on the analysed genetic variation). Yet, we know that by increasing the sample size, a distinction between models can be made (Grusea et al. 2019). The reason for this is that under panmixia and isolation, all individuals are exchangeable and should thus exhibit the same distribution of coalescence times. Under some structured models, individuals are not necessarily exchangeable and will tend to exhibit different coalescence time distributions, particularly in spatially structured models. This is a potentially rich area of scientific research that is still in its infancy. And we should stress that most studies in theoretical population genetics are usually very clear about the assumptions they make.

10.3.7 Hybridization and Population Structure

The problem of hybridization is different from that of population size changes because it is invoked for models that now involve at least two species or subspecies.

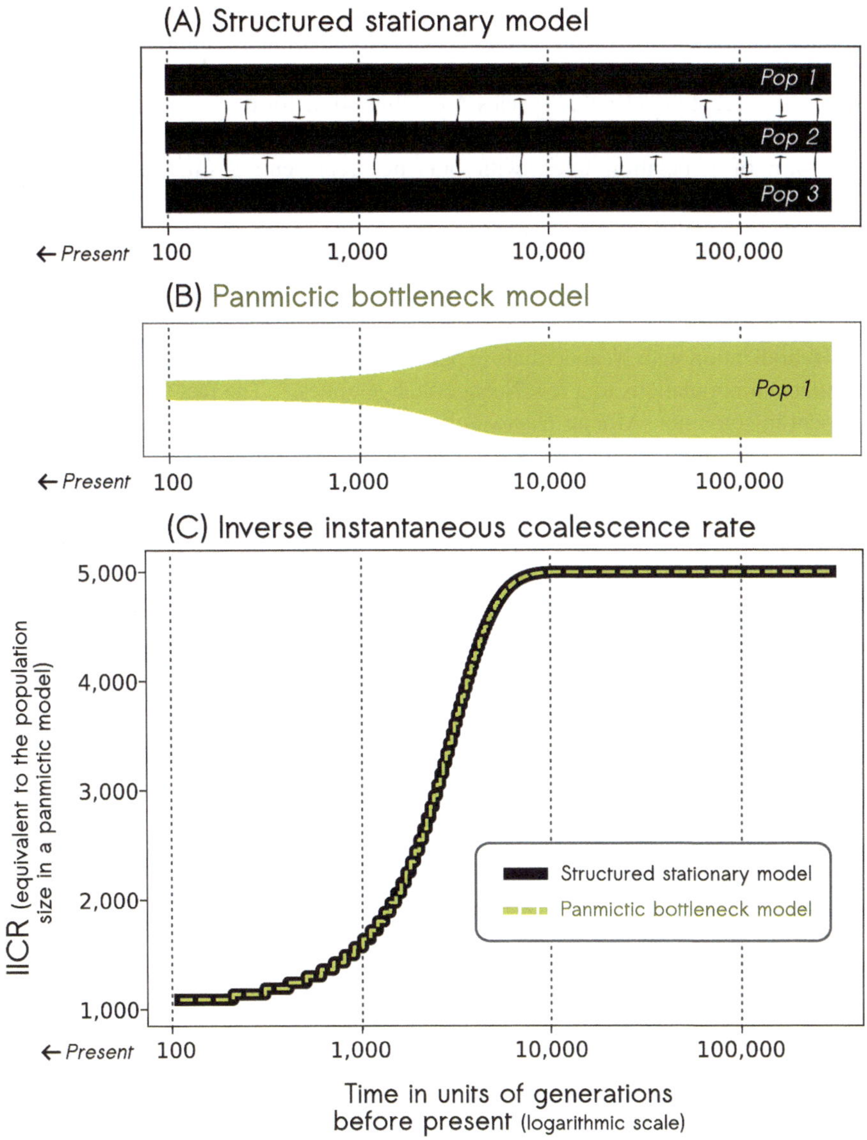

Fig. 10.3 Expected inverse instantaneous coalescence rate (IICR) curve under two demographic models. This figure shows three panels with the x-axis representing time, where time flows from the right to the left. Note that the x-axis represents time in a logarithmic scale (base 10). Panel (**A**) shows a stationary structured model with three constant-sized populations exchanging genes. Its corresponding IICR curve is represented by the black solid line in panel (**C**). Panel (**B**) represents a panmictic model with a continuous population size change from an ancient large population in the past to a small population in the recent past. The figure looks like a horizontal bottle, and the population decrease to a bottleneck. The corresponding IICR curve is represented as a green dashed curve in panel (**C**). Panel (**C**) shows the IICR curves for one diploid individual under the two

The authors of such studies are interested in determining whether hybridization of species A (say, Sapiens) with species B (say, Neanderthals) occurred in some areas of the distribution of species A, and possibly not in others.[31] The authors thus implicitly accept the idea that species A exhibits some form of population structure. However, in the case of *Homo sapiens* at least, this structure is typically modelled in the form of population trees, with branches representing different continental populations. They do not consider that continents can be modelled as populations connected by gene flow, which are themselves connected to other continents by gene flow. Rather, such tree models assume that phylogenetic trees, similar to species trees, are a reasonably good way of summarizing the relationships between human populations, each representing a whole continent such as Africa, Europe and Asia.[32]

Hybridization with Neanderthals in particular is thus inferred by comparing these continental populations to a few Neanderthal genomes.[33] The fact that some alleles present in some non-African *Homo sapiens* populations are similar to Neanderthal alleles is then interpreted as being the result of a hybridization event (Fig. 10.4). By assuming that continental populations have been panmictic (and isolated, in many human demographic models) since they split from each other (typically tens of thousands of years ago), these models assume that the only possible form of gene exchange is admixture with Neanderthals. It is then not necessarily surprising that adding discrete hybridization events will improve the fit between the observed genetic data (for real *Homo sapiens* and Neanderthal populations) and the expectations, compared to models without admixture.

One can see the problems above through different lenses. From a statistical point of view, if geneticists ignore something that is known to exist (population structure and gene flow within continents and between continents) to infer something that

Fig. 10.3 (continued) models represented in the other two panels. The black curve corresponds to a diploid individual sampled in one of the three populations, whereas the dashed green curve corresponds to an individual sampled in the bottlenecked population. More precisely, the "structured stationary model" is composed of three populations (n-island model, where $n = 3$) of size 1,000 diploids each, all interconnected by symmetric gene flow with a per-generation migration rate of 10^{-4} between demes. Population sizes and gene flow do not vary through time. In contrast, the "panmictic bottleneck model" is composed of a single panmictic population that goes through a steady exponential size reduction towards the present, changing from a diploid size of 5,000 to a size of 1,000. The two IICR curves are indistinguishable. With genomic data sampled from a single individual, population geneticists typically use the PSMC method of Li and Durbin (2011) to produce an estimated or inferred IICR curve that they generally interpret in terms of population size changes (as in Panel A), even when the sampled population is actually structured (as in Panel B), as is the case for most great apes (Mazet et al. 2016; Steux et al. 2025)

[31] We do not consider here the cases where a whole species could result from a hybridization event.

[32] With the reification of Europe as a continent equivalent to Africa or Asia, a reification that is rarely discussed or questioned.

[33] Often with the (strong) assumption that sub-Saharan African genomes are devoid of Neanderthal ancestry.

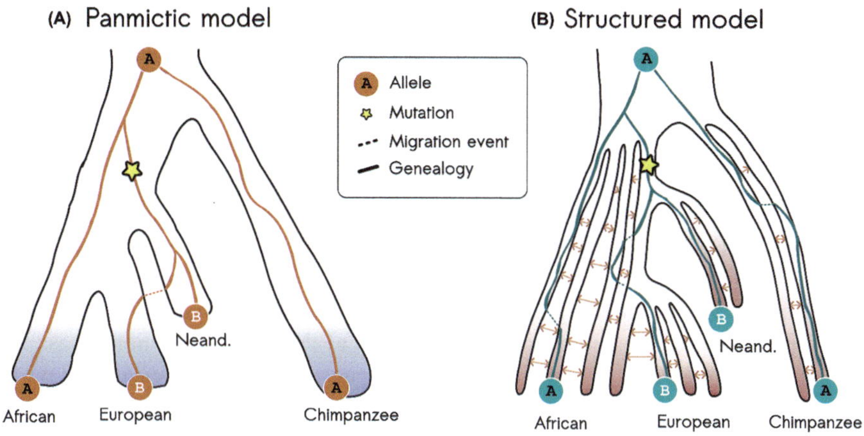

Fig. 10.4 Illustration of the ABBA-BABA hybridization test. This test and the corresponding *D*-statistic are classically used to detect admixture or hybridization. The general idea is that we can identify two patterns in the DNA that can be symbolized by ABBA or BABA, where A represents the state observed in chimpanzees at a particular site and B the state in Neanderthals at the same site, with B different from A. Among these sites we will then focus on those that are polymorphic in humans such that we may observe an A or a B in Europeans, Asians or Africans. If we focus now on an individual from Europe and another from Africa, and if we assume that a tree is a good representation of the evolution of these four entities (chimpanzee, Neanderthal, European, African) then we expect that the number of cases where Europeans and Africans (or any other pair of interest) exhibit an A or a B should be random and thus identical, on average. We can thus compute the difference between the number of DNA sites that have the pattern "ABBA" vs. "BABA", $D = n_{ABBA} - n_{BABA}$. It should be zero when the Neanderthal admixture is similar (or absent) in African and European populations. This test was developed under the assumption of a panmictic model represented on the left, where the relationship between African, European, Neanderthal and Chimpanzee populations is represented by a simple bifurcating tree. This model assumes that if an admixture event happened between Neanderthal and either one of the ancestors of Europeans or Africans, the numbers of ABBA or BABA patterns would change. Since Neanderthals have a B, we expect more ABBA patterns if the admixture is with Europeans as in the left figure, and more BABA if Neanderthals admixed with Africans. If we can identify many of the DNA patterns in the genome and compute the *D* statistic, we can test if it is statistically different from zero. In the figure, we represent this admixture event with a dotted orange segment between Neanderthals and Europeans. Under such an assumption, a coalescent tree for a locus in the genome is represented in orange line, where the European lineage coalesces with the Neanderthal lineage first (backward in time). A mutation (star) occurs on the lineage ancestral to this coalescence event, changing the ancestral allele “A” to “B”. This generates the “ABBA” allelic pattern. However, this pattern can also be observed in the case of a spatially structured population model (right). Here, all the populations are subdivided and their subpopulations are connected by gene flow. Because of this spatial subdivision, a few European lineages may by chance coalesce with Neanderthals first, because of the spatial distribution of diversity before the origin of Neanderthals and present-day *sapiens*. Thus, the expected value of *D* under no admixture is not zero anymore. The expected *D* value will depend on the details of the spatial processes. Consequently, the ABBA-BABA test cannot be used directly to decide whether an observed allelic pattern results from admixture or population structure as was shown by Eriksson and Manica (2012), Tournebize and Chikhi (2025) and Steux et al. (2025)

cannot be easily detected nor quantified or observed directly,[34] they will very likely infer non-zero values in the free parameters compensating for the absence of population structure in the underlying model. Another way to understand it is to imagine the spatial distribution across Africa of the ancestors of *Homo sapiens* and Neanderthals before they expanded into Eurasia. If we see these ancestors as a spatially organized meta-population in Africa, we can imagine that when some of these African ancestors expanded into Eurasia to later become what we identify as the Neanderthals, they brought a subset of the genetic diversity from the African regions from which they expanded. While the ancestral African meta-population evolved into what we identify as *Homo sapiens* in Africa for hundreds of thousands of years, the local populations maintained some of the spatial variation across the metapopulation. A similar spatial process of expansion likely happened later when some ancient *Homo sapiens* populations expanded from the regions where ancestral Neanderthals originated. It is thus not absurd to consider that some shared ancient genetic variation was geographically maintained. This shared ancestral structure with the ancestors of Neanderthals could have left a signature which is now interpreted as a recent admixture event. Indeed, these signatures could have equally likely been the result of common ancestors expanding from the same African regions. This scenario is not only verbally reasonable, but possible as was shown more than a decade ago by Eriksson and Manica (2012, 2014) and more recently by Tournebize and Chikhi (2025) (see Fig. 10.4). In other words, reasonable levels of population structure such as those observed in humans today are actually able to explain why we detect signals of Neanderthal hybridization in humans and genetic bottlenecks. This does not mean that hybridization did not take place, but perhaps that available genetic data and statistical methods are insufficient to separate signal from noise and to unambiguously infer the underlying scenario. And if this is true, why do some authors insist on stating that admixture is indisputable when they sometimes stated the opposite with the same vigour 20 years ago.

10.3.8 Storytelling, Inference and Genetic Data

To summarize, the standard coalescent theory was developed in the 1980s and 1990s, and it has significantly increased our understanding of the theoretical distribution of genetic diversity through space and time under a large variety of models and evolutionary histories. We can thus ask and answer many questions and compare alternative scenarios.

Population geneticists have known for at least two and a half decades that when populations are structured, interpreting genetic data while ignoring that structure will likely and often necessarily lead to the inference of events or parameters that may

[34] Ancient population size changes or hybridization events are generally inferred not observed. A few studies have identified admixed individuals, but these individuals actually have surprising patterns of hybridization (Tournebize and Chikhi 2025) and are often seen as "dead-ends" that did not contribute to present-day populations.

correspond to nothing real, akin to the apophenias of Marks (2019). Whenever we analyse observed genetic data by constructing a phylogenetic tree between human populations, treating them like "species" that do not exchange genes (when we *know* they do), the fact that we infer significant horizontal gene exchange (admixture) with Neanderthals may not be that surprising, although such exchange would be artefactual (i.e. a by-product of the choices made when we constructed the models). Similarly, if we construct a model where gene flow (between populations) and hybridization (with another species) are not allowed and where population size is the only parameter of the underlying models, it might not come as a surprise if we detect significant size changes.

Even though we did not discuss the issue of selection in this chapter, a similar case could probably be made when selection is detected under panmictic models, as population structure, admixture and non-equilibrium dynamics could mimic selection signals. Genes exhibiting a behaviour that is not expected under neutrality and panmixia may be interpreted as being under selection when perhaps other parameters are at work. We wish to add that some authors take the view that selection is actually pervasive in the genome and that reconstructing the demographic history of species by assuming neutrality of the genome could be very misleading. These are complex evolutionary and statistical issues, where population geneticists generally do their best and have to make simplifying assumptions. But these are also issues related to the sociology of science: Why are some parameters considered important while others are seen as unimportant?

These issues have been discussed in the literature, but to our knowledge, and based on what we understand of structured models (n-island or stepping-stone models with a reasonable number of populations), only three studies have tested for the robustness of admixture statistics to population structure, and all three have concluded that the statistics used as evidence in favour of admixture in human evolution were not robust (Eriksson and Manica 2012, 2014; Tournebize and Chikhi 2025).[35] There is thus to this date no indisputable evidence of admixture between Neanderthals and *Homo sapiens* in the genomes of present-day individuals. In the next section, we try to develop our reflection with elements from the philosophy of science.

10.4 Philosophical Considerations on Indeterminacy in Population Genetics

10.4.1 Plurality of Positions

10.4.1.1 The Notion of Pluralism

The two case studies presented in the previous section (that is, bottlenecks during hominin evolution and hybridization with Neanderthals) each confront us with a situation of explanatory plurality. The question we are interested in here is whether

[35] An additional one (Steux et al. 2025) recently concluded that admixture between bonobos and chimpanzees was also unwarranted for similar reasons.

this plurality should translate into some kind of pluralism,[36] a theme that has given rise to numerous analyses in the philosophy of science over recent decades.

Authors such as Mitchell have proposed to distinguish several forms of pluralism, such as "anything goes pluralism", "moderate pluralism", "isolationist pluralism" and, finally, her own position, described as "integrative pluralism" (Mitchell 2003; see Ludwig and Ruphy 2021; Van Bouwel 2014). To understand the significance of these distinctions, we may start with the following basic observation. The existence of a plurality of competing explanations is a perfectly normal situation in science, insofar as it is part of the norms of scientific activity: that a hypothesis or theory is discussed and opposed by another theory. But it has long been thought that this situation should be temporary, and should ideally lead to a resolution. This is what Mitchell calls "moderate pluralism", precisely because of the temporary nature of that pluralism. The idea of a crucial experiment, an experiment that can decide once and for all whether a theory is true or not, occupies an important place in this conception of science. This idea suffered devastating criticism from Duhem (2016), but it was not until the 1970s, with Suppes' article in particular (Suppes 1978), that a more general criticism of an ideal described as "monistic" began to emerge. The monist asserts that (in the ideal) "the ultimate aim of a science is to establish a single, complete, and comprehensive account of the natural world (or the part of the world investigated by the science) based on a single set of fundamental principles" (Kellert et al. 2006). This is not a position that concerns only the goal of science, but an ideal that includes an ontological aspect, insofar as it simultaneously asserts that "the nature of the world is such that it can, at least in principle, be completely described or explained by such an account". For Suppes, it is not a question of renouncing this ideal simply because it would be inaccessible (which is a characteristic of any ideal), but rather of "constructing a different metaphysics or general epistemology" (Suppes 1985). In such an epistemology, the plurality of explanations is an unproblematic phenomenon. This pluralism is more radical because it is not intended to be superseded. Thus, there are indeed several forms of pluralism, and we propose to assess the type of pluralism manifested in our case

[36]Readers unfamiliar with this area of philosophy of science may wonder about the relationship between "plurality" and "pluralism". Kellert et al. (2006) offer the following clarification: "It is useful to distinguish between plurality in the sciences and pluralism about the sciences. The former is a feature of the present state of inquiry in a number of areas of scientific research [...]. These are characterized by multiple approaches, each revealing different facets of a phenomenon. There can be plurality of representational or classificatory schemes, of explanatory strategies, of models and theories, and of investigative questions and the strategies appropriate for answering them. Pluralism is a view about this state of affairs: that plurality in science possibly represents an ineliminable character of scientific inquiry and knowledge (about at least some phenomena), that it represents a deficiency in knowledge only from a certain point of view, and that analysis of metascientific concepts (like theory, explanation, evidence) should reflect the possibility that the explanatory and investigative aims of science can be best achieved by sciences that are pluralistic, even in the long run. [...] Philosophers who advocate pluralism can and do differ as to the extent of the plurality they attribute to the sciences, the strength of the pluralism they adopt, and the broader philosophical implications they draw from it".

studies: is it radical, or are we dealing with different explanations that are reducible to some form of unity?

10.4.1.2 The Modesty of Models Versus the Ambition of Paradigm Shifts

During school disputes, it is common to see two argumentative strategies confront each other: the challengers invoke the need for a change of "paradigm" (even if it means recruiting Kuhn's notion in a less than rigorous sense) while the supporters of the dominant position tend to downplay the differences and incompatibilities between the positions. Assessing the radicality of the break between positions is made more difficult by the fact that the proponents of a minority position do not necessarily question all the basic principles and concepts of a discipline (Racine 2024).[37] This is particularly the case when explanatory pluralism concerns models.[38] While a theory can be characterized as a set of principles posited as true—which makes, when confronting theories, any pluralism with respect to putatively contradictory principles difficult to maintain—building a model[39] is characterized by an assumed activity of idealization, simplification and approximation[40] that can differ across models (Godfrey-Smith 2007; Levins 1966; Odenbaugh 2007; Orzack 2005). Explanatory pluralism seems thus less difficult to maintain if we are speaking about models. But even in this case, we need to measure how ontologically different two models can be.

10.4.1.3 Ontological Plurality of Model Families

In the two cases that interest us here, there is indeed an ontological break between two families of models (structured vs. panmictic), and not simply a pragmatic difference. These models are not dealing with the same objects. The addition of a structure dimension makes certain objects of the structured model non-describable (or incommensurable) in the terms of a panmictic model (e.g. in a model with $n > 1$ islands, the identity of an island, its size or the rate of migration have no meaningful sense in a panmictic Wright-Fisher model). This ontological break is accompanied

[37] Kuhn's examples to illustrate a paradigm shift are sometimes misleading: a central concept of pre-Lavoisian chemistry, such as phlogiston, disappeared completely with the revolution that took place at the end of the eighteenth century. But a paradigm shift in explanations of cancer, for example, does not require us to call into question the whole of cellular theory (Montévil and Pocheville 2017).

[38] In the philosophy of biology, a very active current defends the thesis that the work of abstraction is mainly based on models rather than on theories, to the point that some philosophers question the existence of theories in certain disciplinary fields (Pradeu 2014).

[39] See Varenne (2016) and Vorms (2011) on the place of models in the syntactic conception of theories. The notion of model becomes central in the so-called semantic conception, but in a sense quite different from the one we are interested in.

[40] A position classically summarized by the biologists themselves, who readily quote George Box: "All models are wrong, but some are useful". So much so that the idea that models are fictions is increasingly acknowledged. If they are fictions, then it is obviously not problematic to have more than one. See for example Frigg 2010 and Godfrey-Smith 2009.

by an epistemic break: the models do not pose the same questions. A structured model may prompt us to question the temporal variation of connectivity among human populations, issues that do not make sense in a panmictic model where the only degree of freedom is population size variation (or admixture, or selection). Similarly, imposing a tree model invites us to calculate divergence times, a question that does not arise in the same way with a population network model (Goldstein and Chikhi 2002; Scerri et al. 2019). At the end of this section, we shall see the effect of this ontological break when it comes to inferring the narrative of human history.

10.4.1.4 Asymptotic Congruence of Models

There is, however, beyond this ontological break, a formal articulation between these two families of models: the predictions of the structured model tend asymptotically—without ever, however, being equal to them—towards those of a panmictic model when the structure becomes negligible (everything else being equal between the two models), that is to say when the migration rate between subpopulations increases—in the manner, to take an image, in which special relativity tends towards Newtonian physics when speed decreases.

10.4.2 (Un-)Decidability of Empirical Inferences Under Different Model Families

Let us acknowledge the disagreement between the ontologies of the two families of models considered. Is this disagreement decidable from the point of view of empirical inference? If it is not, are we asking the right questions? And if it is, can we identify the conditions under which it would be? We will examine the first question from the point of view of epistemic values (fit to observations, consistency with the rest of knowledge), then from the point of view of pragmatic values (simplicity, calculability).[41] We will return to the other two questions in the conclusion.

10.4.2.1 Epistemic Values

A strong and widespread thesis in philosophy of science indicates that it is impossible, in principle, to make an empirical inference with certainty: theoretical inference is said to be underdetermined by empirical data (Nola and Sankey 2014). Duhem, for example, early noticed that there can be no experiment that can definitively decide between two theories, each theory being able to be saved from empirical embarrassment by the *ad hoc* construction of auxiliary hypotheses (Duhem 2016). The same

[41] Kuhn (1977) represents an important moment in the reflection on the place of values in science. Values play a crucial role when it comes to choosing between different theories. He distinguishes between epistemic and non-epistemic values, a distinction that has become classic and has given rise to much discussion. Among the epistemic values, he cites precision or empirical adequacy, internal and external coherence, scope, simplicity and fruitfulness. We feel it is appropriate to make a new distinction and call a value such as simplicity "pragmatic" insofar as the manipulability of models is an important issue, but different from their relationship to data (Nola and Sankey 2014).

underdetermination weighs on the choice of models: for a given dataset, an indefinite number of models explaining the data are possible. Faced with this strict underdetermination of abstractions (theories and models) by experience, we will adopt a position that is both modest and practical: we will ask ourselves whether these models are equivalent in terms of plausibility (or likelihood), that is to say both in terms of fit to observations and *a priori* plausibility of their hypotheses (that is, their *a priori* compatibility with the rest of human knowledge).

Regarding the fit to observations, we saw in Sect. 10.2 that, even if the evolutionary history of a population is very complex, some methods will always reconstruct a curve that will be interpretable as a history of population size changes in a panmictic population (see Fig. 10.3). This was formally demonstrated in the case where we only have access to the genome of a single individual. The reason for this is that there will always be a panmictic model that will generate the same coalescence time distribution and therefore the same patterns of genetic diversity as any more complex model. We mentioned it as a problem of identifiability, but this can also be seen as a problem of undecidability in this specific case, as formally demonstrated by Mazet et al. (2016) for the distribution of coalescence times along the genome of a single diploid individual, assuming that these times are independent. This apparently purely theoretical model has important consequences because Li and Durbin (2011) developed a method called the PSMC that allows to infer such a curve from the genome of a single diploid individual (see Box).

Box: How Does Population Structure Lead to the Inference of Bottlenecks that May Never Have Occurred? The PSMC Case

The relationship between coalescence times and N_e, the effective population size, underlies a large body of methods used for demographic inference. In 2011, Li and Durbin introduced a new method called the "pairwise sequential Markovian coalescent" (PSMC) which allows inferring changes in N_e from the genome of a single individual. From the measured density of heterozygous sites along the genome, with some assumptions on the ratio between mutation and recombination rates, the PSMC infers the underlying distribution of coalescence times for all pairs of non-recombining loci and uses an optimization algorithm to infer the trajectory of N_e that best fits this coalescence time distribution. The method assumes that population structure is negligible and thus infers changes in N_e under the panmictic coalescent with population size change. The PSMC produces a histogram or smoothed curve, where the x-axis represents time (often represented in a logarithmic scale) and the y-axis represents N_e. Yet, as discussed in the main text, structured population models can exhibit distributions of coalescence times for two copies of the same loci across the whole genome that are similar or identical to those of panmictic populations changing size, despite both models being fundamentally different in the way they represent populations and their history. Mazet et al. (2016)

(continued)

introduced the concept of IICR (inverse instantaneous coalescence rate) and showed that the PSMC is actually an estimator of the *IICR*. They showed that under panmixia, the IICR is equivalent to N_e, but under structured models, the IICR is related in a complicated way to changes in connectivity and population size. We illustrate this issue in Fig. 10.3, where we calculated the IICR curves (i.e. the theoretical expectation of the PSMC) of two models. The first one is a structured model with three interconnected demes and is stationary with no size or connectivity change. The second one represents a single panmictic population undergoing a population reduction towards the present (a bottleneck). The IICRs obtained from both models are identical. Thus, in the absence of any knowledge about the actual underlying population model (panmictic or structured), one cannot decide whether the genomic data were obtained from a stationary structured population or from a bottlenecked population (panmictic model). This leads to an identifiability problem that cannot be solved by increasing the number of independent loci, making the interpretation of the PSMC results difficult. The PSMC method is generally used and interpreted under the assumption that population structure is negligible. The PSMC curves are thus assumed to represent a "trajectory of effective population size changes", when they could actually indicate changes in connectivity, or both. Interestingly, the panmictic interpretation dominates population genetic studies, even when the same study may use other methods that assume population structure, creating self-contradictory narratives that are rarely discussed. As a final note, the identifiability problem discussed by Mazet et al. (2016) is valid under the assumption that one has access to the distribution of coalescent times. A recent unpublished study suggests that the PSMC method could still contain subtle information that would help separate structured and panmictic models despite their having similar PSMC curves.

However, this undecidability is true only relatively to the type of statistic employed, here the PSMC curve. As soon as we have several genomes, and therefore several PSMC curves, we can show that for some structured models at least, differences will appear between the curves, differences that are not theoretically possible in panmixia (because of exchangeability, see Sect. 10.2 and Box). Conversely, significant differences between IICRs (the theoretical object that the PSMC method estimates, as defined by Mazet et al. (2016)) may be real but not visible or detectable from the point of view of the PSMC estimator, because the PSMC is not a perfect estimator of the IICR. There is thus a real theoretical problem of identifiability to which is added an equally real problem linked to the limits of the information extractable from real genomes.

As for the plausibility of the hypotheses (i.e. structure or not), we can note that the human species appears, today, structured and that there are good reasons, both archeological and paleontological, to think that it has been so in the past. Furthermore, species close to humans such as great apes and other primates, as well as the majority of mammal species, are spatially structured. This makes it rather unlikely that there was a period when humans or their ancestors were panmictic over vast regions. The question is therefore whether this structure, actualized in history, has had *negligible* effects, or not, on the variables that interest geneticists, like the distribution of coalescence times. The question could be posed from two opposite angles. It might seem reasonable that the burden of proof for the non-negligibility of structure should fall on those who advocate it. It would be up to them to show that the conclusions change when the known structure of human populations is taken into account. But it might also be reasonable to say that those who claim that structure is negligible must show that their conclusions are robust to the introduction of population or spatial structure reasonably close to that observed in humans. As we have shown in the previous section, it has been known for at least 25 years that when populations are structured, inferences related to changes in size will most likely not be robust. Wakeley (1999) even observed bottlenecks in structured populations where all populations were actually growing and increasingly connected to each other.

10.4.2.2 Pragmatic Values of Each Family of Models

In contrast to the epistemic reasons mentioned above that would lead us to favour structured models, there are pragmatic reasons to prefer panmictic models, which could explain why they remain so popular in population genetics. Panmictic models can be simpler as they do not require several parameters to be specified—like the number of subpopulations, the rates of gene flow between subpopulations or the sampling schemes. Their more limited number of parameters can increase the statistical power of inferential approaches. Secondly, their mathematical tractability has spurred more developments in theoretical population genetics, leading to a greater number of tools and methods to perform genetic simulations and demographic inference. These tools may also be easier to use than equivalent methods based on structured models. In addition, panmictic models can be less computationally costly than structured models because they do not require modeling migration events, which can represent a heavy simulation load. For instance, simulating genetic data under a panmictic stationary model of $N = 3{,}000$ individuals can be up to 40 times faster than an equivalent structured stationary model composed of 3 demes of 1,000 individuals each (migration rate of 10^{-6} per generation from one deme to another), where half of the samples are taken from one population and the other half from another population.

The simplicity, mathematical tractability and computational efficiency of panmictic models, along with the greater number of methodological resources, could thus represent pragmatic and understandable reasons for a certain disciplinary conservatism. This is reminiscent of the coalescent theory that we introduced earlier: it remains very popular, despite the great diversity of other frameworks that are available to researchers to model genetic evolution. Since the coalescent theory focuses on the properties of the *sample* as a function of the *population* parameters, it is computationally very efficient. This is not the case when population geneticists use forward simulations, where all the individuals (and their genomes) must be effectively simulated. The efficiency (and for some, mathematical beauty) of the coalescent theory explains its success, despite the fact that the coalescent theory, through its general reliance on the Wright-Fisher population model, can pose several limitations and can be too simplistic to represent actual evolution in real life.

10.4.3 Pragmatic Pluralism

This conflict between epistemic and pragmatic values of our two families of models (panmictic vs. structured) invites us to consider a pragmatic pluralism of the use of these models (panmictic models can be used as very first approximations, to address a specific question) and/or a pluralism of *explananda* (of questions to explore). For example, panmictic models could be used not to infer the demographic past in all its complexity, but to detect signatures of putative selection events, or to identify and date a particular demographic event (with the *proviso* that structure has sufficiently negligible effects on the detection of traces of selection).

However, this pragmatic pluralism in no way means an acceptance of the narrative constructed under the assumption of negligible structure. Rather, as for any model, this narrative must be presented and understood as a hypothesis that must be tested, and a model whose *robustness* must be assessed. The heuristic value of this pragmatic use of models will be reinforced if their predictions (or retrodictions) are robust to uncertainty[42]: that is, if the predictions are preserved when the values of the parameters are varied in their likelihood interval (e.g. migration rate), when the structure of the model is varied in a set of plausible structures (e.g. number of islands, equidistance or not between the islands), and, finally, when the predictions are consistent with the results of other disciplines. We can therefore see here that, in such a framework of pragmatic pluralism, multidisciplinarity cannot be limited to affirming that such an ancient event inferred by genetics is consistent with a particular model, but to discussing this consistency with or against alternative models recognized as potentially plausible.

10.4.4 On Monism and Agnosticism in Historical Inference: Raison D'être for Models

If, as we have seen, the structured and panmictic models are ontologically different, and if they can be used pragmatically on different points, it is reasonable to ask whether there is sense in putting them in competition in order to choose, locally, one family over another.

Here we must return to the *raison d'être* for these models. They are used to infer an aspect of human history in order to draw a reasonable narrative. This narrative should, hopefully, correspond to real events that took place at a time close to those

[42] On robustness, see: Lesne (2008, 2012) and Soler et al. (2012).

inferred. Our position here will be that of a monist historical realism: we assume the existence of a single past.[43,44] If, for a given past, an indefinite number of narratives is possible (by focusing on this or that aspect or statistic), these narratives must, in a monist position, be at least compatible with each other: a human population cannot *altogether* experience a continuous demographic growth *and* one or more demographic bottlenecks (and not just genetic ones) during the same period.

Here, we note that our two families of models (structured and panmictic) can be used to infer stories of human demography that are not only different, but also incompatible: the structured model in Fig. 10.3 does not tell the same story as its panmictic counterpart. Assuming a structured model of constant size, and in equilibrium, the interpretation of the corresponding genetic data under the assumption of panmixia, however, implies the existence of a recent bottleneck and a departure from equilibrium. Thus, interpreting the data under one of the models might lead us to identify events that never occurred and a disequilibrium in a population that may be at equilibrium—the type of apophenia mentioned by Marks (2019).

10.4.5 Towards a Prudent Agnosticism

Rather than choosing, for the epistemic reasons mentioned above, the structured model as a more plausible representation of the past, we consider that our monist historical realism invites us, in this precise case, to a prudent agnosticism, that is to say to alert on the fragility (or the non-robustness) of certain narratives drawn from standard methods based on hypotheses that are ultimately fragile. Thus, defending the relevance of population structure does not amount to defending a narrative of structure and rejecting bottlenecks and hybridization events, but to alert on the fragility of a narrative of hybridization or of a bottleneck, in a pluralism of perspectives that can strengthen our objectivity. For instance, in the same way as population structure is a reasonable hypothesis to explain some patterns of human genetic diversity, adding variance in reproductive success to population geneticists' models could change some conclusions obtained under classical structured models, and this possibility should reinforce our cautious agnosticism. In other words, comparing models from different families makes more salient the interpretative shift that occurs when a model is used to make an inference that leads to an interpretation that the model does not strictly allow us to make—for example, the genetic bottlenecks in panmictic models do not necessarily mean real demographic bottlenecks. Similarly, a population number inferred under an n-island model does not necessarily correspond to the actual number of past human populations. For instance, the n-island model ignores the spatial structure (in the sense of geographic distance) that we know is indisputable in the human species.

[43] Other positions are possible, see for instance Turner (2016).

[44] Note that the relationship between pluralism and anti-realism is itself underdetermined: one can be both a pluralist and a realist(e.g. a perspectivist Giere 2010), or an anti-realist without referring to pluralism.

10.4.6 Some Sociological and Ethical Considerations

We have focused, in our discussion, on the epistemic and pragmatic aspects of historical inference, but it would be interesting, in future work, to examine its sociological aspects. Due to the social structuring of science, there may be sociological reasons for conservatism in the activity of modelling, or, on the contrary, for the adoption of challenging positions. We will note here that our prudent agnosticism is not without consequences for the ethics of the structuring of the human population genetics field: if the question between bottleneck and structure narratives is undecidable, is it ethical for some research groups to monopolize public research funds to tell stories whose objectivity seems questionable (Chikhi 2023)? Other scientific disciplines may offer competing, alternative or corroborating insights in the search for a better understanding of human evolution. Yet genetic data seem to always be taken more seriously than other types of data even when their contribution is either not necessarily relevant or even less valid than data from other fields (Källén 2025). Several epistemologists have raised the issue of the so-called "truth effect" that genetics is currently benefitting from, at the expense of other disciplines, like archeology or paleoanthropology. In this situation, discourses from a specific field (genetics) are generally perceived as possessing a stronger epistemic status than others because of historical, institutional or social reasons (Cartwright 2010), i.e. a form of "prestige bias". Beyond genetics, there is a general problem with historical sciences and their use for ideological purposes (Arnold 1990; El-Haj 2001; Gable and Handler 1993; Marks 2017a, b, 2019, 2024). Our duty as scientists should probably not be to replace one dominant discourse with another dominant discourse but to clarify the limits of genetic data and their interpretations because of the strong symbolic power attributed to genetics in many discourses.

When it comes to human history, we expect the narratives presented to society to amplify the sociological factors constraining epistemic choices, favouring simple narratives and fragile assertions, even if denials are subsequently published.[45] For example, the story of an admixture between *Homo sapiens* and Neanderthals (Green et al. 2006), or of a near-extinction of the human lineage (Hu et al. 2023), seems more exciting for the (general and scientific) public than apparently dull stories revolving around the subtle effects of population structure. Beyond the excitement that these stories produce, we can also wonder about the lightness with which such stories are brought to the public, given the history of the use of archeology and human genetics in the political arena.

[45] As noted above recent work (Deng et al. 2025) suggests that the article published less than 2 years ago (Hu et al. 2023) claiming that the human lineage went through a bottleneck 900,000 years before present is the result of an artefact of the method used. Another article suggests another interpretation (Cousins and Durvasula 2025). Note also that Mazet et al. 2016 had shown that under a simple structured model, the genetic data, in the form of a *PSMC* curve, could be rather explained with a change in connectivity around 900 kya.

10.5 Towards a Pluralistic Research Programme

Throughout this chapter, we have tried to show that there may be reasonable reasons to question the reality of events that have been inferred, dated and quantified in the field of human population genetics. We have tried to explain why genetic and genomic data may be more difficult to interpret than what is perhaps believed. We framed our general approach in relation to several problems arising from the philosophy, history or sociology of science. The questions addressed in this work are connected to the problem of modelling in science, of the under-determination of models, of the robustness of models to departures from assumptions, of pluralism, monism and realism. We could not develop all these philosophical issues as much as they deserve, but tried at least to identify questions that are too rarely asked in human evolutionary genetics. For instance, if, as we argue, there was no unambiguously identified bottleneck during the history of humankind, why or how did a whole scientific community act as if there was, discussing such events and their possible causes and consequences across hundreds of scientific articles? Similarly, the idea that *sapiens* and Neanderthals hybridized is not only assumed to be indisputable today, but articles about the consequences of such hybridization events on several phenotypic traits in modern humans have been published in several high-profile journals, e.g. related to autism, depression, diabetes 2, severe COVID, autoimmune diseases (Dannemann 2021; Gregory et al. 2021; Pauly et al. 2024; Simonti et al. 2016; Zeberg et al. 2020; Zeberg and Pääbo 2020). Beyond their claimed effects, the idea that Neanderthal genes are only found in some populations but not in Africa should perhaps have rung some bells in the scientists' ears, given the historical implication of biological anthropology and human genetics in racism and exclusion. One could ask: how likely is it that the Neanderthal contribution to Europeans will not be used by racist groups as a "proof" of the superiority of Europeans (and Asians) over Africans.[46] While we realize the complexity of such issues and the difficulty to address them in scientific articles, shouldn't we also acknowledge that too often too little care is taken when presenting some genetic results as indisputable in the scientific community, and to a non-specialist audience?

We would like to go further and propose a research programme which we believe could help make progress in a field (human population genetics) that remains highly controversial—a field where strong claims can be made, and where careers and funding may depend on such claims. It would be, for instance, interesting to quantify the funding that has been given to ancient DNA and genomic studies in relation to other research fields interested in the human past.

We suggest defending a research programme which is open to uncertainty and to the use of significantly different models, not as a means to propose a final scenario,

[46]Chou, V. 2017. How Science and Genetics are Reshaping the Race Debate of the Twenty-First Century. Science in the News, Harvard. Available at: https://sites.harvard.edu/sitn/2017/04/17/science-genetics-reshaping-race-debate-21st-century/

but rather as a way to identify evolutionary events of potential importance. We defend this as a form of methodological pluralism.

To clarify this position, we will use the case of structured and panmictic models, which we have discussed throughout this chapter. We have defended the view that it is not possible to understand human evolutionary history outside structured models, for the simple reason that human and primate populations are factually structured, and because we have now hopefully demonstrated that population structure can significantly impact our understanding of the evolutionary history of species. Population structure appears thus as an important requirement for models of human evolution. However, we also realize that structured models represent an infinitely large family of models that are impossible to infer exhaustively from a necessarily finite and limited amount of data and computing resources. This means that we will need to clarify the parts of the model that we wish to understand and determine how our conclusions on that aspect of human evolution are robust to changes in the models and assumptions used. This could be used as a first step to then increase the complexity of our models.

However, it also means that we could alternatively fall into a dangerous modelling trap: identifying a highly complex metapopulation model which could possibly fit any type of data by simply tweaking some parameters. This may not necessarily improve our understanding of human evolution, and may be misleading because many complex models could be equally able to explain the data available. Model validation through the prediction of independent statistics may then be a way to make progress. The idea would be to use some statistics to infer a model and then to predict another set of statistics (not used in the process of model fitting) with the inferred model. Such an approach could convey a stronger epistemic value than inferring the best model explaining all available statistics. For instance, Tournebize and Chikhi (2025) recently showed that several published models invoking Neanderthal admixture were poor predictors of human genetic diversity, a statistic which was not used in the original model-fitting procedure.

At the same time, we defend pluralism and argue that simpler models could also be used in an exploratory way to identify selection, admixture or population size changes, in a lucid, temporary suspension of disbelief, and a more critical suspension of belief. The uncertain balance between belief and disbelief should be maintained at all times, and should live with us when interpreting the results derived from such complex and sometimes contradictory analyses. The belief in the real existence of such events should probably be kept at a low level until it can be demonstrated that they are consistent with structured models, and/or can be validated by independent interdisciplinary research. We can thus use such simple models as means to provide explicit hypotheses that can be later tested using more complex and realistic models.

Acknowledgements We thank Jonathan Marks, João Zilhão and Guido Barbujani for comments on the first version of the manuscript. We are very grateful to Paul-Antoine Miquel, original philosopher and interdisciplinary researcher, who brought us all together and was at the origin of meetings during which we discussed ideas around the philosophy of biology, ideas that led to this chapter. We are thus grateful to early members of the group, Mathilde Tahar and Alexis Boisseau, for their comments during these meetings. We also thank the editors for their patience and support throughout the writing process, and the referees for important and critical comments that helped us improve our manuscript. While we may not have always followed their suggestions, we appreciated them. Funding: LC's research was supported by the DevOCGen project, funded by the Occitanie Regional Council's "Key challenges BiodivOc" programme. This work was also supported by the LABEX entitled TULIP (ANR-10-558 LABX-41 and ANR-11IDEX-0002-02) as well as the Investissement d'Avenir grant of the Agence Nationale de la Recherche (CEBA: ANR-10-LABX-25-01). We thank the IRP BEEG-B (International Research Project Bioinformatics, Ecology, Evolution, Genomics and Behaviour) funded by CNRS and University of Toulouse for facilitating travels and collaboration between Toulouse (CRBE, IMT and INSA) and Lisbon (IGC and cE3c).

References

Abbott RJ, Barton NH, Good JM (2016) Genomics of hybridization and its evolutionary consequences. Mol Ecol 25:2325–2332. https://doi.org/10.1111/mec.13685

Alves I, Arenas M, Currat M et al (2016) Long-distance dispersal shaped patterns of human genetic diversity in Eurasia. Mol Biol Evol 33:946–958. https://doi.org/10.1093/molbev/msv332

Ambrose SH (1998) Late Pleistocene human population bottlenecks, volcanic winter, and differentiation of modern humans. J Hum Evol 34:623–651. https://doi.org/10.1006/jhev.1998.0219

Amos W (2019) Flanking heterozygosity influences the relative probability of different base substitutions in humans. R Soc Open Sci 6:191018. https://doi.org/10.1098/rsos.191018

Amos W (2020) Statistics used to infer inter-breeding between humans and Neanderthals are strongly predicted by flanking sequence heterozygosity. bioRxiv. 2020–10

Amos W, Elhaik E (2025) Unexpected D-tour ahead: why the D-statistic, applied to humans, measures mutation rate variation not Neanderthal introgression. bioRxiv. 2024–12

Arnold B (1990) The past as propaganda: totalitarian archaeology in Nazi Germany. Antiquity 64: 464–478. https://doi.org/10.1017/S0003598X00078376

Arredondo A, Mourato B, Nguyen K et al (2021) Inferring number of populations and changes in connectivity under the n-Island model. Heredity 126:896–912. https://doi.org/10.1038/s41437-021-00426-9

Auton A, The 1000 Genomes Project Consortium et al (2015) A global reference for human genetic variation. Nature 526:68–74. https://doi.org/10.1038/nature15393

Barbujani G, Chikhi L (2007) Human genetic diversity and its history. In: Balding DJ, Bishop M, Cannings C (eds) Handbook of statistical genetics, 1st edn. Wiley, pp 1067–1108

Barbujani G, Magagni A, Minch E, Cavalli-Sforza LL (1997) An apportionment of human DNA diversity. Proc Natl Acad Sci U S A 94:4516–4519. https://doi.org/10.1073/pnas.94.9.4516

Barbujani G, Bertorelle G, Chikhi L (1998) Evidence for paleolithic and neolithic gene flow in Europe. Am J Hum Genet 62:488–491. https://doi.org/10.1086/301719

Beaumont MA (1999) Detecting population expansion and decline using microsatellites. Genetics 153:2013–2029. https://doi.org/10.1093/genetics/153.4.2013

Beaumont MA (2004) Recent developments in genetic data analysis: what can they tell us about human demographic history? Heredity 92:365–379. https://doi.org/10.1038/sj.hdy.6800447

Beerli P, Felsenstein J (1999) Maximum-likelihood estimation of migration rates and effective population numbers in two populations using a coalescent approach. Genetics 152:763–773. https://doi.org/10.1093/genetics/152.2.763

Boitard S, Rodríguez W, Jay F et al (2016) Inferring population size history from large samples of genome-wide molecular data - an approximate Bayesian computation approach. PLoS Genet 12: e1005877. https://doi.org/10.1371/journal.pgen.1005877

Brand CM, White FJ, Rogers AR, Webster TH (2022) Estimating bonobo (Pan paniscus) and chimpanzee (Pan troglodytes) evolutionary history from nucleotide site patterns. Proc Natl Acad Sci U S A 119:e2200858119. https://doi.org/10.1073/pnas.2200858119

Bräuer G, Stringer C (1997) Models, polarization, and perspectives on modern human origins. In: Clark GA, Willermet CM (eds) Conceptual issues in modern human origins research. Aldine de Gruyter, New York, pp 191–201

Cartwright N (2010) The dappled world: a study of the boundaries of science. Cambridge University Press, Cambridge

Charlesworth B (1994) The effect of background selection against deleterious mutations on weakly selected, linked variants. Genet Res 63:213–227. https://doi.org/10.1017/S0016672300032365

Charlesworth B (2009) Effective population size and patterns of molecular evolution and variation. Nat Rev Genet 10:195–205. https://doi.org/10.1038/nrg2526

Chen H, Huang S (2020) Modern alleles in archaic human Y chromosomes support origin of modern human paternal lineages in Asia rather than Africa. arXiv preprint arXiv:2003.13473

Chikhi L (2009) Genetic data and storytelling—from Archaeogenetics to Astrologenetics? Hum Biol 81:639–643. https://doi.org/10.3378/027.081.0612

Chikhi L (2023) Basato Su Una Storia Vera: Come Fanno i Genetisti Delle Popolazioni Umane a Conoscere Le Storie Che Raccontano? In: Evoluzione umana e origini di Homo sapiens. Accad Lincei Coll, Roma, pp 181–211

Chikhi L, Destro-Bisol G, Bertorelle G et al (1998) Clines of nuclear DNA markers suggest a largely Neolithic ancestry of the European gene pool. Proc Natl Acad Sci U S A 95:9053–9058. https://doi.org/10.1073/pnas.95.15.9053

Chikhi L, Bruford MW, Beaumont MA (2001) Estimation of admixture proportions: a likelihood-based approach using Markov chain Monte Carlo. Genetics 158:1347–1362. https://doi.org/10.1093/genetics/158.3.1347

Chikhi L, Nichols RA, Barbujani G, Beaumont MA (2002) Y genetic data support the Neolithic demic diffusion model. Proc Natl Acad Sci U S A 99:11008–11013. https://doi.org/10.1073/pnas.162158799

Chikhi L, Sousa VC, Luisi P et al (2010) The confounding effects of population structure, genetic diversity and the sampling scheme on the detection and quantification of population size changes. Genetics 186:983–995. https://doi.org/10.1534/genetics.110.118661

Chikhi L, Rodríguez W, Grusea S et al (2018) The IICR (inverse instantaneous coalescence rate) as a summary of genomic diversity: insights into demographic inference and model choice. Heredity 120:13–24. https://doi.org/10.1038/s41437-017-0005-6

Cousins T, Durvasula A (2025) Insufficient evidence for a severe bottleneck in humans during the early to middle Pleistocene transition. Mol Biol Evol 42:msaf041. https://doi.org/10.1093/molbev/msaf041

Crow JF, Kimura M (1970) An introduction to population genetics theory. Harper & Row, New York, Evanston, London

Currat M, Excoffier L (2004) Modern humans did not admix with Neanderthals during their range expansion into Europe. PLoS Biol 2:e421. https://doi.org/10.1371/journal.pbio.0020421

Dannemann M (2021) The population-specific impact of Neandertal introgression on human disease. Genome Biol Evol 13:evaa250. https://doi.org/10.1093/gbe/evaa250

Darwin C (1872) The descent of man, and selection in relation to sex, vol 2. D. Appleton

Deng Y, Nielsen R, Song YS (2025) A previously reported bottleneck in human ancestry 900 kya is likely a statistical artifact. Genetics 229:iyae192. https://doi.org/10.1093/genetics/iyae192

Donnelly P, Tavaré S (1995) Coalescents and genealogical structure under neutrality. Annu Rev Genet 29:401–421. https://doi.org/10.1146/annurev.ge.29.120195.002153

Duhem P (2016) La théorie physique. Son objet, sa structure (1906). ENS Éditions

Durand EY, Patterson N, Reich D, Slatkin M (2011) Testing for ancient admixture between closely related populations. Mol Biol Evol 28:2239–2252. https://doi.org/10.1093/molbev/msr048

Edmonds CA, Lillie AS, Cavalli-Sforza LL (2004) Mutations arising in the wave front of an expanding population. Proc Natl Acad Sci U S A 101:975–979. https://doi.org/10.1073/pnas.0308064100

El-Haj NA (2001) Facts on the ground: archaeological practice and territorial self-fashioning in Israeli society. University of Chicago Press

Eriksson A, Manica A (2012) Effect of ancient population structure on the degree of polymorphism shared between modern human populations and ancient hominins. Proc Natl Acad Sci U S A 109:13956–13960. https://doi.org/10.1073/pnas.1200567109

Eriksson A, Manica A (2014) The doubly conditioned frequency spectrum does not distinguish between ancient population structure and hybridization. Mol Biol Evol 31:1618–1621. https://doi.org/10.1093/molbev/msu103

Frigg R (2010) Models and fiction. Synthese 172:251–268. https://doi.org/10.1007/s11229-009-9505-0

Fu Y-X (1997) Statistical tests of neutrality of mutations against population growth, hitchhiking and background selection. Genetics 147:915–925. https://doi.org/10.1093/genetics/147.2.915

Fujimura JH, Bolnick DA, Rajagopalan R et al (2014) Clines without classes: how to make sense of human variation. Sociol Theory 32:208–227. https://doi.org/10.1177/0735275114551611

Gable E, Handler R (1993) Colonialist anthropology at colonial Williamsburg. Mus Anthropol 17: 26–31. https://doi.org/10.1525/mua.1993.17.3.26

Gerbault P, Allaby RG, Boivin N et al (2014) Storytelling and story testing in domestication. Proc Natl Acad Sci 111:6159–6164. https://doi.org/10.1073/pnas.1400425111

Giere RN (2010) Scientific perspectivism. University of Chicago Press

Godfrey-Smith P (2007) The strategy of model-based science. Biol Philos 21:725–740. https://doi.org/10.1007/s10539-006-9054-6

Godfrey-Smith P (2009) Models and fictions in science. Philos Stud 143:101–116. https://doi.org/10.1007/s11098-008-9313-2

Goldstein DB, Chikhi L (2002) Human migrations and population structure: what we know and why it matters. Annu Rev Genomics Hum Genet 3(1):129–152. https://doi.org/10.1146/annurev.genom.3.022502.103200

Green RE, Krause J, Ptak SE et al (2006) Analysis of one million base pairs of Neanderthal DNA. Nature 444:330–336. https://doi.org/10.1038/nature05336

Gregory MD, Eisenberg DP, Hamborg M et al (2021) Neanderthal-derived genetic variation in living humans relates to schizophrenia diagnosis, to psychotic symptom severity, and to dopamine synthesis. Am J Med Genet B Neuropsychiatr Genet 186:329–338. https://doi.org/10.1002/ajmg.b.32872

Griffiths RC, Tavaré S (1994) Sampling theory for neutral alleles in a varying environment. Philos Trans R Soc Lond B Biol Sci 344:403–410. https://doi.org/10.1098/rstb.1994.0079

Grusea S, Rodríguez W, Pinchon D et al (2019) Coalescence times for three genes provide sufficient information to distinguish population structure from population size changes. J Math Biol 78: 189–224. https://doi.org/10.1007/s00285-018-1272-4

Haigh J, Maynard Smith J (1972) Population size and protein variation in man. Genet Res 19:73–89. https://doi.org/10.1017/S0016672300014282

Hajdinjak M, Mafessoni F, Skov L et al (2021) Initial Upper Palaeolithic humans in Europe had recent Neanderthal ancestry. Nature 592:253–257. https://doi.org/10.1038/s41586-021-03335-3

Harpending HC, Sherry ST, Rogers AR, Stoneking M (1993) The genetic structure of ancient human populations. Curr Anthropol 34:483–496. https://doi.org/10.1086/204195

Heller R, Chikhi L, Siegismund HR (2013) The confounding effect of population structure on Bayesian skyline plot inferences of demographic history. PLoS One 8:e62992. https://doi.org/10.1371/journal.pone.0062992

Herbots, HMJD (1994) Stochastic models in population genetics: genealogy and genetic differentiation in structured populations. PhD Thesis, Queen Mary University of London

Hey J, Nielsen R (2004) Multilocus methods for estimating population sizes, migration rates and divergence time, with applications to the divergence of Drosophila pseudoobscura and D. Persimilis. Genetics 167:747–760. https://doi.org/10.1534/genetics.103.024182
Howell N (1976) Toward a uniformitarian theory of human paleodemography. J Hum Evol 5:25–40. https://doi.org/10.1016/0047-2484(76)90097-X
Hu W, Hao Z, Du P et al (2023) Genomic inference of a severe human bottleneck during the early to middle Pleistocene transition. Science 381:979–984. https://doi.org/10.1126/science.abq7487
Hudson RR (1990) Gene genealogies and the coalescent process. In: Futuyma D, Antonovics J (eds) Oxford surveys in evolutionary biology, vol 7. Oxford University Pres, New York, pp 1–44
Jones JS, Rouhani S (1986) How small was the bottleneck? Nature 319:449–450. https://doi.org/10.1038/319449b0
Källén A (2025) The trouble with ancient DNA: telling stories of the past with genomic science. University of Chicago Press, Chicago
Kellert SH, Longino HE, Waters CK (eds) (2006) Scientific pluralism. University of Minnesota Press, Minneapolis
Kennett D, Timpson A, Balding D, Thomas M (2018) The rise and fall of BritainsDNA: a tale of misleading claims, media manipulation and threats to academic freedom. Genealogy 2:47. https://doi.org/10.3390/genealogy2040047
Kingman JFC (1982) The coalescent. Stoch Process Appl 13:235–248. https://doi.org/10.1016/0304-4149(82)90011-4
Krings M, Stone A, Schmitz RW et al (1997) Neandertal DNA sequences and the origin of modern humans. Cell 90:19–30. https://doi.org/10.1016/S0092-8674(00)80310-4
Kuhlwilm M, Gronau I, Hubisz MJ et al (2016) Ancient gene flow from early modern humans into eastern Neanderthals. Nature 530:429–433. https://doi.org/10.1038/nature16544
Kuhn TS (1977) Objectivity, value judgment, and theory choice. In: Kuhn TS (ed) The essential tension—selected studies in scientific tradition and change. The University of Chicago Press, Chicago, pp 320-339
Kuhner MK, Yamato J, Felsenstein J (1995) Estimating effective population size and mutation rate from sequence data using Metropolis-Hastings sampling. Genetics 140:1421–1430. https://doi.org/10.1093/genetics/140.4.1421
Lapierre M, Lambert A, Achaz G (2017) Accuracy of demographic inferences from the site frequency Spectrum: the case of the Yoruba population. Genetics 206:439–449. https://doi.org/10.1534/genetics.116.192708
Lesne A (2008) Robustness: confronting lessons from physics and biology. Biol Rev 83:509–532. https://doi.org/10.1111/j.1469-185X.2008.00052.x
Lesne A (2012) Robust modeling in natural sciences. Annales de l'ISUP 56(2–3):109–118
Lester JD, Vigilant L, Gratton P et al (2021) Recent genetic connectivity and clinal variation in chimpanzees. Commun Biol 4:283. https://doi.org/10.1038/s42003-021-01806-x
Levins R (1966) The strategy of model building in population biology. Am Sci 54(4):421–431
Lewontin RC (1972) The apportionment of human diversity. In: Dobzhansky T, Hecht MK, Steere WC (eds) Evolutionary biology. Springer US, New York, pp 381–398
Li H, Durbin R (2011) Inference of human population history from individual whole-genome sequences. Nature 475:493–496. https://doi.org/10.1038/nature10231
Liu H, Prugnolle F, Manica A, Balloux F (2006) A geographically explicit genetic model of worldwide human-settlement history. Am J Hum Genet 79:230–237. https://doi.org/10.1086/505436
Ludwig D, Ruphy S (2021) Scientific pluralism. In: Zalta EN (ed) The stanford encyclopedia of philosophy. https://plato.stanford.edu/archives/win2021/entries/scientific-pluralism/
Marks J (2017a) Human biodiversity: genes, race, and history. Routledge
Marks J (2017b) Is science racist? Polity Press
Marks J (2019) Naming the sacred ancestors: 294Taxonomic reification and Pleistocene genomic narratives. In: Porr M, Matthews J (eds) Interrogating human origins. Routledge, pp 295–309

Marks J (2024) Understanding human diversity. Cambridge University Press, Cambridge
Mazet O, Rodríguez W, Grusea S et al (2016) On the importance of being structured: instantaneous coalescence rates and human evolution—lessons for ancestral population size inference? Heredity 116:362–371. https://doi.org/10.1038/hdy.2015.104
Mitchell SD (2003) Biological complexity and integrative pluralism. Cambridge University Press
Montévil M, Pocheville A (2017) The hitchhiker's guide to the cancer galaxy. How two critics missed their destination. Organisms J Biol Sci 1(2):37–48. https://doi.org/10.13133/2532-5876_2.9
Nei M, Maruyama T, Chakraborty R (1975) The bottleneck effect and genetic variability in populations. Evolution 29:1. https://doi.org/10.2307/2407137
Nielsen R (2001) Statistical tests of selective neutrality in the age of genomics. Heredity 86:641–647. https://doi.org/10.1046/j.1365-2540.2001.00895.x
Nielsen R, Beaumont MA (2009) Statistical inferences in phylogeography. Mol Ecol 18:1034–1047. https://doi.org/10.1111/j.1365-294X.2008.04059.x
Nielsen R, Akey JM, Jakobsson M et al (2017) Tracing the peopling of the world through genomics. Nature 541:302–310. https://doi.org/10.1038/nature21347
Nola R, Sankey H (2014) Theories of scientific method: an introduction. Routledge, London
Nordborg M (1998) On the probability of Neanderthal ancestry. Am J Hum Genet 63:1237–1240. https://doi.org/10.1086/302052
Nordborg M (2019) Coalescent theory. In: Balding D, Moltke I, Marioni J (eds) Handbook of statistical genomics, 1st edn. Wiley, pp 145–130
Notohara M (1990) The coalescent and the genealogical process in geographically structured population. J Math Biol 29. https://doi.org/10.1007/BF00173909
Odenbaugh J (2007) The strategy of "The strategy of model building in population biology". Biol Philos 21:607–621. https://doi.org/10.1007/s10539-006-9049-3
Orzack SH (2005) Discussion: what, if anything, is "the strategy of model building in population biology?" a comment on Levins (1966) and Odenbaugh (2003). Philos Sci 72:479–485. https://doi.org/10.1086/498475
Parreira BR, Chikhi L (2015) On some genetic consequences of social structure, mating systems, dispersal, and sampling. Proc Natl Acad Sci U S A 112. https://doi.org/10.1073/pnas.1414463112
Parreira BR, Gopalakrishnan S, Chikhi L (2025) Effects of social structure on effective population size change estimates. Evol Appl 18:e70063. https://doi.org/10.1111/eva.70063
Pauly R, Johnson L, Feltus FA, Casanova EL (2024) Enrichment of a subset of Neanderthal polymorphisms in autistic probands and siblings. Mol Psychiatry 29:3452–3461. https://doi.org/10.1038/s41380-024-02593-7
Pradeu T (2014) Regenerating theories in developmental biology. In: Minelli A, Pradeu T (eds) Towards a theory of development. Oxford University Press, pp 15–32
Prado-Martinez J, Sudmant PH, Kidd JM et al (2013) Great ape genetic diversity and population history. Nature 499:471–475. https://doi.org/10.1038/nature12228
Prüfer K, Racimo F, Patterson N et al (2014) The complete genome sequence of a Neanderthal from the Altai Mountains. Nature 505:43–49. https://doi.org/10.1038/nature12886
Prugnolle F, Manica A, Balloux F (2005) Geography predicts neutral genetic diversity of human populations. Curr Biol 15:R159–R160. https://doi.org/10.1016/j.cub.2005.02.038
Racine J (2024) Médecine Personnalisée et Oncogènes: Un Paradigme En Question. Doctorat, Université Toulouse 2 Jean Jaurès, Toulouse
Ragsdale AP, Weaver TD, Atkinson EG et al (2023) A weakly structured stem for human origins in Africa. Nature 617:755–763. https://doi.org/10.1038/s41586-023-06055-y
Ramachandran S, Deshpande O, Roseman CC et al (2005) Support from the relationship of genetic and geographic distance in human populations for a serial founder effect originating in Africa. Proc Natl Acad Sci U S A 102:15942–15947. https://doi.org/10.1073/pnas.0507611102

Rodríguez W, Mazet O, Grusea S et al (2018) The IICR and the non-stationary structured coalescent: towards demographic inference with arbitrary changes in population structure. Heredity 121:663–678. https://doi.org/10.1038/s41437-018-0148-0

Rogers AR, Harpending H (1992) Population growth makes waves in the distribution of pairwise genetic differences. Mol Biol Evol. https://doi.org/10.1093/oxfordjournals.molbev.a040727

Scerri EML, Thomas MG, Manica A et al (2018) Did our species evolve in subdivided populations across Africa, and why does it matter? Trends Ecol Evol 33:582–594. https://doi.org/10.1016/j.tree.2018.05.005

Scerri EML, Chikhi L, Thomas MG (2019) Beyond multiregional and simple out-of-Africa models of human evolution. Nat Ecol Evol 3:1370–1372. https://doi.org/10.1038/s41559-019-0992-1

Simonti CN, Vernot B, Bastarache L et al (2016) The phenotypic legacy of admixture between modern humans and Neandertals. Science 351:737–741. https://doi.org/10.1126/science.aad2149

Sjödin P, Kaj I, Krone S et al (2005) On the meaning and existence of an effective population size. Genetics 169:1061–1070. https://doi.org/10.1534/genetics.104.026799

Slatkin M, Hudson RR (1991) Pairwise comparisons of mitochondrial DNA sequences in stable and exponentially growing populations. Genetics 129:555–562. https://doi.org/10.1093/genetics/129.2.555

Smith JM, Haigh J (1974) The hitch-hiking effect of a favourable gene. Genet Res 23:23–35. https://doi.org/10.1017/S0016672300014634

Soler L, Trizio E, Nickles T, Wimsatt WC (eds) (2012) Characterizing the robustness of science: after the practice turn in philosophy of science. Springer Verlag, Dordrecht, New York

Steinrücken M, Kamm J, Spence JP, Song YS (2019) Inference of complex population histories using whole-genome sequences from multiple populations. Proc Natl Acad Sci U S A 116: 17115–17120. https://doi.org/10.1073/pnas.1905060116

Steux C, Couloigner C, Arredondo A et al (2025) On the demographic history of chimpanzees and some consequences of integrating population structure in chimpanzees and other great apes. Hum Popul Genet Genomics 5(2). https://doi.org/10.47248/hpgg2505020004

Stringer C (2014) Why we are not all multiregionalists now. Trends Ecol Evol 29:248–251. https://doi.org/10.1016/j.tree.2014.03.001

Suppes P (1978) The plurality of science. In: PSA: Proceedings of the Biennial Meeting of the Philosophy of Science Association, vol 1978. Cambridge University Press, pp 2–16. https://doi.org/10.1086/psaprocbienmeetp.1978.2.192459

Suppes P (1985) Probabilistic metaphysics. B. Blackwell, Oxford

Tajima F (1983) Evolutionary relationship of DNA sequences in finite populations. Genetics 105: 437–460. https://doi.org/10.1093/genetics/105.2.437

Tajima F (1989a) Statistical method for testing the neutral mutation hypothesis by DNA polymorphism. Genetics 123:585–595. https://doi.org/10.1093/genetics/123.3.585

Tajima F (1989b) The effect of change in population size on DNA polymorphism. Genetics 123: 597–601. https://doi.org/10.1093/genetics/123.3.597

Tibshirani RJ, Efron B (1993) An introduction to the bootstrap. Chapman & Hall, New York, London

Tournebize R, Chikhi L (2025) Ignoring population structure in hominin evolutionary models can lead to the inference of spurious admixture events. Nat Ecol Evol 9:225–236. https://doi.org/10.1038/s41559-024-02591-6

Turner DD (2016) A second look at the colors of the dinosaurs. Stud Hist Philos Sci A 55:60–68. https://doi.org/10.1016/j.shpsa.2015.08.012

Van Bouwel J (2014) Pluralists about pluralism? Different versions of explanatory pluralism in psychiatry. In: Galavotti (ed) New directions in the philosophy of science. Springer. https://doi.org/10.1007/978-3-319-04382-1_8

Varenne F (2016) Théorie, réalité, modèle: épistémologie des théories et des modèles face au réalisme dans les sciences. Matériologiques

Villanea FA, Schraiber JG (2019) Multiple episodes of interbreeding between Neanderthal and modern humans. Nat Ecol Evol 3:39–44. https://doi.org/10.1038/s41559-018-0735-8
Vorms M (2011) Qu'est-Ce Qu'une Théorie Scientifique ? Vuibert
Wakeley J (1999) Nonequilibrium migration in human history. Genetics 153:1863–1871. https://doi.org/10.1093/genetics/153.4.1863
Wang K, Mathieson I, O'Connell J, Schiffels S (2020) Tracking human population structure through time from whole genome sequences. PLoS Genet 16:e1008552. https://doi.org/10.1371/journal.pgen.1008552
Wilkinson-Herbots HM (1998) Genealogy and subpopulation differentiation under various models of population structure. J Math Biol 37:535–585. https://doi.org/10.1007/s002850050140
Wolpoff MH (1989) Multiregional evolution: the fossil alternative to Eden. In: Mellars P, Stringer CB (eds) The human revolution: behavioural and biological perspectives on the origins of modern humans. Princeton University Press, Princeton, pp 62–108
Wolpoff MH (2020) Human evolution: multiregional origins. In: Smith C (ed) Encyclopedia of global archaeology. Springer International Publishing, Cham, pp 5312–5317
Wright S (1931) Evolution in Mendelian populations. Genetics 16:97–159. https://doi.org/10.1093/genetics/16.2.97
Wright S (1943) Isolation by distance. Genetics 28:114–138. https://doi.org/10.1093/genetics/28.2.114
Zeberg H, Pääbo S (2020) The major genetic risk factor for severe COVID-19 is inherited from Neanderthals. Nature 587:610–612. https://doi.org/10.1038/s41586-020-2818-3
Zeberg H, Dannemann M, Sahlholm K et al (2020) A Neanderthal sodium channel increases pain sensitivity in present-day humans. Curr Biol 30:3465–3469.e4. https://doi.org/10.1016/j.cub.2020.06.045
Zilhão J (2006) Neandertals and moderns mixed, and it matters. Evol Anthropol Issues News Rev 15:183–195. https://doi.org/10.1002/evan.20110
Zilhão J (2010) Neanderthals are us: genes and culture. Radical Anthropol 4:4–15

Human Behavior Genetics and Human Evolution

11

Stephen M. Downes

Abstract

I assess the potential for human behavior genetics research in advancing our understanding of human evolution. I argue that, despite such work in genetics producing alternate measures of heritability, it is not set up to provide evolutionary answers to questions about human variation or diversity. I pursue this by comparing and contrasting the way in which human behavior geneticists, on the one hand, and human evolutionary geneticists, on the other, pursue genetics research. Each area is guided by different research questions, different theoretical assumptions, and different uses of or understandings of key theoretical concepts. In Sect. 11.2, I outline the type of research included under the heading "human behavior genetics" and consider some examples of work on human height and educational attainment. In Sect. 11.3, I characterize human evolutionary genetics via a brief introduction to the shared theoretical assumptions and key concepts in the field and introduce a number of examples of human evolutionary genetics research. In Sect. 11.4, I compare and contrast the deployment of evolutionary concepts in human evolutionary genetics and human behavior genetics, focusing on several key concepts: trait and phenotype; genetic load; and additivity. I argue that in each of these cases, human behavior geneticists adapt evolutionary concepts to the context of their research, which is the establishment of relations between genotype and phenotype. I conclude that this move undercuts any potential for GWAS-based genetics to contribute to our understanding of human evolution.

S. M. Downes (✉)
Department of Philosophy, University of Utah, Salt Lake City, UT, USA
e-mail: s.downes@utah.edu

M. Lequin, J. M. Rodríguez Caso (eds.), *The Field of Human Evolution: Critical Perspectives from History and Epistemology*, Evolutionary Biology – New Perspectives on Its Development 11, https://doi.org/10.1007/978-3-032-13925-2_11

Keywords

Genetics · GWAS · Human behavior genetics · Human evolution · Human evolutionary genetics

11.1 Introduction

Early twin studies researchers (see Holzinger 1929) claimed that not only were genes responsible for variation in human physical characteristics like height, but genes also impacted upon psychological characteristics. The message from the contemporary human behavior genetics community is the same, but many more traits are implicated, and the confidence of researchers in finding the genetic underpinning of these traits is much higher than that of early twin researchers (see Plomin 2019). These genetic findings are also offered as transformative, or at least a boost, to social science research on a wide range of human behavior (Harden and Koellinger 2020). Twin studies researchers (see Clark 1956) also noted the relevance of their findings for human evolution. Contemporary human behavior geneticists don't generally promote the relevance of their work for understanding human evolution but do rely on key evolutionary concepts in their work.

Here I will explore examples of human behavior genetics research on educational attainment, which has led to claims of not only establishing heritability measures for higher or lower educational attainment but the pinpointing of specific genes, or regions of the genome, that lead to higher or lower educational attainment (Lee et al. 2018; see also Harden (2021) for an overview of recent work in behavior genetics on educational attainment; Okbay et al. 2022). I compare and contrast behavior genetics work on educational attainment with work on the genetics of human height, as very similar techniques and theoretical assumptions apply in each case (see Mills et al. (2020) for an overview). Human height is highly heritable (but see Turkheimer (2000) for cautionary notes on high heritability measures), and this is taken by some in human genetics research to mean that genes are much more influential on variation in human height than relevant environments. With the introduction of genome-wide association studies (GWAS) and associated techniques, starting around 2005 (see Visscher et al. (2017) for some history here), the hope was that the relevant regions of the genome that impacted height would be discovered. The application of GWAS to human height initially returned very low values for heritability compared to values obtained by traditional methods, dubbed the "missing heritability problem" for height (Matthews and Turkheimer 2021). Researchers found that they could close the gap between traditional heritability measures and those derived from GWAS and associated techniques by increasing the number of sites looked at in the genome and increasing their sample sizes to hundreds of thousands or even millions of people. A recent study used whole genome sequences from a huge sample of people in order to establish very high heritability measures for height (and body mass index (BMI)) (Wainschtein et al. 2019).

Enthusiasm about finding the genetic underpinnings of educational attainment has also been fueled by using huge sample sizes of people and sampling large areas of the genome, although to date no whole genome sequencing has been used in this context (Lee et al. 2018; Okbay et al. 2022). Several people have argued that the highly impressive data analysis techniques and statistics developed and applied within GWAS-related research do not amount to the right approach to ascertain regions of the genome causally responsible for a person's level of educational attainment or even their height (see Turkheimer 2012; Longino 2013; Dupré 2014; Schaffner 2016; Woodward and Kendler 2023). Rather than focusing on this, albeit important, issue of discovering genetic causation here, I turn my attention to the potential for human behavior genetics research in advancing our understanding of human evolution. I argue that, despite such work in genetics producing alternate measures of heritability, it is not set up to provide evolutionary answers to questions about human variation or diversity. I pursue this by comparing and contrasting the way in which human behavior geneticists, on the one hand, and human evolutionary geneticists, on the other, pursue genetics research. Each area is guided by different research questions, different theoretical assumptions, and different uses of or understandings of key theoretical concepts.

Here's how I will proceed: in Sect. 11.2, I outline the type of research included under the heading "human behavior genetics." I also consider some examples of work on human height and educational attainment by these researchers. In Sect. 11.3, I characterize human evolutionary genetics via a brief introduction to the shared theoretical assumptions and key concepts in the field. Here, I also introduce a number of examples of human evolutionary genetics research. In Sect. 11.4, I turn to the characterization and deployment of evolutionary concepts in human evolutionary genetics and human behavior genetics. Here, I focus on several key concepts: trait and phenotype; allele; genetic load; and additivity. In each of these cases, I introduce the version of the concept deployed by human behavior geneticists and then compare and contrast this with standard usage in (human) evolutionary genetics. I argue that in each of these cases, human behavior geneticists adapt evolutionary concepts to the context of their research, which is the establishment of relations between genotype and phenotype. I conclude that this move undercuts any potential for GWAS-based genetics to contribute to our understanding of human evolution.

11.2 Human Behavior Genetics

In this chapter, I use the phrase "human behavior genetics" as a placeholder to cover a huge area of research. All researchers in this broad area share the use of several key approaches centered around genome-wide association studies (GWAS). It seems odd to say that height, for example, is a topic of research for human behavior geneticists, but this oddity is resolved by a brief look at both the history of human behavior genetics and the shared research approach and methods of human behavior geneticists and a large number of other human geneticists. First, I'll consider history.

Early twin studies researchers (see Newman et al. 1937) were intent on uncovering the contribution our genes made to a wide variety of human traits. Human traits were divided up into physical and behavioral, and early expectations were that, though genetics was expected to contribute to both, genetics would play a larger role for physical traits. Early twin study researcher Holzinger put it this way: "nature is much more effective than nurture for physical traits" (Holzinger 1929). Human traits studied by twin researchers included height and fingerprints on the one hand and IQ and various personality traits on the other. Most traits studied by these researchers are measurable, height in the obvious way and IQ via agreed upon IQ scales (see Gould (1996) for a critical evaluation of approaches to human trait measurement). Identical or monozygotic twins were assumed to be genetically identical, and this was a reasonable assumption based on the genetics of the 1930s up until the genome sequencing era of the 2000s (see Jonsson et al. (2021) for an overview of genetic differences between monozygotic twins). The key output from twin studies is a number, a heritability coefficient, that twin study researchers take to measure the relative contribution of genes to a trait as opposed to the environment (see Downes and Turkheimer (2022) for some history of the heritability concept). Heritability coefficients are numbers between 0 and 1. Some take heritability coefficients to be estimates of "genetic influence" on a trait (see Plomin 2019), whereas others adopt a more cautious view: "a heritability estimate tells you how much of the total phenotypic variation in a trait (i.e., the variation due to both genetics and environment) can be explained by genetic variation alone" (Schaffner 2016). Heritability coefficients are derived from equations that combine information from monozygotic twins, dizygotic twins, and their shared and non-shared environments (See Downes and Matthews (2024) for details).[1] Holzinger's idea that "nature is more effective than nurture" can be expressed numerically via heritability coefficients. Twin study derived heritability for human height runs at over 0.8 whereas IQ comes in at around 0.6. There is no difference in approach between establishing heritability for height or for IQ. Both are measurable human traits and both can be addressed by the methods outlined here. Human behavior geneticists agree that twin studies have been surpassed by studies based on GWAS (see Plomin 2019), and we turn to this approach next (see Turkheimer (2024) for an account of the transition from twin studies to GWAS).

GWAS is applied in a wide range of studies of human traits under variation. GWAS-based approaches provide a shared methodology along with dizzying array of statistical methods (Mills et al. (2020) provide a thorough introduction to all of these methods) for researchers interested in both physical and behavioral human traits. Sociogenomics applies GWAS to socio-behavioral traits, medical genomics applies GWAS to a wide range of diseases, and human behavior genetics applies GWAS to human behavioral traits such as IQ and educational attainment (the level

[1] There is a huge critical literature on the role of heritability coefficients and their proper interpretation. Much of this literature stems from Lewontin (1974). Downes and Matthews (2024) summarize the state of play in this area of philosophy of science.

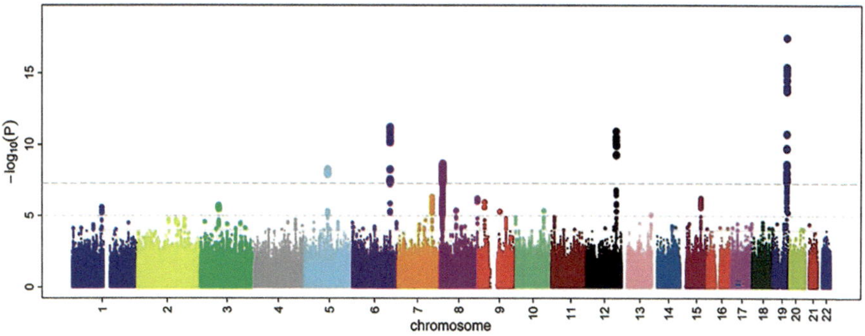

Fig. 11.1 Example of a Manhattan plot on which the *y*-axis is −log10 of the *p*-value for the association between a SNP location and a trait. (From Ikram et al. (2010))

of education an individual achieves in their lifetime). GWAS-based approaches are claimed to be superior to twin studies (and related approaches) because they rely on molecular biology or molecular information (see Plomin 2019). We will see why this claim is made when we outline some of the key features of GWAS. It is worth noting that others are suspicious of this appeal to molecular biology and consider GWAS, with its strong reliance on statistical models, to be more in line with quantitative social sciences than molecular biology (see Turkheimer 2012, 2024). Either way, from here on out, I treat both GWAS-based research and twin studies research as alternate types of genetics research.

There are two main types of GWAS research: Case/control studies and studies of continuously varied quantitative traits, like height. I will deal with each in turn. Case control studies are familiar in medical contexts including psychiatry. Traits of interest here are schizophrenia, congestive heart disease, and so on. The aim of such studies is to assess whether any areas of the genome are associated with the trait of interest. Genetic samples are taken both from the control group and the group with the trait of interest. These samples are taken from a large number of chromosomes, but the samples focus on areas of the genome at which there are single-nucleotide polymorphisms (SNPs). These are genome sites which present variation in at least 1% of the population. In GWAS, SNP hits are sites that are associated with the trait of interest.

GWAS results are presented via Manhattan Plots (see Fig. 11.1 for an illustration). The *x*-axis has SNP sites on each chromosome. The *y*-axis records the significance, *p*-value, of the association between each SNP location and the trait of interest, where the null hypothesis is that there is no association. Minus log10 of all *p*-values are presented so that the smallest *p*-values recorded are the highest values on the plot. Also, significance levels are very different to those familiar to social science researchers, where the significance value is usually set at 0.05. For GWAS, significance is set at between 10^{-5} and 10^{-8}, as in Fig. 11.1. The significance level is shown in Fig. 11.1 by the heavy dashed line. Figure 11.1 reveals 5 SNP hits, reaching the relevant significance value, which many researchers interpret as

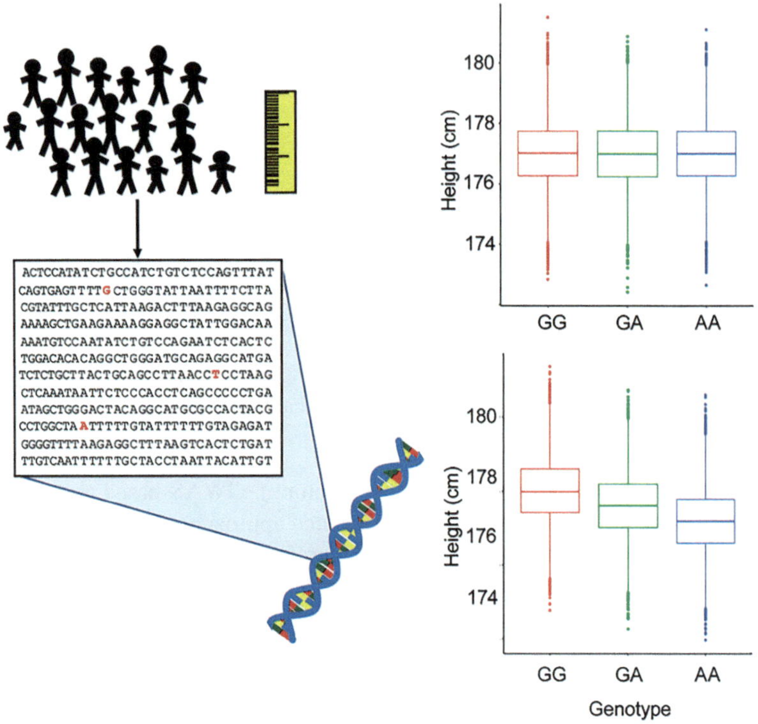

Fig. 11.2 Illustration of a population GWAS, an alternative approach to case/control studies, applied to cases such as height and BMI. (From Guo et al. (2018))

indicating locations of genes of interest for the relevant trait in the genome. There may be many SNPs associated with a given trait, which comports with a background assumption in this area of genetics: many small genetic differences contribute to trait differences. Further information can be derived from GWAS, including polygenic scores (PGS) or polygenic risk scores in the case of diseases. Here the relevant measures are not population wide but rather measures for individuals. I am at risk for a disease to the extent that I have the relevant collection of SNP variants identified via GWAS (see Mills et al. (2020) for details of the methods for deriving PGSs.). As I noted above, there is considerable debate about whether claims about SNPs associated with a trait are straightforwardly causal claims about the role of the SNPs in producing the trait (Woodward and Kendler 2023; Turkheimer 2024). This is not our focus here.

GWAS are also applied to continuous varying quantitative traits like height. A somewhat different procedure than the case/control approach is adopted here. In these studies, every subject is genotyped and their height is also measured (see Visscher 2008). The left-hand side of Fig. 11.2 illustrates the different sampling procedures for population studies. The top graphs illustrate the case in which alternate genotypes do not vary with change in height and the bottom graph illustrate those that do. Studies such as these can show an association between a number of SNPs and variation in height. Guo et al. make this point as follows: "Understanding the genetic architecture of height—the number of genetic variants that contribute to variation in height, their frequencies in the population, and the magnitude of each genetic variant's effect (i.e., effect sizes)—can teach us lessons about how to study the genetic underpinnings of other complex genetic traits and disease" (Guo et al. 2018).

GWAS researchers are in the business of attempting to establish the genes underlying our many and varied traits. They start out by parsing variation in a sample population in terms of genetic differences in that population. Then, via PGSs or similar measures, they propose a genetic basis for traits within individuals. Helen Longino (2013) notes that behavioral geneticists look for the genetic causes of aggressive behavior, whereas sociologists look for the social causes of such behavior. This point is echoed by Elisabeth Lloyd (2015) who talks of the research questions guiding scientists in different fields. Borrowing from both Lloyd and Longino, a way to understand what behavior geneticists relying on GWAS are doing is to understand their research question as "what are the genes underlying a trait?". Let's look at some examples of this approach.

Visscher et al. have conducted a series of studies on the genetics of human height using GWAS and related techniques. Twin studies assessed the heritability of human height at 0.8, which is very high. This measure is interpreted by human behavior genetics to reveal that height is under a great degree of genetic control (see Plomin 2019). Earlier GWAS work on height did not produce figures anywhere close to these earlier heritabilities (see Visscher et al. 2010). However, researchers were undaunted and proceeded to roll out further studies vastly increasing both the number of participants and the number of SNP sites sampled in each participant. Yengo et al. (2018) used a sample of 250,000 people to generate the result that ~700 SNPs are associated with height (this paper presents both height and body mass index results together). And more recently (Yengo et al. 2022), this same research group used a sample of 5.4 million people to conclude that over 12,000 SNPs were associated with height to the relevant significance level, leading some observers to say that they had discovered "nearly all genetic variants linked to height" (DiCorato 2022). The way that the question "what are the genes underlying the trait of height?" is answered in this research tradition is by offering up a seemingly ever-increasing number of SNPs that associate with height via the stringent significance measures adopted in GWAS. The assumption these researchers operate on is that height is a complex or polygenic trait (see Sect. 11.4) and that the impact of all of these SNPs on the trait is additive in some sense (see also Sect. 11.4).

Educational Attainment (EA) is treated very similarly to height by human behavior geneticists. However, the results, couched in contemporary versions of heritabilities, are nowhere near as impressive as those for height. The trajectory of ever-increasing sample sizes for each iteration of EA studies tracks that of those for height. Okbay et al. (2016) sampled close to 300,000 individuals and concluded that 74 SNPs were associated with EA. Lee et al. (2018) sampled 1.1 million individuals revealing 1271 relevant SNPs but also concluding that genes accounted for 11–13%

of the variance in EA. This number is very roughly like a heritability coefficient for EA. Heritability coefficients from twin studies are the additive genetic variance component of the phenotypic variance of a trait (see Downes and Matthews (2024) for details), and here the relevant SNPs account for a specific amount of the total variance. However, such comparisons should be made cautiously and only used as a very rough guide (Downes and Matthews 2024; Turkheimer 2024). Finally, Okbay et al. (2022) used a sample of three million people finding, a perhaps not surprising increase in the number of associated SNPs, 3952, but a minimal increase in the figure for how much variance in EA was accounted for by genes: 12–16%. Conclusions about what genes underlie EA are derived from the application of very similar methods to those used in the case of height. Also, as in the height studies, EA geneticists assume that EA is polygenic and that the trait is additive in some sense. We turn to a critical appraisal of these assumptions in Sect. 11.4. Now we turn to human evolutionary genetics noting the alternate research question(s) that these scientists are guided by.

11.3 (Human) Evolutionary Genetics

In evolutionary genetics, the main idea is to represent evolution via changing gene frequencies. Evolutionary geneticists, whether they are population geneticists, quantitative geneticists, molecular geneticists, or combine methods from all of these approaches, focus on changing gene frequencies over time. From a modeling perspective, human evolutionary genetics is just like the evolutionary genetics for any other organism. However, we do no breeding experiments, knock-out/knock-in experiments, etc. on humans. The starting point for all accounting of evolutionary change in these terms are one locus, two allele models. These models provide a good first pass at representing the patterns of variation we see.

The Hardy-Weinberg model is a good starting point for this discussion: $p^2 + 2pq + q^2 = 1$, where $p =$ frequency of allele A_1 and $q =$ frequency of allele A_2, giving p^2 for A_1A_1, q^2 for A_2A_2 and $2pq$ for A_1A_2. Any shift from this equilibrium state indicates that various evolutionary influences are at play, e.g., selection vs drift and so on. The initial model also assumes random mating, discrete generations, an infinite (or near infinite) population, and so on. There are numerous ways in which population genetics models have been expanded and elaborated upon including non-dominance (such as additivity or epistasis), multiple loci, shifting population parameters, modeling genes as nucleotide sequences, and so on (see Lloyd 1988). There are also numerous extensions of our starting set of theoretical concepts. For example, there are different types of selection, including directional selection and balancing selection. We may also expand our models to capture neutral evolution, drift, draft, hitchhiking, and so on. The overall aim remains the same: to model evolutionary change via models of changes in gene frequency. Lloyd (2015) calls this the "evolutionary factors" approach, which is guided by the research question "What evolutionary factors account for the form and distribution of this trait?" Evolutionary genetics models are supported by all manner of empirical work including various gene discovery techniques in molecular genetics and more recent innovations such as ancient DNA (aDNA) sequencing (see Downes (2021) for an introduction to aDNA research). Some examples help illustrate the approach.

Here are three examples of human evolutionary genetics in practice: lactase persistence, sickle cell disease, and (ancient human) population bottlenecks. We look at each in turn. Lactase persistence is a widely distributed trait in humans marked by adults being able to digest milk via the enzyme lactase. However, not all adult humans have lactase, there is variation, and this variation is what evolutionary geneticists seek to explain. A directional or positive selection account has been given for lactase persistence in humans. On this account, the allele for lactase has been driven to fixation in some human populations. The idea here is that retaining lactase in adulthood is an adaptive response to increased dairy in some human diets (see Bersaglieri et al. 2004). This provides an example of a fairly recent evolutionary change in some human populations. Early models of this evolutionary process relied on simple models of the type outlined above, but modeling now invokes the molecular structure of the allele for lactase and its alternates.

Sickle cell disease is a condition in which red blood cells are misshaped and seriously undermine the ability to process oxygen. Sickle cell disease was originally modeled on the assumption that disease sufferers had two copies of a recessive allele. However, this account introduced a puzzle: why does the recessive allele persist if the disease is so debilitating and often fatal? Shouldn't there be a selective sweep removing the offending allele from the population? Early work on the genetics of sickle cell disease, in combination with epidemiology work, revealed that while sickle cell disease is indeed debilitating, heterozygotes for sickle cell have immunity from malaria (see Allison 1954). The standard approach to sickle cell disease now is to treat it as an example of balancing selection—selection that acts to maintain genetic diversity (see Piel et al. 2010). In parts of the world where malaria is prevalent, malaria is part of the selectively relevant environment. The heterozygote state leads to higher fitness in these environments and so the recessive, and deleterious, allele persists in the population.

Bottlenecks in ancestral human populations are situations when population sizes dropped. Some estimates have ancient human populations dropping to around 1280[2] breeding individuals (see Hu et al. 2023). Population bottlenecks are not traits expressed in individuals, like sickle cell disease and lactase persistence, rather they are features of our evolutionary past that can be detected by understanding and modeling available genetic data (see Rogers (2022) for an overview of some ways of doing this). Genetic signatures of population bottlenecks include less genetic diversity and an accumulation of deleterious alleles. Gene frequency change in small populations is largely attributed to drift, and this type of change is relatively slow. Severe population bottlenecks in ancestral human populations could explain our relatively low genetic diversity.

[2]This very precise number will doubtless be challenged in subsequent work and fluctuate.

I include this final example to emphasize the point that human evolutionary genetics can both model the persistence or lack thereof of specific traits through time but also model broader patterns of variation relevant to evolution, such as the patterns indicating population bottlenecks. Human evolutionary genetics is founded on the assumption that we best model evolutionary change in terms of genes, alleles, nucleotide sequences, and so on. However, in contrast to human behavior genetics, the main aim of human evolutionary genetics is not to find the genes that contribute to our various traits. In the case of lactase persistence and sickle cell disease, we have a very good understanding of the relationship between alleles and traits but we did not need this understanding to model the patterns in variation of the relevant traits in terms of gene frequencies. Such models were developed before any knowledge of the molecular structure and mode of expression of the relevant genes. With this molecular knowledge in hand, models in human evolutionary genetics can be enhanced for specific traits, but this is not necessary for differentiating between alternate patterns of variation.

11.4 Some Evolutionary Concepts Deployed by Human Behavior Geneticists

In what follows, I take a look at the way human behavior geneticists deploy some evolutionary concepts. For each concept, I compare and contrast human behavior geneticists' use and definition with those of evolutionary geneticists. The way in which each of the two groups of researchers uses and deploys these concepts appears to be in tension. This provides the basis for my argument that human behavior genetics, practiced via GWAS, cannot contribute to our understanding of human evolution in any straightforward way. First, I look at the population genetics concept of additivity.

Here is Peter Visscher and collaborators in an early GWAS paper on human height: "Despite the recent suggestion that variation due to epistasis is too-often neglected in complex-trait studies, we found no evidence of any nonadditive genetic variance for height, the complex trait we studied. Our results imply, at least for the quantitative trait height in humans, that genetic variation can be explained by many loci distributed over all the autosomes with an additive mode of gene action" (Visscher et al. 2007). Here is a similar point made in a much more recent paper on human height: "our study has been able to show empirically that the combined additive effects of tens of thousands of individual variants, detectable with a large enough experimental sample size, can explain substantial variation in a human phenotype" (Yengo et al. 2022). There is a lot going on here, but the main point to flag is that the researchers claim to have discovered that additive gene action is responsible for human height. Let's now look more carefully at how additivity is understood by evolutionary geneticists.

There are two main senses of additivity identified by evolutionary geneticists: First, additivity of gene action: genes/alleles act additively to produce a trait. For example, the more relevant alleles you have, the taller you are (see Fig. 11.3). This is

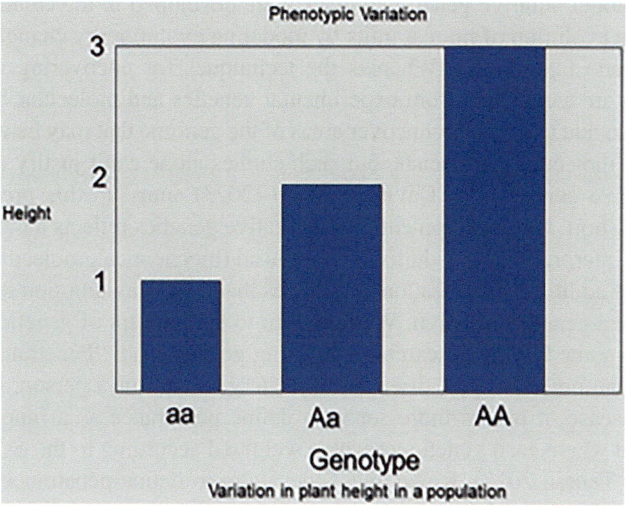

Fig. 11.3 An illustration of the pattern in variation in a population attributed to additivity. This pattern is distinct from a situation in which dominance is at play, where both aA and AA would be expressed as the same phenotype, here height 3

a notion that best applies in the context of gene expression or development. Second, additive genetic variance: phenotypic variance is attributable to genetic variance, which includes additive variance, dominance variance, and epistatic variance (Fisher 1918). This notion of additivity involves the idea that the type of phenotypic variation in the population is best accounted for by additive genetic variance. Dominance variance accounts for characteristic patterns in variation due to the presence of dominant and recessive alleles for traits. Epistatic variance accounts for variance due to interactions between genes at different loci. This notion of additivity is trickier to grasp than additivity of gene action can be conflated with it. Here Falconer lays this out this worry: "The concept of additive variance does not carry with it the assumption of additive gene action; and the existence of additive variance is not an indication that any of the genes act additively (i.e. show neither dominance nor epistasis)" (1960).

GWAS, as we noted, supports conclusions about associations between collections of SNPs and selected phenotypes. Visscher et al.'s claim above summarizes results from an early GWAS of human height and transitions from a claim about additive genetic variance to a claim about additive gene action. As Falconer points out, additivity of variance does not allow us to infer additivity of gene action. Yengo et al., quoted above, make a more ambiguous claim than Visscher et al. They claim that their approach, using much larger samples of people and many more areas of each genome sampled, allows them to explain variation in height in terms of the additive effects of thousands in individual variants. This is neither straightforwardly a claim about additivity of variance nor additivity of gene action. Further, the statistical techniques in quantitative genetics that support inferences about additive genetic variance were developed in the context of understanding the evolution of human traits by modeling evolutionary change in terms of changing gene frequencies. Whereas the techniques for uncovering additivity of gene action are techniques from experimental genetics and molecular biology. It is certainly true that GWAS can uncover areas of the genome that may be worth further scrutiny by molecular geneticists, but such studies alone can't justify a conclusion about additive gene action. Davide Serpico (2023) sums up this predicament as follows "In short, the development of quantitative genetics reflects a transition from a statistical interpretation of additivity to a causal (biochemical, molecular) one." He calls this the additivity idealization and argues that it is an assumption that underlies GWAS-based genetics research. We now turn to the concept of genetic load.

Here are some GWAS researchers invoking genetic load: "Penetrance is usually defined as the probability of disease given the genotype of a person, however for complex disease, it makes more sense to define penetrance as a function of total genetic load where each genetic variant is weighted according to the risk it confers" (Rowe and Tenesa 2012). Rowe and Tenesa aim to define penetrance in terms of genetic load. They say that their notion of penetrance is the usual definition, but this depends crucially on context. In evolutionary contexts, both penetrance and genetic load are handled quite differently. A more standard version of penetrance used in evolutionary contexts is something like this: penetrance is expressed as a number which gives the percentage of individuals in a population who have both an allele and the trait usually associated with that allele. Rowe and Tenesa's definition of penetrance is narrower than this but not too different. Their notion of genetic load, however, is quite different from its evolutionary genetics counterpart.

There has been plenty of controversy about genetic load in the history of genetics (see Teicher 2018), but one thing all participants in the debate agree on is that genetic load is a concept understood via natural selection. Crow's (1958) formulation of genetic load is "the proportional decrease in the mean fitness of a population, compared to an ideal, optimal genotype" (Teicher 2018). Crow introduced a formal treatment of Haldane and Muller's thinking about the action of selection on mutations. The concept is understood via (one locus) population genetics models of evolutionary change.

$$L = \frac{W_{\max}\overline{W}}{W_{\max}}$$

where L is the genetic load, $W_{\max}$ is the fitness of the fittest possible genotype in the population, and $\overline{W}$ is the mean fitness in the population. For evolutionary geneticists, genetic load is understood in terms of genotype fitness in populations, whereas in Rowe and Tenesa's hands, genetic load appears to be a concept involving the risk of disease a particular allele confers on an individual. This appears to be a different notion of genetic load than the one developed by evolutionary geneticists. This idea of genetic load is clearly related to Crow et al.'s notion, but the latter notion only makes sense in the context of genetic models of evolutionary change. Now let's turn to traits and phenotypes.

Here are some GWAS researchers invoking traits and phenotypes: "Educational attainment (EduYears), a heritable trait often used as a proxy for cognitive ability, is associated with various health and social outcomes" (Chen et al. 2024). "Educational attainment (EduYears: years of education as a continuous phenotype) is a behavioural trait that has been studied extensively and linked to various social, economic and health-related outcomes" (Chen et al. 2024). "Human height is an anthropometric trait that varies considerably within human populations as well as across the globe" (Muthuirulan and Capellini 2019). "Among the most commonly studied human traits are extreme tall or short height phenotypes, defined as heights greater than two standard deviations above and below the mean, respectively" (Muthuirulan and Capellini 2019). Educational attainment (EA) is the number of years of education an individual attained in their lifetime. It is characterized as a behavioral trait by these human behavior geneticists and all others working in this research tradition. EA is certainly measurable quite straightforwardly and as such allows for a GWAS approach similar in line with the one outlined for human height in Sect. 11.2. This is the approach adopted in the studies we looked at in Sect. 11.2. Height is a reasonably uncontroversial candidate for a phenotype, but "extreme tall or short height" is more controversial. These are defined statistically here, and this way of addressing phenotype appears at odds with that of evolutionary geneticists.

Here is Douglas Futuyma's definition of phenotype: "The phenotype (observed physical characteristics) is different from the genotype (the set of genes carried by an individual), and the phenotypic differences among organisms can be due partly to genetic differences and partly to direct effects of the environment" (1998). Futuyuma's definition doesn't rule out behavioral phenotypic traits. Bird mating displays, for example, are observable characteristics that vary considerably between species of birds. The key is that we understand an organism's phenotypes to differ as a result of differences in genes or environment and interactions between genes and environment. However, in population genetics we model trait variation via genotypic variation. For example, assuming dominance and representing one locus, we say all AA and Aa genotypes are tall plants and all aa genotypes are short plants. These genotypes stand for the relevant phenotypes. The notion of gene at play here is called either gene-p (for phenotype) (Moss 2003) or Mendelian gene (Griffiths and Stotz 2013). This simple example includes a hypothesis that the observed differences in plant height are due to genetic differences. The way in which we test this hypothesis is via breeding experiments on our plants.

In human behavior genetics, the relation between modeling, on the one hand, and characterizing phenotypes, on the other, appears messier than it is in population genetics. Plus, in human genetics, we can't test our hypotheses about what produces variation in phenotypes via breeding experiments. Here is an illustrative example: Human behavior geneticists and medical geneticists pay a great deal of attention to what they call complex traits. This is a tricky concept and is defined in a number of ways. For example, "Complex traits are determined by the interplay of multiple genetic and environmental factors" (Rowe and Tenesa 2012). Being determined by multiple genetic and environmental factors does not distinguish any subset of traits as, according to Futuyma, all phenotypes just are the product of genes, environment, and their interaction. However, Rowe and Tenesa have something different in mind. Complex traits are to be distinguished from monogenic traits, such as Sickle Cell Disease, our example from Sect. 11.3. So, complex traits are traits whose assumed genetic underpinnings are different from monogenic traits. Complex phenotypic traits now appear to be traits modeled as polygenic. Height, for example, is a complex trait in this sense. This notion of phenotypic traits appears to owe very little to Futuyma's observed physical characteristics account and more to the assumed model of the genetic underpinning of the trait in question. This situation perhaps helps us understand how EA can be a behavioral trait. EA is measurable and is modeled as a polygenic trait and hence is a complex phenotypic trait (Serpico (2020) suggests that headway can be made here by adopting character/character state terminology.).

Human behavior genetics researchers using GWAS and related approaches invoke key concepts from evolutionary genetics but define and deploy them differently than evolutionary geneticists. I have just surveyed a few sample concepts here but there are more to explore (See e.g. Remington (2015) on alleles; and Coop and Przeworski (2022) on population structure). These mismatches between human behavior genetics and human evolutionary genetics do not bode well for the application of GWAS methods to the study of human evolution.

11.5 Conclusion

In the previous section, I appealed to some examples to illustrate the point that evolutionary concepts appear to be differently defined and applied in human behavior genetics. My hope would be for more consistency and more work on sustaining clear conceptual overlap between the various genetics practices. Specifically, my hope is that human behavior geneticists bring their key concepts more closely in line with the evolutionary genetics concepts they are derived from. This would enable continued, enriching theoretical discussion between researchers from both areas to inform future research (see Veller et al. (2024) for an example of critical work on GWAS research from the perspective of population genetics). I am not optimistic about such discussions developing in the near future. My optimism is tempered by what I have presented here and also by statements in a key training manual for GWAS practitioners (Mills et al. 2020) that researchers need no background in molecular biology or genetics to engage in this kind of genetics research; this is put forth as a selling point for them. If GWAS practitioners need no background in biology, evolutionary or otherwise, it does not put them in a good position to work towards achieving more conceptual overlap between their work and that of evolutionary geneticists. This will result in human behavior genetics and other areas of genetics based on GWAS moving further away from evolutionary genetics and evolutionary biology in general. This situation compounds that of human behavior geneticists being guided by starkly different research questions than (human) evolutionary geneticists. Models and theoretical concepts developed under the guidance of the question "What evolutionary factors account for the form and distribution of this trait?" have, on my account, been adapted and reworked to fit genetics practice guided by this question: "What are the genes underlying a trait?" All these considerations undercut any potential for GWAS-based genetics to contribute to our understanding of human evolution.

An alternate way of thinking about this situation is to consider the possibility that the alternate versions of evolutionary genetics concepts produced by GWAS researchers provide a welcome new look at the problems at hand or conceptual diversification.[3] However, conflation of additivity of gene action and additivity of genetic variance, for example, appears to be a case of conceptual confusion, rather than conceptual diversification. Ironing out such confusion requires discussion by both evolutionary geneticists and human behavior geneticists, and such a discussion relies on a shared understanding of the relevant concepts and how they guide research. If GWAS results were only presented as suggestive of places to look in the human genome for sites of interest, GWAS research could be a helpful supplement to human evolutionary biology. How I see the situation is that this is not the norm; rather, the norm is for GWAS researchers to use population-based, statistical methods to uncover relations between specific genotypes and phenotypes. This does not seem to be a productive way forward for cooperation between evolutionary geneticists and human geneticists who deploy GWAS and related methods.

References

Allison AC (1954) The distribution of the sickle-cell trait in East Africa and elsewhere, and its apparent relationship to the incidence of subtertian malaria. Trans R Soc Trop Med Hyg 48:312–318. https://doi.org/10.1016/0035-9203(54)90101-7

Bersaglieri T et al. (2004) Genetic Signatures of Strong Recent Positive Selection at the Lactase Gene. Am J Hum Genet 74(6):1111–1120

Chen T-T, Kim J, Lam M et al (2024) Shared genetic architectures of educational attainment in East Asian and European populations. Nat Hum Behav 8:562–575. https://doi.org/10.1038/s41562-023-01781-9

Clark PJ (1956) The heritability of certain anthropometric characters as ascertained from measurements of twins. Am J Hum Genet 8(1):49–54

Coop G, Przeworski M (2022) Lottery, luck, or legacy. A review of "The genetic lottery: why DNA matters for social equality". Evolution 76(4):846–853. https://doi.org/10.1111/evo.14449

Crow JF (1958) Some possibilities for measuring selection intensities in man. Hum Biol 30(1):1–13

DiCorato A (2022) Largest genome-wide association study ever uncovers nearly all genetic variants linked to height. Broad Institute. https://www.broadinstitute.org/news/largest-genome-wide-association-study-ever-uncovers-nearly-all-genetic-variants-linked-height. Accessed 29 Mar 2023

Downes SM (2021) The role of ancient DNA research in archaeology. Topoi 40(1):285–293. https://doi.org/10.1007/s11245-019-09665-2

Downes SM, Matthews LJ (2024) Heritability. In: Zalta EN (ed) Stanford encyclopedia of philosophy, Spring 2024. Stanford University. https://plato.stanford.edu/entries/heritability/

[3]Both Marshall Abrams and an anonymous referee for this volume pressed this point. The referee offered the term "conceptual diversification."

Downes SM, Turkheimer E (2022) An early history of the heritability coefficient applied to humans (1918–1960). Biol Theory 17(2):126–137. https://doi.org/10.1007/s13752-021-00392-9

Dupré J (2014) Processes of life: essays in the philosophy of biology. Oxford University Press, Oxford

Falconer DS (1960) Introduction to quantitative genetics. Ronald Press Co., New York

Fisher RA (1918) The correlation between relatives on the supposition of Mendelian inheritance. Earth Environ Sci Trans R Soc Edinb 52(2):399–433. https://doi.org/10.1017/S0080456800012163

Futuyma DJ (1998) Evolutionary biology. Sinauer Associates, Sunderland

Gould SJ (1996) The mismeasure of man (Revised). W.W. Norton and Co., New York

Griffiths P, Stotz K (2013) Genetics and philosophy: an introduction. Cambridge University Press, Cambridge

Guo MH, Hirschhorn JN, Dauber A (2018) Insights and implications of genome-wide association studies of height. J Clin Endocrinol Metabol 103(9):3155–3168. https://doi.org/10.1210/jc.2018-01126

Harden KP (2021) The genetic lottery: why DNA matters for social equality. Princeton University Press, Princeton

Harden KP, Koellinger PD (2020) Using genetics for social science. Nat Hum Behav 4:567–576. https://doi.org/10.1038/s41562-020-0862-5

Holzinger KJ (1929) The relative effect of nature and nurture influences on twin differences. J Educ Psychol 20(4):241–248. https://doi.org/10.1037/h0072484

Hu W, Hao Z, Du P, Di Vincenzo F, Manzi G, Cui J, Fu Y-X, Pan Y-H, Li H (2023) Genomic inference of a severe human bottleneck during the Early to Middle Pleistocene transition. Science 381(6661):979–984. https://doi.org/10.1126/science.abq7487

Ikram MK, Xueling S, Jensen RA et al (2010) Four novel loci (19q13, 6q24, 12q24, and 5q14) influence the microcirculation in vivo. PLoS Genet 6:e1001184. https://doi.org/10.1371/journal.pgen.1001184

Jonsson H, Magnusdottir E, Eggertsson HP et al (2021) Differences between germline genomes of monozygotic twins. Nat Genet 53:27–34. https://doi.org/10.1038/s41588-020-00755-1

Lewontin RC (1974) The analysis of variance and the analysis of causes. Am J Hum Genet 26:400–411

Lee JJ et al. (2018) Gene discovery and polygenic prediction from a genome-wide association study of educational attainment in 1.1 million individuals. Nature Genetics 50(8):1112–1121. https://doi.org/10.1038/s41588-018-0147-3

Lloyd EA (1988) The structure and confirmation of evolutionary theory. Princeton University Press, Princeton

Lloyd EA (2015) Adaptationism and the logic of research questions: how to think clearly about evolutionary causes. Biol Theory 10(4):343–362. https://doi.org/10.1007/s13752-015-0214-2

Longino HE (2013) Studying human behavior: how scientists investigate aggression and sexuality. University of Chicago Press, Chicago

Matthews LJ, Turkheimer E (2021) Across the great divide: pluralism and the hunt for missing heritability. Synthese 198:2297–2311. https://doi.org/10.1007/s11229-019-02205-w

Mills M, Barban N, Tropf FC (2020) An introduction to statistical genetic data analysis. The MIT Press, Cambridge, MA

Moss L (2003) What genes can't do. MIT Press, Cambridge

Muthuirulan P, Capellini TD (2019) Complex phenotypes: mechanisms underlying variation in human stature. Curr Osteoporos Rep 17(5):301–323. https://doi.org/10.1007/s11914-019-00527-9

Newman HH, Freeman FN, Holzinger KJ (1937) Twins: a study of heredity and environment. University of Chicago Press, Oxford, England

Okbay A, Beauchamp JP, Fontana MA et al (2016) Genome-wide association study identifies 74 loci associated with educational attainment. Nature 533:539–542. https://doi.org/10.1038/nature17671

Okbay A, Wu Y, Wang N et al (2022) Polygenic prediction of educational attainment within and between families from genome-wide association analyses in 3 million individuals. Nat Genet 54:437–449. https://doi.org/10.1038/s41588-022-01016-z

Piel FB, Patil AP, Howes RE et al (2010) Global distribution of the sickle cell gene and geographical confirmation of the malaria hypothesis. Nat Commun 1:104. https://doi.org/10.1038/ncomms1104

Plomin R (2019) Blueprint: how DNA makes us who we are. Penguin Books, London

Remington DL (2015) Alleles versus mutations: understanding the evolution of genetic architecture requires a molecular perspective on allelic origins: perspective. Evolution 69(12):3025–3038. https://doi.org/10.1111/evo.12775

Rogers AR (2022) Using genetic data to build intuition about population history. arXiv:2201.02668. https://doi.org/10.48550/arXiv.2201.02668

Rowe SJ, Tenesa A (2012) Human complex trait genetics: lifting the lid of the genomics toolbox – from pathways to prediction. Curr Genomics 13(3):213–224. https://doi.org/10.2174/138920212800543101

Schaffner KF (2016) Behaving: what's genetic, what's not, and why should we care? Oxford University Press, Oxford/New York

Serpico D (2020) Beyond quantitative and qualitative traits: three telling cases in the life sciences. Biol Philos 35(3):34. https://doi.org/10.1007/s10539-020-09750-6

Serpico D (2023) A wolf in sheep's clothing: idealisations and the aims of polygenic scores. Stud Hist Phil Sci 102:72–83. https://doi.org/10.1016/j.shpsa.2023.10.006

Teicher A (2018) Caution, overload: the troubled past of genetic load. Genetics 210(3):747–755. https://doi.org/10.1534/genetics.118.301093

Turkheimer E (2000) Three laws of behavior genetics and what they mean. Curr Dir Psychol Sci 9(5):160–164. https://doi.org/10.1111/1467-8721.00084

Turkheimer E (2012) Genome wide association studies of behavior are social science. In: Plaisance KS, Reydon TAC (eds) Philosophy of behavioral biology. Springer Netherlands, Dordrecht, pp 43–64

Turkheimer E (2024) Understanding the nature–nurture debate. Cambridge University Press, Cambridge

Veller C, Przeworski M, Coop G (2024) Causal interpretations of family GWAS in the presence of heterogeneous effects. Proc Natl Acad Sci U S A 121(38):e2401379121. https://doi.org/10.1073/pnas.2401379121

Visscher PM (2008) Sizing up human height variation. Nat Genet 40(5):489–490. https://doi.org/10.1038/ng0508-489

Visscher PM, Macgregor S, Benyamin B, Zhu G, Gordon S, Medland S, Hill WG, Hottenga J-J, Willemsen G, Boomsma DI, Liu Y-Z, Deng H-W, Montgomery GW, Martin NG (2007) Genome partitioning of genetic variation for height from 11,214 sibling pairs. Am J Hum Genet 81(5):1104–1110

Visscher PM, McEvoy B, Yang J (2010) From Galton to GWAS: quantitative genetics of human height. Genet Res 92(5–6):371–379. https://doi.org/10.1017/S0016672310000571

Visscher PM, Wray NR, Zhang Q, Sklar P, McCarthy MI, Brown MA, Yang J (2017) 10 years of GWAS discovery: biology, function, and translation. Am J Hum Genet 101(1):5–22. https://doi.org/10.1016/j.ajhg.2017.06.005

Wainschtein P, Jain D, Zheng Z et al (2019) Recovery of trait heritability from whole genome sequence data. bioRxiv. https://doi.org/10.1101/588020

Woodward J, Kendler K (2023) Polygene risk scores: a philosophical exploration. Philos Med 4(1): 1–21. https://doi.org/10.5195/pom.2023.156

Yengo L, Sidorenko J, Kemper KE et al (2018) Meta-analysis of genome-wide association studies for height and body mass index in ~700000 individuals of European ancestry. Hum Mol Genet 27:3641–3649. https://doi.org/10.1093/hmg/ddy271

Yengo L, Vedantam S, Marouli E et al (2022) A saturated map of common genetic variants associated with human height. Nature 610:704–712. https://doi.org/10.1038/s41586-022-05275-y

Part IV

Otherness and Diversity in Human Evolution: Anthropological and Political Perspectives

Critical Hominin Theory

12

Jonathan Marks

Abstract

In the last few decades, biological anthropologists came to realize that thinking taxonomically about human variation obscured, rather than clarified, the evolutionary patterns that actually exist in our species. Approaching hominin ancestry taxonomically can be especially vexing, for the simple anthropological reason that dead ancestors are universally special in human societies, even in those societies that are composed of scientists. Although paleoanthropology seems to have been riddled with perpetually corrupted biological taxonomy, perhaps the problem is not incompetent biology, but a misapplication of eighteenth-century biological principles. Here, I discuss how the species in paleoanthropology transcends biology, and I suggest the possible merit in trying to make sense of recent human ancestry anthropologically and non-taxonomically.

Keywords

Taxonomy · Classification · Systematics · Ontology · Species

12.1 Introduction: Physical Anthropology as Pseudo-taxonomy

At one time, the field of physical anthropology was occupied primarily with establishing the number and natures of the elementary natural units of the human species. Radical biologists, like Ernst Haeckel (1868), W. C. Osman Hill (1940), and Reginald R. Ruggles Gates (1944), even held the different varieties of living peoples to be themselves species, rejecting the familiar interbreeding criterion. Nevertheless,

J. Marks (✉)
Department of Anthropology, University of North Carolina, Charlotte, NC, USA
e-mail: jmarks@uncc.edu

M. Lequin, J. M. Rodríguez Caso (eds.), *The Field of Human Evolution: Critical Perspectives from History and Epistemology*, Evolutionary Biology – New Perspectives on Its Development 11, https://doi.org/10.1007/978-3-032-13925-2_12

once that question was resolved to everyone's satisfaction, in the early 1960s (e.g., Barnicot 1963), the question still remained about how to understand the elementary units of the human species taxonomically below the species level. Ever since Linnaeus (1758), human taxonomy has been a starting point for understanding physical anthropology.

Two centuries later, it was still a major issue in the leading scientific journals. In *Nature*, Campbell (1962) formally identified *H. sapiens sapiens*, *H. sapiens afer*, *H. sapiens asiaticus*, *H. sapiens americanus*, *H. sapiens australasicus*, and *H. sapiens neptunianus*. Concurrently, based on serological genetic considerations, Boyd (1963) proposed 13 human "races," clustered into seven "groups" in *Science*. Garn (1962) identified 9 "geographical races" and 32 "local races." Coon (1962) identified five subspecies of living *Homo sapiens* (ignoring the additional racist baggage about their having evolved in parallel from five subspecies of *Homo erectus*, and its possible bearing on school integration), namely, Caucasoids, Mongoloids, Negroids, Capoids, and Australoids.

In the intervening decades, we stopped thinking about and teaching about human variation that way in physical anthropology. We have learned to talk about human variation without formal biological taxonomy (Marks 2020). The actual empirical biological structure of our species is "regarded as constituting a widespread network of more or less interrelated, ecologically adapted and functional entities" (Weiner 1957: 80), or in contemporary language, a "structured metapopulation" that defies taxonomic labelling; a population of populations, rich in local idiosyncrasies and connected genetically, socially, and economically to others. The groups that compose our species are created for particular purposes, and while they may correlate with some biological patterns, they do not represent natural divisions, but political identities. Classifying our species as Linnaeus did turned out to be a square peg—round hole problem (Livingstone 1962). Race can thus be modeled on the classic anthropological example of African witchcraft (Evans-Pritchard 1937; Fields and Fields 2012), which structures people's lives in spite of not having any material, naturalistic existence.

Indeed, partly as a result of the historical baggage associated with treating the human species in such a pseudo-taxonomic fashion, we terminated physical anthropology altogether and rebranded ourselves formally as biological anthropology. Yet even as biological anthropology, and without formal taxonomy describing present-day humans, we retain the urge to try and make sense of the ostensible lineages in our ancestry taxonomically (e.g., Wood and Boyle 2016; Reed et al. 2023; Martin et al. 2024), a daunting task that even Linnaeus himself never faced back in the eighteenth century.

A popular science bestseller, Harari's (2014) *Sapiens*, told readers on its back cover that "[o]ne hundred thousand years ago, at least six different species of humans inhabited Earth." That, of course, might indeed be true. What are the 6 species he identified 100,000 years ago? One, *Homo sapiens*, modern people. That is fair enough, even if they were a bit on the robust side. Two, *Homo neanderthalensis*. Indeed, according to the best genetic data 30 years ago (Krings et al. 1997); but now the geneticists say that I have 2% Neanderthal DNA, which presumably changes the status of Neanderthals, or the status of species, or both (Harvati and Ackermann 2022; Weasel 2022). Three, *Homo Denisova*. Yet even the geneticists, who are the only people to whom it is visible, say this is a genomic subgroup or sister group of Neanderthals (Meyer et al. 2012; Zeberg et al. 2024)—who, as just noted, may not even be their own species in the first place. Four, *Homo erectus*. That is largely uncontroversial as a taxon, but not at 100,000 years ago, unless perhaps we regard species #6 as *H. erectus* (Rizal et al. 2020). Five, *Homo floresiensis*. Clearly, the mysterious and isolated *H. floresiensis* was something (Madison 2023). And six, *Homo soloensis*, named for a different set of Indonesian fossils than *H. floresiensis*, and which are often regarded as archaic *H. sapiens* (Swisher et al. 1996) on account of their anatomical continuity with both *H. erectus* and *H. sapiens*.

All of a sudden, we are down from "at least six different species" to humans, the hobbit, and possibly Neanderthals. This is not to single out Harari (2014) for criticism, but rather to note that biological anthropologists know that all of these species are vexatious at best (e.g., Athreya and Hopkins 2021; Delson and Stringer 2022; Bae et al. 2024), while outside of biological anthropology, the nuanced winks that generally accompany these taxa tend to get lost.

The problem is not a new one. Reviewing mammalian systematics in 1945, G. G. Simpson was frustrated by the difference he encountered between paleontological and paleoanthropological taxonomy. Since humans are mammals, it stands to reason that an expert on mammalian species should be able to make sense of extinct hominin[1] species. Like any other biological taxonomic enterprise, although there may be minor disagreements, there is a proper taxonomic scheme, which reflects a proper understanding of the fossil species in that particular evolutionary lineage. It is simply a matter of finding it. Some people see too few species (lumpers), and some people see too many species (splitters), while a competent paleontologist should produce results that are as Goldilocks found the baby bear's porridge: just right.

The problem, however, is not simply that everybody fancies themselves to be the baby bear, but rather lies more fundamentally in the assumption that the elementary units in paleoanthropology and the elementary units of paleontology are equivalent. I think it is time to call that assumption into question. The units of paleontology, and of biology more generally, are different from the units of paleoanthropology, in that the latter are *units in a story of our ancestors, and the ancestors are invariably sacred*. Paleoanthropology is ultimately about a very specific program: the construction of narratives, bound by the rules of science, to explain our descent and relations—in a word, the oldest anthropological subject of all, kinship.

[1] Although I reject the two premises that led to the adoption of "hominin" (only naming clades, and privileging our genetic intimacy with the apes over our ecological difference from them), I use the term for its hipness.

12.2 What Are Primate Species?

A species, like a culture or a gene, is infamously difficult to define precisely and satisfactorily. Animal species appear to be units of nature of some sort, but different kinds of units than species among plants or bacteria (Godfrey and Marks 1991; Barraclough 2019), and with boundaries somewhat more porous than thought a few decades ago (Walker et al. 2024). Nevertheless, for multicellular animals at least, it appears to be an elemental unit of ecology, as a cell is an elemental unit of physiology, a genotype is an elemental unit of population genetics, and an organism is an elemental unit of society. If we restrict ourselves just to the more familiar mammalian and primate species, we can ideally identify several properties possessed by a species: (1) it is composed of organisms related to one another as either potential mates or competitors for mates, thus bounded by the limits of a gene pool; (2) it constructs and is adapted to a niche, filling a role in a dynamic ecosystem; (3) it has a locus in space and a duration in time as an evolutionary lineage; and (4) it can replicate, making more species, each slightly different from itself. Different definitions of species, from Buffon (1753; Farber 1972) through Mayr (1942) and beyond, highlight one or another of these features.[2]

A more recently appreciated feature of species is that they are also units of conservation legislation, and thus partly cultural as well. That is why the number of extant primate species has risen so dramatically in recent decades, from about 170 in the mid-1980s (Richard 1985) to over 500 today (Strier 2021). It is not simply that the number of species is being taxonomically inflated, and is thus *less* accurate than it used to be (Rosenberger 2012); or that new species are finally being recognized, and their number is thus *more* accurate than it used to be (Groves 2014). In a sense, comparing the number of living primate species 35 years ago and today is simply unfair, because they are actually tabulating different things than they used to (Rylands and Mittermeier 2014).

However entropic the universe of extant primate species seems to have become, it is disconnected from extinct primate species, which are of course unaffected by conservation legislation. They nevertheless have their own issues, and their status is constantly being negotiated by the community of primate paleontologists. Perhaps unsurprisingly, the closer we get to ourselves, the more complex things seem to become (Kimbel and Martin 1993).

Hominin species seem just as troublesome today as they did to Simpson back in 1945. Today public intellectuals reiterate uncritically the extreme taxonomic diversity that they hear about. Simpson's explanation for the taxonomic ambiguity was that the people doing the paleoanthropology were generally coming from medical anatomy, and simply did not understand taxonomy well enough: "much of the work on primates has been done by students who had no experience in taxonomy and who

[2] I regard a species to be a group of organisms composed of potential mates and competitors for mates, and representing a temporarily stable state in a biopolitical field of evolution, ecological relations, capitalism, conservation science, and governmental action.

were completely incompetent to enter this field, however competent they may have been in other respects" (Simpson 1945:181). In particular, he noted, "Dart's placing of *Australopithecus* in a family 'Homo-simiadae' (1925) only served to exemplify the total ignorance of zoology so common among the special students of these higher primates (although, of course, Dart's work is excellent in his own field)" (Simpson 1945:188). Tim White (2000) added "careerism" to the causes of ostensibly poor paleoanthropological taxonomy.

12.3 The Ontology of Hominin Species

There is a diverse literature on what a species is, or ought to be, generally centering on the gene pool (Mayr 1942), which is directly inaccessible paleontologically; or on the possession of derived characters (cf. Tattersall 2014), which would have significant implications for European blondes, who are monophyletic and possess just such a derived character (Eiberg et al. 2008; Guenther et al. 2014). The underlying presupposition is that hominin biological history is composed of zoological units that are equivalent to species; however, species may be defined biologically. Thus, *Homo naledi*, *Homo longi*, *Homo heidelbergensis*, *Homo luzonensis*, *Homo bodoensis*, *Homo juluensis*, etc., either are or are not distinct phylogenetic lineages, and only a proper taxonomy will resolve and describe them.

Suppose, however, that there are no recoverable biological species in the history of our lineage, perhaps because much of that history is occupied biologically by a structured meta-population rather than by distinct species (Pääbo 2015; Scerri et al. 2019; van Holstein and Foley 2022). And perhaps because there is more at stake in identifying our own ancestors than there is in identifying the ancestors of other species, because ancestries are linked to personal identities. And perhaps because the names themselves represent not so much biological ancestors, as anthropological ancestors.

What do I mean by an anthropological ancestor? I mean the subject of a sacred narrative about our past, a story intended to explain who we are and how we came to be here. The diverse manners by which peoples construct relations of (horizontal) kinship and (vertical) ancestry were among the earliest discoveries of anthropology (Morgan 1871; Franklin and McKinnon 2001). Relatives are made and named culturally, and never in precisely the way that scientists conceptualize them. And while we have talked for decades about looking at science as a culture (Snow 1959), we are only more recently actually getting around to doing it (Franklin 1995). I want to suggest that the reason that hominin taxonomy is, and has always been, so vexed is because its elements are *not* biological species. The species that compose our own lineage are mythic ancestors, which are intended to resemble biological species. But they are actually elements in a story, the *bricolage* (Lévi-Strauss 1962) out of which scientific narratives of our origins (Landau 1991) are constructed.

In one story, the actor in a creation narrative might be Mother Corn Spirit. In another, it might be *Homo luzonensis*. In still another, it might be *Homo soloensis*. And they might have precisely the same ontological status: as nonexistent entities in

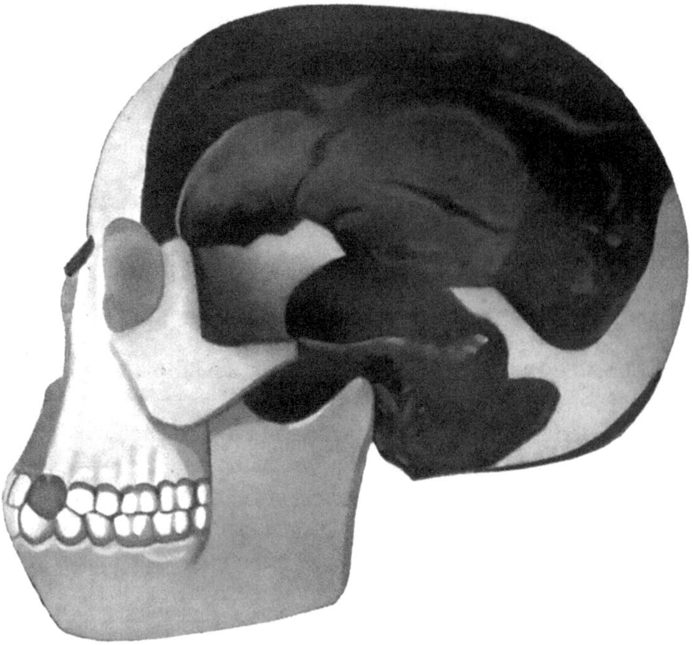

Fig. 12.1 A notoriously sacred ancestor, *Eoanthropus dawsoni*, or Piltdown Man. (Public Domain)

the naturalistic ancestry of the human species. As some other kind of entities, namely, as characters in an origin narrative, they may have a very different kind of existence.

In other words, the persistent confusion over hominin taxonomy is a result of misunderstanding the nature and existence of hominin species. It is not that there is a correct taxonomic interpretation of the fossils, which will reveal itself under the proper analytic technique. It is rather that (1) this taxonomic practice is different from other ostensibly zoological taxonomic practices, by virtue of being reflexive; (2) the origin of this difference lies in the fact that it is naming and describing our own ancestors; and (3) ancestors are invariably sacred (in the broad anthropological sense of "special," rather than in the narrower sense of "holy," although the ancestors may of course sometimes be that too; Zerubavel 2012).

Anthropologically speaking, hominin species are sacred ancestors. In some cases, this is more obvious than others. The role of *Eoanthropus dawsoni* as a sacred British ancestor (Fig. 12.1), which effectively precluded its exposure as a hoax for decades, is well known (Weiner 1955). Another notable example is *Sinanthropus pekinensis*. We often retell the story of the loss of the Peking Man fossils on Pearl Harbor Day. They were being transported for safekeeping to the naval base in Hawaii to protect them from the Japanese, who were very interested in taking control of the fossils (Roberts et al. 2021). We never tell a parallel story in Europe. After all, there were hominin fossils scattered around Europe in 1941, but they did not require

Fig. 12.2 An ancestor (STS-5 or Mrs. Ples) on a postage stamp. (Collection of the author)

special protection from the Germans. Why did Peking Man require protection from the Japanese? Presumably, because the fossils were regarded as ancestors of the Chinese (Sautman 2001; Schmalzer 2008), and thus possessed symbolic power and value to East Asians, which was quite different from the cultural meanings of European fossils.

And it was not even *Homo erectus* at the time, but *Sinanthropus pekinensis*. That is what all of these names represent: not elements of biological history, but elements of stories *about* biological history. They are attempts to create formal biological characters and distinctive elements in order to convey a story of human origins, where there may be, in fact, no such distinctive biological elements in nature. This situation is familiar from the neontological end of physical anthropology, in the form of two centuries of misguided racial classification. I am suggesting that the problem with paleoanthropological taxonomy today is the same problem faced by racial taxonomy decades ago, namely, a fundamental confusion of categories, mistaking units of symbolic culture for units of biological nature.

Some famous fossil ancestors have appeared on national postage stamps (Fig. 12.2), an unusual location for scientific data. Even from a philatelic standpoint, studying and narrating one's own ancestry is a culturally powerful activity. The elements of a scientific story of our ancestors are species, and those species are not like other species, by virtue of being our ancestors (Walker et al. 2021). We are humans studying human ancestors; the reflexivity is built into the system. In order to break out of the loop (Huxley 1863:69), we can pretend to be space aliens ("scientific Saturnians, if you will"), but that is hardly a scientific solution to any problem.

We no longer teach the pattern of extant human variation as being taxonomically structured, because we have come to realize that the taxonomies of race did not in fact represent what they purported to: the existence of a fairly small number of fairly discrete and fairly natural clusters of people (Marks 2024). Quite possibly the most pernicious popular idea today involves reifying the "Denisovans," who are known from DNA sequences from teeth in a Siberian cave around 50,000 years ago, and are only visible to geneticists. One popular interpretation has the Denisovans as Asian counterparts of European Neanderthals and African "moderns" (e.g., Zeberg et al. 2024), based on projecting three continental forms of *Homo sapiens* back into the Pleistocene. Their particular genetic idiosyncrasies have turned up in other fossils, and even to a limited extent in living peoples, although not in Siberians, but rather in the genomes of Australians and New Guineans. What that means is anybody's guess, but there is certainly a lot of curious prehistoric biogeographical genetic information to be made sense of, which formal Linnaean taxonomic labels, based on an eighteenth-century view of nature, only confuse further. Indeed, the idea of continental boundaries dividing a multi-partite Linnaean human species has been unpopular since the 1960s, and for good reason: It is empirically false and historically biblical, not microevolutionary. Simply imagining Europe, Asia, and Africa to be distinct from one another was foolish even to Herodotus (*The Histories*, IV, 45), who did not have to worry about mapping human or hominin taxa onto them.

Our named human taxa are place-holders, mere samples of the genetic and anatomical diversity in our lineage at any time. They represent, at best, local populations and local gene pools. The mistake here is to confuse the scientific contributions of Linnaeus with those of Darwin; a static eighteenth-century view of Nature full of pigeonholes to be labeled and organized, versus a dynamic nineteenth-century view of Nature full of pigeon breeders filling holes they have created.

By shifting the ontological status of race from the biological to the cultural (obviously, sometimes correlated with a bit of biology), race has become no longer a subject of formal biological taxonomy. Consequently, whether the International Code of Zoological Nomenclature recognizes *Homo sapiens afer* as a valid subspecies is irrelevant, for it is not actually a biological thing. It is simply one of the many cultural ways that humans form politically salient groups, and consequently one of the diverse identities that are available for people to adopt, or to be assigned, in a particular time and place.

Similar reasoning can be applied to human ancestry. We may begin to regard *Homo heidelbergensis*, for example, as ontologically equivalent to *Homo sapiens asiaticus*. That is to say, as an attempt to apply scientific taxonomic rigor where it is inapplicable, by naming something and thus willing it into existence. This is the fallacy of reification, or misplaced concreteness. There is no *Homo sapiens asiaticus*, although we can certainly discuss the indigenous peoples of East Asia, their gene pools, their adaptations, and their descendants. But we do not recognize Linnaean subspecies taxa as describing our species at present. Are the Linnaean species taxa that ostensibly describe our ancestry any realer? Or are they primarily serving to conceal the actual structure of human prehistoric biology? It is, after all, at least conceivable that the human lineage was more like a structured metapopulation than like a macroevolutionary clade, all the way down. If that were true, then gene flow would presumably be occurring, which becomes a mechanism for the horizontal transmission of derived characters. If derived characters are transmitted horizontally by gene flow, rather than vertically by common ancestry, that would render any ostensible phylogenetic analyses untenable; in which case the goal of establishing a proper taxonomy and phylogeny for recent hominin fossils (e.g., Bae et al. 2024) would be impossible.

12.4 Conclusion

We talk today about a species *Homo sapiens*, and even about a subspecies, *Homo sapiens sapiens*, without further formal taxonomic structure, although with real and studiable patterns of biological diversity below that level. But instead of characterizing *Homo sapiens europaeus* as if it were a naturalistic taxon, today we ask instead why scientists thought it existed as such for two or three centuries when it actually does not (Fig. 12.3). Instead of asking how many and what the human races are, today we ask how and why they get made, and what work they do. We may look similarly at the question in paleoanthropology of how hominin species get made and manipulated—without necessarily assuming that hominin species are biological things, for that is what we think our prehistory is supposed be composed of. It may be that the quest for a proper and correct scientific biological taxonomy of hominin species is itself a vain one.

The eighteenth-century Linnaean tree of classification does not map perfectly onto the nineteenth-century Darwinian tree of phylogeny. Why not? Because adaptive Darwinian divergence creates paraphyletic Linnaean categories—such as invertebrates (minus vertebrates), fish (minus tetrapods), reptiles (minus birds), monkeys (minus hominoids), whales (minus dolphins), and apes (minus humans). Consequently, twenty-first-century cladistic classifications often have weird components, as they attempt to make all recognized taxa monophyletic (Withgott 2000). Nevertheless, even Linnaeus was not faced with trying to incorporate extinct taxa into his work; he did not even believe in extinction. Moreover, Linnaeus did not know about macro- and micro-evolution when he formulated his system, and the distinction has vexed systematists ever since, since speciation is a process and not an event, and thus reproductive isolation (if that is your criterion of species) is often incomplete (Dobzhansky 1937; Walker et al. 2024).

A recent paper by Meneganzin and Bernardi (2023) about Neanderthal specieshood underscores the problem. With imprecise definitions of what "human," "Neanderthal," and "species" mean, the Linnaean system simply breaks down here. Neanderthals may indeed be a "good" species, assuming that species actually mean something real and biological in this context. In the 1990s, the Neanderthals became firmly embedded within the genus *Homo* as a separate species from *Homo sapiens*, given the apparent distinction of their gene pool from "ours"; and the assumption that a separate gene pool was what distinguished one species from another. But today neither of those conditions can be taken for granted. Interbreeding between our lineage and Neanderthals is now generally acknowledged; so whatever the nature of their gene pools may have been, they are no longer seen as quite so distinct from one another. Moreover, with the classic concept of species itself being increasingly questioned, the discovery of widespread genetic introgression among primates and the privilege now accorded to conservation issues in primate taxonomy has resulted in an apparent explosion of extant primate diversity, in spite of their dwindling actual numbers. In brief, the taxonomic status of Neanderthals is as unclear as ever.

Fig. 12.3 Five official kinds of living people, from 1911. (Public Domain)

And they are the extinct large-brained hominins that we know best. But I would suggest that Neanderthals are neither a "good" species nor a "bad" species; the imposition of biological species labels where the species are at best biologically ambiguous is the problem. That would be a problem anywhere on the "tree" of life, but Neanderthals have the added burden of sitting on the boundary between the human and the nonhuman, which is about as singularly symbolically charged as any place you can imagine.

Scientific taxonomy is purposeful and is based on specific criteria, as all folk taxonomy is. Here I have tried to make sense of paleoanthropological science anthropologically, which may enable us to harmonize it more faithfully to the empirical record of our ancestry, and to relativize it among systems of knowledge. Perhaps hominin species do more harm than good at present, when it comes to understanding the bio-historical processes and patterns that got us here. Perhaps hominin species are examples of scientific pareidolia: projecting faces onto objects like clouds and tortillas. If you really want them to be there, you can always find them. The problem lies in convincing everyone else that they are there.

References

Athreya S, Hopkins A (2021) Conceptual issues in hominin taxonomy: *Homo heidelbergensis* and an ethnobiological reframing of species. Yearb Phys Anthropol 175(Suppl. 72):4–26

Bae CJ, Aiello LC, Hawks J et al (2024) Moving away from "the Muddle in the Middle" toward solving the Chibanian puzzle. Evol Anthropol 33:e22011. https://doi.org/10.1002/evan.22011

Barnicot NA (1963) Racial taxonomy. Curr Anthropol 4:117–118

Barraclough TG (2019) The evolutionary biology of species. Oxford University Press, New York

Boyd WC (1963) Genetics and the human race. Science 140:1057–1065

Buffon C d (1753) L'Asne. In: Histoire Naturelle, Générale et Particuliére, vol IV. Imprimerie Royale, Paris, pp 337–436

Campbell B (1962) The systematics of man. Nature 194:225–232

Coon CS (1962) The origin of races. Knopf, New York

Delson E, Stringer C (2022) The naming of *Homo bodoensis* by Roksandic and colleagues does not resolve issues surrounding Middle Pleistocene human evolution. Evol Anthropol 31:233–236. https://doi.org/10.1002/evan.21950

Dobzhansky, T. (1937) *Genetics and the Origin of Species*. Columbia University Press, New York

Eiberg H, Troelsen J, Nielsen M et al (2008) Blue eye color in humans may be caused by a perfectly associated founder mutation in a regulatory element located within the HERC2 gene inhibiting OCA2 expression. Hum Genet 123:177–187. https://doi.org/10.1007/s00439-007-0460-x

Evans-Pritchard EE (1937) Witchcraft, oracles, and magic among the Azande. Oxford University Pres, Oxford

Farber PL (1972) Buffon and the concept of species. J Hist Biol 5:259–284

Fields B, Fields K (2012) Racecraft: the soul of inequality in American life. Verso, New York

Franklin S (1995) Science as culture, cultures of science. Annu Rev Anthropol 24:163–184

Franklin S, McKinnon S (2001) Relative values: reconfiguring kinship studies. Duke University Press, North Carolina

Garn S (1962) Races. Charles C. Thomas, Springfield

Gates, R. R. (1944) Phylogeny and Classification of Hominids and Anthropoids. *American Journal of Physical Anthropology*, 2: 279–292.

Godfrey L, Marks J (1991) The nature and origins of primate species. Yearb Phys Anthropol 34:39–68

Groves CP (2014) Primate taxonomy: inflation or real? Annu Rev Anthropol 43:27–36

Guenther CA, Tasic B, Luo L et al (2014) A molecular basis for classic blond hair color in Europeans. Nat Genet 46:748–752. https://doi.org/10.1038/ng.2991

Sautman, B. (2001) Peking man and the politics of paleoanthropological nationalism in China. *The Journal of Asian Studies*, 60: 95–124
Haeckel, Ernst (1868) *Natürliche Schöpfungsgeschichte*. Reimer, Berlin
Harari YN (2014) Sapiens: a brief history of humankind. Random House, New York
Harvati K, Ackermann RR (2022) Merging morphological and genetic evidence to assess hybridization in Western Eurasian late Pleistocene hominins. Nat Ecol Evol 6:1573–1585. https://doi.org/10.1038/s41559-022-01875-z
Huxley T (1863) Man's place in nature. Williams and Norgate, London
Kimbel WH, Martin LB (1993) Species, species concepts and primate evolution. Plenum, New York
Krings M, Stone A, Schmitz RW et al (1997) Neandertal DNA sequences and the origin of modern humans. Cell 90:19–30. https://doi.org/10.1016/S0092-8674(00)80310-4
Landau M (1991) Narratives of human evolution. Yale University Press, New Haven
Lévi-Strauss C (1962) The savage mind. University of Chicago Press, Chicago
Livingstone FB (1962) On the non-existence of human races. Curr Anthropol 3:279–281
Madison P (2023) Tug-of-War: bones and stones as scientific objects in postcolonial Indonesia. Isis 114:77–98
Marks J (2020) Naming the sacred ancestors: taxonomic reification and Pleistocene genomic narratives. In: Porr M, Matthews J (eds) Interrogating human origins. Routledge, New York, pp 295–309
Marks J (2024) Understanding human diversity. Cambridge University Press, New York
Martin JM, Leece AB, Baker SE et al (2024) A lineage perspective on hominin taxonomy and evolution. Evol Anthropol 33:e22018. https://doi.org/10.1002/evan.22018
Mayr E (1942) Systematics and the origin of species. Columbia University Press, New York
Meneganzin A, Bernardi M (2023) Were Neanderthals and Homo sapiens 'good species'? Quat Sci Rev 303:107975. https://doi.org/10.1016/j.quascirev.2023.107975
Meyer M, Kircher M, Gansauge M-T et al (2012) A high-coverage genome sequence from an Archaic Denisovan individual. Science 338:222–226. https://doi.org/10.1126/science.1224344
Morgan LH (1871) Systems of consanguinity and affinity of the human family. Smithsonian contributions to knowledge, vol 17. Smithsonian Institution, Washington, DC
Osman Hill WC (1940) Classification of Hominidæ. Nature 146:402–403. https://doi.org/10.1038/146402b0
Pääbo S (2015) The diverse origins of the human gene pool. Nat Rev Genet 16:313–314. https://doi.org/10.1038/nrg3954
Reed DN, Raney E, Johnson J et al (2023) Hominin nomenclature and the importance of information systems for managing complexity in paleoanthropology. J Hum Evol 175:103308. https://doi.org/10.1016/j.jhevol.2022.103308
Richard AF (1985) Primates in nature. W. H. Freeman, San Francisco
Rizal Y, Westaway KE, Zaim Y et al (2020) Last appearance of Homo erectus at Ngandong, Java, 117,000–108,000 years ago. Nature 577:381–385. https://doi.org/10.1038/s41586-019-1863-2
Roberts MF, DeVisser EM, Marrant JP (2021) Treacherous evidence: archival documents and the search for Peking Man. PaleoAnthropology 2021. https://doi.org/10.48738/2021.ISS1.72
Rosenberger AL (2012) New world monkey nightmares: science, art, use, and abuse (?) in platyrrhine taxonomic nomenclature. Am J Primatol 74:692–695. https://doi.org/10.1002/ajp.22037
Rylands AB, Mittermeier RA (2014) Primate taxonomy: species and conservation. Evol Anthropol 23:8–10. https://doi.org/10.1002/evan.21387
Scerri EML, Chikhi L, Thomas MG (2019) Beyond multiregional and simple out-of-Africa models of human evolution. Nat Ecol Evol 3:1370–1372. https://doi.org/10.1038/s41559-019-0992-1
Schmalzer S (2008) The people's Peking Man: popular science and human identity in twentieth-century China. University of Chicago Press, Chicago
Simpson GG (1945) The principles of classification and a classification of mammals. Bull Am Mus Nat Hist 85:1–350

Snow CP (1959) The two cultures and the scientific revolution. Cambridge University Press, London
Strier KB (2021) Primate behavioral ecology, 6th edn. Routledge, London
Swisher CC, Rink WJ, Antón SC et al (1996) Latest *Homo erectus* of java: potential contemporaneity with *Homo sapiens* in Southeast Asia. Science 274:1870–1874. https://doi.org/10.1126/science.274.5294.1870
Tattersall I (2014) Recognizing species, present and past. Evol Anthropol 23:5–7. https://doi.org/10.1002/evan.21385
Van Holstein LA, Foley RA (2022) A process-based approach to hominin taxonomy provides new perspectives on hominin speciation. Evol Anthropol 31:166–174. https://doi.org/10.1002/evan.21946
Walker J, Clinnick D, White M (2021) We are not alone: William King and the naming of the Neanderthals. Am Anthropol 123:805–818. https://doi.org/10.1111/aman.13654
Walker JM, Van Der Heijden ESM, Maulana A et al (2024) Common misconceptions of speciation. Evol J Linnean Soc 3:kzae029. https://doi.org/10.1093/evolinnean/kzae029
Weasel L (2022) How Neanderthals became white: the introgression of race into contemporary human evolutionary genetics. Am Nat 200:129–139. https://doi.org/10.1086/720130
Weiner JS (1955) The Piltdown forgery. Oxford University Press, London
Weiner JS (1957) Physical anthropology: an appraisal. Am Sci 45:79–87
White TD (2000) A view on the science: physical anthropology at the millennium. Am J Phys Anthropol 113:287–292. https://doi.org/10.1002/1096-8644(200011)113:3<287::AID-AJPA1>3.0.CO;2-8
Withgott J (2000) Is it "So Long, Linnaeus"? Bioscience 50:646. https://doi.org/10.1641/0006-3568(2000)050[0646:IISLL]2.0.CO;2
Wood B, Boyle EK (2016) Hominin taxic diversity: fact or fantasy? Am J Phys Anthropol 159:37–78. https://doi.org/10.1002/ajpa.22902
Zeberg H, Jakobsson M, Pääbo S (2024) The genetic changes that shaped Neandertals, Denisovans, and modern humans. Cell 187:1047–1058. https://doi.org/10.1016/j.cell.2023.12.029
Zerubavel E (2012) Ancestors and relatives: genealogy, identity, and community. Oxford University Press, New York

Is and Ought? Problems with Deriving "Natural" and Prescribing "Normal" via the Human Evolutionary Record

13

Agustín Fuentes

Abstract

There is a strong entanglement between the evolutionary history of any given lineage of organisms and assumptions about the "nature" of those organisms. Humans evolved, and continue evolving, and thus our evolutionary history shapes much of who and how we are. However, humans are also diverse, complex, fascinating, and biocultural, and there is no one specific ecology, society, or behavior that epitomizes all of humanity. Thus, interpretations of the human past, and assumptions about the human present, often straddle the boundaries between "is," a descriptive approach of the available data and analyses on human evolution, and "ought," a proscriptive assertion about how humans truly are and thus should be. Many biases in understandings of the past and of evolutionary processes lead to flawed reconstructions of human evolutionary history and promote misconceptions, "oughts," about human behavior and societies. Highlighting a suite of data and analyses from the study of Pleistocene *Homo* that illustrate a range of variation, including complexity in reproducing and caretaking, and multifarious, malleable intergroup relations across space and time, can offer a corrective to common misconceptions. This chapter reviews current knowledge of Pleistocene *Homo* and the human niche, offering two examples of what the human evolutionary record can and cannot inform us about becoming and being human. The examples are: intergroup relations and xenophobia, and the core social structures around child-rearing.

A. Fuentes (✉)
Department of Anthropology, Princeton University, Princeton, NJ, USA
e-mail: afuentes2@princeton.edu

M. Lequin, J. M. Rodríguez Caso (eds.), *The Field of Human Evolution: Critical Perspectives from History and Epistemology*, Evolutionary Biology – New Perspectives on Its Development 11, https://doi.org/10.1007/978-3-032-13925-2_13

Keywords

Homo · Pleistocene · Family · Xenophobia · Human nature · Caretaking · Human niche

13.1 Introduction

Today humans are diverse, complex, fascinating, and biocultural. There is not one specific ecology, society, or behavior that epitomizes all of humanity, but there are key patterns and some typical biological parameters. Humans evolved, and continue evolving, and our evolutionary history shapes much of who and how we are. When it comes to specific aspects of being human, a diverse array of data and analyses from the study of Pleistocene *Homo* suggest a range of variation, including complexity in reproducing and caretaking, and multifarious, malleable intergroup relations across space and time.

There is a strong entanglement between the evolutionary history of any given lineage of organisms and assumptions about the "nature" of those organisms. The genus *Homo* developed a multifaceted suite of morphological/physiological, cognitive, behavioral, and ecological structures and processes across the Pleistocene, resulting in a distinctive human niche. The emergence of this niche and its particular biocultural dynamics facilitated the dramatic demographic, technological, ecological, and socio-cultural changes in Holocene and in contemporary humans with broad ecological impacts on the planet. Contemporary humans display social, economic, political, and cultural diversity that is more complex and larger in scope and impact than that of other organisms. Clearly many aspects of contemporary human bodies and behavior are structured by, rooted in, constrained, and/or facilitated by the dynamics of Pleistocene *Homo* evolution. However, it is common for a wide range of scholars across multiple disciplines, and much of the lay public, to point to specific interpretations of Pleistocene *Homo* data as evidence for the "naturalness" of a particular contemporary experience. This is especially true for aspects of sex and sexuality (e.g., that differential evolutionary pressures created different male and female *Homo* "natures" reflected in contemporary sex and sexuality dynamics), gender/sex roles (e.g., that males are primary provisioners/protectors and females are primary caretakers), family/kin structures (e.g., the nuclear family is the basal feature of human social organization), and intergroup relations (e.g., humans exhibit a natural xenophobia and thus contemporary modes of inter-group conflict are "normal"). These interpretations of the human past, and their concomitant assumptions about the human present, often straddle the boundaries between "is," a descriptive approach of the available data and analyses on human evolution, and "ought," a proscriptive assertion about how humans truly are and thus should be. This mode of discourse about humanity is used by a wide swath of disciplines and institutions to explain and justify particular views about contemporary human societies. This chapter reviews current knowledge regarding Pleistocene *Homo* and the development of a human niche offering two examples to assess what the human evolutionary record can and cannot inform us about becoming and being human. The examples are: intergroup relations and xenophobia, and the core social structures around child-rearing.

13.2 Stories Matter

The stories about human evolution structure the ways in which people think about a "human nature," and there is no shortage of scholars producing such stories in the current moment (e.g., Fuentes 2017a, b; Wengrow and Graber 2021; Harari 2015; Henrich 2016; Hrdy 2024; Pinker 2011; Wrangham 2019a, b). This connection, human evolution as human nature, is not limited to the influence of academics writing popular books, but is a shared cultural dynamic widespread across much of the public and the academy. Take, for example, the often-used phrase: "it's in our genes." This assertion argues that whatever "it" is, a behavior, a belief, a habit, perceptions or ideology, is rooted in the very core of our bodies, our biology, our DNA, our evolution. That biology and DNA is the product of some degree of evolutionary history and thus the "it is in our genes" assertion reflects the perspective that humans evolved the bodies and minds to do this particular thing. One need not hold the antiquated "selfish gene" model of evolution to use this phrase, or even have a robust grasp of genetics or evolutionary biology, rather all one needs is a basic belief that certain elements of being human, and thus human society, are produced and/or constrained by specific events or processes in the human evolutionary past. In such a context, what one thinks happened during human evolution is deeply relevant to how one perceives and envisions something (whatever that is) akin to a "human nature" (e.g., Midgley 2002; see also Lock and Palsson 2016).

For the most part, few members of the public and only a smattering of academics care about the more banal incarnations of human evolutionary patterns and their structuring effects on human bodies and lives. For example, humans walk on two legs, and that is the product of our specific evolutionary history. This aspect of what one might term our "nature" often flies under the radar, despite the fact that the evolution of bipedality (obligatory locomotion on the hind legs) has massive implications for our bodies, our behavior, our senses and perceptions, and the way we "do" the thing we call human (DeSilva 2021). In truth, there is an enormous amount of relatively complex, multifaceted, substantive aspects of our bodies and lives that draw on, that emerge from, are facilitated by, or incorporate aspects of our evolutionary histories and trajectories; and this is the main focus of much academic study of human evolution. But only some of these myriad features and processes draw significant attention from the broader public and those in the academy outside of the fields most closely tied to human evolutionary studies. It is those aspects of human bodies and related behaviors that speak to contemporary lives in particularly culturally salient, and often political and contested, manners that draw the most interest.

When one veers into conversations about gender, sex, violence, power, greed, inequality, etc., the role of a "human nature" becomes central and the interest in potentially related aspects of human evolution is pronounced. Here the evolutionary past, and its assumed "proof" of the biological (read "genetic" and "natural") development of a particular pattern or experiential reality becomes a critical factor in the debate about "is" and "ought." If something evolved a certain way, it is assumed to be "natural" and thus intrinsic to the human being (it "is"). Many would then argue any individual or society can make the decision about whether to support or oppose specific aspects of human nature on cultural/political/philosophical grounds (a form of individual or societal "free will") but most would still maintain the position that one cannot deny that the data demonstrate a particular "is" and that this is the way humans naturally are, and thus, many argue, should be ("ought"). This is, superficially, similar to the fictional philosopher Pangloss' pronouncements that all is as it should be in Voltaire's Candide…but this superficial similarity should not be dismissed as a literary flight of fancy because the conviction that something emerging from "human nature" (a particular interpretation of human evolution) reflects how things *should be* remains quite prominent in both public and academic narratives.

Importantly, and to the point of this chapter, before one even gets to assessing the is/ought-descriptive/proscriptive debate and discussion, it is critical to note that much in what is asserted and assumed as "fact" about human evolution, and used to support popular (and much academic) assertion about "human nature," misrepresents, misunderstands, or even falsifies the actual published data, analyses, and inferences from the scientific investigation into human evolution. Two salient examples of this are in the assertion that humans are naturally ("evolved to be") xenophobic and that the nuclear family (the heterosexual pair and their offspring) is the core unit of human reproductive biology and society. This chapter will illustrate how in these two cases many basal assumptions/assertions about the data and analyses are incorrect and lead to substantive misrepresentation of the human past and mistaken declarations about causal relationships undergirding the human present. These misrepresentations can result from a lack of familiarity with the datasets and analyses, misreading or selective reading of original work in the area, and/or active distortion of data and analyses by scholars translating to broader audiences or to academic disciplines far afield of human evolution (such as economic and political sciences, e.g., Ashraf and Galor 2013).

There are indeed many scholars and scientists of human evolution who write for audiences beyond their peers (e.g., Henrich 2020; Hrdy 2024) and many scholars who do not study human evolution professionally but who write about human evolution as intellectual "experts" on the topic (e.g., Harari 2015). Many of these books, and even a number of peer-reviewed articles (e.g., Benenson et al. 2022; Wrangham 2019a, b), come from scholars who are making a human evolution = human nature argument by drawing on some aspects of human evolutionary studies and placing them in a form of causal dialogue with contemporary aspects of the human experience. These stories are especially relevant as they are seen as validated by "experts" and thus carry particularly robust explanatory weight (at least in many circles). These "experts" are usually from a very limited demographic[1] in the academy (and in society) and particular national, linguistic, and intellectual frames. I should know as I, in many but not all ways, am one of them (Fuentes 2017a, b, 2019). All stories matter, and multiple different ones can be honestly and accurately told, but when it comes to narratives about human evolution that are used to validate, justify, contest, or describe "human nature," these stories told by "experts" matter in particularly powerful ways (see Athreya and Ackermann 2020; Kissel 2020; Schroeder and Ackermann 2023; Ackermann et al. 2024; Menéndez and Yañez 2025, this volume).

13.3 "Normal" and the "Is" and "Ought" That Comes from the Study of Human Evolution

Something has "normativity" when it entails that some action, attitude, or mental state is justified, is an action one ought to do, or a state one ought to be in. The view of "what is normal=what is natural" creates a normative framing for/in human society and is deeply interrelated to the is/ought-descriptive/proscriptive discourse. There are multiple aspects of contemporary human social, economic, and political experience that are represented as "normative" for humans based on assumptions and assertions about events and dynamics of the human evolutionary record. But "normal" is often a poor categorical tool for describing human bodies, lives, and experiences (Darwall 2016; Nelson 2021a, b; Wiley 2021).

Variation in any human trait, morphological or behavioral, comes with a set of ideas about what constitutes the "normal" portion of that variation. And the idea of "normal" is usually accompanied by specific opinions about what that trait *should* be like. For example, it is "normal" to expect a height difference between human males and females,[2] with males being taller. However, in any given population of humans, the overlap between these two categories (male and female) in height is approximately 78%, meaning that the "normal" expectation is breached more often than not in random selections of individuals from any given population. Humans as a species also vary in multiple morphological and behavioral categories across the geographic and cultural range of the species with multiple fully successful sizes, shapes, and behaviors. It turns out that human biology and behavior are highly variable in many aspects, and this results in substantive diversity and variation of *Homo sapiens*, past and present. This reality creates substantive difficulties and errors in the assignation of the description of "normal" to many aspects of human morphological and behavioral variation (Fuentes 2022, 2025; Wiley 2021). The descriptor of "normal"

[1] Primarily white men (occasionally white women) from the United States, the United Kingdom, or certain European nations, elite academic institutions, and particular economic and cultural backgrounds.

[2] Here I am referring to the commonly used definition of male and female associated with a pattern of concordance between specific genitals, gonads, and 23rd chromosomes, fully aware that this categorization does not accurately describe all humans.

is generally a poor category and framing to "think with" when discussing human evolution.

Contemporary humans display social, economic, political, and cultural diversity that is more complex and larger in scope than that of other organisms. It is common for a wide range of scholars across multiple disciplines, and much of the lay public, to point to specific interpretations of Pleistocene *Homo* data as evidence for the "naturalness" of a particular contemporary experience. These interpretations of the human past, and their concomitant assumptions, often straddle the boundaries between "is" and "ought" in discourse about humanity and are used by a wide swath of disciplines and institutions to explain and justify contemporary human inequities and disparities. And, many of these assertions/assumptions about is and ought, normal, and natural, are wrong. Here I illustrate two particularly powerful and influential examples. The nuclear family and its concomitant gender/sex roles and xenophobia/us versus them models of human nature are experiential realities for many in the contemporary world, or "is"s, but not always in the manners most people think. Within the label/category "nuclear family" and that of "in-group versus out-group" conflict, there is substantive variation in structure, dynamics, and outcomes. Even these assumed "core" categories are in fact quite variable in their specific instantiations. There is a deep history of scholars, and the public, pointing to human evolution (and assumptions about human biology) to assert that the nuclear family and intrinsic xenophobia are both natural and normative, an "is" and an "ought," due to their emergence as natural processes across human evolution. But such assertions beg the question: what do we know about human evolution, and how does that knowledge bear directly on these topics?

13.4 Nuclear Family

The nuclear family, defined as the unit consisting of one male and one female and their biological offspring, was a key focal point for much theorizing about humans in the past (Lee and Devore 1968; Lovejoy1981; Parsons and Bales 1956; Wilson 1978) and remains so in the present (Becker 2009; Chapais 2010; Hawkes et al. 2001; Lovejoy 2009; Wrangham 2019a, b). The concept, and idealization, of the nuclear family is also at the heart of many current social, political, and religious movements (Dowland 2015) and used as a core premise for arguments against LGBTQ rights (Family Research Council 2004).

Interestingly, there is a solid argument that anthropology created the term. The anthropologist Bronislaw Malinowski (1924) coined the term "nuclear family complex" in an essay discussing and critiquing Freud's, and the broader psycho-analytical field's, contribution to the sociology of the family. In the essay, Malinowski, while acknowledging the possible insights and contributions of the psycho-analytical approach, notes that anthropological research demonstrates that humans have "great variations in the conditions of the family" across culture and space, past and present. In that same essay, Malinowski also notes the power and influence of the "nuclear family complex" on the formation of myth, legend, and fairytale in certain human societies. So, more than a century ago, the anthropological literature demonstrated that (a) the nuclear family is neither ubiquitous nor typical across humans (debunking the "is") and (b) in certain societies in "the west" the concept of the nuclear family held much sway in the development of mythos and societal expectations (developing the "ought"). In short, by 1924, it was clear that in humans, nuclear families are not normative, or typical, but that the notion of the "nuclear family complex" developed into a powerful trope in particular societies' visualizations and expectations of normal. And it is still a substantive force in both academic and public (especially political) discourse about how humans are and should be (see also Nelson 2021a; Connally 2023; Sear 2016, 2021).

The "nuclear family complex" and its belief to be central in the human experience is a testable scientific premise (a hypothesis) undergirded by a suite of assumptions about the evolution of specific behavior and physiology in humans. These are: (a) that human females are highly constrained by the cost of bearing and raising young, (b) that humans males act a primary provisioners and/or protectors to support those females and their offspring, (c) that pair-bonds are both central to human evolution and primarily related to sexual reproduction, and (d) that social structures in humans are centered around reproductive units. The current data and analyses of the human evolutionary record either refutes or complexifies each of these assumptions and supports a rejection of the assertion of one normative format for human families and reproductive units (Sear 2016, 2021; Nelson 2021a; Fuentes 2025).

The concept that human females are highly constrained by the cost of bearing and raising young rests on core biological assumptions derived from anisogamy (different sized gametes) and its purported downstream evolutionary implications for differences between females and males. The core assumption is that females invest much more in reproduction than do males, via the costs of producing large gametes (the ova) versus small gametes (sperm). The fruit fly biologist Angus Bateman reportedly demonstrated that differences in gamete size led to large differences in mating investment and thus behavior and ecology between the sexes. And because humans are mammals many scholars then asserted that these differences are even more extreme in humans due to the "costs" of gestation and lactation. Basically, the concept is that reproduction costs females a lot in an energetic and evolutionary sense and costs males little.

This concept initiates with Charles Darwin's (1871) assertions about evolved differences in males and females and the assumption that they are based on different investment in reproduction. The work on anisogamy by Bateman (1948) was subsequently mathematically modeled by Trivers (1972) and codified by Wilson (1975) and others into the assertion that one should base evolutionary explanations on the "fact" that individual females pay very high costs for reproduction and thus should behave and invest their efforts in very specific ways. This established the contemporary biological framing that females and males have differential reproductive investment costs and thus different evolutionary strategies, with the females being the more constrained sex. However, much in Darwin's (and subsequent) characterization of uniformity and structures of constraint on females was/is incorrect (Cooke 2022). There are numerous and diverse sets of options for reproduction across the animal kingdom, and there is no single consistent physiological and behavioral pattern for female bodies and behavior regarding reproductive efforts (Cooke 2022; Fuentes 2025; Lents 2025). Also, Bateman's pioneering work on anisogamy was challenged and complexified, demonstrating that simplistic large versus small gamete assumptions are neither ubiquitously correct nor always useful (Tang-Martinez 2005; Drea 2005; Gowaty et al. 2012). In short, the implications of reproduction regarding behavior and bodies, based on the concept of extreme constraints on females, are not as clear-cut as they are often assumed to be.

This is especially true for humans. Work on human physiology, human evolution, and human reproductive dynamics (Bogin et al. 2014; Brown et al. 2009; Hrdy 2009, 2024; Rosenberg 2021; Spikins et al. 2019, 2021; Burkart et al. 2009; Sear 2016; Gettler 2010, 2016) demonstrates that the genus *Homo* evolved a system of cooperative and biocultural parenting and broad-scale caretaking by at least the middle Pleistocene. Paleoanthropological, archeological, and contemporary work on *Homo* behavioral ecology and reproduction rejects a simple calculus based on the assumptions that the burden (and the costs) of reproduction are best modeled as falling exclusively on the gestating female, but rather includes a caretaking niche where the mother is one of many caretakers and much of the physiological, behavioral, and energetic costs associated with raising an infant are redistributed across members of the social group. While the energetic costs of reproduction are indeed high for the genus *Homo*, the best current data and analyses suggest that they are socially and physiologically mediated by cooperative actions, and physiologies, of group members. There is strong evidence to support the argument that during the Pleistocene, the genus *Homo* evolved a reproductive system that involved multiple individuals, enabling the development of an extended childhood connected to substantial neurobiological and social development post-utero (Fuentes 2025; Hrdy 2009, 2024; Rosenberg 2021).

While there is substantial popular buy-in to the "man the hunter/provider" mythos, there is abundant evidence from archeological, paleoanthropological, and contemporary study that *Homo sapiens* did not evolve a social system and ecological dynamic in which males are always primary provisioners for females and their offspring. In fact, such a system is not at all common across mammals and is non-existent in the primates. Despite a historical focus on "man the hunter" and "man the provider" in modeling human evolution (e.g., Washburn and Lancaster 2017; Lovejoy1981, 2009), most work in the last few decades recognizes complexity in both reproductive (Hrdy 2009, 2024; Gettler 2016; Rosenberg 2021) and foraging/hunting (Anderson et al. 2023; Haas et al. 2020; Ocobock and Lacy 2023; Lacy and Ocobock 2014; Venkataraman et al. 2024) physiology, behavior, and dynamics in the genus *Homo* (see also Morbeck et al. 1996). Prior to terminal Pleistocene and the Holocene, there is little fossil or archeological evidence of contemporary-type gender distinction patterns or specializations in behavior or technology related to hunting/foraging, ranging, tool creation and manufacture, and an assortment of other relevant factors (e.g., Arthur 2010; Fuentes 2017a, b, 2021, 2025; Wragg Sykes 2020, but also see Kuhn and Stiner 2006). Specifically, there is no evidence to support the assertion that males in the genus *Homo* were the primary providers of caloric inputs or protection for females and young across the Pleistocene.

The idea that pair-bonds are both central to human evolution and primarily related to sexual reproduction (Chapais 2010; Lovejoy 2009) remains a common assumption, especially outside of the human evolution research community. However, substantive data demonstrate that pair bonding does not always co-occur with monogamy (socially and/or reproductively) or a specific caretaking investment pattern, and there is no evidence across the primates and other social mammals that exhibit pair bonding that such behavior is necessarily connected to a form of nuclear family structure (e.g., Fuentes 1998, 2002; Bales et al. 2021). Clearly, humans do engage in a wide range of pair-bonding behaviors, but these are not limited to heterosexual reproductive dyads nor are they exclusively (or sometimes at all) tied to parenting dynamics (Fuentes 2022).

The final assumption in the "nuclear family as basal for *Homo*," that social structures in humans are centered around reproductive units, has basic support. But not as it is often framed. For organisms with post-birth care of young (mammals and birds, plus a selection of other animals), the patterns and overall system of such care usually play a significant role in the formation of the social structure of the species. This is true for all primates and assumed to be the case for the hominins. However, in the case of the genus *Homo*, it is also clear that there is a diversity in demographic, behavioral, and ecological patterns associated with social structure and that both inter- and intragroup dynamics are quite complex relative to other primates (Bogin et al. 2014; Fuentes 2016; Pisor and Surbeck 2019). So, yes, human social structure is related to our system of reproduction and caretaking, but neither of these two facets are necessarily tied to, or constrained by, a nuclear family structure (Sear 2016).

13.5 Xenophobia

Intergroup competition, warfare, and racism are all present, and often common, patterns in contemporary human societies. This reality is often invoked to undergird an assumption of xenophobia as a basal human state. In many evolutionary analyses, the guiding set of assumptions undergirding investigation and explanation of the evolution of intergroup relations is one of a basal nature to, an evolved pattern of, intragroup cooperation and intergroup competition (Darwin 1871; Wilson 1975; Wrangham 2019a, b). The contemporary evolutionary theory infrastructure for these assumptions emerges from the models of W.D. Hamilton (1964) and their continuation and expansion by E.O. Wilson (1975, 1978) and their robust explanatory reliance on kin selection and altruism theory and the array of work on competition and cooperation in organisms. The "selfish-gene" motif (Dawkins 1976, 1982), while generally acknowledged as overly simplistic, still forms a nexus for the understanding of in-group cooperation and out-group competition. Changes to the modeling of the foci of selection (e.g., individual and group, Wilson and Wilson 2007) and challenges to the basal selection arguments about altruism and cooperation (Nowak et al. 2010) still frame intergroup relations as the locus for conflict and that conflict as a central evolutionary dynamic. In short, the assumption is that xenophobia, an evolved tendency to be fearful of strangers, seems to be an outcome of general patterns of social animal evolution, and that hominin and human evolutionary trajectories fall into this pattern, and thus it is rooted in a "human nature." The bottom line for much of this theoretical positioning is that there is a "strong human universal toward parochial altruism—in-group favoritism at out-group cost" (see Pisor and Surbeck 2019). Recently, one can also see the specific frame of Pleistocene *Homo* populations and their evolution of intense intragroup cooperation as emerging because "…competition among bands, clans, and tribes drove the cultural spread of cooperative norms that permitted groups to survive both violent conflicts with other groups and natural disasters" (Henrich 2020).

It is a common assumption that humans are characterized by an evolved tendency towards xenophobia because Pleistocene *Homo* populations were characterized by competition between groups that drove the evolution (biological and cultural) of cooperative norms that enabled the current success of *Homo sapiens* (Henrich 2016; Wrangham 2019a, b). In much academic discussion of human evolution, intergroup competition and warfare are seen as common patterns guiding a set of assumptions of a basal nature to intragroup cooperation and intergroup competition in the genus *Homo* (Wilson 1975, 1978; Wrangham and Peterson 1996; Wrangham 2019a, b; Pinker 2002, 2011). The bottom line that a strong human universal toward parochial altruism, individual sacrifice to benefit the in-group and harm the out-group (see Pisor and Surbeck 2019), is both evidenced by intergroup conflict and related characteristics of the Pleistocene fossil and archeological record. But this is not true.

The range of relationships between groups and populations of the genus *Homo* across the Pleistocene is not wholly clear, but some key patterns emerge when reviewing the available data. First, there is little evidence that interpersonal violence was a more common form of death or injury than other processes of mortality. Second, there is no evidence that intergroup conflict was a common lethal pattern across the majority of the Pleistocene. There are indeed instances of lethal violence evidenced in the fossil record (e.g., Kissel and Kim 2019; Sala et al. 2021; Glowacki 2024), but they are neither consistent nor patterned across sites, times, and taxa for most of the Pleistocene. Starting around 14,000 years ago, there is increasing evidence for intergroup lethal violence (Kim and Kissel 2018; Kissel and Kim 2019; Ferguson 2008; Fry 2013; Walker 2001). This is not to say that interpersonal violence did not happen earlier in the Pleistocene and that it was not important as a factor in *Homo* behavioral ecology; it likely did and was. But fossil evidence in the form of lethal injury and archeological evidence in the form of specific weaponry and evidence of organization of intergroup violence are not common (or even present) in most of the Pleistocene data currently available. This is significant because if intergroup lethal violence was a significant/consistent enough evolutionary pressure to cause psychological and behavioral adaptive shifts in *Homo*, one would expect there to be a sufficient degree of fossil and archeological evidence associated with that pattern. And there is not. The absence of evidence does not wholly affirm that such an ecology or set of selection pressures was not present, but it does suggest that we cannot at this time assume that it was.

Human evolutionary history is not chock-full of either warfare or other intergroup violence. Nor are we simply peace-loving primates. However, the material record shifts in the terminal Pleistocene/early Holocene, and by 4000–7000 years ago, evidence of warfare and mass killings becomes increasingly present. The key to the emergence of warfare and large-scale and temporally extended intergroup conflict is not simply greater stakes in resource competition, but the capacity to turn such conflicts into ideological struggles (Glowacki 2024; Kim and Kissel 2018; Kissel and Kim 2019). War is not *the* legacy of our evolutionary history, but it is obviously a major part of our potential.

There is now substantial evidence that rather than avoidance or direct conflict, various groups and populations of the genus *Homo* across the Pleistocene, even those that are sometimes argued to be distinct species, exchanged both genetic and cultural materials (Ahlquist et al. 2021; Hajdinjak et al. 2021; Gopalan et al. 2021). While we cannot know how each of such interfaces was characterized, we do not have archeological or fossil evidence suggesting that these interfaces were predominantly characterized by aggressive interactions/conflict. A recent overview of demographic processes in Pleistocene *Homo* concluded, "For most time periods and in most regions it is no longer reasonable to make the default assumption of no gene flow between human populations" (Scally 2021).

It is becoming increasingly evident that rather than perpetual or regular conflict or avoidance between groups, the exchange of material culture and the likelihood of dynamic social relationships between groups of the genus *Homo* in populations across Africa and Eurasia was common in the last 300–400,000 years of the Pleistocene (Ashton and Davis 2021; Princeton Brooks et al. 2018; MacDonald et al. 2021). By the last 100,000 years or so, there is robust evidence of wide-scale exchange networks across groups and even populations (e.g., Ostrich egg shell bead networks, Miller and Wang 2022). Models drawing on the fossil and archeological evidence arguing for substantial social tolerance and affiliative relationships between groups as a major process in the genus *Homo* are becoming more commonplace in assessing the Pleistocene (e.g., Spikins et al. 2021) and are supported by increasing recognition of the multiscalar relational inter-community dynamics of contemporary human "small-scale" society populations (Bird et al. 2019). Even primate-wide comparisons have increasingly come to the conclusion that the genus *Homo* has a particularly high intensity of prosocial intergroup relations relative to other primates, with a recent overview concluding "The preponderance of between-community relationships in humans suggest that humans are an outlier in the Primate order with respect to our intergroup behavior, even relative to more distantly related primates that, like humans, live in multilevel societies" (Pisor and Surbeck 2019).

The neuroscience of intergroup behavior in contemporary humans offers a particularly relevant final insight into assumptions of a normativity of xenophobia. While it is indeed common, if not universal, for humans to develop in-group and out-group recognition criteria, "the biological bases of these processes does not suggest that specific in-group and out-group effects are 'hard-wired.' On the contrary, one of the major contributions of this recent work is the demonstration of how flexibly these biological systems represent the self and others, carving up the social world into 'us' and 'them' at the toss of a coin" (Cikara and Van Bavel 2014). Such neurobiological patterns suggest an evolutionary history of dynamic social landscapes of relations between groups and across communities rather than a consistent set of selection pressures favoring a deep-seated xenophobia and an evolved antagonistic set of responses to humans with whom one is not directly familiar.

I am not asserting that intergroup conflict never was/is not evolutionarily significant. This is both absurd and false. Intergroup conflict is and has been a substantive element in human evolutionary histories. However, the evidence from the fossil and archeological record, plus comparative work on contemporary human and other primate societies, robustly suggests that intergroup conflict, and violence, was not the main driving evolutionary pressure or selective context for the genus *Homo* across much of our history. The available data from Pleistocene *Homo*, humans as compared to other primates, and contemporary human neurobiology neither naturalizes, or fully explains, contemporary patterns and contexts of xenophobia (Fry 2013; Pisor and Surbeck 2019; Spikins et al. 2021). In fact, such data challenge current assertions for ubiquity of intergroup conflict as a central evolutionary force in the genus *Homo*, at least across the Pleistocene.

13.6 Human Evolution and Human Nature(s): Against Myopic Assertions of Is and Ought

Philosophers, biologists, psychologists, anthropologists, theologians, many other scholars, and much of the lay public opine and assert, frequently, about what constitutes human nature. A common theme for such argumentation is that there is indeed one true nature, one way of truly and originally being human, the one that arose as a result of our evolutionary history. But what if there is no single way of being, in most cases, for the species *Homo sapiens*? What if the evolutionary history of the genus *Homo* facilitated many successful ways to be human?

Over the last two and a half million years, members of the genus *Homo* underwent significant changes in brains, bodies, and behavior as they created and were shaped by a new niche, a distinctly human niche. Across the Pleistocene morphological changes alongside behavioral and cognitive shifts facilitated the emergence of certain behavioral and physiological patterns in *Homo*, including hyper-cooperation and complex collaboration in social interactions and the creation and use of material technologies, substantially extended childhoods and complex parenting/caretaking behavior involving multiple individuals, diverse foraging and hunting patterns involving multifaceted technologies, novel and dynamic material and representational cultures eventually resulting in complex cognitive and material meaning-making processes, emergence of exchange networks and increasingly dynamic intergroup relations, and increasingly complex communication and information sharing eventually resulting in language (Foley 2016; Fuentes 2017a, b,

2018; Galway-Witham et al. 2019; Kissel and Fuentes 2021). Given the morphological and ecological diversity, and the multiple social and ecological structures, material technologies, and lifeways evident across the Pleistocene, it appears that there were indeed multiple successful ways to navigate and participate in the human niche and that contemporary *Homo sapiens* are inheritors of a diversity of biological and cultural histories from diverse populations of *Homo* across the Pleistocene (Kissel and Fuentes 2021; Ragsdale et al. 2023; Scerri et al. 2018; Schroeder and Ackermann 2023).

I, drawing on many others, have argued elsewhere (Fuentes 2015, 2016, 2017a, b, 2021, 2022, 2025) that human processes, in all their complexity, cannot be treated as a social, material, historical, and perceptual veneer laid over a basal set of physiological capabilities and biological structures. In humans, the biological and sociocultural should not be envisioned as distinct entities (Marks 2012, 2015; Nelson 2020, 2021a, b). Human experiences and actions are not best understood via patterns of *either* biological or cultural processes; rather, they are developed, constrained, and afforded by behavior, material technologies, socioeconomic and behavioral contexts, ecologies, physiologies, evolutionary histories, and more. The webs of action and perception, memory and history, bodies and behaviors, items and ideas that humans are entangled in is the dynamic and fundamental constituent of a human niche that is simultaneously constructed by and constructing of the human experience (Fuentes 2016). Such a tangled biocultural web is highly relevant and central to both evolutionary inquiry and the ways in which such inquiry is used. An integrative biocultural approach to human evolution offers as robust and diverse toolkit to examine the patterns and processes at play in humans who evolved in, shaped, and are shaped by complex dynamic niches and who continue to do so at a rapid pace.

Rather than presenting a single pattern, an "is," that validates a particular feature of contemporary humanity, creating an "ought," most data and analyses emerging from the study of the human deep past offer ranges and possibilities for human lives. The more accurate, but less specific, interpretations of such study result in a complicated, less linear and more messy picture than many arguing for a "human nature" idealize. Combining information from the human evolutionary record with contemporary human biology and society can offer general insights into why humans do what they do but does not, and likely cannot, provide a specific template for one right way to be human.

Biases in understandings of the past and of evolutionary processes lead to flawed reconstructions of human evolutionary history and promoting misconceptions about human behavior and societies (Athreya and Ackermann 2020; Kissel 2020; Marks 2012; Schroeder and Ackermann 2023). This is why diverse voices, with a multitude of lived experiences from across multiple human societies, need to provide input for the interpretation of human evolution and its connections to contemporary human lives (Ackermann et al. 2024; Athreya 2019; Menéndez and Yañez 2025, this volume; Nelson 2021b). Diversity is not simply about social justice and equal opportunity—although it serves these ends—but about building the best possible programs, doing the best possible research, and training the best students (Antón et al. 2018). This is why who constructs and who tells the stories about humans is so important and why scholars and the public should be extremely careful about definitive and exclusivist assertions about human nature.

References

Ackermann R, Athreya S, Black W et al (2024) Paleoanthropology: decolonizing and building for the future. In: Callan H, Coleman S (eds) The international encyclopedia of anthropology, 1st edn. Wiley, pp 1–9

Ahlquist KD, Bañuelos MM, Funk A et al (2021) Our tangled family tree: new genomic methods offer insight into the legacy of archaic admixture. Genome Biol Evol 13:evab115. https://doi.org/10.1093/gbe/evab115

Anderson A, Chilczuk S, Nelson K et al (2023) The myth of man the hunter: women's contribution to the hunt across ethnographic contexts. PLoS One 18:e0287101. https://doi.org/10.1371/journal.pone.0287101

Antón SC, Kuzawa CW (2017) Early *Homo*, plasticity and the extended evolutionary synthesis. Interface Focus 7:20170004. https://doi.org/10.1098/rsfs.2017.0004

Antón SC, Malhi RS, Fuentes A (2018) Race and diversity in U.S. biological anthropology: a decade of AAPA initiatives. Am J Phys Anthropol 165:158–180. https://doi.org/10.1002/ajpa.233822

Arthur KW (2010) Feminine knowledge and skill reconsidered: women and flaked stone tools. Am Anthropol 112:228–243. https://doi.org/10.1111/j.1548-1433.2010.01222.x

Ashraf Q, Galor O (2013) The "out of Africa" hypothesis, human genetic diversity, and comparative economic development. Am Econ Rev 103:1–46. https://doi.org/10.1257/aer.103.1.1

Ashton N, Davis R (2021) Cultural mosaics, social structure, and identity: the Acheulean threshold in Europe. J Hum Evol 156:103011. https://doi.org/10.1016/j.jhevol.2021.103011

Athreya S (2019) "But you're not a *real* minority": the marginalization of Asian voices in paleoanthropology. Am Anthropol 121:472–474. https://doi.org/10.1111/aman.13216

Athreya S, Ackermann RR (2020) Colonialism and narratives of human origins in Asia and Africa. In: Porr M, J. Matthews J (eds) Interrogating human origins: decolonisation and the deep past. Routledge, London, pp 72–95

Bales KL, Ardekani CS, Baxter A et al (2021) What is a pair bond? Horm Behav 136:105062. https://doi.org/10.1016/j.yhbeh.2021.105062

Bateman AJ (1948) Intra-sexual selection in drosophila. Heredity 2(3):349–368

Becker GS (2009) A treatise on the family. Harvard University Press, Cambridge

Benenson JF, Webb CE, Wrangham RW (2022) Self-protection as an adaptive female strategy. Behav Brain Sci 45:e128. https://doi.org/10.1017/S0140525X21002417

Bird DW, Bird RB, Codding BF, Zeanah DW (2019) Variability in the organization and size of hunter-gatherer groups: foragers do not live in small-scale societies. J Hum Evol 131:96–108. https://doi.org/10.1016/j.jhevol.2019.03.005

Bogin B, Bragg J, Kuzawa C (2014) Humans are not cooperative breeders but practice biocultural reproduction. Ann Hum Biol 41:368–380. https://doi.org/10.3109/03014460.2014.923938

Boyd R (2018) A different kind of animal. Princeton University Press

Brown GR, Laland KN, Mulder MB (2009) Bateman's principles and human sex roles. Trends Ecol Evol 24:297–304. https://doi.org/10.1016/j.tree.2009.02.005

Burkart JM, Hrdy SB, Van Schaik CP (2009) Cooperative breeding and human cognitive evolution. Evol Anthropol Issues News Rev 18:175–186. https://doi.org/10.1002/evan.20222

Chapais B (2010) Primeval kinship: how pair-bonding gave birth to human society. Harvard University Press, Cambridge

Cikara M, Van Bavel JJ (2014) The neuroscience of intergroup relations: an integrative review. Perspect Psychol Sci 9:245–274. https://doi.org/10.1177/1745691614527464

Connally B (2023) The nuclear family: how a complex became a norm. Parapraxis. https://www.parapraxismagazine.com/articles/the-nuclear-family
Cooke L (2022) Bitch: on the female of the species. Basic Books, New York
Dapschauskas R, Göden MB, Sommer C, Kandel AW (2022) The emergence of habitual ochre use in Africa and its significance for the development of ritual behavior during the middle stone age. J World Prehist 35:233–319. https://doi.org/10.1007/s10963-022-09170-2
Darwall S (2016) Normativity. In: Routledge encyclopedia of philosophy, 1st edn. Routledge, London. https://www.rep.routledge.com/articles/thematic/normativity/v-1
Darwin C (1871) The descent of man, and selection in relation to sex, 1st edn. John Murray, London
Dawkins CR (1976) The selfish gene. Oxford University Press, Oxford
Dawkins CR (1982) The extended phenotype. Oxford University Press, Oxford
DeSilva J (2021) A most interesting problem. Human evolution 150 years after Darwin's descent of man. Princeton University Press, Princeton
Dowland S (2015) Family values and the rise of the Christian right. University of Pennsylvania Press, Philadelphia
Drea CM (2005) Bateman revisited: the reproductive tactics of female primates. Integr Comp Biol 45:915–923. https://doi.org/10.1093/icb/45.5.915
Family Research Council (2004) Ten arguments from social science against same-sex marriage. In: FRC. https://www.frc.org/issuebrief/ten-arguments-from-social-science-against-same-sex-marriage. Accessed 16 June 2024
Ferguson BR (2008) Ten points on war. Soc Anal 52:32–49. https://doi.org/10.3167/sa.2008.520203
Foley RA (2016) Mosaic evolution and the pattern of transitions in the hominin lineage. Philos Trans R Soc Lond B Biol Sci 371:20150244. https://doi.org/10.1098/rstb.2015.0244
Fry D (2006) Beyond war: the human potential for peace. Oxford University Press, Oxford
Fry DP (2013) War, peace, and human nature: the convergence of evolutionary and cultural views. Oxford University Press, New York
Fuentes A (1998) Re-evaluating primate monogamy. Am Anthropol 100:890–907. https://doi.org/10.1525/aa.1998.100.4.890
Fuentes A (2002) Patterns and trends in primate pair bonds. Int J Primatol 23:953–978. https://doi.org/10.1023/A:1019647514080
Fuentes A (2015) Integrative anthropology and the human niche: toward a contemporary approach to human evolution: integrative anthropology and the human niche. Am Anthropol 117:302–315. https://doi.org/10.1111/aman.12248
Fuentes A (2016) The extended evolutionary synthesis, ethnography, and the human niche: toward an integrated anthropology. Curr Anthropol 57:S13–S26. https://doi.org/10.1086/685684
Fuentes A (2017a) Human niche, human behaviour, human nature. Interface Focus 7:20160136. https://doi.org/10.1098/rsfs.2016.0136
Fuentes A (2017b) The creative spark: how imagination made humans exceptional. Dutton/Penguin, New York
Fuentes, A. (2018) How humans and apes are different, and why it matters Journal of Anthropological Research 74(2):151–167. https://doi.org/10.1086/697150
Fuentes A (2021) Searching for the "roots" of masculinity in primates and the human evolutionary past. Curr Anthropol 62:S13–S25. https://doi.org/10.1086/711582
Fuentes A (2022) Race, monogamy and other lies they told you: busting myths about human evolution. University of California Press, Berkeley
Fuentes A (2025) Sex is a spectrum: the biological limits of the binary. Princeton University Press, Princeton
Galway-Witham J, Cole J, Stringer C (2019) Aspects of human physical and behavioural evolution during the last 1 million years. J Quat Sci 34:355–378. https://doi.org/10.1002/jqs.3137
Gettler LT (2010) Direct male care and hominin evolution: why male–child interaction is more than a nice social idea. Am Anthropol 112:7–21. https://doi.org/10.1111/j.1548-1433.2009.01193.x

Gettler LT (2016) Becoming DADS: considering the role of cultural context and developmental plasticity for paternal socioendocrinology. Curr Anthropol 57:S38–S51. https://doi.org/10.1086/686149

Glowacki L (2024) The evolution of peace. Behav Brain Sci 47:e1. https://doi.org/10.1017/S0140525X22002862

Gopalan S, Atkinson EG, Buck LT et al (2021) Inferring archaic introgression from hominin genetic data. Evol Anthropol 30:199–220. https://doi.org/10.1002/evan.21895

Gowaty PA, Kim Y-K, Anderson WW (2012) No evidence of sexual selection in a repetition of Bateman's classic study of *Drosophila melanogaster*. Proc Natl Acad Sci USA 109:11740–11745. https://doi.org/10.1073/pnas.1207851109

Haas R, Watson J, Buonasera T et al (2020) Female hunters of the early Americas. Sci Adv 6: eabd0310. https://doi.org/10.1126/sciadv.abd0310

Hajdinjak M, Mafessoni F, Skov L et al (2021) Initial Upper Palaeolithic humans in Europe had recent Neanderthal ancestry. Nature 592:253–257. https://doi.org/10.1038/s41586-021-03335-3

Hamilton WD (1964) The genetical evolution of social behaviour. I. J Theor Biol 7:1–16. https://doi.org/10.1016/0022-5193(64)90038-4

Harari YN (2015) Sapiens. Harper, New York

Hawkes K, O'Connell JF, Blurton Jones NG (2001) Hunting and nuclear families: some lessons from the Hadza about mens work. Curr Anthropol 42:681–709. https://doi.org/10.1086/322559

Henrich J (2016) The secret of our success: how culture is driving human evolution, domesticating our species, and making us smarter. Princeton University Press, Princeton

Henrich J (2020) WEIRD: how westerners became psychologically peculiar and particularly prosperous. Farrar, Straus and Giroux, New York

Hrdy SB (2009) Mother and others: the evolutionary origins of mutual understanding. Belknap, New York

Hrdy SB (2024) Father time: a natural history of men and babies. Princeton University Press, Princeton

Kim NC, Kissel M (2018) Emergent warfare in our evolutionary past. Routledge, New York

Kissel M (2020) Anthropology now: how popular science (mis)characterizes human evolution. In: Willermet C, Lee S-H (eds) Evaluating evidence in biological anthropology: the strange and the familiar. Cambridge University Press, Cambridge, pp 55–70

Kissel M, Fuentes A (2021) The ripples of modernity: how we can extend paleoanthropology with the extended evolutionary synthesis. Evol Anthropol 30:84–98. https://doi.org/10.1002/evan.21883

Kissel M, Kim NC (2019) The emergence of human warfare: current perspectives. Am J Phys Anthropol 168:141–163. https://doi.org/10.1002/ajpa.23751

Kramer KL (2010) Cooperative breeding and its significance to the demographic success of humans. Annu Rev Anthropol 39:417–436. https://doi.org/10.1146/annurev.anthro.012809.105054

Kropotkin KPA (1902) Mutual aid: a factor of evolution. Project Guttenberg. https://gutenberg.org/ebooks/4341

Kuhn SL, Stiner MC (2006) What's a mother to do?: the division of labor among Neandertals and modern humans in Eurasia. Curr Anthropol 47:953–981. https://doi.org/10.1086/507197

Lacy S, Ocobock C (2023) Woman the hunter: The archaeological evidence. American Anthropologist. https://doi.org/10.1111/aman.13914

Laland KN (2017) Darwin's unfinished symphony: how culture made the human mind. Princeton University Press, Princeton

Lee RB, DeVore I (1968) Man the hunter. Aldine Publishing Company, Chicago

Lents N (2025) The sexual evolution: a provocative look at sexual behavior through the lens of evolution, expand your understanding. Mariner Books

Lock M, Palsson G (2016) Can science resolve the nature/nurture debate? Polity Press, Cambridge

MacDonald K, Scherjon F, Van Veen E et al (2021) Middle Pleistocene fire use: the first signal of widespread cultural diffusion in human evolution. Proc Natl Acad Sci USA 118:e2101108118. https://doi.org/10.1073/pnas.2101108118
Malinowski B (1924) Psychoanalysis and anthropology. Psyche 4:293–332
Marks J (2012) The biological myth of human evolution. Contemp Soc Sci 7:139–157. https://doi.org/10.1080/21582041.2012.691989
Marks J (2015) Tales of the ex-apes: how we think about human evolution. University of California Press, Berkeley
Menéndez LP, Yañez B (2025) A geopolitical and multilayered analysis on the silencing of Latin American biological anthropology: the case of "the peopling of the Americas" research. In: Lequin M, Rodriguez Caso JM (eds) The field of human evolution viewed from critical history and epistemology. Springer
Midgley M (2002) Evolution as a religion. Routledge, New York
Miller JM, Wang YV (2022) Ostrich eggshell beads reveal 50,000-year-old social network in Africa. Nature 601:234–239. https://doi.org/10.1038/s41586-021-04227-2
Morbeck ME, Galloway A, Zihlman A (1996) The evolving female: a life history perspective. Princeton University Press, Princeton
Mulder MB (2004) Are men and women really so different? Trends Ecol Evol 19:3–6. https://doi.org/10.1016/j.tree.2003.11.002
Nelson R (2020) Methods without meaning: moving beyond body counts in research on behavior and health. In: Willermet C, Lee S-H (eds) Evaluating evidence in biological anthropology: the strange and the familiar. Cambridge University Press, Cambridge, pp 86–100
Nelson RG (2021a) Theoretical constraints: science, caretaking, and the creation of normative ideals. Am J Hum Biol 33:e23650. https://doi.org/10.1002/ajhb.23650
Nelson RG (2021b) The sex in your violence: patriarchy and power in anthropological world building and everyday life. Curr Anthropol 62:S92–S102. https://doi.org/10.1086/711605
Nowak MA, Tarnita CE, Wilson EO (2010) The evolution of eusociality. Nature 466:1057–1062. https://doi.org/10.1038/nature09205
Ocobock, C. and Lacy, S. (2023) Woman the hunter: the physiological evidence. American Anthropologist. https://doi.org/10.1111/aman.13915
Parsons T, Bales RF (1956) Family socialization and interaction process. Routledge, Abingdon
Pinker S (2002) The blank slate: the modern denial of human nature. Viking, New York
Pinker S (2011) The better angels of our nature: why violence has declined. Penguin, New York
Pisor AC, Surbeck M (2019) The evolution of intergroup tolerance in nonhuman primates and humans. Evol Anthropol 28:210–223. https://doi.org/10.1002/evan.21793
Princeton Brooks AS, Yellen JE, Potts R et al (2018) Long-distance stone transport and pigment use in the earliest middle stone age. Science 360:90–94. https://doi.org/10.1126/science.aao2646
Ragsdale AP, Weaver TD, Atkinson EG, Hoal EG, Möller M, Henn BM, Gravel S. A weakly structured stem for human origins in Africa. Nature. 2023 May;617(7962):755–763. 10.1038/s41586-023-06055-y. Epub 2023 May 17. Erratum in: Nature. 2023 Aug;620(7972):E11. https://doi.org/10.1038/s41586-023-06433-6
Read D (2011) How culture makes us human: primate social evolution and the formation of human societies. Left Coast, Walnut Creek
Rosenberg KR (2021) The evolution of human infancy: why it helps to be helpless. Annu Rev Anthropol 50:423–440. https://doi.org/10.1146/annurev-anthro-111819-105454
Scally A (2021) Molecular anthropology is dead, long live molecular anthropology. J Anthropol Sci:179–182. https://doi.org/10.4436/JASS.99009
Scerri EML, Thomas MG, Manica A et al (2018) Did our species evolve in subdivided populations across Africa, and why does it matter? Trends Ecol Evol 33:582–594. https://doi.org/10.1016/j.tree.2018.05.005
Schroeder L, Ackermann RR (2023) Moving beyond the adaptationist paradigm for human evolution, and why it matters. J Hum Evol 174:103296. https://doi.org/10.1016/j.jhevol.2022.103296

Sear R (2016) Beyond the nuclear family: an evolutionary perspective on parenting. Curr Opin Psychol 7:98–103. https://doi.org/10.1016/j.copsyc.2015.08.013
Sear R (2021) The male breadwinner nuclear family is not the 'traditional' human family, and promotion of this myth may have adverse health consequences. Philos Trans R Soc Lond B Biol Sci 376:rstb.2020.0020, 20200020. https://doi.org/10.1098/rstb.2020.0020
Smuts B (1995) The evolutionary origins of patriarchy. Hum Nat 6:1–32. https://doi.org/10.1007/BF02734133
Spikins P, Needham A, Wright B et al (2019) Living to fight another day: the ecological and evolutionary significance of Neanderthal healthcare. Quat Sci Rev 217:98–118. https://doi.org/10.1016/j.quascirev.2018.08.011
Spikins P, French JC, John-Wood S, Dytham C (2021) Theoretical and methodological approaches to ecological changes, social behaviour and human intergroup tolerance 300,000 to 30,000 BP. J Archaeol Method Theory 28:53–75. https://doi.org/10.1007/s10816-020-09503-5
Sterelny K (2007) Social intelligence, human intelligence and niche construction. Philos Trans R Soc Lond B Biol Sci 362:719–730. https://doi.org/10.1098/rstb.2006.2006
Sterelny K (2021) The Pleistocene social contract: culture and cooperation in human evolution. Oxford University Press, Oxford
Stout D (2011) Stone toolmaking and the evolution of human culture and cognition. Philos Trans R Soc Lond B Biol Sci 366:1050–1059. https://doi.org/10.1098/rstb.2010.0369
Stout D, Khreisheh N (2015) Skill learning and human brain evolution: an experimental approach. Camb Archaeol J 25:867–875. https://doi.org/10.1017/S0959774315000359
Tang-Martinez Z (2005) The problem with paradigms: Bateman's worldview as a case study. Integr Comp Biol 45:821–830. https://doi.org/10.1093/icb/45.5.821
Tomasello M (2014) A natural history of human thinking. Harvard University Press, Cambridge
Trivers RL (1972) Parental investment and sexual selection. In: Campbell B (ed) Sexual selection and the descent of man, 1871–1971. Aldine, Chicago, pp 136–179
Venkataraman VV, Hoffman J, Farquharson K et al (2024) Female foragers sometimes hunt, yet gendered divisions of labor are real: a comment on Anderson et al. (2023) The myth of man the hunter. Evol Hum Behav 45:106586. https://doi.org/10.1016/j.evolhumbehav.2024.04.014
Walker PL (2001) A bioarchaeological perspective on the history of violence. Annu Rev Anthropol 30:573–596. https://doi.org/10.1146/annurev.anthro.30.1.573
Washburn SL, Lancaster GS (2017) The evolution of hunting. In: Lee RB, Devore I (eds) Man the hunter, 1st edn. Routledge, pp 293–303. https://doi.org/10.4324/9780203786567-39
Wengrow D, Graber D (2021) The Dawn of everything: a new history of humanity. Farrar, Straus and Giroux, New York
Wiley AS (2021) Pearl lecture: biological normalcy: a new framework for biocultural analysis of human population variation. Am J Hum Biol 33:e23563. https://doi.org/10.1002/ajhb.23563
Wilson EO (1975) Sociobiology: the new synthesis. Harvard University Press, Cambridge
Wilson EO (1978) On human nature. Harvard University Press, Cambridge
Wilson EO (1979) Sociobiology: sex and human nature. The Wilson Quarterly (1976-) 3(4):92–105
Wilson DS, Wilson EO (2007) Rethinking the theoretical foundation of sociobiology. Q Rev Biol 82:327–348. https://doi.org/10.1086/522809
Wragg Sykes R (2020) Kindred: Neanderthal life, love, death and art. Bloomsbury Sigma, London
Wrangham RW (2019a) The goodness paradox: the strange relationship between virtue and violence in human evolution. Pantheon, New York
Wrangham RW (2019b) Hypotheses for the evolution of reduced reactive aggression in the context of human self-domestication. Front Psychol 10:1914. https://doi.org/10.3389/fpsyg.2019.01914
Wrangham R, Peterson D (1996) Demonic males: apes and the origin of human violence. Mariner Books, New York

14 When Neanderthals and Modern Humans Coexisted: How Philosophy, Archeology and Prehistory Novels Are Working Towards a More Inclusive Understanding of Deep Human History

Hub Zwart, Susan Peeters, Karel Kuipers, and Marie Soressi

Abstract

Is there something exceptional about modern humans compared to early human ancestors? Until recently, Neanderthal research focused on human uniqueness: what made us decisively different? Although Neanderthals are increasingly recognised as similar to us, researchers tended to ask whether there was a decisive difference that enabled modern humans to move from "nature" to "culture" and develop complex technology, language, art and social organisation. This question shaped both scientific debates and Neanderthal novels, where encounters between "superior" humans and "inferior" Neanderthals often explained the latter's demise. Today, however, a shift is visible both in paleo-archeology and in Neanderthal novels. The binary opposition of "them" versus "us" is giving way to a more dynamic, interactive and inclusive view of early human history, a change that affects how we see ourselves. We address the question of whether and how a non-binary, non-hierarchical perspective on the Neanderthal–human encounter is taking shape in science, philosophy and literature. We approach this question interdisciplinarily, combining philosophy, archeology and literary studies. In research on the Châtelperronian period, when Neanderthals and modern humans coexisted, the focus is moving from competition to interaction and exchange. Likewise, Neanderthal novels show a transition from depictions of violence and superiority to narratives of connection and coexistence. For example, Jack London's *Before Adam* contrasts with recent works that imagine shared contemporaneity in Pleistocene Europe. This reorientation matters today, as it

H. Zwart (✉) · S. Peeters · K. Kuipers
Erasmus School of Philosophy, Erasmus University Rotterdam, Rotterdam, Netherlands
e-mail: zwart@esphil.eur.nl

M. Soressi
Faculty of Archaeology, Leiden University, Leiden, Netherlands

M. Lequin, J. M. Rodríguez Caso (eds.), *The Field of Human Evolution: Critical Perspectives from History and Epistemology*, Evolutionary Biology – New Perspectives on Its Development 11, https://doi.org/10.1007/978-3-032-13925-2_14

fosters a more dynamic and inclusive conception of humanity—one that emphasises interaction, embeddedness and shared belonging over hierarchy and separation.

Keywords

Philosophy of deep history · Neanderthal research · Neanderthal novels · Modern human Exceptionalism · Jack London · The human revolution · Philosophical anthropology

14.1 Introduction

Reflections on the planetary place and role of human beings often tend towards exceptionalism, i.e. the conviction that there is something unique about humans compared to other living beings: a uniqueness which notably involves our capacity to think (e.g. to think about ourselves: self-consciousness) which is closely connected with language and the use of symbols, thereby distancing human existence from the rest of nature. As philosopher Martin Heidegger phrased it, whereas animals dwell in an *Umwelt* [environment], human beings have entered a *Lichtung* [clearing], a world of language. If we adopt an evolutionary perspective on human existence, however, the question emerges whether, when and how this clearing, this *leap* towards self-consciousness came about? And research into early human history is evidently highly relevant here. The question even becomes more specific: is there something exceptional about *(anatomically) modern humans* compared to other early human ancestors? And Neanderthal research becomes notably interesting in this respect because Neanderthals are seen as our closest relatives, and they have been the object of extensive research during the past three decades, while Neanderthals and modern humans co-existed in what is now Europe roughly between 50,000 and 40,000 years ago.[1] To address this question, we will argue, a *philosophy of deep history* is required which builds on interdisciplinary dialogue between philosophy and archeology, but also engages with literature studies and genres of the imagination, e.g. novels dedicated to the encounter between modern humans and Neanderthals (Table 14.1).

Until recently, the question concerning the uniqueness of modern humans also guided Neanderthal research: what made "us" (modern humans) decisively different? Although Neanderthals are increasingly seen as similar to us in many ways, the

[1]We view (anatomically) modern humans, Neanderthals and Denisovans as part of a "metapopulation" (Bergström et al. 2021; Pääbo 2015) and "anatomically modern humans" as the direct ancestors of all humans alive today. To avoid taking a stance in the species debate, we will consistently use the term "modern humans" from now on and refrain from using terms like *Homo sapiens* (Meneganzin and Bernardi 2023; Langdon 2022; Pearson 2008). For us, the term "modern humans" refers to a population originated in Africa and later colonised the rest of the planet in multiple waves. We try to avoid normative connotations (e.g. modern human as "more" human) but acknowledge that neutral or unquestionable terms are not available.

Table 14.1 Overview of temporal markers for a philosophy of deep history

Years ago	Event	Practices	Worldview
~250,000	Emergence of *Homo sapiens*	Hunting and gathering; Promethean revolution (pyro-technology)	Birth of shamanism?
~100,000	Migration "Out of Africa"	Stone tools, ornaments	
~50,000	Modern humans–Neanderthals encounter	Stone tool industries and art	
~10,000	Neolithic revolution: first "biotechnological" revolution	Emergence of agriculture at various locations around the globe	Polytheism
~5000	Urban revolution (emergence of big cities)	Agriculture reaches what is now Western Europe	Invention of writing
~2500	Beginnings of empire building	Class societies	Birth of philosophical cosmologies
~1500	Medieval epoch	Monasticism	Metaphysical cosmologies
~500	Modernity	Colonisation: migration of farmers, animals and crops	Modern science
~250	"Industrial" revolution	Emergence of modern cities	Political ideologies
~50	Environmental concerns	Biotechnological (or "green") revolution	Globalisation, bioethics

focus often tends to be on detecting the so-called factor X, the decisive difference that makes us truly human (Peeters and Zwart 2020). What made modern humans take the decisive step from natural habitats towards increasingly self-designed cultural ambiances, and capable of developing complex forms of technology, language, art and social organisation?

This question (why are "we" so similar to Neanderthals, and yet different) not only guided much of Neanderthal research but was also explored in Neanderthal novels. Here, the encounter between (culturally and genetically) "superior" humans and "inferior" Neanderthals, resulting in the demise of the latter, often served as a core literary thematic.

Currently, however, we notice a change of perspective, both in paleo-archeology[2] as a scientific practice and in Neanderthal novels as a literary genre. The binary logic of "them" versus "us" is giving way to a more dynamic, interactive and inclusive view of early human history, a shift, we will argue, which has repercussions for how

[2]Different terms are employed to refer to the research field in question, often reflecting the disciplinary background or national research traditions of the authors (archeology, paleoanthropology, human origins research, etc.). We opted for the term paleo-archeology here, i.e. the archeology of deep history, notably the Pleistocene. We see paleo-archeology as an interdisciplinary scientific field that builds on methods and concepts from archeology, anthropology, cultural anthropology, ethology, gender studies and other fields.

we see ourselves as human beings today. This, therefore, is the core question of our contribution: whether and how a non-binary, non-hierarchical, more inclusive view of the relationship between Neanderthals and modern humans is taking shape, both in scientific research and in Neanderthal novels? And, subsequently, what implications does this entail for human self-understanding today?

To explore these issues in depth, a collaboration across disciplines is important, notably involving philosophical anthropology (devoted to address the question "What makes us human?") and paleo-archeology, but other fields, such as gender studies, epistemology, art history and literature studies, prove highly relevant as well. The structure of our interdisciplinary contribution (a dialogue between philosophy, paleo-archeology and literature studies) is as follows. First, we will discuss the importance of broadening philosophy's temporal horizon by developing a philosophy of deep history. This requires reflection on the results of contemporary paleo-archeological research, notably the work that focusses on a period of technological innovation often referred to as the Châtelperronian, when Neanderthals and humans co-existed in what is now France for thousands of years. Subsequently, we will indicate how the tendency towards a more inclusive view on early human history is noticeable in Neanderthal novels as well. Here, we will compare Jack London's Neanderthal classic *Before Adam* with more recent literary enactments of the encounter between Neanderthals and modern humans. Finally, we analyse how these shifting views challenge the exceptionalist tradition in human self-understanding.

14.2 The Emergence of Historical Thinking

Although archeology in the broader sense can boast an impressive track record, including for instance the antiquarianism that flourished during the Song dynasty (960–1279) in China or during the Renaissance in Europe, the birth of archeology in the modern sense (of archeology as a systematic science) is often situated in the nineteenth century and associated with the work of scholars such as Augustus Henry Lane Fox Pitt Rivers (1827–1900), who adopted the methodological principle that *all* artefacts (not only the beautiful and exceptional ones, the collectors' items if you like) are relevant when it comes to studying the evolution of human craftmanship during various epochs as they are unearthed in archeological stratums (Bowden 1991).

This connection with the nineteenth century is no coincidence because it is not only the century when the word "scientist" was introduced for the first time, but also the century of *historisation* (cf. Sera-Shriar 2018), both in the natural sciences (*Naturwissenschaften*), notably the life sciences (e.g. Darwin's theory of evolution), and in the humanities (*Geisteswissenschaften*), e.g. the rise of critical philology. In philosophy, this tendency towards historisation is noticeable in the writings of historically minded authors such as Hegel, Marx and Nietzsche. Hegel's thinking for instance builds on the conviction that humanity's present condition can only be understood against the backdrop of a profound understanding of our history: the present must be grasped as a historical *result.*

At the same time, a difference can be pointed out between the life sciences on the one hand and the humanities on the other in terms of scale. Philosophically speaking, a very important aspect of Darwin's theory was the tremendous expansion of our temporal horizon it entailed, for which the geological work of Charles Lyell and others prepared the ground. Until then, natural history was often discussed in terms of thousands of years, using *Genesis* as a frame of reference, but Darwin convincingly argued that the concept of evolution only becomes credible when we realise that it may take millions of years to sculpt new species and that nature has "incomprehensibly vast periods of time" available for "the work of natural selection" (Darwin 1859/1985, p. 147, p. 293). In philosophy, however, given its dependence on written documents, reflections on the historical nature of human thinking tended to stay within a much more limited timeframe. Here, the so-called "axial age" (twenty-five centuries ago) was often taken as point of departure. In his concise world history of philosophy, for instance, Hans Joachim Störig (1961) argued that philosophy as we now know it began around 550 BC at different locations around the globe. In China and India for instance, but also along the shores of Asia Minor and Southern Italy, the world suddenly began to think in the philosophical sense of the term, so that mythical worldviews (i.e. the tendency to attribute changes and events to supernatural powers) were replaced by maxims such as "for every change, there is a natural cause". What was remarkable, however, was that this awakening was a worldwide phenomenon. Störig emphasises the global scope of the transition. The rise of Greek thinking was not a stand-alone event. Buddha, Confucius and Lao-Tse were contemporaries of Parmenides and Heraclitus, as were Jeremiah and Zarathustra. Karl Jaspers likewise argued that, during the axial period, thinking *as such* suddenly came into existence (Jaspers 1949). This idea, that at several places on earth, simultaneously more or less, but independently and at great distances from one another, human self-consciousness suddenly seemed to make a leap, also implied that systematic reflection, once spurred into action, managed to reach impressive heights in a relatively short period of time (e.g. the miraculous birth of Greek mathematics and philosophy, seemingly ex nihilo). We notice the same conviction in Oswald Spengler's comprehensive conception of history (Spengler 1918/1923).

Although the axial epoch represents a remarkable event in human history, there is a drawback to this emphasis on the axial revolution as well, because it tends to obfuscate everything that must have occurred before that time (cf. Grosos 2023). Authors such as Hegel and Spengler do not have much to say about what came before: about "archaic" culture as they call it. Spengler does mention that archaic cultural forms emerging in "immense natural landscapes" (art, dance, music, poetry) were meant to adjure natural forces or deities ("*Beschwörung*"), but of the songs and dances of archaic cultures, hardly anything remains, he argues, and the little that remains concerns mostly the ornamental side. Hegel likewise surmises how, with the help of particular words and gestures, shamans once tried to implore the inscrutable powers of nature and to move them to pliability, hypnotising nature as it were (Hegel 1969), but the empirical basis for his reflections was quite limited at the time. In other words, "archaic" culture was considered a lost world, beyond the scope of philosophical reflection.

14.3 Broadening Philosophy's Temporal Horizon

This has evidently changed during recent decades, due to important developments in archeological research. Philosophers may try to keep track of archeological investigations so as to broaden their perspective, not only geographically (aspiring to supersede their Eurocentric bias) but also temporally, trying to become more inclusive, for instance by reflecting on the importance of the neolithic revolution (the emergence of agriculture), again in various parts of the globe, independently from one another, around ten millennia ago. The shift from a nomadic to a sedentary existence, and from hunter-gatherer to farming cultures, must have resonated with a different mindset or style of thinking as well, as humans began to drastically reshape their environment and human existence settled in self-supportive village communities—a relatively stable agricultural way of life that lasted for millennia across the globe, until it was disrupted by the industrial revolution in the eighteenth and nineteenth century (Zwart 2009; Grosos 2023).

In other words, while authors such as Hegel focussed on what was traditionally known as "history" (characterised by the presence of written documents), later authors began to push the starting point of their reflections increasingly deeper into pre-history, e.g. authors such as Arnold Gehlen (1964) in *Urmensch und Spätkultur* ("Primordial man and late culture") or Hans Blumenberg, author of *Höhlenausgänge* (*Coming out of the Cave*) and multiple other writings (Velasco 2023). More recently, Philipe Grosos argued that the "birth" of philosophy in ancient Greece, twenty-five centuries ago, was actually the culmination of extensive prior processes (Grosos 2023; cf. Geneste et al. 2023) so that Heraclitus, Parmenides, Plato and Aristotle (inhabitants of large urban centres of agricultural societies dominated by males), was not the beginning, but rather a late culmination. Using paleolithic and neolithic art for guidance, Grosos aims to reconstruct the obfuscated origins of philosophy in pre-history, emphasising for instance how in paleolithic and neolithic art the human way of being in the world shifts from an ontology of *participation* to an ontology of anthropocentric presence, and while in neolithic art the vast majority of humans represented were male, Paleolithic artworks mainly depicted female figures. Thus, while in earlier days pre-history was equated with mythology, and rational philosophy with ancient Greek philosophy, we are now challenged to include "pre-historic" ontologies as well, so that archeology as the study of prehistorical remains becomes increasingly important as a source of information for philosophical reflection. *Philosophy of archeology* (literally: the logic of beginnings) thus becomes an *archeology of philosophy*.

The prefix "pre-" does not seem very functional in this regard, however, not only because archeology often combines and compares written sources with material findings, but also because the vast period referred to as "pre-history" is evidently much more expanded than "history",—a relatively recent epoch in the journey of human evolution, if we consider that modern humans entered the scene around 200,000 years ago,—230,000 years ago according to a recent *Nature* publication (Vidal et al. 2022). Therefore, in this contribution we will prefer to use the term "deep history" rather than "pre-history" (cf. Zwart and Soressi 2025).

French philosopher Michel Foucault argued in the 1960s that we should reverse our perspective by also studying written sources in an archeological manner (Foucault 1969). In the history of written discourse, Foucault argues, stratums can likewise be discerned. For an archeological philosopher, *all* documents are equally relevant, not only the exceptional ones, written by "great thinkers". The focus should shift to the enormous layers of (often anonymous) "discourse" accumulated in libraries and archives. Adopting an archeological approach, Foucault argues, allows us to determine the typical manner of speaking, observing and thinking in vogue during a particular epoch: the logic, the basic vocabulary all its authors have in common (rather than emphasising their theoretical differences). Again, however, by focussing on written sources, the starting point of his efforts, gargantuan as they may seem, is fairly late.

In order to gain access to deep history, and to enter worlds of human experience where written sources (the usual window of access for philosophical reflection on particular historical ambiances) cannot be found, intense forms of collaboration between archeology and philosophy are evidently pivotal, at least from the philosophical point of view. Whereas hermeneutics (the traditional method of the *Geisteswissenschaften*) presupposes the availability of texts, a philosophy of deep history requires the adoption of different sets of tools and is only possible as an intrinsically interdisciplinary endeavour, involving collaborations with archeology first and foremost, but also with evolutionary theory, semiotics and art history.

Although a philosophy of deep history faces many obstacles (such as the absence of written source materials, as indicated), it at the same time entails a number of important potential benefits. During the past decades, philosophers began to realise that, from Plato and Aristotle onwards, philosophy has been subject to a series of biases (Corbey 2005; Porr and Matthews 2019). Increasingly, philosophy (notably: modern Western philosophy) is now seen as *egocentric* (obfuscating the dependence of humans on social networks), as *anthropocentric* (obfuscating the embeddedness and connectedness of human existence within the web of life), as *Eurocentric* (tending to see European concepts and categories as the standard) and as *androcentric* (identifying "human" with "male", and humanity with masculinity). These biases are far from being a typical philosophical predicament. Rather, they are discernible in other research fields as well, including archeology, evolutionary theory and art history. Precisely here, an interdisciplinary approach to deep history may broaden our perspective and spur us to overcome these implicit biases, for instance by situating human symbolism within the broader field of biology, while replacing a view of human history based on exceptionality with a more inclusive understanding. In the next section, we will describe how in paleo-archeological research a similar strive towards a more dynamical and inclusive understanding of deep history is emerging.

14.4 Paleo-archeology: Challenging the Human Revolution Model

Until the 1990s, the *human revolution model* was the dominant paradigm in paleo-archeological research. Its proponents (Mellars 1989; Donald 1993; Noble and Davidson 1991; Klein 1995) argued that, at a certain point in human evolution, a leap-like shift occurred in human cognition, resulting in the birth of language, art and symbolism. In their book *In Search of the Neanderthals: Solving the puzzle of human origins*, for instance, Chris Stringer and Clive Gamble (1993) claimed that Neanderthals did not have culture in the sense that we understand it today and still recognise it in the remains of early modern humans such as cave art, thereby solidifying the "them vs. us" dichotomy. This has changed. As Andra Meneganzin and Adrian Currie (2022) argue, the archeological expectation of discovering a "Rubicon" event, i.e. evidence of the sudden emergence of "modern behaviour as an integrated package", is giving way to a view which sees the origin of behavioural modernity as a spatially and temporally distributed process. Yet, although the Rubicon model has been criticised and debunked (Cartmill 1990; King 1994; McBrearty and Brooks 2000; Roebroeks and Corbey 2001; Nowell 2010), and even Stringer and Gamble themselves adapted their views considerably since the publication of their book, the human revolution model was not very favourable for the Neanderthal image and its impact on Neanderthal discourse is still noticeable today (Meneganzin and Currie 2022; Nowell 2010; Shea 2011). "Policing boundaries" (i.e. defining a number of attributes that are unique and qualitatively exceptional for humans) is still an important trend, not only between humans and other primates (Cartmill 1990, p. 178), but also between modern humans and other early human populations including Neanderthals. Whenever a particular trait is under threat of not being uniquely human, the trait is often redefined, thereby effectively moving "the goal posts" to ensure that the exceptionality of humans remains intact.

A telling example of this logic of policing is the debate concerning symbolic behaviour (Kuipers et al. 2024). Many scholars agree that *symbolic behaviour* is a core defining trait of *modern behaviour*, i.e. the behavioural repertoire of modern humans (Nowell 2010). Therefore, symbolic behaviour became the new synonym for humanness for many archeologists (Bushozi 2020; Finlayson et al. 2012; Rodríguez-Vidal et al. 2014). Yet, again, the claim that symbolic behaviour is exceptional proves problematic (Kuipers et al. 2024). How to determine that a series or notches found on bones or antlers should be considered as symbols? And if symbolic behaviour is attributed exclusively to (modern) humans, how to assess evidence of symbolic behaviour in other early human populations? Indeed, evidence of symbolic behaviour can be found in many animals as well (Hersh et al. 2022; Pardo et al. 2023). The focus on some "essential" human trait which allegedly sets modern humans apart from others proves counterproductive. Commonalities are at least as interesting and allow us to re-embed early human history in a more comprehensive ecological and evolutionary perspective.

Therefore, to answer our questions, our aim is to reconnect the philosophical question "What makes us human?" with the evolutionary framework of archeological research into human origins ("How did we become what we are?"). Philosophers must familiarise themselves with the advancements made in human evolution research, while archeologists may benefit from close collaboration and dialogue between disciplines when it comes to reconsidering their conceptual frameworks. An additional advantage of interdisciplinary work is that we can critically consider the biases outline above (exceptionalism, anthropocentrism, androcentrism) across research fields. This is evidently an ambitious programme. Therefore, we will focus on one particular case study, namely, Neanderthal research.

We can summarise our argument (already outlined above) as follows. Until recently, the logic of binary and hierarchical thinking (humans vs. animals, males vs. women, Greeks vs. barbarians, masters vs. slaves, etc.) was not only noticeable in philosophy, but in Neanderthals research as well. While modern humans were seen as exceptional, Neanderthals tended to be seen as inferior, in the sense of insufficiently human, and the focus was on detecting decisive behavioural, technological and genetic differences between "them" and "us" (Peeters and Zwart 2020). Moreover, whereas modern humans were seen as taking the decisive revolutionary step from nature to culture (with the help of technology, language, art, complex social organisation, etc.), Neanderthals were basically seen as remaining embedded in nature, and as not yet having taken that decisive step. An important decisive factor, defining our humanity, was symbolism, moreover. Although Neanderthals evidently used tools, the difference between them and us was allegedly noticeable in symbolic behaviour (defining symbolic behaviour as that which makes us human, an argument which can already be found in Aristotle, e.g. in the first book of his *Politics*). Last but not least, human exceptionalism was connected with masculinity, based on the idea that the traits that supposedly made us human (or more human than Neanderthals) were notably to be found in males (innovation, competition, intelligence, cooperation, art, craftmanship, etc., cf. Peeters and Zwart 2026). We will argue that, building on recent developments in Neanderthal research, this binary way of thinking has become increasingly untenable at all levels.

First of all, for many of the differences posited (language use, music, burial practices, art, ornaments, etc.), there is mounting evidence of Neanderthals having developed such practices as well. Moreover, in an era of global environmental crisis, we are becoming increasingly conscious of our embeddedness and dependence on nature, notwithstanding our technological prowess, while the idea of human history as an exceptional evolutionary "success story" is under siege. In addition, basic conceptual markers that have framed this discussion so far (e.g. "symbolic behaviour", "modern behaviour", "humanity", and their numerous interconnections) are in need of thorough reconsideration. Also, we are becoming aware of how our understanding of Neanderthals (both in scholarly discourse and in the literary domain) is tainted by gender stereotypes, as indicated for instance by how gender roles are imagined and enacted in deep history novels. Concrete archeological evidence of gender roles in the deep past is sparse (Coltofean-Arizancu et al.

2021), but the discovery of female big-game hunters highlighted uncritical assumptions about past gender roles (Haas et al. 2020) that echoed the "paleo-fantasy" of Man the Hunter (Lacy and Ocobock 2023). That gender roles in history may diverge considerably from persistent assumptions was confirmed by the discovery that the individual buried in a high-status Viking warrior grave was not male—as had been assumed since its excavation in 1878—but female (Hedenstierna-Jonson et al. 2017; Price et al. 2019; Brown 2021).

In the following sections, we will zoom in on the modern humans–Neanderthals encounter: did it take place at all, was it necessarily a violent and competitive event or did it rather involve mutual learning and technological and cultural exchange? We will compare how the modern humans–Neanderthal encounter was envisioned by novelists such as the American writer Jack London (who wrote his Neanderthal classic *Before Adam* more than a century ago) and the Neanderthal best-sellers by Jean Auel, and compare these fictional enactments with emerging insights into the extensive period of co-existence of modern humans and Neanderthals in contemporary archeological research.

14.5 The Beginnings of Neanderthal Fiction

The first deep history novel, entitled *Paris avant les hommes* ("Paris before the arrival of humans"), was written by the French botanist and geologist Pierre Boitard (1789–1859) and published posthumously in 1861 (Boitard 1861). It is a time-travelling story. The narrator, after reading Cuvier, dreams that he is taken back 12,000 years in time by a demon who warns him to leave his preconceptions about human prehistory at home. In the *Grotte de Souvignargues* in southern France, he is confronted with a revolting creature resembling an orangutan. Boitard's intention was to show that our ancestors were not Adam and Eve but ape-like cave dwellers. His book contained a notorious illustration of a hideous caveman standing in front of his cave, loosely dressed in animal skin and wielding a stone axe to protect a frightened cavewoman with child hiding inside.

Dreams also play an important role in the novels of American novelist Jack London (1876–1916) who suggests that there is an alternative pathway to access lost Pleistocene worlds of experience apart from science. Besides archeological research based on excavations, unconscious ancestral memories offer an alternative route. Jack London's novels can be regarded as paleo-archeological experiments with literary means (Zwart 2019). This notably applies to two of his most remarkable novels, namely, *Before Adam* (1907) and *The Star Rover* (1915). Both novels rely on a technique that can be referred to as "active imagination". In *The Star Rover*, the protagonist Darrell Standing introduces himself as a former scientist, a "precise laboratory scientist" employed by the University of California in Berkeley as a "laboratory slave" (p. 16). Running into conflict with a colleague from another university, he kills his competitor in a surge of anger and is sent to the California State Prison at San Quentin for life. Here he proves to be an impossible and "incorrigible" prisoner (p. 15), rebelling against the prison regime and subjected to stringent disciplinary measures, such as spending long hours in isolation, wrapped in a straitjacket. In response, he develops a mental technique, putting himself in a state of trance, which allows him to wander off into distant times and places, entering the consciousness of alter egos who lived in a distant past, exploring vanished worlds of experience in which these alter egos once dwelled.

Initially, Darrell experiences this as a liberating "orgy of the imagination" (p. 83), but gradually his experiences become more concentrated, directed and controlled. He discovers inside himself an "awareness" of other times and places (p. 9), of other persons, of distant memories and previous selves and, in the course of the novel, he becomes increasingly apt at conducting time-travelling "experiments" (p. 79, p. 84, p. 156). His technique is reminiscent of what C.G. Jung refers to as "active imagination" (Jung 1968, p. 96; 1959, p. 356). All human beings have the potential to enter other worlds, Darrell Standing argues, and to experience vistas of other times and places (p. 9), but in normal life we tend to suffer from "forgetfulness" (p. 10). It is only in our dreams that we see "glimpses of other worlds and previous manners of life" (p. 9).

In *Before Adam* the narrator focusses on a particular dreamscape, which he explores systematically and in detail. As long as he can remember, the narrator is tormented by nightmares of fear, exposing him to "a fear of long ago", the fear that "reigned supreme in that period known as the mid-Pleistocene" (p. 9). Initially, he experiences these dreams as a "screaming incoherence" (p. 10), but gradually he discovers how to explore these nocturnal visions in a more systematic manner. During the day, he is a modern city dweller, but in his dreams, he wanders through interminable oak forests. These dream images originate in a previous existence, and he gradually recognises that his dream personality, his alter ego, "lived in the long ago", before humans as we now know them came to be (p. 18). Somehow, our cerebral cells still retain memories of the life once lived by remote ancestors. These reminiscences are "stamped into the stuff of our heredity" (p. 20).

As indicated, the narrator's alter ego lived during "the Mid-Pleistocene" (p. 21), the period of "our becoming" (p. 25), and somehow his experiences are still accessible, although going back to that early world of experience via what seems to be a jumble of nocturnal images requires considerable training and exercise. He decides to take his dreams seriously and discovers that his "uncanny" (p. 21) dream experiences guide him back to a transition stage in human evolution, with early humans dwelling in caves. He discovers that he retains a vision of his ancestor mother, formidable of build, with stout legs, swift, swinging muscles and a furious temper (p. 30), and that he somehow remembers the language spoken in those days, when the vocabulary was limited to a minimal set of sounds, modifiable by intonation, pitch and pantomime (p. 39). But he also remembers how his "horde" of cave dwellers was tyrannised by an enormous aggressive male. Although the horde buried its dead, they had not yet developed any form of cooperative action. They had a capacity for "great laughter" (p. 65), however, and "played along through life" (p. 79). There was no seriousness in them, and they had not yet become sufficiently human-like to work out a problem or to innovate, so that little or no progress was made. They did enjoy collective singing in the sombre twilight of their caves, and they would strike a unanimous rhythm for a moment, but soon the "art nascent" of their primordial chorus would disintegrate into pandemonium (p. 144).

At a certain point, the horde is attacked by what the narrator refers to as the "fire people" (i.e. modern humans), who not only know how to use fire, but are much more advanced socially, while their faces are more pleasing because "their nose-orifices open downward and the bridges are more developed, so that they look less squat". They have speech and co-operation, while their bows and arrows are "enormous extension of their leaping and striking muscles" (p. 154). Indeed, these artificial fangs turn them into "the most terrible of all the hunting animals that ranged the primeval world" (p. 155). They rapidly exterminate the horde, although a small number manages to survive the cataclysm (p. 183). Somehow the narrator's genetic stuff must have mingled with theirs, stamping into "the cerebral constitution of one of his progeny all the impressions of his life", so indelibly that the enormous laps of time failed to obliterate them (p. 187).

Evidently, Jack London's novels, and the genetic and neurophysiological theories on which they build, raise a host of questions, but if we are willing to *suspend our disbelief* for a moment, it is not difficult to read *Before Adam* as a description of an imaginary encounter between early humans, Neanderthals perhaps, apparently not fully human, and modern humans. Thus, we might see *Before Adam* as an early example of a Neanderthal novel, a Neanderthal *classic* if you like. If we subject Jack London's novel to a quick analysis, some features are fairly typical for Neanderthal novels as a genre. First of all, the novel stages an *encounter* between Neanderthals and modern humans and this event (the first encounter as a primal scene) is a basic component in many Neanderthal novels (Peeters et al. 2025). Also, we notice that the Neanderthals as depicted by London as having developed a culture of their own, which includes language, music (cave singing) and burial practices. Yet, compared to the culture of modern humans, Neanderthal culture is presented as deficient. Neanderthals use language, but their vocabulary is limited, while their social organisation is not very advanced (they are a horde rather than a tribe). Their faces are not quite human, moreover, and although they developed primitive tools, advanced human artifacts (e.g. arrows) are unknown to them. The encounter is sudden, violent and short-lived, moreover. The Neanderthals are eliminated or expelled, although some interbreeding must have taken place. The Clan People (Neanderthals) were in the process of becoming human, but the Fire people (modern humans) are depicted as far superior in terms of language, tool-use and cooperation, while Neanderthals represent an evolutionary dead end (Drell 2000).

This is how Jack London imagined the Neanderthals–modern humans encounter, more than a century ago. Let us now briefly compare his version with the image of this encounter as it emerges from paleo-archeological research, notably as it has evolved during recent years. First of all, although archeologists still devote multiple debates to distinctive differences between Neanderthals and modern humans, these differences tend to be seen as considerably less dramatic nowadays than London depicted them. In addition, increasingly often, they are no longer framed in such a way that Neanderthals are seen as deficient, as not yet human. And a third striking tension between London's early twentieth-century fantasy and contemporary scholarly discourse concerns the encounter as such. Instead of seeing it as a single, dramatic and violent event, with our ancestors in the role of invading predators and Neanderthals as inferior victims, scientific research suggests that Neanderthals and modern humans co-existed in what is now Europe for thousands of years, while during this period, various kinds of interaction and exchange (ranging from technologies to genes) took place. Let us have a closer look.

14.6 The Encounter According to Current Paleo-archeological Research

Neanderthals were Pleistocene hunter-gatherers who lived in (western) Eurasia between ~400,000 and 40,000 years ago (Roebroeks and Soressi 2016). Hundreds of thousands of years of relative separation between African and Eurasian humans led to the emergence of different phenotypes, yet Roebroeks and Soressi argued that genetic evidence and archeological data suggest that the biological and cultural gaps between Neanderthals and modern humans were smaller than previously thought. The cultural capacities of the late Middle and early Late Pleistocene populations in Africa and Eurasia were remarkably comparable (Villa and Roebroeks 2014). It is only around 40 ka, tens of thousands of years after the modern human phenotype appeared outside of Africa, and thousands of years after the first interbreeding between modern humans, Neanderthals and Denisovans occurred, that we see a major change in the archeological record. Explaining this transformation solely in linear terms, in the sense that innovative and advanced technocomplexes are consistently attributed to invading modern humans, must give way to a more fluid and interactive understanding of this period of transition.

Although the disappearance of Neanderthals during the Upper Paleolithic coincided with the spread of modern humans, Neanderthal presence persisted in what is now Europe for several millennia after modern humans entered Europe from the East. And rather than a clear Neanderthal–modern humans divide, research suggests a "patchwork" of human populations and technocomplexes in Europe during a transitional period which lasted for thousands of years (Mylopotamitaki et al. 2024), although debates concerning whether Neanderthals were "replaced" or rather "assimilated" by modern humans are still ongoing (Zilhão et al. 2024). The dispersal of modern humans occurred during a "transition" period which lasted a few thousand years, eventually resulting in the replacement of indigenous Neanderthals around 40,000 years ago (Abrams 2023). We emphasise, however, that, although his transition period, during which Neanderthals and modern humans coexisted in Europe, coincided with an epoch of technological transformation, it would be a simplification to say that this was solely due to innovations introduced by modern humans. Archeological evidence suggests a more complex dynamics of transition.

The Châtelperronian industry (emerging at the end of the Middle or the beginning of the Upper Paleolithic) was especially interesting because, whereas traditionally Neanderthals were often seen as "conservative" compared to "innovative" modern humans, Châtelperronian industries demonstrate Neanderthal capacity for creativity, innovation and change (Soressi and Roussel 2014). The traditional view among researchers was more or less "London-like" in the sense that the arrival of modern humans in Europe allegedly resulted in the demise of Neanderthals due to modern humans' technical superiority (e.g. London's 'arrows'), but according to Roussel and Soressi almost every behaviour previously thought to be unique to modern humans was shared, at least occasionally, by Neanderthals—although they also notice a tendency to interpret this period in terms of a contrast between Neanderthals and modern humans, rather than as a unique Neanderthal culture, a tendency which concurs with how the Neanderthal–modern humans encounter is often depicted in popular sources (p. 2680). Since 2014, however, more findings and interpretations have become available that allow us to further develop our understanding of this intriguing epoch in early human history. While concrete evidence of interaction between Neanderthals and modern humans is sparse, the scenario of a sudden emergence of a "modernity package" is giving way to seeing early human history as an intricate mosaic of multiple and temporally variable pathways of techno-cultural practice (Scerri and Will 2023). Or, as Finlayson et al. (2023) formulate it: traditional narratives about human migrations and the biological and/or cultural advantages of our own species over the Neanderthals are now giving way to the study of the biological and cultural dynamics of past human populations and the nature of their interactions in time and space.

14.7 Shifting Scenarios

As indicated earlier, during the 1980s, the arrival of modern humans in Europe and the concurrent demise of Neanderthals was viewed as a "revolutionary" event, characterised by the rapid and straightforward replacement of one species (Neanderthals and their Middle Paleolithic stone tools) by another (modern humans and their Upper Paleolithic tool kit) (Mellars and Stringer 1989). However, already by the late 1970s, the discovery of Neanderthal remains in a layer with Upper Paleolithic stone tools (Lévêque and Vandermeersch 1980) suggested a much more complex history, possibly pointing toward "acculturation" of Neanderthals under the influence of modern humans before their demise (Demars and Hublin 1989). In the early 1990s, new dating techniques revealed that some of the latest Neanderthals were indeed contemporaneous to the first modern humans in Europe (Mercier et al. 1991). On the basis of these data, scientists cautiously concluded that "the possibility of contact between the West European Neanderthals and the intrusive modern humans who replaced them [could not] therefore be excluded" (Mercier et al. 1991: 737).

Since then, efforts have continued to further our understanding of the duration of coexistence of Neanderthals and modern humans and to identify biological and cultural changes in the last Neanderthals that might reflect hybridisation and interactions (Stringer and Grün 1991). In 1996, the attribution of a piece of skull found in a layer with personal ornaments (perforated and grooved mammal teeth) to Neanderthals confirmed that they were using decorative objects (Hublin et al. 1996), possibly due to trading or imitation. Although this could be interpreted as empirical evidence for contact and acculturation, Neanderthals could also have independently invented and manufactured these beads (d'Errico et al. 2005; Hublin 2007). In the early 2000s, the analysis of one of the oldest modern human skulls found in Europe revealed a mosaic of morphological traits suggesting possible interbreeding with Neanderthal, though not conclusively (Trinkaus et al. 2003).

14.8 DNA Sequencing: How Currently Living People Inherited Genes from Long-Demised Neanderthals and Denisovans

In 2010, the advent of high-throughput sequencing enabled the extraction and successful sequencing of nuclear DNA from Neanderthal bones (Green et al. 2010). This breakthrough showed that genomes of modern humans living outside of Africa contain a few per cent Neanderthal DNA, indicating that modern humans interbred with Neanderthals more than 40,000 years ago. Science journalist Rex Dalton concluded in *Nature* that this discovery significantly changed our views on humanity (Dalton 2010). Subsequent years provided further confirmation through in-depth analyses of living humans (Vernot and Akey 2014; Sankararaman et al. 2014) and the sequencing of early modern human remains (Hajdinjak et al. 2021), including the remains of a 40,000-year-old modern human with Neanderthal ancestry, possibly as close as a great-great-grandparent (Fu et al. 2014). Science journalist Ewen Callaway summarised the results by stating that "our inner Neanderthal was revealed" (Callaway 2014). In 2018, the sequencing of DNA from a hybrid individual with a Neanderthal mother and a Denisovan father (Slon et al. 2018) highlighted the interbreeding between Neanderthals and other hominins besides modern humans as well. Denisovans are a population of extinct humans who lived in Asia from approximately 300,000 to circa 25,000 years ago (Brown et al. 2022). Recently, it was confirmed that intermingling between modern humans and Neanderthals may have occurred as early as 200,000 years ago, with another major period of mixing around 100,000 years ago (Li et al. 2024).

The past fifteen years have seen numerous discoveries that have reshaped our view on the origin of our genetic makeup. These discoveries were triggered by technical advancements—specifically the development of high-throughput sequencing—and the dedicated work of scholars like Svante Pääbo and his team at the Max Plank Institute in Leipzig, Germany. Pääbo received the Nobel Prize in Physiology and Medicine in 2022 for his work, which highlighted the lasting impact of Neanderthal (and other fossil hominin) DNA on the immune systems of modern humans. Initially, genetic studies focused on Neanderthal bones, and then broadened to early human remains older than 40,000 years (Peyrégne et al. 2024). Yet, while evidence for the mixing of different hominin populations in the deep past has become abundant over the past fifteen years, evidence for cultural interactions remains elusive. Chu et al. (2024) for instance report that in present Romania, bones of modern humans with recent Neanderthal ancestry have been found, while there is no evidence of interaction in tool use.

14.9 Cultural Legacies and Exchanges

Over the past decade, improvements in radiometric dating have enabled more accurate dating of layers attributed to the last Neanderthals and the earliest modern humans in Europe. The latest evaluations suggest that Neanderthal and modern human presence in western Europe overlapped between one thousand and three thousand years before the Neanderthals disappeared (Djakovic et al. 2022). Recent discoveries also indicate that some modern humans could have reached France as early as 50,000 years ago (Slimak et al. 2022) and Greece as early as 200,000 years ago (Harvati et al. 2019). Despite this temporal potential for cultural interactions, evidence for interactions remains limited so far.

In 2013, one of us (M.S.) and colleagues published the discovery of bone tools known as *lissoirs* and used to soften hides in layers associated with Neanderthals and dated to circa 50,000 years ago (Soressi et al. 2013). Interestingly, similar bone tools are still used today by humans to work leather, representing some of the oldest evidence of a technology invented long ago that remains in use (Callaway 2014). These *lissoirs* were the most ancient ever discovered. Since they had not been found in Africa or in the Near-East but became common in western Europe by 35 ka, the authors suggested they might be the first evidence of Neanderthal technology transmitted to modern humans. However, similar objects, albeit broader and *spatula-like*, were later discovered in North Africa in layers dated to 120,000 to 90,000 years ago (Hallett et al. 2021) and in Eastern Europe in layers younger than 50,000 years (Martisius et al. 2022), suggesting that similar tools may also have been independently invented by modern humans. The hypothesis of Neanderthals inventing lissoirs and transmitting them to modern humans has neither been contradicted nor reinforced in recent years, indicating the need for further study to flesh out a more complex scenario than the one originally suggested. Still, we notice a shift of emphasis from the late 1980s view of Neanderthals being "acculturated" by modern humans to considering how modern humans may also have been influenced by Neanderthals and how population contact was a driver of cultural change at the time of the Neanderthal demise (Greenbaum et al. 2019; Creanza et al. 2017).

In 2016, Marie Soressi and colleagues also published stone tools found in layers then associated with the latest Neanderthals in France. These tools mimicked the shape of those used by modern humans in the same area during that time period. However, the manufacturing process for these Neanderthal tools differed from that of the modern human tools, suggesting superficial contact between the two groups (Roussel et al. 2016). Despite the extended interaction period in Western Europe, more evidence of interactions has yet to be found. One possibility is that it has not been researched extensively enough—the theoretical framework for identifying cultural interaction in the Pleistocene archeological record is still in its infancy (Tostevin 2000, 2007; Gravina et al. 2022). To further complicate matters, several scholars have suggested that these types of assemblages may not have been produced by Neandertals, but by Homo sapiens instead or a mixture of both (see Djakovic et al. 2022). Yet, compared to research in the past, the impact of

Neanderthals–modern humans encounters is now symmetrically considered in both archeology and paleo-genomics, reflecting a more balanced view of the cultural exchanges between these ancient populations. Another difference is that the perception of Neanderthals has evolved from that of an extinct and inferior species to the view of Neanderthals as humans with behaviours as sophisticated and unique as their modern human counterparts (Villa and Roebroeks 2014; Roebroeks and Soressi 2016; Zilhão et al. 2010; Hoffmann et al. 2018). For instance, Neanderthals likely invented the first stone strike-a-lights (Sorensen et al. 2018), discovered the fire-starting chemical properties of manganese dioxide powder (Heyes et al. 2016) and operated in large social units for the butchering of large animals as early as 125,000 years ago (Gaudzinski-Windheuser et al. 2023). Thus, instead of a one-sided cultural transmission from modern humans towards Neanderthals, there is some evidence demonstrating independent innovation by Neanderthals or cultural diffusion from Neanderthals to modern humans. Last but not least, the collaboration between paleo-genomicists and archeologists is intensifying, which may allow for a more refined understanding of interactions in the deep past.

If we compare current paleo-archeological research with previous approaches focussing on modern human superiority (a focus that was amplified in *Before Adam* as an early Neanderthal novel), we may conclude that the traditional understanding of the Neanderthal–modern humans encounter is being revisited. Rather than a single, revolutionary event, resulting in the elimination of "inferior" Neanderthals by "superior" modern humans, communities of early humans co-existed and interacted with one another in what is now Europe for thousands of years. And rather than seeing their interaction as unidirectional (explaining the technological revolution that occurred during this period solely in terms of Neanderthals adopting "superior" industries developed by modern humans), there is ample evidence for Neanderthal dexterity and creativity as well, and the traffic of ideas may also have occurred the other way around.

14.10 More Recent Literary Versions of the Encounter

A similar trend, towards a more dynamic and interactive view of the coexistence of Neanderthals and modern humans, can be discerned in Neanderthal novels as well, although hierarchical and binary scripts (pitting an inferior "them" against a "superior" us) still prove quite tenacious. As indicated earlier, both in research and in literary fiction, the tendency had been to depict Neanderthals as "not yet us", which meant that they offered a way of defining ourselves by contrasting "us" with "them" (Peeters and Zwart 2020). This also tended to be the focus of attention in literary visions of the encounter of (late) Neanderthals and (early) modern humans. The idea that our ancestors once were face to face with a more "primitive" human species has long captured the imagination of archeologists, novelists, filmmakers and broader audiences alike, although this fascination tended to reflect a progressivist view on history (the encounter occurred when the Neanderthal chapter was about to be closed while the advance of *Homo sapiens* would continue). Many works of fiction revolving around the encounter are stories of competition, of winners and losers, where modern humans emerge as heroic victors (Hackett and Dennell 2003). This is a common narrative in stories about human evolution (Landau 1993). Apparently, they appeal to our sense of progress and affirm our sense of exceptionality and superiority.

This scenario is also quite evident for instance in the 2018 BBC documentary *Neanderthals: Meet Your Ancestors*, staging a *fight club* where a Neanderthal male has to compete with a modern human male in a life-or-death combat. The documentary starts by explaining that Neanderthals were not the brutish ape-men of popular imagination, but "surprisingly" like us. The combat that follows is nonetheless a combat between unequally equipped males, between brawns (them) and brains (us), where our cognitive advantage allows the victorious human male to deliver the knock-out blow. This example also emphasises how the alleged superiority of modern humans is closely connected with stereotypically masculine traits (e.g. innovation, expansion, competition, aggression). This heroic identity is not only problematic from a scientific point of view, where archeology offers evidence of coexistence, interaction and admixture, but it also reinforces a one-sided and stereotypical view of what it means to be human.

More recent Neanderthal novels often try to offer an alternative view. Interestingly, although the first articles providing evidence for long-term coexistence and interactions between early Homo sapiens and late Neanderthals in Europe were published during the 1990s (e.g. Hublin et al. 1996), already in the 1980s two novels appeared that explored this coexistence, namely, *Dance of the Tiger* by paleontologist Bjorn Kurtén (Kurtén 1980/1995) and Jean Auel's world-famous *The Clan of the Cave Bear* (Auel 1980). In both novels, Neanderthals are comparable to modern humans, although Auel's Neanderthals are doomed to extinction because of their conservatism.

With the modern human protagonist Ayla, Jean Auel created a heroine who has many of the qualities traditionally associated with male heroes, e.g. independence, physical strength, a gift for innovation and intelligence.[3] However, precisely these qualities are also the reason why she finds it difficult to adapt to the traditionalist expectations of her adoptive Neanderthal clan. Physically and intellectually, Ayla (an athletic beauty, both feminine and strong) is more evolved than the Neanderthal Clan people and struggles to live up to the Clan's rigid gender norms. Being more emancipated than the Neanderthal clan, she experiences the Neanderthal clan culture as claustrophobic and oppressive. Neanderthals have a different brain capacity and

[3] In an interview, Auel (1986) explains that she wanted to write about "a heroine that I didn't have when I was growing up, because when I read books very often, if there were woman, or girls in them, they were kind of sitting around in the background being Heroines! Waiting for someone to come and save them." Likewise in books about cave men, feminine characters "were always in the background". The androcentricity of Western culture entails that the masculine (with a focus on autonomy and rationality) is seen as universal, as the standard, while the feminine (with a focus on emotion, the body and relationships) is traditionally constructed in a hierarchical opposition to masculinity, also in novels.

find it difficult to learn new things. Males are in charge, and women are submissive by nature. Contrary to Jack London's version, Neanderthals neither laugh nor cry. They do not understand "abstractions" and can hardly count. Cave males dislike Ayla (the rebel) because she is seen as one of the "Others" (the Cave term for modern humans), no longer constrained by the traditions wired into the Neanderthal brain ("a brain that was nearly all memory"). Her forward-thinking frontal brain provides her with insight and allows her to forge ideas which Clan members never even dreamt of. She represents a new manner of being woman and being human, destined to supplant "the ancient, dying race", as is the way of nature (p. 162). Although her first novel in many ways continues to reflect the "conservative Neanderthals" versus "innovative human" scenario, the contemporaneity of Neanderthals and modern humans remains a key ingredient of Auel's later novels, which are informed by more recent archeological findings and interpretations.

14.11 Jean Auel and the Châtelperronian Novel

Like other Jean Auel Neanderthal novels, *The Shelters of Stone* is set in a Châtelperronian landscape precisely during this archeological period, when the presence of protective limestone shelters made this part of present-day France "among the most heavily populated in all that cold ancient world" (Auel 2002/2011, p. 136). Rather than on active imagination, as in the case of Jack London, Jean Auel relies on paleo-archeological expertise to gain access into the Châtelperronian world where modern humans and Neanderthals encountered one another. She mentions the support of several archeologists in her *Acknowledgements,* and she also mentions *The Shamans of Prehistory* by Jean Clottes and David Lewis-Williams (1998), who argued that polychrome Cro-Magnon cave paintings were part of Shamanistic rituals. After reading papers and manuscripts by professional archeologists and visiting sites, Auel relied on her imagination to fill in the blanks.

To a considerable extent, Auel's description of modern humans (the Zelandonii) and Neanderthals (the Clan people) continues to adopt a "them" versus "us" approach similar to what happens in her first novel and in archeology during the reign of the human revolution model. Zelandonii (modern humans) have fair skins and blue eyes, and the women decorate their bodies and faces with charcoal and ochres in front of polished wooden mirrors (2002/2011, p. 117), while Neanderthals are brown-eyed and dark, with prominent brow ridges and noses, and are referred to by the Zelandonii as "flatheads". While Neanderthals live in small groups (clans), the Zelandonii dwellings are much more populated, and their crafts include "the production of small sculptures [that] seemed to be made for themselves" (p. 15), resulting in figurines and necklaces, a novelty for Ayla (the modern human female protagonist who was raised by a Neanderthal clan), while modern human tools and sculptures are "carved with symbols" of animals and plants. Whereas the Zelandonii worship the Great Earth Mother, Neanderthals are devoted to Ursus, a mythical Cave Bear, supposedly a more primitive cult. And while Zelandonii women play a prominent role in the daily life of the tribe, Clan women "quietly fade into the background and seem to disappear" (p. 113). Finally, while Zelandonii communicate by means of language, Neanderthal Clan people only use a limited number of "very ancient words" (throaty, guttural sounds pronounced with a low-pitched voice), combining them with onomatopoeic phrases, signs, gestures, facial expressions and other forms of non-verbal communication (p. 182 and elsewhere).

Gradually, however, the difference becomes less hierarchical. Throughout the novel, there is a discussion among the modern humans whether "flatheads" are animals or humans, and the consensus gradually slides towards the latter. Neanderthals bury their dead and their storytelling actually proves to be quite intricate and advanced. Their signs and gestures evolve into a form of dance (p. 71, p. 80). Their culture is different (they "talk with their hands", p. 71), but not necessarily inferior to modern human communication. Whereas Neanderthal traditions change slowly (p. 79), they do change, and while both modern human and Neanderthal women have acquired medicinal knowledge, only modern humans (notably males) are given to a rather bad habit: brewing and drinking alcoholic beverages. Still, modern humans definitely have an advantage when it comes to the use of art (paintings, carvings, decorations) and the development of new crafts and skills. Contrary to humans, the Clan people create "no legacy of art" (p. 201).

The most decisive difference between the Pleistocene worlds as depicted by Jack London and Jean Auel, however, is the temporal dimension. Modern humans and Neanderthals share the same landscape for extended periods of time. They co-exist and interact, via trading, occasional encounters, mutual learning and sexual intercourse.

In the final book of the series, *The Land of the Painted Caves* (Auel 2011), Ayla (who already had a reputation as a horse, wolf and cave lion whisperer) becomes the Zelandonii shaman and in this role, she embarks on a pilgrimage to visit sacred caves (Cro Magnon painted caves in Southern France) as part of her training to become a spiritual leader. A hallucinogenic liquid sends her on a three-day psychedelic trip that ends in one of these caves, where she experiences her calling by the Great Earth Mother. This gift of knowledge changes the scene, however, because due to Ayla's insights, human males become aware of their role in procreation, which leads to the collapse of matriarchy and the rise of patriarchy, while Neanderthals (Clan people) end up at the wrong side of history and are bound for extinction.

Auel's novels could be seen as intermediary novels, still sharing much of the "them" versus "us" logic so characteristic of London's classic, while on the other hand depicting Neanderthals as humans. More recent novels move further away from the logic of human (masculine) exceptionalism and tend to focus even more on issues of connection, interaction and vulnerability. An interesting example is *The Last Neanderthal* by Claire Cameron (2017), a story of two women separated by millennia (a Pleistocene Neanderthal woman and a modern archeologist) but connected by shared experiences of early motherhood. The story follows Girl and her Neanderthal family as they struggle to survive, and archeologist Rosamund Gale who, living in the present, excavates Neanderthal remains while being pregnant. Rather than reinforcing the "them" versus "us" divide, the novel enacts communalities and encounters between women then and now, bypassing fantasies of sex and war while emphasising empathy and vulnerability.

Ghost Species by James Bradley (2020) likewise questions long-standing assumptions about human exceptionalism. Here, de-extinction of Neanderthals with the help of paleo-DNA becomes a core theme. What does de-extinction imply for our existential loneliness as modern humans, in a world where so much is lost? Set against the backdrop of a rapidly escalating climate catastrophe, scientists and partners Kate Larkin and Jay Gunesekera are recruited by tech billionaire Davis Hucken to explore the forests of Tasmania. Davis has ambitious plans to save the planet by reintroducing long extinct creatures such as thylacines and mammoths, thus reversing the damage done by modern humans in the course of deep history, and this includes a secret program to recreate Neanderthals with the help of paleo-DNA. By focussing on Kate and Eve, the child that is born, the novel focusses on compassion and connectedness. In short, both in contemporary scientific discourse and in recent Neanderthal novels, we notice a shift from reinforcing the "Neanderthal versus modern humans" divide towards a more inclusive emphasis on connectedness and interaction.

14.12 Conclusion

Our understanding of early human evolution is changing, and this has repercussions for how we see ourselves as human beings. Although we are now the only humans on earth, our story is part of a broader history, where the extended period of contemporaneity of Neanderthals and modern humans emerges as an epoch of special significance. As we argued in a previous publication (Peeters and Zwart 2020), Neanderthals are more like us than other early humans, but the guiding conviction was that they somehow missed something "we" (our early ancestors) allegedly had. Precisely this (genetic, behavioural and cultural) difference allegedly confirmed our exceptionality as modern humans. The overall trend was to frame modern humans as favoured evolutionary "winners", whose alleged "superiority" led to the demise of Neanderthals (Villa and Roebroeks 2014). The human revolution was attributed exclusively to the innovativeness of modern humans, and especially of males.

As indicated, research in paleo-archeology is rapidly questioning this scenario. Neanderthals and modern humans co-existed in what is now Europe for an extended period of time, and more data concerning interactions and transmissions are becoming available. Whereas in traditional novels encounters between Neanderthals and modern humans tended to be depicted as violent and brutal, resulting in the elimination of the culturally and biologically "inferior" Neanderthals by "superior" modern humans, more recent novels put more emphasis on contemporaneity, interaction, creativity and mutual exchange, although in the Neanderthal novels by Jean Auel (the most popular Neanderthal novels perhaps), Neanderthals are still depicted as people with limited brain capacity, conservative and narrow-minded rather than innovative and emancipatory, and therefore bound to extinction.

Images of Neanderthals in novels and in scholarly discourse evolve in parallel, but in an interactive manner. While novelist Jean Auel (2002/2011) for instance acknowledges the support received from several archeologists in her *Acknowledgements*, Rebecca Wragg Sykes emphasises in her book *Kindred* (2020) that it is precisely the interaction between imaginative and academic interpretations that has revolutionised our understanding of the vanished world of Neanderthal culture, convincing us that Neanderthal culture wasn't a "dead end", even though the question whether apparently artistic Neanderthal remnants (Bruniquel etc.) can be considered as art the way *we* understand it continues to be a matter of debate.

Both in archeology and in Neanderthal novels, we notice the need for a more inclusive perspective on early human history, where cultural, technical and behavioural similarities and differences are studied in a non-hierarchical manner. Rather than on Neanderthal extinction (caused by their alleged cognitive disabilities and behavioural deficits, compared to modern humans), the focus now centres on their extended history and presence: on how Neanderthals *lived*; on their manual dexterity, technical competence and cognitive versatility (Carrión et al. 2019). This will foster a non-hierarchical conception of difference and a more inclusive and fluid conception of human identity, no longer dependent on the disqualification of otherness. In this manner, paleo-archeology becomes cultural anthropology of the past, sensitive to diversity and difference, but refraining from assessing such differences as "inferior" or "superior" on the basis of resemblance with modern human behavioural patterns. The emergence of modern symbolic behaviour is a gradual and multifaceted process involving a broad variety of expressions across space and time. Stability or a slower pace of change has often been interpreted as a lack of progress or capacity for innovation and has even been equated with backwardness (Montón-Subías 2024), but stability may well have been a strength under Pleistocene circumstances. The predilection for innovation reflects assumptions about masculinity and Enlightenment (suggesting that "progress", i.e. change, is good by definition) that are open for reconsideration, while putting more emphasis on the importance of cooperation, interdependence, bonding and care (Montón-Subías and Hernando 2018) rather than competitiveness.

This shift from exceptionalism towards a more inclusive view has conceptual repercussions as well, notably for concepts such as humaneness. Neanderthals were humans. Their manner of being in the world, their language and their crafts may have differed from those of modern human communities, but not in the sense of being deficient, given that Neanderthals were able to maintain themselves in challenging Pleistocene environments for extended periods of time, relying on their cultural and technical dexterity. Moving away from the idea of sudden and brutal encounters, more attention is now given to the *contemporaneity* of Neanderthals and modern humans in the immense natural landscapes of the Pleistocene (Conard 2006). This will also allow us to move away from the traditional androcentric focus on "masculine" characteristics, e.g. violence, competition and expansion. The title *Before Adam* as such is already symptomatic of the focus on masculinity in London's novel, seeing the Pleistocene world as dominated by aggressive males. Research into what actually was an extended period of co-existence (referred to here as the Châtelperronian period) opens up a different cultural landscape, with diverse but coexisting and interacting technocomplexes.

There is an additional reason, albeit a somewhat speculative one, why Neanderthal research and Neanderthal novels draw so much attention among scholars and the broader public alike. Both the late Pleistocene past and the Anthropocene present are precarious periods of crisis and transition. The current interest in Neanderthal extinction in deep history may resonate with concerns about our own future. The possibility of modern humans becoming extinct is widely discussed from multiple perspectives, and perhaps we would like to consult them. A much stronger awareness of our human embeddedness in and dependence on nature is something we might learn from them.

While we increasingly see Neanderthals as human, for us the question is not whether Neanderthals should be "out" or "in". Rather, we advocate a more fluid and dynamic understanding of the concept of humanity and related concepts such as symbolic behaviour and modern behaviour, encompassing various manners of being human evolving in a Pleistocene ambiance. Thus, Neanderthal research provides us with an opportunity to reconsider our basic categories and frameworks for self-reflection. Besides a shift of focus from competitiveness to cooperation and dependence, future research will need to reconsider our definitions of humanness (which doesn't necessarily mean being comparable to contemporary humans). And while most Neanderthal findings come from what is now Europe, the abundance of insights from Africa and Asia must be incorporated more systematically as well, as the focus on Europe as the culmination point of history echoes Enlightenment thinking. The Châtelperronian for instance refers to a relatively well-studied but specific area within the immense Neandertal world that reached from Gibraltar up to the Altai Mountains.

Acknowledgements This publication is part of the project entitled Neanderthals and "us": how the golden age of Neanderthal research challenges human self-understanding with file number 406.21. FHR.011 which is financed by the Dutch Research Council (NWO).

References

Abrams G (2023) Close encounters of the third kind? Neanderthals and modern humans in Belgium, a bone story. Retrieved from https://hdl.handle.net/1887/3618716
Auel J (1980) The clan of the cave bear. Crown, New York
Auel J (1986) Jean Auel – interviews and memories. Don's Maps. https://www.donsmaps.com/auel.html. Accessed 13 Oct 2023
Auel J (2002) The shelters of stone. Bantam Books, New York
Auel J (2011) The land of the painted caves. Crown, New York
Bergström A, Stringer C, Hajdinjak M et al (2021) Origins of modern human ancestry. Nature 590: 229–237. https://doi.org/10.1038/s41586-021-03244-5
Boitard P (1861) Etudes antédiluviennes. Paris avant les hommes: l'Homme fossile, etc. Passard
Bowden M (1991) Pitt rivers: the life and archeological work of Lieutenant-General Augustus Henry Lane Fox Pitt Rivers. Cambridge University Press

Bradley J (2020) Ghost species. Hodder Studio, London
Brown NM (2021) The real Valkyrie: the hidden history of Viking warrior women. St. Martin's Press, New York
Brown S, Massilani D, Kozlikin MB et al (2022) The earliest Denisovans and their cultural adaptation. Nat Ecol Evol 6:28–35. https://doi.org/10.1038/s41559-021-01581-2
Bushozi PM (2020) Middle and later stone age symbolism: stone beads from Mumba rock-shelter in northern Tanzania. Utafiti 15:1–27. https://doi.org/10.1163/26836408-15010020
Callaway E (2013) Neanderthals made leather-working tools like those in use today. Nature. https://doi.org/10.1038/nature.2013.13542
Callaway E (2014) Modern human genomes reveal our inner Neanderthal. Nature. https://doi.org/10.1038/nature.2014.14615
Cameron C (2017) The last Neanderthal. Little, Brown and Company, New York
Carrión JS, Lalueza-Fox C, Stewart J (2019) Neanderthals: ecology and evolution. Q Sci Rev 217: 1–6. https://doi.org/10.1016/j.quascirev.2019.05.008
Cartmill M (1990) Human uniqueness and theoretical content in paleoanthropology. Int J Primatol 11:173–192. https://doi.org/10.1007/BF02192868
Chu W, Doboş A, Soressi M (2024) Interactions in bones but not stone: anomalous cultural transmission gaps in Romania's Middle to Upper Paleolithic Transition. Q Sci Rev 329: 108546. https://doi.org/10.1016/j.quascirev.2024.108546
Coltofean-Arizancu L, Gaydarska B, Matić U (eds) (2021) Gender stereotypes in archeology. A short reflection in image and text. Sidestone Press, Leiden
Conard NJ (ed) (2006) When Neanderthals and modern humans met. Kerns Verlag, Tübingen
Corbey R (2005) The metaphysics of apes: negotiating the animal-human boundary. Cambridge University Press, Cambridge
Creanza N, Kolodny O, Feldman MW (2017) Greater than the sum of its parts? Modelling population contact and interaction of cultural repertoires. J R Soc Interface 14:20170171. https://doi.org/10.1098/rsif.2017.0171
d'Errico F, Henshilwood C, Vanhaeren M, van Niekerk K (2005) Nassarius kraussianus shell beads from Blombos Cave: evidence for symbolic behaviour in the Middle Stone Age. J Hum Evol 48: 3–24. https://doi.org/10.1016/j.jhevol.2004.09.002
Dalton R (2010) Ancient DNA set to rewrite human history. Nature 465:148–148. https://doi.org/10.1038/465148a
Darwin C (1985) [1859] On the origin of species by means of natural selection or the preservation of favoured races in the struggle for life. Penguin, Harmondsworth
Demars P-Y, Hublin J-J (1989) La transition Néandertaliens/Hommes de type moderne en Europe occidentale: aspects paléontologiques et culturels. L'homme de Néandertal 7:23–37
Djakovic I, Key A, Soressi M (2022) Optimal linear estimation models predict 1400–2900 years of overlap between Homo sapiens and Neandertals prior to their disappearance from France and northern Spain. Sci Rep 12:15000. https://doi.org/10.1038/s41598-022-19162-z
Donald M (1993) Origins of the modern mind: three stages in the evolution of culture and cognition. Harvard University Press
Drell JRR (2000) Neanderthals: a history of interpretation. Oxford J Archaeol 19:1–24. https://doi.org/10.1111/1468-0092.00096
Finlayson C, Brown K, Blasco R et al (2012) Birds of a feather: neanderthal exploitation of raptors and corvids. PLoS One 7:e45927. https://doi.org/10.1371/journal.pone.0045927
Finlayson C, Zollikofer C, Ponce de León M et al (2023) Close encounters vs. missed connections? A critical review of the evidence for Late Pleistocene hominin interactions in western Eurasia. Q Sci Rev 319. https://doi.org/10.1016/j.quascirev.2023.108307
Foucault M (1969) L'archéologie du savoir. Gallimard, Paris
Fu Q, Li H, Moorjani P et al (2014) Genome sequence of a 45,000-year-old modern human from western Siberia. Nature 514:445–449. https://doi.org/10.1038/nature13810

Gaudzinski-Windheuser S, Kindler L, MacDonald K, Roebroeks W (2023) Hunting and processing of straight-tusked elephants 125.000 years ago: implications for Neanderthal behavior. Sci Adv 9. https://doi.org/10.1126/sciadv.add8186
Gehlen A (1964) Urmensch und Spatkultur: philosophische Ergebnisse und Aussagen. Athenaum Verlag, Frankfurt am Main
Geneste J-M, Grosos P, Valentin B (eds) (2023) Préhistoire. Nouvelles frontières. Éditions de la Maison des sciences de l'homme, Paris
Gravina P, Ludovisi A, Moroni B et al (2022) Geochemical proxies and mineralogical fingerprints of sedimentary processes in a closed shallow lake basin since 1850. Aquat Geochem 28:43–62. https://doi.org/10.1007/s10498-022-09403-y
Green RE, Krause J, Briggs AW et al (2010) A draft sequence of the Neandertal genome. Science 328:710–722. https://doi.org/10.1126/science.1188021
Greenbaum G, Friesem DE, Hovers E et al (2019) Was inter-population connectivity of Neanderthals and modern humans the driver of the Upper Paleolithic transition rather than its product? Q Sci Rev 217:316–329. https://doi.org/10.1016/j.quascirev.2018.12.011
Grosos P (2023) La philosophie au risque de la préhistoire. Editions du Cerf, Paris
Haas R, Watson J, Buonasera T et al (2020) Female hunters of the early Americas. Sci Adv 6: eabd0310. https://doi.org/10.1126/sciadv.abd0310
Hackett A, Dennell R (2003) Neanderthals as fiction in archeological narrative. Antiquity 77:816–827. https://doi.org/10.1017/S0003598X00061755
Hajdinjak M, Mafessoni F, Skov L et al (2021) Initial Upper paleolithic humans in Europe had recent Neanderthal ancestry. Nature 592:253–257. https://doi.org/10.1038/s41586-021-03335-3
Hallett EY, Marean CW, Steele TE et al (2021) A worked bone assemblage from 120,000–90,000 year old deposits at Contrebandiers Cave, Atlantic Coast, Morocco. iScience 24. https://doi.org/10.1016/j.isci.2021.102988
Harvati K, Röding C, Bosman AM et al (2019) Apidima Cave fossils provide earliest evidence of Homo sapiens in Eurasia. Nature 571:500–504. https://doi.org/10.1038/s41586-019-1376-z
Hedenstierna-Jonson C, Kjellström A, Zachrisson T et al (2017) A female Viking warrior confirmed by genomics. Am J Phys Anthropol 164:853–860. https://doi.org/10.1002/ajpa.23308
Hegel GWF (1969) Vorlesungen über die Philosophie der Religion I, Werke 16. Suhrkamp, Frankfurt am Main
Hersh TA, Gero S, Rendell L et al (2022) Evidence from sperm whale clans of symbolic marking in non-human cultures. Proc Natl Acad Sci 119:e2201692119. https://doi.org/10.1073/pnas.2201692119
Heyes PJ, Anastasakis K, de Jong W et al (2016) Selection and use of manganese dioxide by Neanderthals. Sci Rep 6:22159. https://doi.org/10.1038/srep22159
Hoffmann DL, Angelucci DE, Villaverde V et al (2018) Symbolic use of marine shells and mineral pigments by Iberian Neandertals 115,000 years ago. Sci Adv 4:eaar5255. https://doi.org/10.1126/sciadv.aar5255
Hublin J-J (2007) What can Neanderthals tell us about modern human origins. In: Mellars P, Boyle K, Bar-Yosef O, Stringer C (eds) Rethinking the human revolution: new behavioural and biological perspectives on the origin and dispersal of modern humans. McDonald Institute Monographs, Cambridge, pp 235–248
Hublin J-J, Spoor F, Braun M et al (1996) A late Neanderthal associated with Upper paleolithic artefacts. Nature 381:224–226. https://doi.org/10.1038/381224a0
Jaspers K (1949) Vom Ursprung und Ziel der Geschichte. Piper, München
King BJ (1994) Evolutionism, essentialism, and an evolutionary perspective on language: moving beyond a human standard. Lang Commun 14:1–13. https://doi.org/10.1016/0271-5309(94)90016-7
Klein RG (1995) Anatomy, behavior, and modern human origins. J World Prehist 9:167–198. https://doi.org/10.1007/BF02221838
Kuipers K, Kloeg J, Soressi M (2024) What is 'human' in human evolution? Jahrbuch für philosophische Anthropologie 14. (forthcoming)

Kurtén B (1995) Dance of the tiger: a novel of the Ice Age. University of California Press

Lacy S, Ocobock C (2023) Woman the hunter: the archeological evidence. Am Anthropol 126:19–31. https://doi.org/10.1111/aman.13914

Landau M (1993) Narratives of human evolution. Yale University Press, New Haven

Langdon JH (2022) Late Pleistocene Homo and the emergence of modern humans. In: Langdon JH (ed) Human evolution: bones, cultures, and genes. Springer, Cham, pp 497–537

Lévêque F, Vandermeersch B (1980) Découverte de restes humains dans un niveau castelperronien à Saint-Césaire (Charente-Maritime). Bulletin de la Société Préhistorique Française Comptes Rendus des Séances Mensuelles Paris 77(2)

Li L, Comi TJ, Bierman RF, Akey JM (2024) Recurrent gene flow between Neanderthals and modern humans over the past 200,000 years. Science 385:eadi1768. https://doi.org/10.1126/science.adi1768

London J (1907) Before Adam. Mills and Boon, London

London J (1915) The Star Rover. Mills and Boon, London

Martisius NL, Spasov R, Smith GM et al (2022) Initial Upper Paleolithic bone technology and personal ornaments at Bacho Kiro Cave (Bulgaria). J Hum Evol 167:103198. https://doi.org/10.1016/j.jhevol.2022.103198

Mcbrearty S, Brooks AS (2000) The revolution that wasn't: a new interpretation of the origin of modern human behavior. J Hum Evol 39:453–563. https://doi.org/10.1006/jhev.2000.0435

Mellars P (1989) Major issues in the emergence of modern humans. Curr Anthropol 30:349–385. https://doi.org/10.1086/203755

Mellars P, Stringer C (1989) The human revolution: behavioural and biological perspectives on the origins of modern humans. Princeton University Press, Princeton

Meneganzin A, Bernardi M (2023) Were Neanderthals and Homo sapiens 'good species'? Q Sci Rev 303:107975. https://doi.org/10.1016/j.quascirev.2023.107975

Meneganzin A, Currie A (2022) Behavioural modernity, investigative disintegration & Rubicon expectation. Synthese 200:47. https://doi.org/10.1007/s11229-022-03491-7

Mercier N, Valladas H, Joron J-L et al (1991) Thermoluminescence dating of the late Neanderthal remains from Saint-Césaire. Nature 351:737–739. https://doi.org/10.1038/351737a0

Montón-Subías S (2024) Other Pasts. Feminist archeology, maintenance activities, and reconceptualisation. In: Moen M, Pedersen U (eds) The Routledge handbook of gender archeology. Taylor & Francis, pp 383–396

Montón-Subías S, Hernando A (2018) Modern colonialism, eurocentrism and historical archeology: some engendered thoughts. Eur J Archaeol 21:455–471

Mylopotamitaki D, Weiss M, Fewlass H et al (2024) Homo sapiens reached the higher latitudes of Europe by 45,000 years ago. Nature 626:341–346. https://doi.org/10.1038/s41586-023-06923-7

Noble W, Davidson I (1991) The evolutionary emergence of modern human behaviour: language and its archeology. Man 26:223–253. https://doi.org/10.2307/2803830

Nowell A (2010) Defining behavioral modernity in the context of Neandertal and anatomically modern human populations. Annu Rev Anthropol 39:437–452. https://doi.org/10.1146/annurev.anthro.012809.105113

Pääbo S (2015) The diverse origins of the human gene pool. Nat Rev Genet 16:313–314. https://doi.org/10.1038/nrg3954

Pardo MA, Fristrup K, Lolchuragi DS et al (2023) African elephants address one another with individually specific calls. bioRxiv:2023.08.25.554872. https://doi.org/10.1101/2023.08.25.554872

Pearson OM (2008) Statistical and biological definitions of "anatomically modern" humans: suggestions for a unified approach to modern morphology. Evol Anthropol Issues News Rev 17:38–48. https://doi.org/10.1002/evan.20155

Peeters S, Zwart H (2020) Neanderthals as familiar strangers and the human spark: how the 'golden years' of Neanderthal research reopen the question of human uniqueness. Hist Philos Life Sci 42:33. https://doi.org/10.1007/s40656-020-00327-w

Peeters S, Zwart H (2026) Neanderthals and the issue of masculinity. The role of gender in popular Neanderthal discourse. In: Moro Abadía O, Porr M (eds) Alterity and human evolution: deep time and multispecies perspectives on difference and variation. Berghahn Books. (forthcoming)
Peeters S, Soressi M, Jensen S, Zwart H (2025) Close encounters of a human kind. On the need for distinction versus the longing for connection in Neanderthal. Jahrbuch für philosophische Anthropologie 14. (forthcoming)
Peyrégne S, Slon V, Kelso J (2024) More than a decade of genetic research on the Denisovans. Nat Rev Genet 25:83–103. https://doi.org/10.1038/s41576-023-00643-4
Porr M, Matthews JM (eds) (2019) Interrogating human origins: decolonisation and the deep human past. Routledge, Abingdon
Price N, Hedenstierna-Jonson C, Zachrisson T et al (2019) Viking warrior women? Reassessing Birka chamber grave Bj. 581. Antiquity 93:181–198
Rodríguez-Vidal J, d'Errico F, Pacheco FG et al (2014) A rock engraving made by Neanderthals in Gibraltar. Proc Natl Acad Sci 111:13301–13306. https://doi.org/10.1073/pnas.1411529111
Roebroeks W, Corbey R (2001) Biases and double standards in paleoanthropology. In: Corbey R, Roebroeks W (eds) Studying human origins. Amsterdam University Press, Amsterdam, pp 67–76
Roebroeks W, Soressi M (2016) Neandertals revised. Proc Natl Acad Sci 113:6372–6379. https://doi.org/10.1073/pnas.1521269113
Roussel M, Soressi M, Hublin J-J (2016) The Châtelperronian conundrum: Blade and bladelet lithic technologies from Quinçay, France. J Hum Evol 95:13–32. https://doi.org/10.1016/j.jhevol.2016.02.003
Sankararaman S, Mallick S, Dannemann M et al (2014) The genomic landscape of Neanderthal ancestry in present-day humans. Nature 507:354–357. https://doi.org/10.1038/nature12961
Scerri EML, Will M (2023) The revolution that still isn't: the origins of behavioral complexity in Homo sapiens. J Hum Evol 179:103358. https://doi.org/10.1016/j.jhevol.2023.103358
Sera-Shriar E (ed) (2018) Historicizing humans: deep time, evolution, and race in nineteenth-century British Sciences. University of Pittsburgh Press
Shea JJ (2011) Homo sapiens is as Homo sapiens Was. Curr Anthropol 52:1–35. https://doi.org/10.1086/658067
Slimak L, Zanolli C, Higham T et al (2022) Modern human incursion into Neanderthal territories 54,000 years ago at Mandrin, France. Sci Adv 8:eabj9496. https://doi.org/10.1126/sciadv.abj9496
Slon V, Mafessoni F, Vernot B et al (2018) The genome of the offspring of a Neanderthal mother and a Denisovan father. Nature 561:113–116. https://doi.org/10.1038/s41586-018-0455-x
Sorensen AC, Claud E, Soressi M (2018) Neandertal fire-making technology inferred from microwear analysis. Sci Rep 8:10065. https://doi.org/10.1038/s41598-018-28342-9
Soressi M, Roussel M (2014) European Middle to Upper Paleolithic transitional industries: Châtelperronian. In: Smith C (ed) Encyclopedia of global archeology. Springer, New York, pp 2679–2693
Soressi M, McPherron SP, Lenoir M et al (2013) Neandertals made the first specialized bone tools in Europe. Proc Natl Acad Sci 110:14186–14190. https://doi.org/10.1073/pnas.1302730110
Spengler O (1918) Der Untergang des Abendlandes. Umrisse einer Morphologie der Weltgeschichte. Erster Band: Gestalt und Wirklichkeit. Zweiter Band: Welthistorische Perspektiven. Beck, München
Störig HJ (1961) Kleine Weltgeschichte der Philosophie. Kohlhammer, Stuttgart
Stringer C, Gamble C (1993) In search of the Neanderthals: solving the puzzle of human origins. Thames and Hudson, London
Stringer CB, Grün R (1991) Time for the last Neanderthals. Nature 351:701–702. https://doi.org/10.1038/351701a0
Tostevin GB (2000) Behavioral change and regional variation across the Middle to Upper Paleolithic transition in Central Europe, Eastern Europe, and the Levant. Harvard University

Tostevin GB (2007) Social intimacy, artefact visibility, and acculturation models of Neanderthal-modern human interaction. In: Mellars P, Boyle K, Bar-Yosef O, Stringer C (eds) Rethinking the human revolution: new behavioural and biological perspectives on the origins and dispersal of modern humans. McDonald Institute for Archaeological Research, Cambridge, pp 341–357

Trinkaus E, Moldovan O, Milota S et al (2003) An early modern human from the Peştera cu Oase, Romania. Proc Natl Acad Sci 100:11231–11236. https://doi.org/10.1073/pnas.2035108100

Velasco JR (2023) Evolution and paleoanthropology in Hans Blumenberg's Nachlaß. Aisthesis Pratiche, linguaggi e saperi dell'estetico 16:117–132. https://doi.org/10.36253/Aisthesis-13641

Vernot B, Akey JM (2014) Resurrecting surviving Neandertal lineages from modern human genomes. Science 343:1017–1021. https://doi.org/10.1126/science.1245938

Vidal CM, Lane CS, Asrat A et al (2022) Age of the oldest known *Homo sapiens* from eastern Africa. Nature 601:579–583. https://doi.org/10.1038/s41586-021-04275-8

Villa P, Roebroeks W (2014) Neandertal demise: an archeological analysis of the modern human superiority complex. PLoS One 9:e96424. https://doi.org/10.1371/journal.pone.0096424

Wragg Sykes R (2020) Kindred: Neanderthal life, love, death and art. Bloomsbury, London

Zilhão J, Angelucci DE, Badal-García E et al (2010) Symbolic use of marine shells and mineral pigments by Iberian Neandertals. Proc Natl Acad Sci 107:1023–1028

Zilhão J, d'Errico F, Banks WE, Teyssandier N (2024) A data-driven paradigm shift for the Middle-to-Upper paleolithic transition and the Neandertal debate. Q Environ Hum 2:100037. https://doi.org/10.1016/j.qeh.2024.100037

Zwart H (2009) Biotechnology and naturalness in the genomics era: plotting a timetable for the biotechnology debate. J Agric Environ Ethics 22:505–529

Zwart H (2019) Psychoanalysis of technoscience: symbolisation and imagination. LIT Verlag, Berlin/Münster/Zürich

Zwart H, Soressi M (2025) Neanderthals and us: towards a philosophy of deep history. Jahrbuch für philosophische Anthropologie 14. (forthcoming)

A Decolonial and Multilayered Analysis of the Silencing of Latin American Biological Anthropology: The Case of 'the Peopling of the Americas' Research

15

Lumila Paula Menéndez and Bernardo Yáñez

Abstract

The study of human evolution is a broad, multifaceted, and multisituated process that requires a critical approach. In this chapter, we will critically focus on the investigations into the first dispersals of humans in the Americas, which is considered one of the most recent episodes of global human expansions across the world. It consisted of the migration of Paleolithic human populations who moved from somewhere in Asia into the Americas, what has been coined as the "peopling of the Americas." We argue that this research must be tackled not only from a scientific stance, but also through historical, epistemological, and geopolitical lenses. This means that a critical analysis of scientific hypothesis, interpretive methods, and empirical evidence becomes necessary to advancing and refining the science related to the evolutionary process of the human dispersal in the Americas. Therefore, besides the scientific critique of models and hypotheses, here we will discuss specifically a set of biases related to infrastructure, funding, and editorial industry, but overall, we will analyze the tradition of silencing or neglecting the research done by Latin American researchers from the hegemonic scientific mainstream. We claim that, in addition of the core scientific aspects, by taking into account a decolonial, multidisciplinary, multisituated, and polyphonic approach, it is possible to give some light on the differential circumstances of the research done by scholars from the Global North and those from the Global South, which resulted in an asymmetrical treatment to

L. P. Menéndez (✉)
Biological Anthropology Lab, School of Anthropology, University of Costa Rica, Ciudad Universitaria, San José, Costa Rica

B. Yáñez
Dirección de Antropología Física, Instituto Nacional de Antropología e Historia, Ciudad de México, Mexico

M. Lequin, J. M. Rodríguez Caso (eds.), *The Field of Human Evolution: Critical Perspectives from History and Epistemology*, Evolutionary Biology – New Perspectives on Its Development 11, https://doi.org/10.1007/978-3-032-13925-2_15

these different traditions of knowledge in the specialized literature. For instance, in the current literature, there is a misapprehension toward considering the morphometric and genetic uniparental data as obsolete methods for reconstructing human history. Furthermore, in such a context, the research done by scholars from Latin American countries and institutions is frequently neglected due to different reasons, even when it is frequently published in open access journals. Thus, we maintain that this multilayer analysis is important for the integration of scientific, historical, epistemological, and geopolitical elements as relevant issues of the scientific endeavor. In a nutshell, the "peopling of the Americas" is not just a scientific endeavor, but also a geopolitical matter.

Keywords

Biological anthropology · Human evolution · Peopling of the Americas · Geopolitics of knowledge · Silencing · Decolonial · Multidisciplinary · Multisituated · Polyphonic approach

15.1 Introduction

Research on the peopling of the Americas has been shaped by structural biases that extend beyond purely scientific aspects and delve into the geopolitical dynamics of knowledge production. These biases manifest in various ways, including limited access to research funding for Global South and peripheral scholars, unequal opportunities for publication and access to the contents of high-impact journals, and the prioritization of research agendas set by institutions in the Global North. We argue that these dynamics have often led to the marginalization and silencing of academic work produced and published in the Global South, by privileging interpretations and frameworks from scientists based in the Global North (Krotz 1996, 1997). Our perspective is rooted in biological anthropology, a key discipline for both the scientific and humanistic understanding of human evolution (Fuentes 2016, 2018a), and aligns with recent calls in anthropology, archeology, and the history of science to critically assess the structural and epistemic dimensions of knowledge production. As several recent studies have shown—including contributions in the volume *Interrogating Human Origins* (Porr and Matthews 2019a, b; Athreya and Ackermann 2019), recent work on Indigenous global histories (Porr 2021), and disciplinary reflections in paleoanthropology and biological anthropology (Ackermann et al. 2018; Sardi et al. 2024)—the field of human evolution is deeply entangled with colonial legacies, Eurocentric epistemologies, and power asymmetries in research design, authorship, and interpretation. These critiques have prompted scholars to advocate for decolonial, inclusive, and pluralistic approaches that reconceptualize scientific practices as socially situated and geopolitically marked.

Much of this literature has focused on African or Asian contexts, where discussions on human origins or early dispersals are particularly active. In contrast, the Americas—despite their central role in understanding human adaptation, diversity, and mobility—remain comparatively marginalized in global narratives of human evolution. This is partly because the continent is not seen as central to questions about the origins of *Homo sapiens*, nor is it framed as a "cradle of humankind" in the same way Africa is. Whereas Africa remains central to discussions on the origins of modern humans, and Asia has gained visibility due to recent ancient fossil discoveries, the Americas are commonly framed as a late chapter in human history, and often excluded from or downplayed in human evolution global debates. This exclusion is reinforced by dominant models in paleoanthropology that prioritize behavioral "modernity" and its expression in African and Eurasian contexts, leaving little conceptual space for incorporating early human histories in the Americas (Shea 2011). This framing reinforces the invisibilization of local expertise and entrenches dependence on Global North institutions for data validation, interpretation, and prestige.

It is precisely this exclusionary framing that our chapter seeks to challenge. Within this broader context, our focus on the "peopling of the Americas" highlights a region often marginalized—and silenced—in global narratives of human evolution. By focusing on the peopling of the Americas through a geopolitical and epistemological lens, this chapter contributes a perspective that is often absent from broader discussions on scientific inequality and knowledge hierarchies. Our contribution intervenes in this emerging literature by offering a regional, bioanthropological case study that illustrates how geopolitical asymmetries shape not only who gets to formulate research questions, but also whose data are considered valid, and which scientific paradigms or explanations are privileged. For instance, although we situate our argument in dialogue with recent reflections on these structural issues in human evolution, we are compelled to ask how often researchers in core institutions—or working with high-impact topics—reciprocate by considering the epistemological contributions of scholars like us working from and on the Americas.

Here, we address this issue by proposing an analytical framework based on a multilevel approach to examine structural biases related to infrastructure, funding, and editorial practices. We assert that the perpetuation of these biases reinforces a hegemonic narrative predominantly shaped by Global North scientists. Thus, a multilevel analytical approach allows for a more nuanced understanding of how different layers of scientific production—ranging from geopolitical power struggles to editorial policies and disciplinary hierarchies—intersect to reinforce exclusionary dynamics (Harding 1998; Kreimer 2022).

The manuscript begins with a brief historical overview of research on the peopling of the Americas. We then examine structural biases from a geopolitical perspective before presenting our proposal, which advocates for a multilevel analytical framework incorporating decolonial, multidisciplinary, multisituated, and polyphonic approaches. To support our argument, we draw on the concept of "World Anthropologies" (Ribeiro and Escobar 2006, 2008, 2020), which we adopt as an epistemological and methodological framework to critically analyze power dynamics among scientists, institutions, and countries in Global South-Global North scholarly interactions. This concept challenges the universalizing tendencies of Western epistemologies by advocating for epistemic pluralism and the recognition of multiple anthropological traditions shaped by diverse historical and cultural contexts.

In doing so, we seek to contribute to the growing body of work that calls for epistemic justice in science by showing how biological anthropology—despite its interdisciplinary nature—remains structured by extractive dynamics and entrenched hierarchies (de Sousa Santos 2014; Wynter 2003). This approach not only helps trace how knowledge is produced and circulated unevenly across geopolitical contexts, but also provides pathways for reconfiguring collaborations, authorship, and methodological and epistemological inclusion. Recognizing the importance of situated perspectives, we aim to promote analytical frameworks that account for both historical legacies and contemporary conditions shaping the field (Law and Lin 2017).

We further explore the dichotomy between "soft sciences" and "hard sciences," proposing a reconfiguration of this notion based on contributions from various authors who have engaged with this debate (Athreya and Ackermann 2019; Bolnick et al. 2019; Harding 1991). We argue that this dichotomy often serves to devalue qualitative, context-driven research—particularly that produced in the Global South—while reinforcing the dominance of quantitative and laboratory-based methodologies commonly associated with Global North institutions (Porr and Matthews 2019b). Donna Haraway (1989) has shown how these scientific hierarchies are not neutral, but historically tied to political and ideological constructions, particularly masculinist and colonial framings of knowledge in fields like biological anthropology. In this chapter, we extend this reflection to encompass the geopolitical, economic, and technological dimensions that intersect with research activities in the field of human evolution.

Finally, we emphasize the need to incorporate different kinds of data, diverse voices, and knowledge systems to expand the analytical tools available for a more comprehensive and nuanced understanding of the peopling of the Americas. We conclude our reflections by underscoring the significance of adopting a multidisciplinary, multisituated, and polyphonic perspective. Addressing these disparities requires structural changes, including more inclusive funding mechanisms and editorial policies that prioritize both linguistic and geographic diversity, as well as collaborative research frameworks that foster balanced partnerships rather than perpetuating and reproducing extractive and colonial practices.

15.2 A Brief Account on the "Peopling of the Americas" Research

The term "peopling of the Americas" is a relatively recent expression in the scientific literature. It was coined in the 1940s by US scholars to describe a growing field of research focused on when and how the first humans arrived in North, Central, and South America, as well as identifying the ancestral populations to which these early migrants might be linked (Birdsell 1951; Giddings 1954; Hibben 1946; Walter et al. 1953; Willey and Newman 1949). While the earliest inquiries into these questions date back to the late 1400s and early 1500s—coinciding with the arrival of Europeans in the Americas—phrases such as "settlement of the Americas" or "migration to the New World" were more commonly used during that period (Acosta 1590; de las Casas 1877). Nonetheless, "peopling of the Americas" has become a prevalent term in academic literature since at least the 1970s, reflecting ongoing research and debates about the processes of arrival and dispersal of the earliest human inhabitants of the American continents. Today, the term is widely used by researchers worldwide, in its English, Portuguese, and Spanish variants, i.e., *Povamento da América, Poblamiento de América*—appearing in diverse regional and disciplinary contexts (Borrero 2025; Bueno et al. 2020; García-Ortiz et al. 2021; Gruhn 2023; Mazzia et al. 2020; Scott et al. 2023; Wagner 2022; Willerslev and Meltzer 2021).

Beyond the terminology, the earliest scientific accounts of the peopling of the Americas date back to the mid- to late-nineteenth century. These initial investigations primarily focused on determining the timing of the earliest human arrival, often by evaluating the contemporaneity of human skeletons or cultural artifacts with extinct megafauna (Abbott 1878; Ameghino 1880; Burmeister 1891). At that time, the debate centered on whether the first human arrival occurred during preglacial or postglacial periods. Today, however, the discussion is typically framed around whether it took place before or after the Last Glacial Maximum (Becerra-Valdivia and Higham 2020; Bennett et al. 2021). Since those early debates, a long-standing controversy has emerged between researchers who support an early human arrival in the Americas and those who advocate for a more recent one (Hrdlička 1912; Moreno 1882). With the advent of technologies such as radiocarbon dating in the 1950s, the debate expanded to include more refined discussions about the nature of archeological materials, possible migratory routes, their environmental contexts, and the methods used for radiocarbon analysis (Prates et al. 2020; Waters and Stafford 2013).

Additionally, from the 1980s onward, computational advancements in biological analyses, particularly in morphometric and genetic studies, provided new data that complemented findings from the archeological sciences (e.g., Cavalli-Sforza et al. 1988; Greenberg et al. 1986; Neves and Pucciarelli 1989, 1991; Schurr et al. 1990). These studies addressed questions about the identity of the first Americans, the populations from other continents with whom they shared biological affinities, and the biological relationships among Native American groups. Their findings significantly expanded the scope of research by informing interpretations of dispersal routes, timing of arrival, and key population bottlenecks, among other demographic, adaptive, and evolutionary factors.

Currently, the field is dominated by paleogenetic and/or archeogenomic studies, which attract the majority of funding and frequently publish their findings in highly cited general journals such as *Nature*, *Science*, and *Cell* (Moreno-Mayar et al. 2018; Posth et al. 2018; Reich et al. 2012). While genetics is a specialization available in many countries with developed biological anthropology programs, large laboratories in core or Global North countries control much of the processing and publication in ancient DNA (aDNA) research, and often overlooking or devaluing work produced in the periphery. This creates a vicious cycle: researchers from these core countries and institutions publish in top-tier journals, attract greater funding, and continue improving their infrastructure, which in turn enables them to conduct more advanced analyses. As a result, they maintain visibility in high-impact journals, benefit from media coverage, and shape dominant narratives of human evolutionary history. These dynamics contribute to a structurally uneven playing field, where researchers from the Global South face persistent barriers to achieving equivalent visibility and impact, despite conducting rigorous and contextually grounded research.

Recently, there has been a shift in the topics associated with research on the "peopling of the Americas," extending beyond a purely scientific perspective. Over the past decade, Latin American researchers have initiated critical reflections on how science is conducted. Many arguments highlight the importance of actively involving Indigenous communities and developing strategies for capacity building in Global South countries, where the samples are often extracted (Ávila-Arcos et al. 2022; de la Fuente and Figueiro 2024; Nieves-Colón et al. 2021). Other discussions emphasize the need to redefine collaborations among researchers from different specializations and from peripheral and core countries (Argüelles et al. 2022; Ávila-Arcos et al. 2022; di Fabio Rocca et al. 2021; Menéndez et al. 2022; Ruderman 2023; Squires and García-Mancuso 2020; Yáñez et al. 2023). Consequently, calls for ethical, inclusive, and sustainable research practices have gained momentum, emphasizing the need to move beyond solely technical debates and toward a reflexive understanding of the geopolitics of anthropological science. While this criticism is both timely and valuable, it has yet to result in widespread or structural changes within the practices of many Global North institutions and researchers.

In this chapter, we argue that the control and dominance of the field by a small group of researchers—often disconnected from the local knowledge produced by scholars over the past century and by Indigenous communities for millennia—has introduced several biases into the mainstream narrative of human history in the Americas, particularly the peopling of the Americas. While recent advancements have provided critical insights into the theoretical and practical aspects of this field (Argüelles et al. 2022; Ávila-Arcos et al. 2022; de la Fuente and Figueiro 2024; Menéndez et al. 2022; Nieves-Colón et al. 2021; Ruderman 2023; Yáñez et al. 2023), we believe that addressing these biases is essential for fostering decolonial, multidisciplinary, multisituated, and polyphonic narratives about the history of Native American populations. We contend that this topic must be approached not only from a scientific perspective but also through the incorporation of historical, epistemological, and geopolitical dimensions. Without embracing the broader framework we propose, researchers risk perpetuating and reproducing biases that constrain knowledge production, and lead to fragmented and incomplete understandings of the phenomena under study. The continued disregard for scholarship produced and published by local and peripheral researchers effectively crystallizes in the silencing of these perspectives. The issue of the peopling of the

Americas serves as a compelling example of how such structural conditions are reproduced, preventing polyphonic engagement with diverse perspectives on this and related issues. Instead, this dynamic fosters a hegemonic set of interpretations, methods, and questions that dominate the narrative of this bioanthropological topic.

15.3 Geopolitical Biases in "the Peopling of the Americas" Research

We have identified a series of biases related to research on the "peopling of the Americas." These biases primarily stem from neocolonial asymmetries in infrastructure, funding, and the editorial industry (Figueiro et al. 2024). This results in a lack of recognition for the work produced by Latin American researchers, which in turn shapes the dominant narrative surrounding human history in the Americas and the "peopling of the Americas". Among other things, we argue that this dynamic contributes to the silencing of the research conducted and published in the Global South, and more specifically in Latin America. Such marginalization not only excludes valuable perspectives but also sustains an Eurocentric or North Atlantic narrative that overlooks the diversity of scientific contributions from the Global South. Addressing these disparities requires fostering equitable collaborations, increasing investment in local research infrastructure, and advocating for inclusive publishing practices that amplify underrepresented and marginalized voices.

15.3.1 Infrastructure Biases

Researchers from the Global North have access to more advanced equipment and technology due to better resources and easier acquisition opportunities. This disparity creates an uneven playing field in scientific research, where researchers from the Global South often face significant challenges in accessing comparable technology and infrastructure. Even acquiring the same specialized equipment is more expensive in the Global South due to additional taxes and fees imposed by official distributors, who are typically based on other continents, are fewer in number, and face less competition.

Consequently, researchers from the Global South may rely on collaborations with better-funded institutions in the Global North, which can lead to imbalanced partnerships. As a result, instead of playing a relevant role in the project, Global South researchers often become intermediaries, primarily acting as sample providers (Argüelles et al. 2022). They are frequently involved late in the process, sometimes only receiving the manuscript when it is nearly complete and ready for submission, limiting their ability to offer meaningful insights or incorporate their own perspectives. Similarly, they are often invited to collaborate at a later stage, without the opportunity to shape the project's research questions from the outset. These dynamics reinforce existing hierarchies and perpetuate colonial traditions in knowledge production, with researchers from the Global North disproportionately driving the research agenda and receiving the majority of credit for published work and discoveries.

In parallel, some Global North institutions offer visiting fellowships or short-term research stays that appear to support Global South scholars. However, these programs are often structured to serve the strategic interests of the host institutions. Rather than fostering equitable collaborations, they can function as mechanisms to gain access to samples, data, or permits through the invited researchers, thereby reproducing extractive dynamics under the guise of academic exchange. Such practices highlight the need for critical evaluation of so-called capacity-building initiatives, ensuring they do not simply reinforce existing power imbalances, but genuinely support reciprocal and locally driven research agendas. Furthermore, researchers from the Global South who are trained in academic hubs and elite institutions sometimes adopt and reproduce these controversial practices within their own national contexts. This also contributes to whitewashing hierarchical and asymmetric interactions, which accelerate the pace of research and, in turn, hinder the advancement and development of research agendas in the Global South and other peripheral contexts.

Another structural bias addressed further in this chapter is the hegemony of the English language in global scientific endeavors. This affects not only the writing of grant proposals for acquiring funding to access infrastructure and equipment, but also the publication and communication of research findings (see "Publishing and Editorial Biases" section). Consequently, the prevailing imperative for contemporary scientists and graduate students, both from the Global North and the Global South, becomes "publish or perish, but do it in English." Research not published in the hegemonic language is often rendered invisible and uncited, further contributing to the silencing of the scientific production from the Global South and other peripheral contexts. This invisibilization is compounded by structural practices within top-tier journals, which often limit the number of references allowed. When Global South researchers are only included at the final stages of manuscript preparation—as is frequently the case—they have little opportunity to advocate for the inclusion of relevant regional literature. Therefore, work produced and published locally is systematically excluded from citation, even when directly relevant.

15.3.2 Funding Biases

Researchers in core countries benefit from a combination of economic resources, institutional support, technological infrastructure, and political influence, all of which provide them with greater access to research funding and public attention. This is further enabled by the stable and secure living conditions typically found in Global North countries. Thus, the largest research programs and the most influential funding agencies in anthropology are predominantly located in countries from the Global North (e.g., National Geographic, Leakey Foundation, European Commission, and many National Research Councils), and frequently focus their support on core universities, laboratories, and countries. Researchers from the Global North and

South do not share equal starting conditions when applying for funding. Those who are ultimately awarded grants often list affiliations with eminent universities in their CVs, demonstrate previous success in attracting third-party funding, and have a track record of publishing in prestigious journals—all of which reinforce their competitiveness in subsequent applications. In addition, successful applicants are frequently supported—explicitly or implicitly—by senior, well-established scholars whose endorsement, institutional prestige, or co-investigator status can significantly increase the perceived legitimacy and attractiveness of a proposal. This form of academic sponsorship, rarely available to researchers in the Global South, further skews access to international funding opportunities.

The concentration of resources and political influence allows researchers in the Global North to set the priorities and agendas in anthropology and related disciplines. They often dictate which questions are deemed most important, which methodologies are prioritized, and how and where findings are disseminated. This dominance frequently marginalizes perspectives and research interests originating in the Global South, where researchers may face significant obstacles in obtaining funding and institutional support. Moreover, the reliance on Global North funding agencies often comes with conditions that can limit the autonomy of researchers in peripheral countries, pressuring them to align their projects with priorities defined by external institutions or foreign expectations.

This power-imbalanced dynamic becomes even more pronounced when considering the localization of biocultural and patrimonial objects or subjects related to the peopling of the Americas across the continent. From a bioanthropological perspective, the Global South serves as the core region where skeletonized human remains are discovered, recovered, contextualized, and stored. However, the Global North is home to the laboratories, epistemic agendas, academic centers, and funding agencies that drive the research and the editorial industry. These structural asymmetries create an ecosystem where international scientific relationships perpetuate geopolitical and economic biases (Argüelles et al. 2022).

Often, researchers from the Global North justify their geographic focus when applying for funding to conduct research in the Global South by claiming that "there is a complete lack of research in that region." This assertion, however, often overlooks or dismisses the substantial body of knowledge already produced by local researchers and institutions within the Global South. Such claims perpetuate a narrative that frames the Global South as a passive recipient of knowledge rather than an active contributor, thereby reinforcing unequal power dynamics in academic and scientific communities. Furthermore, this approach can lead to the marginalization of local expertise, as funding agencies and global academic audiences prioritize research produced and published by Global North institutions over that of their Southern counterparts. This dynamic not only undermines the recognition and visibility of local scholarship, but also risks erasing culturally and contextually grounded perspectives that are essential for a nuanced understanding of the region.

Some funding agencies, such as the Wenner-Gren Foundation, have begun to address these disparities through bottom-up programs like the Global Initiatives Grant, launched in 2022. This program aims to build capacity by "...encouraging applications from individuals based in countries, regions, and communities where anthropology is disadvantaged or under threat… ." We believe that more targeted programs of this kind should be established. Moreover, when a research project is carried out in, or samples are obtained from, a country in the Global South, funding agencies from the Global North must provide additional support. This supplementary funding should be directed toward capacity building, improving laboratory facilities, and infrastructure in Global South partner institutions, and covering travel costs to facilitate equitable team meetings and sustained collaboration.

15.3.3 Publishing and Editorial Biases

While there is a vast body of literature on the peopling of the Americas, the most visible and frequently cited works are often authored by scholars from the Global North. In contrast, researchers from the Global South or peripheral regions are frequently marginalized, leading to the silencing or devaluation of their contributions (Krotz 1996, 1997). This exclusion is even more pronounced for women and other minority scholars from the Global South, whose perspectives are often overlooked despite their crucial roles in fieldwork and long-term regional research (Hutson 2002; Smith and Garrett-Scott 2021). This silencing perpetuates existing biases and limits the visibility of research produced in the Global South, thereby reinforcing power imbalances within the field. Consequently, key contributions from peripheral researchers are often overlooked or dismissed, despite their potential to provide valuable insights into the diverse histories and cultures of the Americas. Such asymmetries not only reflect broader structural inequalities in academia but also shape which voices are recognized as authoritative and whose knowledge is legitimized in global scientific debates.

A clear example of this marginalization is the limited, or complete absence, of citations of the work published by Global South scholarship in articles led by Global North researchers, even when the work is written in English and available through open access platforms (Menéndez et al. 2025). Frequently, this omission is justified by referencing the project's technological specialization or methodological complexity, while overlooking the unfamiliarity with local knowledge, languages, literature, and scholarly traditions. Yet a meaningful synergy between technical results and regionally grounded expertise, as well as a minimum effort to engage with the local language and cultural context, is essential for advancing the field and producing more balanced, meaningful, and multisituated contributions. Researchers from the Global South not only face limited opportunities to contribute to such articles from the outset, but are also constrained by the structural limitations of high-impact journals—such as strict word counts, limits on the number of references, and expectations of advanced English language proficiency—which further deprioritize their work. For example, *Nature* enforces a maximum of 50 references per article. This creates a vicious cycle in which only research published in top-tier journals is selected for citation, effectively excluding locally produced scholarship from the conversation.

Moreover, researchers from the Global South do not have the same opportunities as their Global North counterparts when it comes to publishing in top-tier journals. Not only are the high publication fees often inaccessible within Global South research budgets—for instance, the cost for publishing in Nature is USD $12.690 while the average 4-year project grant from the Argentinean Research Council was approximately USD $10.000 in 2024. Additionally, those articles must also be written in proficient English and adhere to a particular stylistic register that many researchers from the Global South have not had the privilege or opportunity to master. Furthermore, many articles published in high-impact journals are solicited by invitation from editors-in-chief or associate editors, who are typically based in Global North institutions and tend to invite colleagues within their own networks. This practice facilitates the reproduction of a closed system, whereby a small group of researchers disproportionately publishes in top journals. These influential figures often appear as co-authors on multiple highly cited papers across journals, reinforcing their central role in the field and limiting the visibility of alternative voices. As a result, a few "big fishes" are co-authors of many of the most influential papers in a discipline, both through their authorship and through their gatekeeping positions (Farrera et al. 2024). This concentration of authorship and editorial power narrows the range of voices contributing to the story of human evolution, limiting the field's potential for pluralism and inclusivity.

Another example is the primary language in which articles are written in the Global South. Many researchers in Latin America publish their work in regional scientific journals, where articles are predominantly written in Spanish or Portuguese. Most journals in the region focused on anthropology and archeology—including topics related to the peopling of the Americas—follow a diamond open access model. Examples include *Antípoda, Boletín de Arqueología PUCP, Boletim do Museu Paraense Emílio Goeldi, Chungará, Cuadernos de Antropología, Diario de Campo, Estudios de Antropología Biológica, Estudios Atacameños, Intersecciones en Antropología, Magallania, Relaciones en Antropología, Revista Argentina de Antropología Biológica*, and *Revista del Museo de La Plata*, among others. This model ensures that research articles are freely accessible to all readers without subscription barriers or paywalls, and that authors do not incur publication fees. While the diamond open access model promotes broad accessibility, the use of Spanish or Portuguese as primary languages of publication can limit the visibility and international reach of these articles within the global academic community (Moutinho 2024).

Notably, this dynamic is not reciprocal: researchers from the Global South are still expected to learn, write, and communicate proficiently in English in order to publish their work in high-impact journals and reach broader audiences. Nevertheless, access to English language education is also deeply shaped by socioeconomic inequalities. In many Global South contexts, studying English—especially to the level required for academic writing—is expensive and largely limited to private institutions or urban centers. Thus, English proficiency becomes not only a linguistic barrier but also a class-based one, contributing to the exclusion of researchers who cannot afford such training. This further stratifies who gets to participate in international academic conversations, and who remains relegated to the margins. Unlike many Indigenous peoples in the Global North—where English is often the dominant or official language—Indigenous communities in Latin America face additional language barriers, as limited access to quality language education and broader structural inequalities severely restrict their participation in global academic debates, particularly those concerning their own histories and heritage.

Again, the biases in publication and citation practices are even more pronounced for women scholars from the Global South. Studies have consistently shown that women are underrepresented as lead authors in high-impact journals, particularly in fields such as anthropology and archeology (Fulkerson and Tushingham 2019; Hanscam and Witcher 2023; Hutson et al. 2023; Schroeder and Ackermann 2023; Smith and Garrett-Scott 2021; Turner et al. 2018). For instance, articles authored by women are less likely to be cited compared to those authored by men, a phenomenon known as the "gender citation gap" (Hutson 2002; Smith and Garrett-Scott 2021). As a result of the aforementioned publishing and editorial biases, the dominance of Global North perspectives in mainstream academic discourse perpetuates a narrow and often skewed understanding of the peopling of the Americas, disregarding the complexity and richness of Indigenous knowledge systems, locally developed research, and gender driven perspectives. The reproduction of these intersecting asymmetries—of geography, language, gender, and power—not only silences critical voices but also hinders the possibility of a more plural, inclusive, and organically developed scientific landscape in the Global South.

15.4 The Need of a Decolonial, Multidisciplinary, Multisituated, and Polyphonic Perspective on "the Peopling of the Americas' Research

The peopling of the Americas is a topic of significant relevance to all regions and countries across the American continents and the Caribbean. It represents a scientific inquiry that transcends national borders and extends beyond the Americas itself. Historically, this topic has attracted strong interest from European anthropologists and archeologists, reflected today in the large number of European-led projects funded by bodies like the European Research Council (ERC) and national scientific councils, some of which are led by Latin American scholars, but many still directed and shaped by European logics and researchers. This carries profound philosophical, epistemological, and historical implications for how the subject should be studied and understood by both the scientific community and the diverse human groups inhabiting these lands. We argue that, in addition to the decolonial perspective outlined in previous sections, a multidisciplinary, multisituated, and polyphonic approach is essential for advancing our understanding of the peopling of the Americas. Such an approach enables a more comprehensive and nuanced explanation of this pivotal event in the human evolutionary process, particularly within the broader journey of *Homo sapiens* across the planet. A situated perspective involves examining specific events or phenomena through the lens of their social, historical, political, ethical, and legal contexts. In the case of the peopling of the Americas, this means taking into account the unique circumstances, specific environmental conditions, and particular cultural frameworks in which these processes unfolded.

In the current state of research on the peopling of the Americas, adopting multidisciplinary and multisituated approaches is not merely desirable but essential. Such an integrated framework allows researchers to transcend the limitations of unidimensional analyses, providing a more comprehensive and robust foundation for interpreting this pivotal event in human history (Menéndez et al. 2022; Menéndez and Hubbe 2025). Moreover, it brings to the forefront critical, ethical, and political questions regarding how research is conducted, who participates in it, and how the benefits of the knowledge produced are distributed. In an era of growing global emphasis on digital humanities, open science, and interdisciplinary collaboration, we argue that this multilayered model holds the potential to shape future research that is not only more inclusive, reflexive, and critical, but also transformative in its impact.

15.4.1 World Anthropologies: A Decolonial Epistemological Framework in the Geopolitical Context of Science

The *World Anthropologies* project (Ribeiro and Escobar 2006, 2008, 2020) recognizes that science, like all human endeavors, is profoundly shaped by power structures. This approach goes beyond examining anthropology's theoretical and epistemological dimensions to interrogate the political, economic, and social forces that underpin its academic practices. It seeks to critically analyze power relations within the "world system of anthropology," a multilevel and hierarchical framework that sustains and reproduces hegemonic and subordinate dynamics among individuals, disciplines, institutions, and nations.

In Latin America, this analysis has prompted reflections on the tensions between the anthropologies of the Global North and those of the Global South. Roberto Cardoso de Oliveira (1997), a pioneer in exploring these dynamics, framed this tension as primarily epistemological rather than merely political or social. He regarded it as fertile ground for anthropological innovation, inquiry, and debate. Cardoso de Oliveira introduced the distinction between "metropolitan anthropologies" and "peripheral anthropologies," underscoring the asymmetrical conditions within these different anthropological traditions, on the one hand, and the need to cultivate more equitable relationships across them.

Mexican anthropologist Esteban Krotz further contributed to this discourse with the concept of "anthropologies of the South" (Krotz 1996, 1997, 2006, 2019), which contrasts the anthropological practices of the Global South with those of the industrialized North. He argued that, despite the depth and richness of anthropological contributions in Latin America, these insights are often overlooked or marginalized. This marginalization, he posited, reinforces the notion that "true anthropology" originates in core countries (Krotz 1996, 1997). According to Krotz, the disparities between the North and South are not the result of an intrinsic lag, but rather of historically entrenched asymmetrical pressures imposed by the North.

Building on the work of Krotz, Cardoso de Oliveira, Walter Mignolo, and Anibal Quijano, Gustavo Lins Ribeiro and Arturo Escobar articulated the concept of "World Anthropologies" as part of a decolonial epistemological and methodological project. This framework is founded on two key principles: (i) the proliferation of localized, situated, and context-sensitive anthropological practices, and (ii) the pluralization of anthropology through the adoption of diverse concepts, methodologies, and languages. It emphasizes the critical importance of involving local researchers and amplifying the voices of historically marginalized communities, including Indigenous peoples, Afro-descendants, and other subaltern groups. Ribeiro and Escobar (2008) reject the notion of a single, dominant core for anthropological practice, advocating instead for the emergence of diverse, contextually grounded, engaged, and situated forms of anthropology. This decolonial perspective seeks to foster global dialogue by promoting collaboration and mutual respect among anthropologists from varied traditions, contexts, and cultural frameworks.

Ultimately, the "World Anthropologies" project calls for the recognition and appreciation of the rich and vast anthropological traditions emerging from peripheral contexts, with particular emphasis on Latin American contributions. By critically challenging the discipline's hegemonic hierarchies, it advocates for pathways toward more horizontal, inclusive, and collaborative relationships. As Ribeiro and Escobar (2008) assert, "the locus of enunciation of disciplines is geopolitically marked," prompting a fundamental reconsideration of how anthropological knowledge is constructed, shared, and valued on a global scale.

For instance, a decolonial approach to the study of the peopling of the Americas challenges the hierarchical connotation of evidence that often privileges genetic data over archeological context (Menéndez and Veigl 2026). In many recent studies, paleogenomic findings—though powerful in tracing broad population histories—are often interpreted with limited reference to local archeological records, site formation processes, or regional cultural sequences. This approach tends to reinforce a reductionist narrative of the continent's occupation. This disconnection can lead to overly generalized migration models or misinterpretations of demographic change. By contrast, a decolonial framework insists on assigning equal epistemic weight to archeological, historical, and bioanthropological forms of evidence. It encourages researchers to interpret genetic results in close dialogue with stratigraphic data, radiocarbon chronologies, material culture, and burial contexts, allowing for more fine-grained, culturally informed reconstructions of early human histories. Moreover, this perspective recognizes that archeological expertise—often concentrated in Global South institutions—is crucial for interpreting the significance of ancient individuals beyond their genomes. In doing so, a decolonial approach fosters more balanced and context-sensitive interpretations, while also redressing epistemic asymmetries in collaborative research.

We contend that this decolonial framework is highly relevant for analyzing the power dynamics and epistemic inequalities that structure research on the peopling of the Americas. Although these insights were initially developed within the context of sociocultural anthropology, we argue that they offer a robust epistemological, analytical, and methodological foundation for rethinking contemporary bioanthropological research. Applying a decolonial lens to this field of study, the peopling of the Americas enables a critical examination of how power relations, resource distribution, editorial practices, and academic hierarchies shape both the production and circulation of knowledge about human origins in the Americas. Such an approach is essential not only for exposing epistemic asymmetries but also for envisioning more just and inclusive ways of doing science.

15.4.2 Multidisciplinarity and Asymmetrical Techno-Scientific Relationships: Examining the Divide Between Hard and Soft Sciences, Science and Humanities, and Persistent Dichotomies in Knowledge Production

Multidisciplinarity is now a hallmark of contemporary research, promising to bridge entrenched disciplinary divides and foster holistic approaches to complex global challenges. Yet, this ideal often encounters structural asymmetries rooted in longstanding hierarchies within academia. The division between "hard sciences" (e.g., physics, chemistry, and engineering) and "soft sciences" (e.g., anthropology, sociology, and psychology) reflects entrenched biases that privilege quantitative, empirically rigorous methodologies over qualitative and interpretative ones. These asymmetries influence funding allocation, institutional priorities, and public perceptions of scientific legitimacy. As Klein (1990) highlights, the institutionalization of disciplinary boundaries frequently reinforces power imbalances, thereby limiting the potential for truly integrative research.

The division between hard and soft sciences is not merely methodological, but also historical and epistemological, and thus it is shaped by the evolution of modern science. During the Enlightenment, the emphasis on rationality and empirical observation gave rise to the "scientific revolution," which privileged the natural sciences as the primary producers of legitimate knowledge. Meanwhile, the humanities and social sciences were positioned as subjective and speculative. However, this divide has long hindered intellectual and social progress (Snow 1959), and despite ongoing efforts to foster interdisciplinarity, this dichotomy persists, often framed as a competition for resources and intellectual primacy.

The asymmetry between science and humanities also reflects broader sociopolitical dynamics. Fields like biotechnology are closely tied to economic growth and industrial interests and, as such, receive significant funding and institutional support. In contrast, the humanities often struggle to justify their relevance in market-driven paradigms of education and research. Jasanoff (2004) argues that these asymmetries are not simply a matter of resource distribution, but are deeply embedded in the co-production of scientific knowledge and social order. The privileging of techno-scientific disciplines aligns with neoliberal ideologies that prioritize innovation and productivity over critical inquiry and reflexivity.

While dichotomies between hard and soft sciences, and between science and the humanities, remain entrenched, there are growing efforts to bridge these divides. Latour (1993) critiques the artificial separation between nature and culture, advocating for a more integrated understanding of knowledge production. Similarly, Haraway (1988) emphasizes the value of "situated knowledge," arguing that diverse epistemologies can complement and enrich each other. Scientific pluralism acknowledges the value of multiple, sometimes conflicting, explanatory models, promoting robust and complementary understandings of phenomena (Kellert et al. 2006). Mitchell (2003) further highlights that in domains like biology, where systems are highly complex, pluralism enables the integration of different levels of analysis, such as molecular, ecological, and evolutionary perspectives, to provide a more comprehensive account of life processes.

Krotz also contributes to this critical perspective by challenging traditional divisions between the sciences into two groups: on the one hand, the humanities, and on the other hand, the exact sciences. This division implies that the latter are more solid, reliable, and true, while the former appear vague or inconsistent. Instead, he proposes we could rethink these categories by suggesting that disciplines such as anthropology could be considered as part of the "flexible sciences," while the so-called exact sciences might be more accurately described as "rigid sciences." This shift in terminology challenges hierarchical connotations and opens space to value the adaptability and interpretative breadth that characterize anthropological inquiry (Krotz 1996:37).

Social anthropologists have long challenged such hierarchies by proposing ontological pluralism. Eduardo Viveiros de Castro (1998), for instance, argues that Western metaphysics must be "decentered" in order to recognize radically different ways of knowing and being—insights that are increasingly relevant to debates about pluralism in biological anthropology. Similarly, Jon Marks (2009) emphasizes that biological anthropology's position between the natural and social sciences should not be treated as a liability, but as a strength—an opportunity for critical reflexivity and pluralistic knowledge-making.

Addressing asymmetries in techno-scientific relationships requires more than collaborative frameworks; it demands a critical examination of the underlying power structures that perpetuate these divides. Policymakers, educators, and researchers must advocate for funding models and institutional practices that value diverse forms of knowledge equally. As Jasanoff (2004) suggests, integrating the insights of humanities and social sciences into techno-scientific agendas can foster more inclusive and equitable approaches to global challenges. When pursued equitably, multidisciplinarity holds the potential to transcend disciplinary boundaries and reimagine the very possibilities of knowledge production.

Biological anthropology stands as an exemplary discipline for bridging the traditional divide between the humanities and the sciences, offering a uniquely integrative approach to understanding the human experience. Rooted in the humanities, this field draws on its rich tradition of exploring human culture, history, and behavior, while simultaneously employing scientific methodologies to study the evolutionary history and biological diversity of humans and their primate relatives

(Fuentes and Wiessner 2016). This dual focus enables biological anthropologists to investigate questions that are both deeply contextual and empirically grounded. As biological anthropologists, we are uniquely positioned to foster meaningful dialogue between the humanities and the sciences, combining biological, archeological, and ethnographic data, and to advance more reflexive, collaborative, and inclusive forms of knowledge production.

The study of the peopling of the Americas exemplifies how biological anthropology bridges the humanities and sciences by integrating diverse methodologies and perspectives to address a profoundly interdisciplinary question. This field of research draws on biological data, archeological evidence, and cultural context to reconstruct the origins, timing, routes, and adaptations of the first populations to inhabit the Americas (Menéndez 2024).

While morphometric and genetic data have been foundational to early models of human population history, they are increasingly recognized as limited when interpreted in isolation. For example, despite the advances brought by the genomic revolution, mitochondrial DNA haplogroups remain highly relevant in Latin America, where they continue to be widely used to explore patterns of human affinity, diversity, and population history. This ongoing relevance is due not only to the availability of extensive comparative datasets, but also to the feasibility of conducting mtDNA analyses in local laboratories—making them a more accessible and sustainable resource for regional researchers. However, to fully unlock the interpretive potential of such data, it is essential to situate genetic results within archeological and historical frameworks. The integration of uniparental markers with archeological context, other biological datasets, environmental evidence, and historically informed perspectives enhances the depth and accuracy of population reconstructions (Menéndez and Urban 2025). A multidisciplinary perspective enables this kind of synthesis by bringing together diverse lines of evidence and forms of expertise—often produced in distinct geographic, disciplinary, and epistemic contexts—to revise, complicate, and deepen our understanding of early human histories (Menéndez and Veigl 2026).

By weaving together genetic, archeological, and bioarcheological data, biological anthropology demonstrates the power of an interdisciplinary and integrative approach to address complex questions about the human past (Fuentes 2018b; Fuentes and Wiessner 2016; Menéndez and Veigl 2026). Therefore, the study of the peopling of the Americas not only enhances our understanding of early human migrations but also underscores the value of interweaving scientific and humanistic methods in reconstructing the human experience (for an interesting discussion on this issue from a socio-anthropological perspective, see: Barnard 2016a, b).

However, this multifaceted and multidisciplinary approach often contrasts with the accelerated pace of ancient DNA studies advocating, instead, for a more reflective and slower-paced science, particularly when it comes to the study and analyses of skeletonized human remains (for a discussion on this topic, see Figueiro et al. 2024; Yáñez et al. 2023).

To round up our argument, we, as biological anthropologists, are uniquely positioned to facilitate dialogue between paleogeneticists and archeologists. With our expertise in both the biological and cultural dimensions of human history, we are well equipped to play a decisive role in bridging these fields, ensuring that genetic data are interpreted within the broader context of human adaptation, cultural dynamics, and evolutionary history. The skills we develop through our professional training enable us to synthesize and intertwine diverse data sources, contextualizing findings within broader evolutionary and historical frameworks. This positions us to contribute critically to the construction of comprehensive and multidisciplinary narratives of early human migrations.

15.4.3 The Need for a Multisituated Approach to Contextualize Biological Anthropology

Biological anthropology, with its inherent interdisciplinarity, has increasingly recognized the need to adopt multisituated and polyphonic approaches to better capture the complexities of human and non-human primate experiences (Bourdier et al. 2021; Fuentes 2018a; Little and Haas 1989; Menéndez et al. 2022). These terms reflect a methodological and epistemological shift toward greater inclusivity and contextuality. A multisituated approach emphasizes conducting research across diverse geographic, cultural, and historical settings to account for variability and interconnectedness within a particular subject.

Traditional research in biological anthropology has often been geographically and methodologically siloed, focusing on specific populations, sites, or regions in isolation. These localized studies have generated valuable insights, but they have also tended to reinforce narrow research frameworks that may overlook broader patterns or miss the heterogeneity of human responses across different contexts. Moreover, such siloed approaches can inadvertently replicate colonial structures of knowledge production by privileging data collected under extractive models or by limiting interpretation to external epistemologies. In contrast, a multisituated framework challenges these limitations by encouraging researchers to engage with multiple contexts, populations, and temporal scales to uncover broader transregional patterns and the nuances of local variation. This approach allows scholars to ask more expansive questions while remaining attentive to the complexity and specificity of each case.

Such a multisituated ethnography responds to contemporary anthropology's myriad commitments, as it privileges attention to questions of scale, comparison, and the politics of ethnographic and scientific encounters (Barnard 2016a, 2016b; Cveček 2024; Fuentes and Wiessner 2016; Sunder Rajan 2021). It also reflects an epistemological shift that recognizes the importance of knowledge co-production, sustained regional expertise, and the historical conditions under which data are generated, curated, and interpreted.

An example of the value of this approach can be seen in studies of human adaptation to high-altitude environments. Earlier research often focused exclusively on Andean populations, leading to the assumption that physiological responses to hypoxia were relatively uniform. However, more recent comparative studies have expanded to include Tibetan and Ethiopian groups, revealing that each group has developed distinct behavioral, cultural, genetic, and physiological pathways to cope with similar environmental pressures (Alkorta-Aranburu et al. 2012; Beall 2007; Huerta-Sánchez et al. 2014; Yi et al. 2010). A selective sweep in EPAS1 has been identified in both Tibetans and Andean highlanders (Bigham et al. 2013). Notably, some of these adaptations—particularly among Tibetans—have been linked to introgression from Denisovan hominins, raising intriguing questions about whether similar archaic genetic contributions may have influenced Andean populations as well, not only in relation to high-altitude adaptation but also to other biological traits, especially considering that their ancestors migrated from Asia (Huerta-Sánchez et al. 2014; Zhang et al. 2021). These findings demonstrate that even when selective forces appear constant, the evolutionary outcomes can vary significantly depending on population history, local ecology, and cultural practices. This multisituated lens reveals not only the plasticity of human adaptation but also the limitations of drawing generalizations from single-population studies. In doing so, it underscores the importance of comparative and context-sensitive approaches for producing more comprehensive and globally relevant anthropological knowledge.

Adopting multisituated approaches in biological anthropology is not merely an academic exercise, but a necessity for addressing contemporary challenges. Issues such as health inequities, environmental crises, human history, and global migration demand research that is contextual, inclusive, and collaborative. By grounding research within multiple contexts and, in conjunction with a polyphonic perspective, amplifying diverse voices, biological anthropology can contribute to a more nuanced and socially relevant understanding of human biology and evolutionary history. This paradigm shift not only enriches the discipline but also aligns it with broader efforts to decolonize and democratize scientific inquiry.

15.4.4 The Value of a Polyphonic Perspective for Epistemic Plurality in Biological Anthropology

Complementing the multisituated perspective, a polyphonic approach brings attention to the multiplicity of voices, positionalities, and epistemologies involved in the production of knowledge. A polyphonic perspective advocates for the incorporation of multiple and diverse voices as well as knowledge systems, including those of marginalized communities, to produce a more holistic understanding of human history.

A polyphonic or multivocal biological anthropology prioritizes the integration of diverse perspectives, particularly those from historically underrepresented groups and communities. This includes incorporating Indigenous, Afro-descendant, and other—marginal—knowledge systems, along with feminist critiques and decolonial methodologies, into the discipline. For instance, the inclusion of Indigenous and Afro-descendant oral histories alongside genetic and archeological data has enriched our understanding of the peopling of the Americas. These integrative approaches challenge Eurocentric narratives and emphasize the agency of Indigenous and Afro-descendant communities in shaping their own histories (Balée 2013; Heckenberger et al. 2003; Kuwanwisiwma et al. 2018; Rasmussen et al. 2014; Thornton 2015).

A striking example of the epistemic power of polyphonic approaches is the interpretation of the Inka Qhapaq Hucha ceremony (Capacocha)—ritual sacrifices of children conducted at high-altitude shrines across the Andes. While early interpretations by European-descended scholars often framed these practices through moralistic or sensationalist lenses, recent research has drawn on Indigenous cosmology, Spanish colonial chronicles, archeological context, isotopic data, and ancient DNA to build a more nuanced and respectful understanding. These integrative studies—many led by South American scholars—have revealed complex patterns of child mobility, ceremonial preparation, and sacred geography, highlighting the role of Qhapaq Hucha in Inka statecraft, cosmopolitics, and ancestral renewal (Ceruti 2015; de la Fuente et al. 2024; Santa Gadea 2022). This case exemplifies how a polyphonic approach, rooted in both scientific and Indigenous knowledge systems, can radically transform our understanding of past lifeways and challenge reductive or exoticizing narratives.

Polyphony also extends to collaboration among researchers from diverse disciplinary and cultural backgrounds, fostering dialogue and the co-creation of knowledge. Haraway's (1988) concept of "situated knowledge" is particularly relevant here, as it emphasizes how positionality shapes scientific inquiry and epistemic authority. Embracing polyphony allows biological anthropology to move toward more inclusive and equitable research practices by recognizing the value of diverse epistemologies and lived experiences, including oral traditions and regionally grounded interpretive frameworks.

Another compelling case illustrating the need for a polyphonic approach is the longstanding debate over the morphological differences observed between early South American skeletons—such as those from Lagoa Santa in Brazil and the Pampas in Argentina—and those from the Late Holocene. For decades, these differences were interpreted by Latin American researchers as evidence of two separate migratory events into the continent, a hypothesis grounded in extensive regional work and skeletal analysis (Neves and Pucciarelli 1991). In contrast, recent paleogenomic studies, primarily led by researchers based in the Global North, have proposed a model of long-term genetic continuity, favoring a single major migration followed by regional diversification (Meltzer 2021; Posth et al. 2018; Raff 2022). However, rather than engaging with the morphological divergence, these studies tend to dismiss the two-migration model outright, without addressing the underlying patterns that gave rise to it. Further complicating the picture are documented affinities between certain ancient and contemporary South American individuals and populations from Austro-Melanesia and Polynesia—findings that challenge entrenched models of continental isolation and suggest deeper, possibly shared ancestries (Campelo dos Santos et al. 2022; Castro e Silva et al. 2021; Skoglund et al. 2015). These apparently conflicting lines of evidence—morphological, genomic, archeological, and ethnohistorical—cannot be meaningfully reconciled within a single explanatory paradigm. Instead, they require a pluralistic and polyphonic methodological framework that integrates multiple forms of knowledge and interpretation across epistemic, disciplinary, and geopolitical boundaries (Menéndez and Veigl 2026). Only through such a multivocal approach can we begin to make sense of both the continuity and the complexity that characterize early South American population histories.

Ultimately, embracing and implementing a polyphonic perspective in biological anthropology is not only a methodological choice but an ethical and epistemological commitment. By valuing multiple voices, especially those historically silenced from scientific narratives, we move toward more inclusive, reflective, and socially accountable research. In the context of the peopling of the Americas, such an approach is essential for constructing explanations that are not only scientifically robust but also responsive to the plural realities and lived histories of the populations we seek to understand.

15.5 Final Remarks

The lack of recognition of Latin American researchers not only marginalizes valuable perspectives but also perpetuates a colonialist narrative that overlooks the diversity of scientific contributions from the Global South and contributes to the silencing of the academic work conducted in peripheral or Global South contexts. For instance, studies conducted in Latin America often face challenges in securing funding or gaining visibility in high-impact journals, leading to a skewed representation of archeological and anthropological findings. Addressing these disparities requires fostering balanced collaborations, increasing funding for research infrastructure in the Global South, promoting policies that ensure fair access to advanced equipment and training, empowering local research institutions, creating collaborative frameworks that genuinely value and incorporate diverse perspectives, and advocating for inclusive publishing practices that amplify underrepresented voices. While this chapter focuses on research related to the peopling of the Americas, we believe that the ideas presented here are applicable to other topics and regions where local researchers from the Global South should play a central role—but where narratives are often dominated, and essential resources controlled, by institutions and scholars from the Global North.

We recognize that implementing a polyphonic approach in practice requires more than a commitment to inclusion—it demands intentional structural changes in how research is conceived, organized, and funded. While epistemic inclusion is indeed time-intensive and often constrained by short funding cycles and institutional pressures, meaningful collaboration must begin at the project design stage. This includes co-developing research questions with local scholars, ensuring equitable participation in fieldwork, analysis, and authorship, and allocating time for dialogue rather than treating it as an afterthought. Crucially, this responsibility should not fall on early-career researchers or PhD students alone. It must be assumed by principal investigators, research institutions, and funding agencies, who are in a position to set the terms of collaboration and redistribute epistemic authority. When planned deliberately, such approaches are not only feasible, but also lead to more meaningful and contextually grounded scientific outcomes. As philosopher of science Helen Longino (1990) argues, scientific objectivity is achieved not by silencing diverse and multivocal narratives, but through the inclusion of diverse perspectives and accountability to the social contexts in which knowledge is produced. This perspective aligns with our emphasis on decolonial, multidisciplinary, multisituated, and polyphonic research designs as a way to build more responsible and equitable science.

Implementing a decolonial, multidisciplinary, multisituated, and polyphonic approach is critical for addressing these inequities. This framework emphasizes conducting research across diverse contexts and incorporating a plurality of voices and epistemologies. It ensures that knowledge production is not confined to perspectives rooted in the Global North but instead embraces localized expertise, including Indigenous and Afro-descendant knowledge systems, often essential for interpreting archeological and anthropological data. By situating research in multiple locations, engaging with diverse cultural frameworks, and combining different disciplinary approaches, this model helps dismantle entrenched biases and produces more comprehensive, nuanced narratives. This inclusive model not only promotes equity in knowledge creation but also enriches scientific inquiry by acknowledging and integrating the full spectrum of human experiences and contributions to research on human history.

Acknowledgments We would like to thank Mathilde Lequin and Juan Manuel Rodríguez Caso for inviting us to contribute a chapter to their edited volume. We are also grateful to the other members of our research network, BALANCE (Constanza de la Fuente Castro, Gonzalo Figueiro, Vivette García Deister, Camila Tamburrini, Juan Manuel Argüelles, and Columba González-Duarte), for the fruitful and ongoing discussions that motivate the constant debate that partially inspired the development of this work.

References

Abbott CC (1878) Second report on the Paleolithic implements from the glacial drift: in the valley of the Delaware River, Near Trenton. Salem Press

Ackermann R, Athreya S, Black W, Humphreys R, Pickering R, Sahle Y, Schroeder L (2018) Paleoanthropology: decolonizing and building for the future. In: The international encyclopedia of anthropology, pp 1–9. https://doi.org/10.1002/9781118924396.wbiea2549

Acosta J (1590) Historia natural y moral de las indias. En Casa de Juan de León

Alkorta-Aranburu G, Beall CM, Witonsky DB, Gebremedhin A, Pritchard JK, Di Rienzo A (2012) The genetic architecture of adaptations to high altitude in Ethiopia. PLoS Genet 8(12): e1003110. https://doi.org/10.1371/journal.pgen.1003110

Ameghino F (1880) La antigüedad del hombre en el Plata. Igon Hermanos

Argüelles JM, Fuentes A, Yáñez B (2022) Analyzing asymmetries and praxis in aDNA research: a bioanthropological critique. Am Anthropol 124(1):130–140. https://doi.org/10.1111/aman.13692

Athreya S, Ackermann RR (2019) Colonialism and narratives of human origins in Asia and Africa. In: Porr M, Matthews JM (eds) Interrogating human origins. Routledge, pp 72–95

Ávila-Arcos MC, de la Fuente Castro C, Nieves-Colón MA, Raghavan M (2022) Recommendations for sustainable ancient DNA research in the global south: voices from a new generation of paleogenomicists. Front Genet 13:880170. https://doi.org/10.3389/fgene.2022.880170

Balée W (2013) Cultural forests of the Amazon: a historical ecology of people and their landscapes. University of Alabama Press
Barnard A (2016a) Unity versus interdisciplinarity: a future for anthropology. Curr Anthropol 57 (S13):S145–S153. https://doi.org/10.1086/686022
Barnard A (2016b) Human origins: the case to include social anthropology. Cuicuilco 23(65): 217–226
Beall CM (2007) Two routes to functional adaptation: Tibetan and Andean high-altitude natives. Proc Natl Acad Sci 104(Suppl 1):8655–8660. https://doi.org/10.1073/pnas.0701985104
Becerra-Valdivia L, Higham T (2020) The timing and effect of the earliest human arrivals in North America. Nature 584(7819):93–97. https://doi.org/10.1038/s41586-020-2491-6
Bennett MR, Bustos D, Pigati JS, Springer KB, Urban TM, Holliday VT et al (2021) Evidence of humans in North America during the last glacial maximum. Science 373(6562):1528–1531. https://doi.org/10.1126/science.abg7586
Bigham AW, Wilson MJ, Julian CG, Kiyamu M, Vargas E, Leon-Velarde F, Rivera-Chira M, Rodriquez C, Browne VA, Parra E, Brutsaert TD (2013) Andean and Tibetan patterns of adaptation to high altitude. Am J Hum Biol 25(2):190–197. https://doi.org/10.1002/ajhb.22358
Birdsell JB (1951) The problem of the early peopling of the Americas as viewed from Asia. In: The physical anthropology of the American Indian, vol 1. Viking Fund, New York, p 68
Bolnick DA, Smith RW, Fuentes A (2019) How academic diversity is transforming scientific knowledge in biological anthropology. Am Anthropol 2:464–464. https://doi.org/10.1111/aman.13212
Borrero LA (2025) De Beringia a América del Sur: modelos, asunciones y olvidos. Notas acerca del poblamiento de América. Revista del Museo de La Plata 10(1):22–64. https://doi.org/10.24215/25456377e233
Bourdier P, Duboz P, Macia E, Ka A, Nazare JA, Cohen E, Bergouignan A et al (2021) How interdisciplinary research at the crossroad between socio-cultural anthropology, nutritional and physical activity physiology can help addressing the obesity epidemic. Cahiers de Nutrition et de Diététique 56(1):51–58. https://doi.org/10.1016/j.cnd.2020.11.003
Bueno LDMR, Dias AS, Horta AI (2020) «Poblamientos plurales»: discontinuidades y diversidad cultural en el proceso de poblamiento antiguo del este de América del sur. Boletín Americanista 70(81):39–61. https://doi.org/10.1344/BA2020.81.1002
Burmeister CV (1891) Breves datos sobre una excursión a Patagonia. Revista Museo de La Plata 2: 273–288
Campelo dos Santos AL, Owings A, Sullasi HSL, Gokcumen O, De Giorgio M, Lindo J (2022) Genomic evidence for ancient human migration routes along South America's Atlantic coast. Proc R Soc B 289(1986):20221078. https://doi.org/10.1098/rspb.2022.1078
Cardoso de Oliveira R (1997) Antropologías Periféricas. In V Congreso Argentino de Antropología Social 29 de julio-1 de agosto de 1997 La Plata
Castro e Silva MA, Ferraz T, Bortolini MC, Comas D, Hünemeier T (2021) Deep genetic affinity between coastal Pacific and Amazonian natives evidenced by Australasian ancestry. Proc Natl Acad Sci U S A 118(14):e2025739118. https://doi.org/10.1073/pnas.2025739118
Cavalli-Sforza LL, Piazza A, Menozzi P, Mountain J (1988) Reconstruction of human evolution: bringing together genetic, archaeological, and linguistic data. Proc Natl Acad Sci 85(16): 6002–6006. https://doi.org/10.1073/pnas.85.16.6002
Ceruti MC (2015) Frozen mummies from Andean mountaintop shrines: bioarchaeology and ethnohistory of Inca human sacrifice. Biomed Res Int 2015(1):439428. https://doi.org/10.1155/2015/439428
Cvećek S (2024) Why kinship still needs anthropologists in the 21st century. Anthropol Today 40(1):3–6. https://doi.org/10.1111/1467-8322.12861
David RN, Patterson D, Campbell A, Tandon S, Mazieres N, Ray MV, Parra W, Rojas C, Duque N, Mesa LF, García O, Triana S, Blair A, Maestre JC, Dib CM, Bravi G, Bailliet D, Corach T, Hünemeier M, Bortolini FM, Salzano ML, Petzl-Erler V, Acuña-Alonzo, Aguilar-Salinas S, Canizales-Quinteros T, Tusié-Luna L, Riba M, Rodríguez-Cruz M, Lopez-Alarcón R, Coral-

Vazquez T, Canto-Cetina I, Silva-Zolezzi JC, Fernandez-Lopez V., Contreras G, Jimenez-Sanchez MJ, Gómez-Vázquez J, Molina Á, Carracedo A, Salas C, Gallo G, Poletti DB, Witonsky G, Alkorta-Aranburu RI, Sukernik L, Osipova SA, Fedorova R, Vasquez M, Villena C, Moreau R, Barrantes D, Pauls L, Excoffier G, Bedoya F, Rothhammer J-M, Dugoujon G, Larrouy W, Klitz D, Labuda J, Kidd K, Kidd A, Di RNB, Freimer AL, Price A, Ruiz-Linares (2012) Reconstructing Native American population history Nature 488(7411): 370–374. https://doi.org/10.1038/nature11258

de la Fuente CC, Figueiro G (2024) Human ancient DNA analysis in Latin America: current state and challenges. Archaeometry 67(S2):S4–S26. https://doi.org/10.1111/arcm.12979

de la Fuente CC, Cortés C, Raghavan M, Castillo D, Castro M, Verdugo RA, Moraga M (2024) The genomic and cultural diversity of the Inka Qhapaq hucha ceremony in Chile and Argentina. Genome Biol Evol 16(9):evae196. https://doi.org/10.1093/gbe/evae196

de Las Casas B (1877) Historia de las Indias. Imprenta y litografía de I. Paz

de Sousa Santos B (2014) Epistemologies of the south: justice against epistemicide. Routledge

di Fabio Rocca F, Russo MG, Arencibia V, Seldes V (2021) Ancient DNA studies: use of ethnonyms and collaborative research in South America. J Anthropol Archaeol 61:101265. https://doi.org/10.1016/j.jaa.2020.101265

Eduardo V, de Castro (1998) Cosmological Deixis and Amerindian Perspectivism The Journal of the Royal Anthropological Institute 4(3):469. https://doi.org/10.2307/3034157

Farrera A, Lacoste Jeanson A, Yáñez B (2024) Bioética y antropología biológica: la necesidad de una epistemología para una ciencia humana. Estudios de Antropología Biológica 22(2):9–19. https://doi.org/10.22201/iia.14055066p.2024.90538

Figueiro G, Menéndez LP, Yáñez B (2024) Presentación.(Neo) colonialidades y asimetrías geopolíticas en la bioantropología latinoamericana. Rev Urug Antropología Etnografía 9(2). https://doi.org/10.29112/ruae.v9i2.2430

Fuentes A (2016) The extended evolutionary synthesis, ethnography, and the human niche: toward an integrated anthropology. Curr Anthropol 57(S13):S13–S26. https://doi.org/10.1086/685684

Fuentes A (2018a) How humans and apes are different, and why it matters. J Anthropol Res 74(2): 151–167. https://doi.org/10.1086/697150

Fuentes A (2018b) Towards integrative anthropology again and again: disorderly becomings of a (biological) anthropologist. Interdiscip Sci Rev 43(3–4):333–347. https://doi.org/10.1080/03080188.2018.1524236

Fuentes A, Wiessner P (2016) Reintegrating anthropology: from inside out: an introduction to supplement 13. Curr Anthropol 57(S13):S3–S12. https://doi.org/10.1086/685694

Fulkerson TJ, Tushingham S (2019) Who dominates the discourses of the past? Gender, occupational affiliation, and multivocality in North American archaeology publishing. Am Antiq 84(3): 379–399. https://doi.org/10.1017/aaq.2019.35

García-Ortiz H, Barajas-Olmos F, Contreras-Cubas C, Cid-Soto MÁ, Córdova EJ, Centeno-Cruz F et al (2021) The genomic landscape of Mexican indigenous populations brings insights into the peopling of the Americas. Nat Commun 12(1):5942. https://doi.org/10.1038/s41467-021-26188-w

Giddings JL (1954) Early man in the Arctic. Sci Am 190(6):82–89

Greenberg JH, Turner CG II, Zegura SL (1986) The settlement of the Americas: a comparison of the linguistic, dental, and genetic evidence. Curr Anthropol 27:477–497. https://doi.org/10.1086/203472

Gruhn R (2023) An anthropological conception of the initial peopling of the Americas. PaleoAmerica 9(3):167–173. https://doi.org/10.1080/20555563.2023.2278948

Hanscam E, Witcher R (2023) Women in antiquity: an analysis of gender and publishing in a global archaeology journal. J Field Archaeol 48(2):87–101. https://doi.org/10.1080/00934690.2022.2143896

Haraway D (1988) Situated knowledges: the science question in feminism and the privilege of partial perspective. Fem Stud 14(3):575–599. https://doi.org/10.2307/3178066

Haraway D (1989) Primate visions: gender, race, and nature in the world of modern science. Routledge
Harding S (1991) Whose science? Whose knowledge? Thinking from women's lives. Cornell University Press
Harding S (1998) Is science multicultural? Postcolonialisms, feminisms, and epistemologies. Indiana University Press
Heckenberger MJ, Kuikuro A, Kuikuro UT, Russell JC, Schmidt M, Fausto C, Franchetto B (2003) Amazonia 1492: pristine forest or cultural parkland? Science 301(5640):1710–1714. https://doi.org/10.1126/science.1086112
Hibben FC (1946) The lost Americans. Thomas Y. Crowell Co
Hrdlička A (1912) Early man in South America. Bull Smithsonian Inst Bur Am Ethnol 52:1–405
Huerta-Sánchez E, Jin X, Asan BZ, Peter BM, Vinckenbosch N et al (2014) Altitude adaptation in Tibetans caused by introgression of Denisovan-like DNA. Nature 512(7513):194–197. https://doi.org/10.1038/nature13408
Hutson SR (2002) Gendered citation practices in American antiquity and other archaeology journals. Am Antiq 67(2):331–342. https://doi.org/10.2307/2694570
Hutson SR, Johnson J, Price S, Record D, Rodriguez M, Snow T, Stocking T (2023) Gender, institutional inequality, and institutional diversity in archaeology articles in major journals and sapiens. Am Antiq 88(3):326–343. https://doi.org/10.1017/aaq.2023.36
Jasanoff S (2004) States of knowledge: the co-production of science and social order. Routledge
Kellert SH, Longino HE, Waters CK (2006) Scientific Pluralism. University of Minnesota Press
Klein JT (1990) Interdisciplinarity: history, theory, and practice. Wayne State University Press
Kreimer P (2022) Techno-scientific promises, disciplinary fields, and social issues in peripheral contexts. Sci Cult 31(4):515–536. https://doi.org/10.1080/09505431.2022.2108614
Krotz E (1996) La generación de teoría antropológica en América Latina: silenciamientos, tensiones intrínsecas y puntos de partida. Maguaré 11–12:25–39
Krotz E (1997) Anthropologies of the south: their rise, their silencing, their characteristics. Crit Anthropol 17(3):237–251. https://doi.org/10.1177/0308275X9701700302
Krotz E (2006) La diversificación de la antropología universal a partir de las antropologías del sur. Boletín Antropológico 24(66):7–20
Krotz E (2019) Overseas, continental, and internal colonialism: responses from Latin American anthropologies. In: Shorcowitz D, Chávez J, Schröeder I (eds) Shifting forms of continental colonialism: unfinished struggles and tensions. Springer Nature, pp 71–94
Kuwanwisiwma LJ, Ferguson TJ, Colwell C (eds) (2018) Footprints of Hopi history: Hopihiniwtiput kukveni'at. University of Arizona Press
Latour B (1993) We Have Never Been Modern. Harvard University Press
Law J, Lin W (2017) Provincializing STS: postcoloniality, symmetry, and method. East Asian Sci Technol Soc 11(2):211–227. https://doi.org/10.1215/18752160-3823859
Little MA, Haas JD (eds) (1989) Human population biology: a transdisciplinary science. Oxford University Press
Longino HE (1990) Science as social knowledge: values and objectivity in scientific inquiry. Princeton University Press
Marks J (2009) Why I Am Not a Scientist: Anthropology and Modern Knowledge. Univ of California Press
Menéndez LP, Hubbe M (2025) Toward a Biocultural Synthesis of the Peopling of the Americas: Introduction to the Special Issue Abstract Abstract in Spanish Abstract in Portuguese American Journal of Biological Anthropology 188(3). https://doi.org/10.1002/ajpa.v188.3, https://doi.org/10.1002/ajpa.70160
Mazzia N, Weitzel C, Flegenheimer N (2020) What is new about the early peopling of central and South America? PaleoAmerica 6(3):207–209. https://doi.org/10.1080/20555563.2020.1758548
Meltzer DJ (2021) First peoples in a new world: populating ice age America. Cambridge University Press

Menéndez LP (2024) The antiquity and ancestral origin of humans in the Americas: a five hundred year inquiry from a biological anthropology perspective. J Anthropol Sci 102:7–67. https://doi.org/10.4436/JASS.10201
Menéndez LP, Urban M (2025) Tracing human diversity in South America's southern cone: linguistic, morphometric, and genetic perspectives. Am J Biol Anthropol 187(3):e70077. https://doi.org/10.1002/ajpa.70077
Menéndez LP, Veigl S (2026) It takes two to tango: a pluralist account for building comprehensive explanations in human evolution. Am Anthropol
Menéndez LP, Paul KS, de la Fuente C, Almeida T, Delgado M, Figueiro G, Jorgensen K, Kuzminsky S, López-Sosa MC, Nichols J, Roksandic M, Scott GR, O'Rourke D, Hubbe M (2022) Towards an interdisciplinary perspective for the study of human expansions and biocultural diversity in the Americas. Evol Anthropol 31(2):62–68. https://doi.org/10.1002/evan.21937
Menéndez LP, Figueiro G, de la Fuente Castro C, Yáñez B (2025) Fostering inclusive practices of citation and teamwork in biological anthropology: a commentary on "Petrous Fever" by Källén et al. (2024). Curr Anthropol 66(3):440–441. https://doi.org/10.1086/735687
Mitchell SD (2003) Biological complexity and integrative pluralism. Cambridge University Press
Moreno FP (1882) El origen del hombre sudamericano: Razas y civilizaciones de este continente. Anales de la Sociedad Científica Argentina 14:182–223
Moreno-Mayar JV, Vinner L, de Barros DP, De La Fuente C, Chan J, Spence JP et al (2018) Early human dispersals within the Americas. Science 362(6419):eaav2621. https://doi.org/10.1126/science.aav2621
Moutinho S (2024) Breaking the glass. Science 386(6726):1087–1089. https://doi.org/10.1126/science.adv0401
Neves WA, Pucciarelli HM (1989) Extracontinental biological relationships of early South American human remains: a multivariate analysis. Cienc Cult 41:566–575
Neves WA, Pucciarelli HM (1991) Morphological affinities of the first Americans: an exploratory analysis based on early South American human remains. J Hum Evol 21:261–273. https://doi.org/10.1016/0047-2484(91)90107-7
Nieves-Colón MA, Blevins KE, Contreras-Sieck MÁ, López MB (2021) Paleogenómica y bioarqueología en México. Cuicuilco 28(81):187–223
Porr M (2021) Indigenous global histories and modern human origins. In: McGrath A, Russell L (eds) The Routledge companion to global indigenous history. Routledge, pp 139–164
Porr M, Matthews JM (2019a) Interrogating human origins: decolonisation and the deep past. Routledge
Porr M, Matthews JM (2019b) Interrogating and decolonising the deep human past. In: Porr M, Matthews JM (eds) Interrogating human origins. Routledge, pp 3–32
Posth C, Nakatsuka N, Lazaridis I, Skoglund P, Mallick S, Lamnidis TC et al (2018) Reconstructing the deep population history of central and South America. Cell 175(5):1185–1197. https://doi.org/10.1016/j.cell.2018.10.027
Prates L, Politis GG, Perez SI (2020) Rapid radiation of humans in South America after the last glacial maximum: a radiocarbon-based study. PLoS One 15(7):e0236023. https://doi.org/10.1371/journal.pone.0236023
Raff J (2022) Origin: a genetic history of the Americas. Hachette UK
Rajan KS (2021) Multisituated: ethnography as diasporic praxis. Duke University Press
Rasmussen M, Anzick SL, Waters MR, Skoglund P, DeGiorgio M, Stafford TW Jr et al (2014) The genome of a late Pleistocene human from a Clovis burial site in western Montana. Nature 506(7487):225–229. https://doi.org/10.1038/nature13025
Ribeiro GL, Escobar A (2006) Las antropologías del mundo: transformaciones de la disciplina a través de los sistemas de poder. Univ Humaníst 61:15–50
Ribeiro GL, Escobar A (2008) Antropologías del mundo: transformaciones disciplinarias dentro de sistemas de poder. Envión Editores

Ribeiro GL, Escobar A (2020) World anthropologies: disciplinary transformations within systems of power. In: Ribeiro GL, Escobar A (eds) World anthropologies. Routledge, pp 1–25
Ruderman A (2023) Population diversity and equity in the genomic era: going global to return to the local. J Community Genet 14(6):519–525. https://doi.org/10.1007/s12687-023-00669-5
Santa Gadea LM (2022) La Capacocha o Qhapaq Ucha de los incas the Capacocha or Qhapaq Hucha of the Incas. Revista de Humanidades 3(6):17–41
Sardi ML, Dettler M, Arcidiácono LV (2024) Antropología implacable. La historia de los instrumentos antropológicos del Museo de La Plata en perspectiva global y local. Rev Argent Antropol Biol 26(2):088. https://doi.org/10.24215/18536387e088
Schroeder L, Ackermann RR (2023) Moving beyond the adaptationist paradigm for human evolution, and why it matters. J Hum Evol 174:103296. https://doi.org/10.1016/j.jhevol.2022.103296
Schurr TG, Ballinger SW, Gan YY, Hodge JA, Merriwether DA, Lawrence DN et al (1990) Amerindian mitochondrial DNAs have rare Asian mutations at high frequencies, suggesting they derived from four primary maternal lineages. Am J Hum Genet 46(3):613
Scott GR, Navega D, Vlemincq-Mendieta T, Dern LL, O'Rourke DH, Hlusko LJ, Hoffecker JF (2023) Peopling of the Americas: a new approach to assessing dental morphological variation in Asian and native American populations. Am J Biol Anthropol. https://doi.org/10.1002/ajpa.24878
Shea JJ (2011) Homo sapiens is as Homo sapiens was: behavioral variability versus "behavioral modernity" in paleolithic archaeology. Curr Anthropol 52(1):1–35. https://doi.org/10.1086/658067
Skoglund P, Mallick S, Bortolini MC, Chennagiri N, Hünemeier T, Petzl-Erler ML et al (2015) Genetic evidence for two founding populations of the Americas. Nature 525(7567):104–108. https://doi.org/10.1038/nature14895
Smith CA, Garrett-Scott D (2021) "We are not named": black women and the politics of citation in anthropology. Feminist Anthropol 2(1):18–37. https://doi.org/10.1002/fea2.12038
Snow CP (1959) The two cultures and the scientific revolution. Cambridge University Press
Squires K, García-Mancuso R (2020) Desafíos éticos asociados al estudio y tratamiento de restos humanos en las ciencias antropológicas en el siglo XXI (Ethical challenges associated with the study and treatment of human remains in anthropological sciences in the 21st century). Rev Argent Antropol Biol 23(2). https://doi.org/10.24215/18536387e034
Thornton TF (2015) The ideology and practice of pacific herring cultivation among the Tlingit and Haida. Hum Ecol 43:213–223. https://doi.org/10.1007/s10745-015-9736-2
Turner TR, Bernstein RM, Taylor AB, Asangba A, Bekelman T, Cramer JD et al (2018) Participation, representation, and shared experiences of women scholars in biological anthropology. Am J Phys Anthropol 165:126–157. https://doi.org/10.1002/ajpa.23386
Wagner GP (2022) O povoamento da costa atlântica brasileira, uma abordagem geoarqueológica para a transição entre holoceno inicial e médio. Tessituras: Revista de Antropologia e Arqueologia 10(1):73–111. https://doi.org/10.15210/tes.v10i1.21900
Walter HV, Cathoud A, Mattos A (1953) Theme VI: theories regarding the peopling of the Americas. Origins, Indig Period 51:87
Waters MR, Stafford TW (2013) The first Americans: a review of the evidence for the Late-Pleistocene peopling of the Americas. In: Graf KE, Ketron CV, Waters MR (eds) Paleoamerican Odyssey. Texas A & M University Press, pp 543–562
Willerslev E, Meltzer DJ (2021) Peopling of the Americas as inferred from ancient genomics. Nature 594(7863):356–364. https://doi.org/10.1038/s41586-021-03499-y
Willey GR, Newman MT (1949) Men out of Asia. Am Anthropol 51(1):160–164
Wynter S (2003) Unsettling the coloniality of being/power/truth/freedom: towards the human, after Man, its overrepresentation—an argument. CR: New Centen Rev 3(3):257–337. https://doi.org/10.1353/ncr.2004.0015
Yáñez B, Fuentes A, Silva CP, Figueiro G, Menéndez LP, García-Deister V, de la Fuente-Castro C, González-Duarte C, Tamburrini C, Argüelles JM (2023) Pace and space in the practice of aDNA

research: concerns from the periphery. Am J Biol Anthropol 180(3):417–422. https://doi.org/10.1002/ajpa.24683

Yi X, Liang Y, Huerta-Sanchez E, Jin X, Cuo ZXP, Pool JE et al (2010) Sequencing of 50 human exomes reveals adaptation to high altitude. Science 329(5987):75–78. https://doi.org/10.1126/science.1190371

Zhang X, Witt KE, Bañuelos MM, Ko A, Yuan K, Xu S, Nielsen R, Huerta-Sanchez E (2021) The history and evolution of the Denisovan-EPAS1 haplotype in Tibetans. Proc Natl Acad Sci U S A 118(22):e2020803118. https://doi.org/10.1073/pnas.2020803118

GPSR Compliance

The European Union's (EU) General Product Safety Regulation (GPSR) is a set of rules that requires consumer products to be safe and our obligations to ensure this.

If you have any concerns about our products, you can contact us on ProductSafety@springernature.com

In case Publisher is established outside the EU, the EU authorized representative is:

Springer Nature Customer Service Center GmbH
Europaplatz 3
69115 Heidelberg, Germany

Batch number: 10755214

Printed by Printforce, the Netherlands